The Power of Images in Early Modern Science

Edited by
Wolfgang Lefèvre
Jürgen Renn
Urs Schoepflin

Birkhäuser Verlag
Basel · Boston · Berlin

Editors:

Wolfgang Lefèvre
Jürgen Renn
Urs Schoepflin

Max-Planck-Institut für Wissenschaftsgeschichte
Wilhelmstr. 44
D-10117-Berlin

A CIP catalogue record for this book is available from the Library of Congress, Washington D.C., USA

Bibliographic information published by Die Deutsche Bibliothek
Die Deutsche Bibliothek lists this publication in the Deutsche Nationalbiografie; detailed bibliographic data is available in the internet at http://dnb.ddb.de.

ISBN 3-7643-2434-1 Birkhäuser Verlag, Basel – Boston – Berlin

Printed on acid-free paper produced from chlorine-free pulp. TCF ∞
Cover design: Micha Lotrovsky, CH-4106 Therwil, Switzerland
Cover illustrations: see pp. 113, 148, 84, 17, 110, 190, 229, and 163
Printed in Germany

ISBN 3-7643-2434-1

9 8 7 6 5 4 3 2 1

Contents

IV. Depicting the World at Large and the Hidden Potential of Images

V. Systems of Knowledge and their Representation by Images

Introduction

Wolfgang Lefèvre, Jürgen Renn, and Urs Schoepflin

General

The origin of this volume is a workshop held in 1997 in Berlin as part of a series of workshops organized in the framework of the Network on Science and the Visual Images 1500 - 1800 funded by the European Science Foundation and initiated by William Shea. Meanwhile a selection of contributions was thoroughly revised and prepared for publication together with additionally invited papers for this book. The result is a volume which we hope corresponds to the original intention of the Network to contribute to a historical reconstruction of the role of images in the history of science, still neglected because of the traditional focus of the history of science on texts corresponding to a concentration on scientific theories.

With the widening of the scope of recent history of science to include for example the experimental culture of science, increasing attention has also been paid to images. This widened perspective not only implies the methodological extension to incorporate in particular the methods of the history of art, but also implies a new understanding of scientific knowledge. Paying attention to images increases first of all the awareness of the dependence of scientific knowledge on different external representations and thus deepens the understanding of the historicity of this knowledge. Indeed, external representations such as images are subject to cultural changes and therefore play an important mediatory role between science and its cultural context. Second, the study of images makes it evident that scientific knowledge has a deeper, more complex structure which must be assumed if its analysis is only based on text. In fact, the analysis of the function of images in the early modern period shows that they mediated not only between science and its cultural context, but also between practical knowledge and its theoretical reflection in scientific theories.

The analysis of images thus constitutes an important branch of the history of science that on the one hand is conceived of as part of a more general history of culture and on the other hand as a historical epistemology of knowledge. This book is not a systematic and comprehensive account of scientific images and the early modern period. Nevertheless, it covers a fairly broad spectrum of the domains of scientific knowledge of the period and raises some of the key issues connected with an epistemic history of images.

I. Mechanics Between Practical and Theoretical Knowledge and the Mediatory Function of Images

The first section focuses on mechanical knowledge and illustrates the role of images as mediatory instances between practical and theoretical knowledge, addressing various aspects of their role. The first contribution shows how the practical knowledge of artillerists and its visual representation provided the basis for early modern theories of ballistics, eventually leading to the theories of projectile motion. The second contribution addresses a different type of practical knowledge also relevant to the emergence of mechanical science, that of shipbuilding. It traces the emergence of design techniques and their role in integrating practical knowledge and theoretical context. Marking the passage from medieval culture to Renaissance knowledge, the third contribution describes the evolution of the drawing as an autonomous language for the representation of technical subjects. The fourth contribution analyses the limits of these pictorial representations as mediatory instances and points to the lack of a sophisticated grammar of visual representation in early modern machine drawings. The fifth contribution makes it clear on the other hand, that images of machines did not primarily serve planning purposes, but were intended to convey a specific "image of knowledge" in a social environment in which the inventors of machines depended on patronage.

II. Theories of Matter Between Alchemy and Atomism and the Autonomy of Images

The second section focuses on theories of matter and illustrates the autonomy of images and their history with regard to textual traditions. The first contribution deals with alchemy and the representation of alchemical knowledge by symbolically charged images. It analyses a process in which these images took on an increasing autonomy both with regard to the practice of alchemy and with regard to texts. They rather served the purpose of embedding alchemical knowledge in a universal worldview by drawing on iconographic traditions, making it possible to transcend the limits of the textual traditions. Another transgression of the boundaries of a textually based tradition achieved by images is illustrated by the second contribution dealing with the visual representation of the minimal constituency of matter. It shows the striking independence of this tradition of visual representations from specific theories of matter and thus points to the fact that these theories comprise structures of knowledge invariant with respect to the great conceptual revolutions of science.

III. The Classification of Life and the Interaction Between Images and Texts

The third section focuses on the interaction between images and texts in early modern life sciences. This interaction was driven by the dynamics of the technological development of the production and reproduction of illustrations on the one hand, and the increasing professionalization on the other hand. The latter resulted in the distinction between amateur and professional scientists, each using their own figurative language, also in an increasing canonization of knowledge, which resulted in a decreasing importance of illustrations with respect to texts and schematic representations in professional science. This development of a specific pictorial grammar is similar to that observed in early modern mechanics. Different aspects of this development are highlighted for botany in the first contribution, and for zoology and geology in the second and third contributions.

IV. Depicting the World at Large and the Hidden Potential of Images

The fourth section deals with the knowledge of the world at large and analyses the potential of images to synthesize fragments of knowledge to a global picture, thus providing the fragments with a new meaning. The contributions to this chapter also illustrate the striking power of persistence of images, even if the corresponding textual traditions are broken down or suppressed. This becomes particularly evident from the first contribution that demonstrates, contrary to the received opinion, the presence of astronomical knowledge in early Medieval Europe prior to the advent of the translation movement in the twelfth century. This knowledge was carried by planetary diagrams accompanying

Greek and Roman texts which in themselves provide hardly any detailed information. While the dispersed ancient knowledge of astronomy has been largely reconstituted and substantially augmented in the early modern period, the cultural constraints imposed on science by the catholic church also served to give images a special function making it possible to convey meaning even when textual transmission was impossible. This is illustrated by the second contribution. While the first contribution deals with traditional schematic representation, planetary diagrams, which took on a realistic physical meaning in the early modern period, the third contribution deals with the invention of new schematic representations of the world at large in early modern science. It shows that the representation of the global distribution of winds in the scheme of Halley constituted a new branch of science, climatology.

V. Systems of Knowledge and their Representation by Images

The fifth and final section deals with the organization of knowledge into systems and the use of images to represent such systems, thereby conveying new meaning to them which may even challenge the traditional order. The contributions to this section illustrate that, in the early modern period, visual representations served as an important medium for epistemological reflection. The first contribution takes the example of 17th century encyclopaedic enterprise and shows how the idea of knowledge that drove this enterprise was expressed within allegorical images placing the enterprise in the context of the religious worldview of the time. The second contribution tackles a classical masterpiece of the Renaissance from a new angle, Raphael's School of Athens, which is interpreted as a synthesis of epistemic and political order.

Acknowledgments

This volume would have been impossible without the lively discussions at the Berlin workshop and is thus indebted to all of its participants. It was produced with the help of several collaborators of the Max Planck Institute for the History of Science – in particular Ulrike Burgdorf, Lindy Divarci, and Heinz Reddner. The editors are grateful to the European Science Foundation for supporting a project that contributes to the realization of one of the essential goals of the Max Planck Institute – to pursue the history of science as an interdisciplinary endeavour.

Max Planck Institute for the History of Science

I.
Mechanics Between Practical and Theoretical Knowledge and the Mediatory Function of Images

The Challenging Images of Artillery
Practical Knowledge at the Roots of the Scientific Revolution

Jochen Büttner, Peter Damerow, Jürgen Renn, and Matthias Schemmel

Images and the Material Culture of Science

Princely families of old were in the habit of engaging historians who were charged with producing tailor-made histories, in which the achievements of these families received due attention. For a long time this, remarkably, was exactly what natural scientists also expected from their historians. As a matter of fact, from the point of view of developed science, the knowledge of a discipline is simply represented by the natural laws that define its object. It was accordingly a matter of course for its historians to concentrate solely on the question of who had discovered which of these laws when and in what manner. In this sense, the history of science is a biographically oriented, heroic history of the great discoverers and their discoveries (fig. 1).

To modern historians of science this view represents a naive faith in progress. Recent studies of the social and institutional contexts of scientific discoveries focusing, for instance, on Galileo as a courtier, have broadened the perspective on science in a fundamental way. According to such studies, the credibility of scientists like Galileo is put into perspective by the fact that they were required to position themselves in the patronage system, for they had to praise their knowledge and ability as being especially valuable, to ascribe to their putative discoveries and inventions an outstanding practical significance, to hide from their competitors the actual state of their knowledge, and to conceal their failures.

On the other hand, however, studies of the social context of scientific knowledge run the risk of neglecting its dependence on the material culture of science, and of focusing solely on framing conditions such as patronage that are largely indifferent to scientific objects. But, as a matter of fact, scientific knowledge is not just determined by a sequence of heroic discoveries, or by contingent social constellations indifferent to its objects. It is true that the development of scientific knowledge depends crucially on a social process of the transformation of knowledge systems, but this development also involves a material culture constituted by historically specific forms of material representation as well as specific objects, neither of which are indifferent to the contents of scientific knowledge. An important part of this material culture comprises images.

In the following, the relation between the development of science and material culture will be elaborated with the example of artillery in the early modern period. According to the textbook perspective of a history of science conceived as a sequence of heroic discoveries, the most important event in the history of ballistics is Galileo's discovery of the parabolic shape of the trajectory. But this perspective totally neglects the important role the material culture of representations in the practical knowledge of artillerists played at that time for the development of ballistics. At the dawn of classical physics, visual representations mediated between the immense body of practical experience accumulated by artillerists over centuries and the eventually successful attempts of scientists to overcome the limitations of this knowledge;

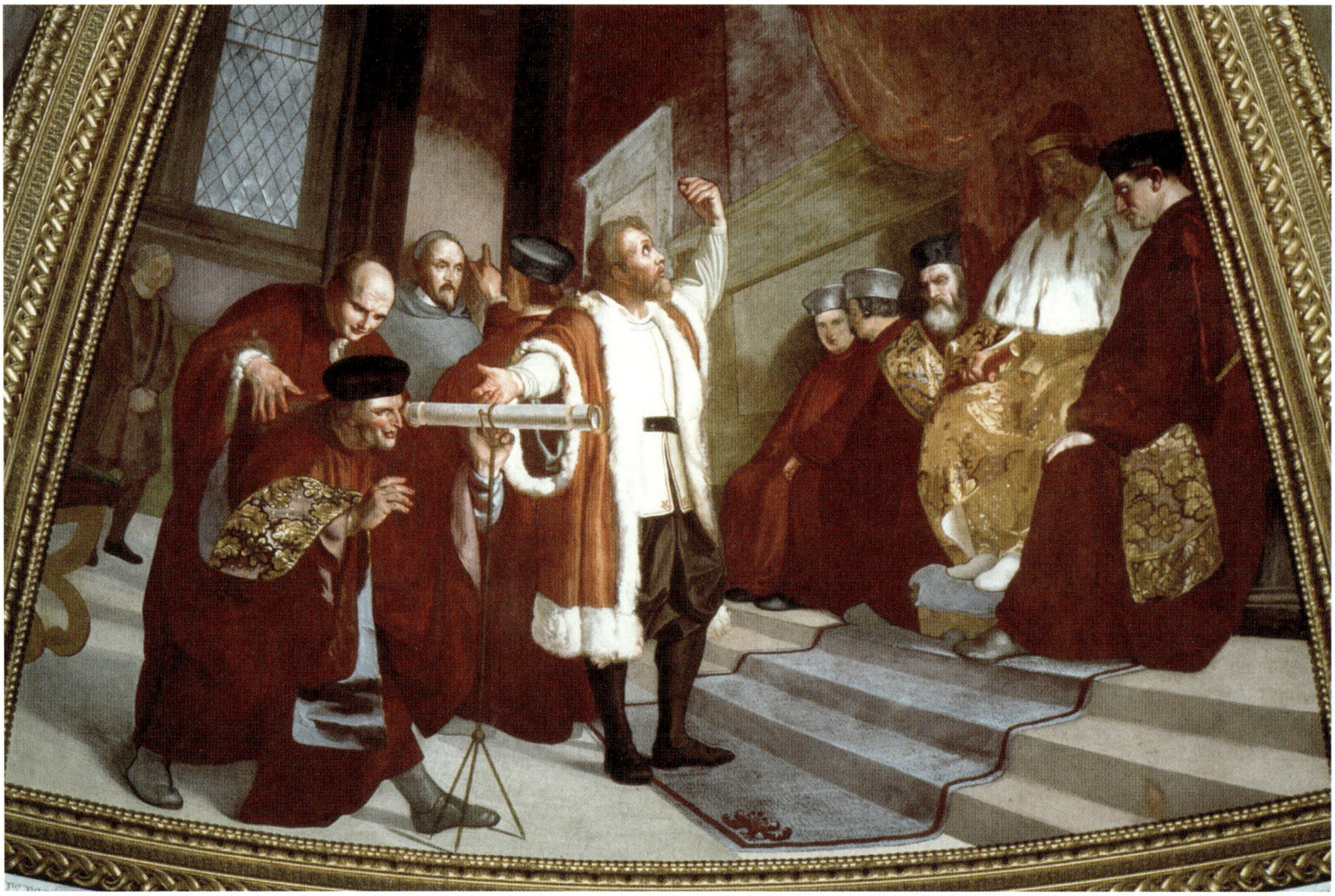

Figure 1
Galileo: hero or courtier? Detail from a fresco in the Tribuna di Galileo, Florence, courtesy of Sezione di Zoologia 'La Specola' del Museo di Storia Naturale dell'Università degli Studi di Firenze.

attempts focusing on theoretical puzzles that would have no practical impact for a long time to come. Images thus turn out to be the backbone of the exchange between the distinct bodies of practical and theoretical knowledge and consequently catalysed the transformation process from medieval natural philosophy to modern science.

This is true, in particular, for the challenging images of artillery. On the one hand, images served as realistic depictions of the practices of the artillerists, representing at the same time the intuitive understanding they had about the physics involved. On the other hand, theoreticians of ballistics produced such images in the early modern period in order to extract theoretical conclusions from the knowledge of the practitioners at that time, conclusions that eventually provided the foundation of later classical mechanics. A specific new type of image based on both geometrical construction and mechanical generation led to major achievements of Renaissance engineers and early modern scientists, among them the discovery of the parabolic trajectory and, as a consequence of the interest in this problem, the law of fall.

Geometry and Construction

The production of images requires tools, theoretical as well as practical. In the early modern period, Euclidean geometry played a key role as such a tool. The role of Euclidean geometry both as a model and as an instrument of the Scientific Revolution has been interpreted as indicating the debt of the new mathematical sciences of the period to an immaterial Platonic conception of nature. As a matter of fact, however, Euclidean geometry, as it was central to efforts of the Renaissance engineers and early modern scientists, was suitable to play this key role because it was itself founded in a specific aspect of material culture, the construction of figures and diagrams with particular devices (fig. 2).

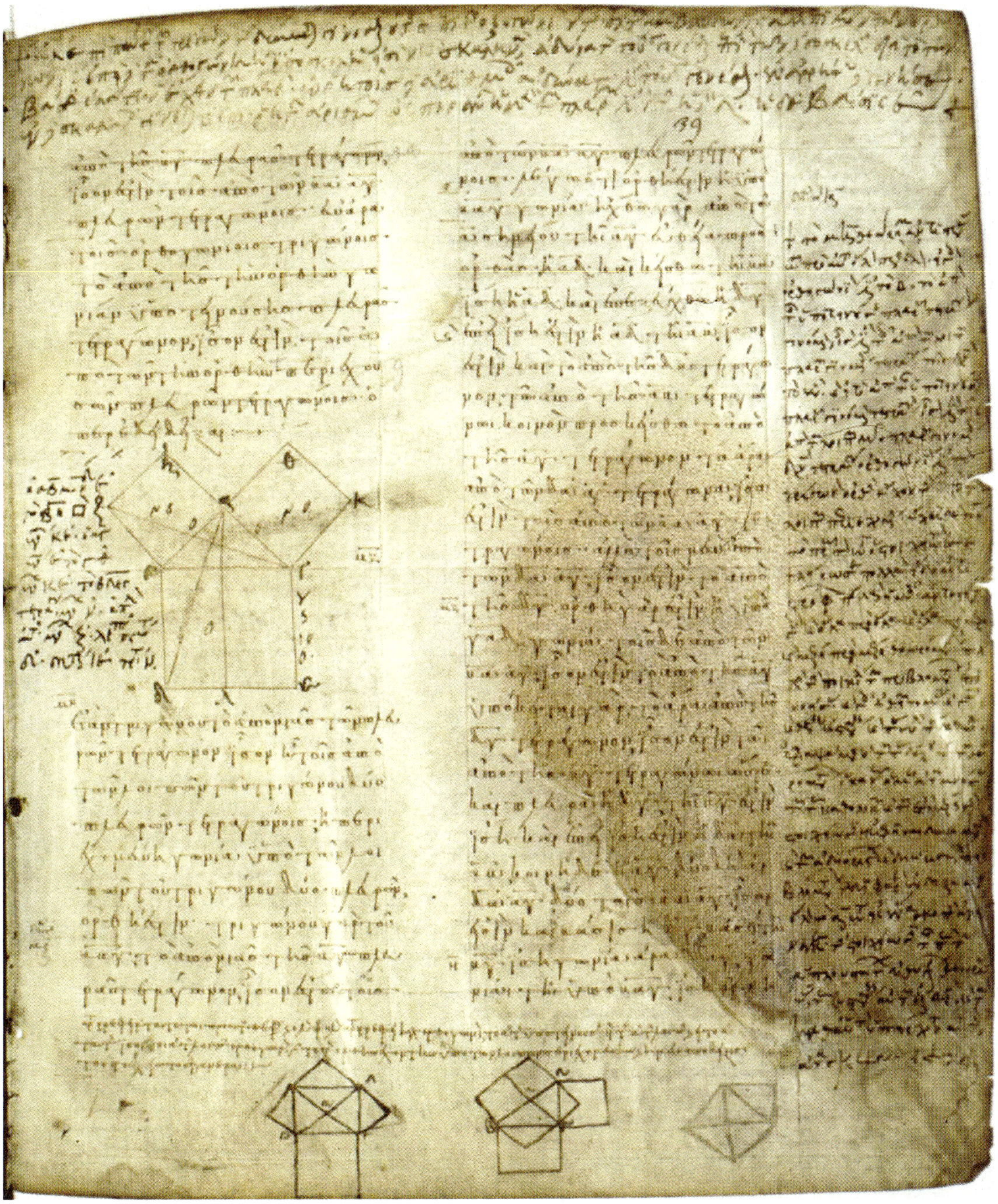

Figure 2
Detail of an ancient manuscript of Euclid's *Elements*.

This becomes clear when one looks at the first three Postulates of Euclid's Elements.[1] The first Postulate states that it is possible to draw a straight line from any one point to any other, the second that it is possible to extend a finite straight line continuously in a straight line, the third that it is possible to describe a circle with any centre and radius. These postulates express the universal applicability of the basic instruments, the straight edge and compasses, and at the same time restrict any construction in Euclidean geometry to the exclusive use of these instruments. More generally, Euclidean geometry can be conceived of as a theory of figures constructed using the straight edge and compasses. This largely determines the internal structure of Euclid's deductive system. The Elements consist, in particular, of "problems" and "propositions." In the "problems" it is shown that figures with certain properties can be constructed and how this can be done. In the "propositions" it is shown that certain constructed figures have certain properties.

Euclidean geometry turned out to be a powerful instrument for gaining new knowl-

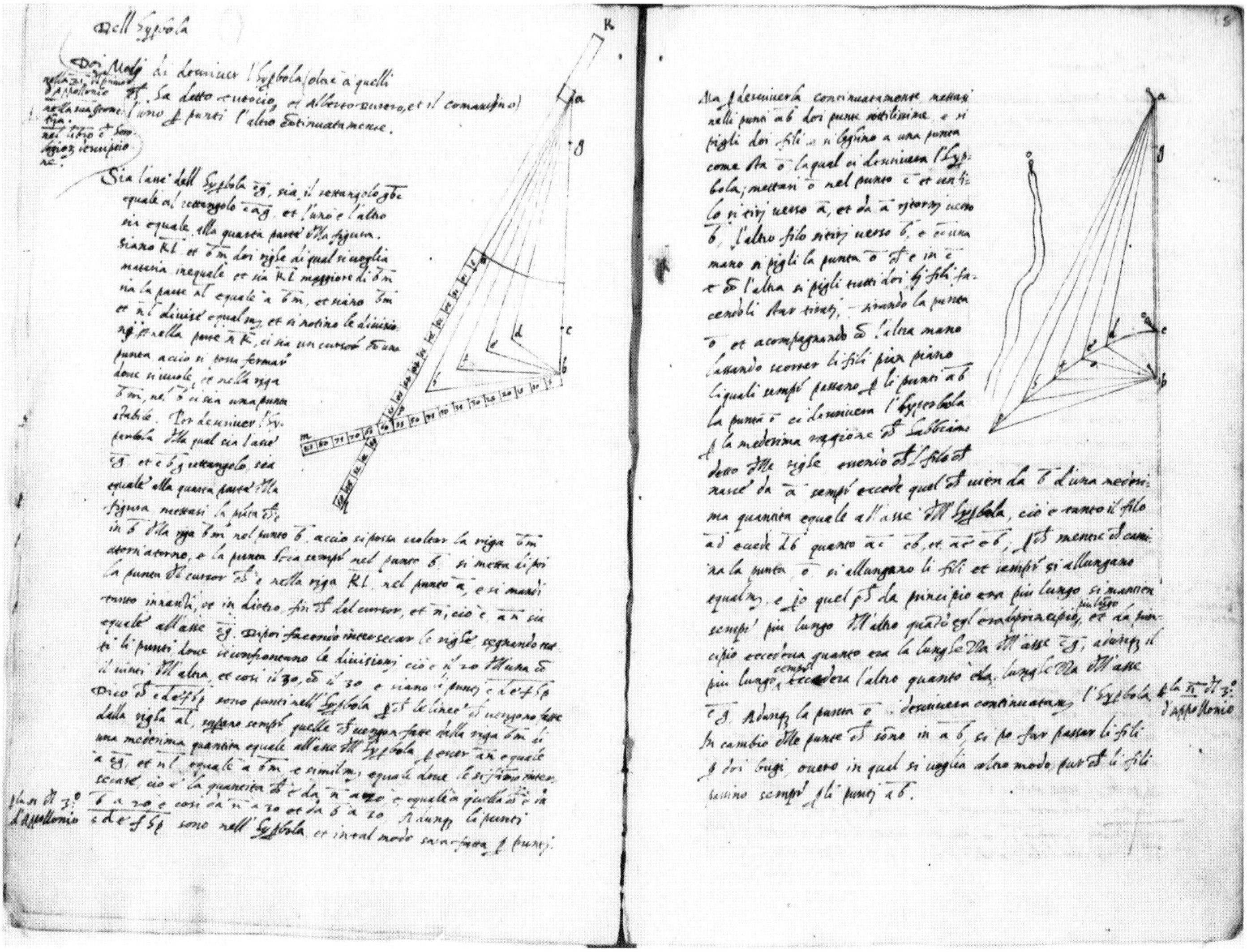

Figure 3
Instrument for drawing hyperbolas in a manuscript by Guidobaldo del Monte [ca. 1587–1592].

edge. One example, which turned out to be most influential, is the derivation of properties of higher-order curves like the parabola that can be constructed only pointwise. The Greeks, however, had also already realized the limitations inherent in being restricted to constructions made with the straight edge and compasses. They identified, in particular, the famous seemingly "unsolvable" problems of squaring the circle, doubling the cube, and trisecting the angle. But nevertheless, the principal restriction to constructions made with the straight edge and compasses remained unchallenged in antiquity.

In the early modern period, on the other hand, the situation was completely different. The development of devices for mechanically generating complex curves did question this restriction of Euclidean geometry. Descartes, for instance, wrote in his famous geometry:[2]

I am unable to understand why they [the Ancients] called these curves [the higher-order curves] mechanical, rather than geometrical. For if we say that they are so called because it is necessary to use a certain instrument to describe them, it is necessary to reject, for the same reason, circles and straight lines, seeing that they can only be described on paper with a compass and ruler, which we can also call instruments. Nor is it because the instruments that are used to trace them, being more complex than the ruler and compass, are therefore not as exact, for were this the case, it would be necessary to exclude them from mechanics, where exactness of works made by hand is desired, rather than from Geometry, where we seek only exactitude of reasoning, which can certainly be as exact with reference to these lines as to simpler ones.

In antiquity the study of conic sections such as the parabola was merely an esoteric special discipline of Euclidean geometry. In early modern times, however, conic sections took on a central practical relevance. In accordance with this increased relevance of conic sections, the engineer-scientists of the time searched for practicable methods to draw these curves. For instance, in a manuscript by the engineer-scientist Guidobaldo del Monte,[3] one of Galileo's closest collaborators and correspondents during the latter's early research on mechanics, one finds two such practical methods for constructing a hyperbola, one with an instrument, the other using threads (fig. 3). This example is not unique. The early modern period saw the development of ever more sophisticated drawing and construction techniques in order to meet the challenges of the time such as the construction of sun dials, the design of ships, the planning of architecture, and, in particular, the theoretical reconstruction of the trajectory of projectiles.

Blending the Knowledge of Practitioners with Aristotelian Physics

The prevailing theory of motion at that time was based on Aristotelian physics with its distinction between violent and natural motion. This theory provided a canonical interpretation for the motion of a projectile. According to this interpretation, the beginning of the projectile trajectory represents the violent part, while the end represents the natural motion tending downwards to the centre of the earth.[4] The transition from the violent to the natural part was sometimes considered as a motion mixed of both kinds (fig. 4), although such an interpretation is not necessarily implied by Aristotelian physics in its dogmatic form.

Even rough sketches of the trajectory illustrate the efforts of contemporary practitioners and engineer-scientists to use geometrical means for interpreting characteristics of the trajectory as known from practical experience in terms of Aristotelian physics. In the following, three examples will be considered.

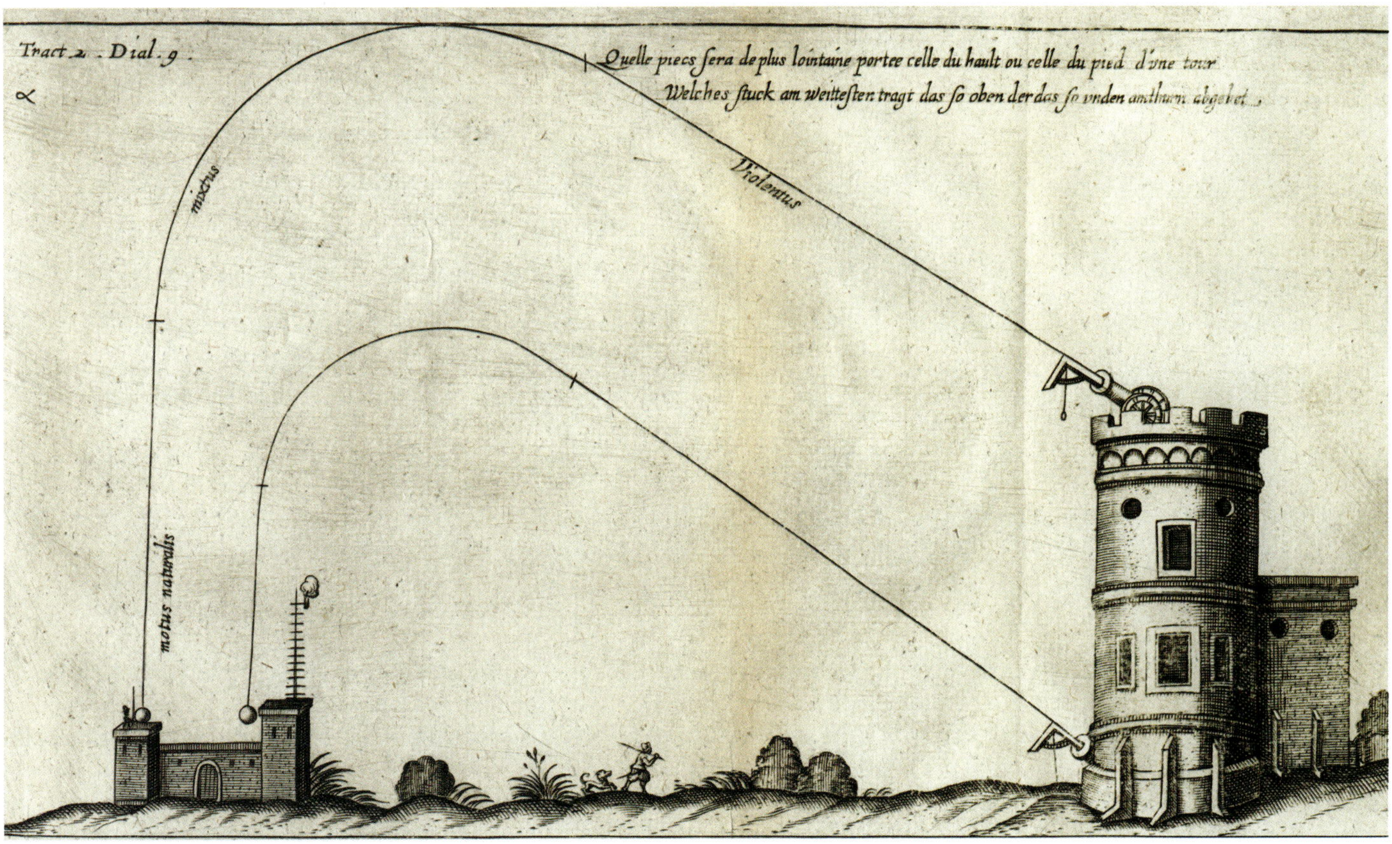

Figure 4
Aristotelian interpretation of the motion of a projectile. Ufano, D. [1928].

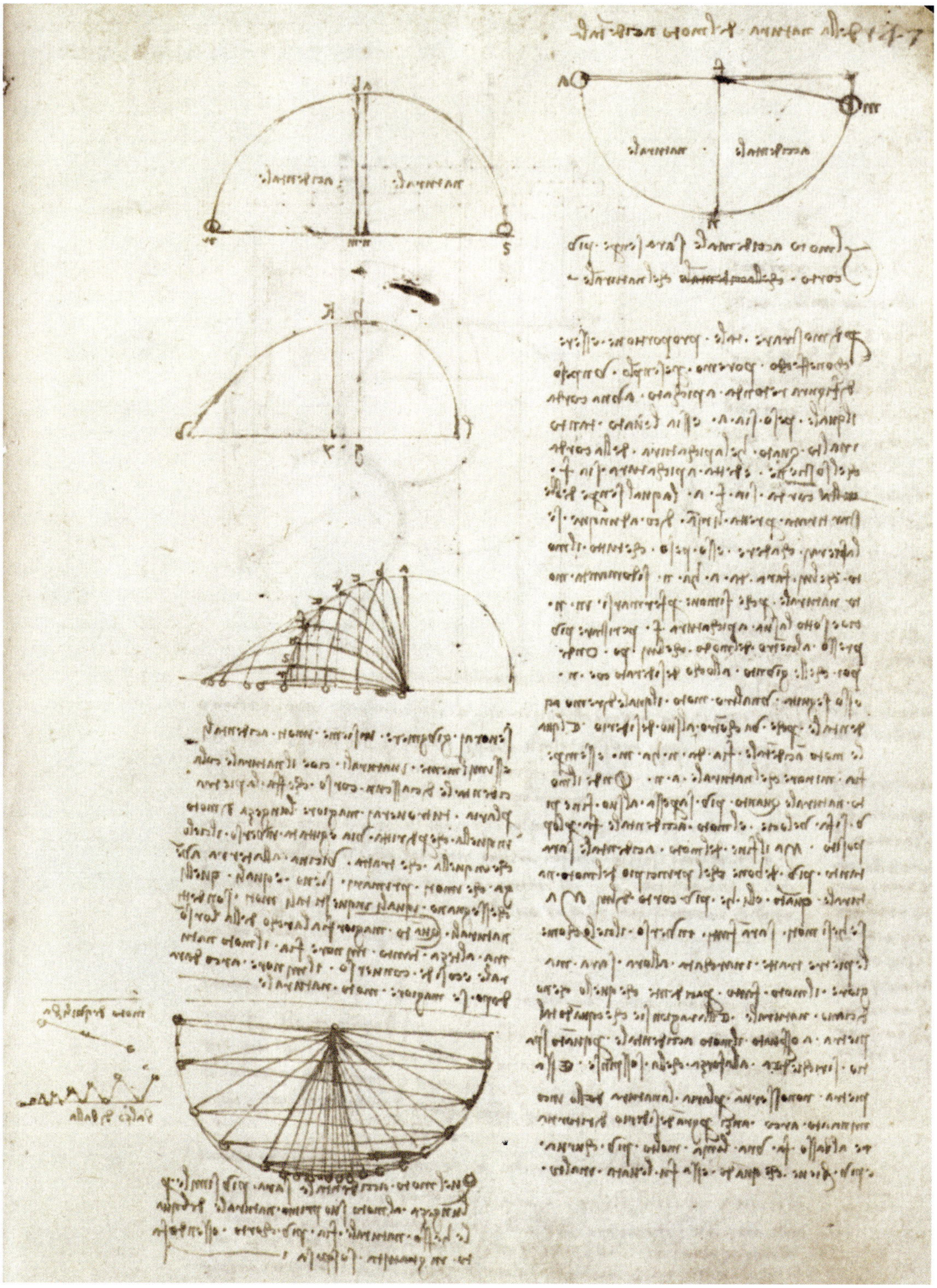

Figure 5
Geometrical rule for the variation of a trajectory with the elevation of the shot in Leonardo da Vinci's *Codex Madrid.*
Leonardo da Vinci [1974].

An early effort to reconstruct essential features of the trajectory dating from the last decade of the 15th century is found in Leonardo da Vinci's Codex Madrid.[5] Leonardo drew a diagram showing a construction of the overall characteristics of a number of trajectories, accompanied by a brief dynamical explanation of the asymmetrical nature of projectile trajectories (fig. 5). He attempted to capture the available practitioners' knowledge on projectile motion in a theoretical framework rooted in the Aristotelian dynamics of natural and violent ("accidental") motions and hoped to be able to determine essential features of the projectile trajectory on this basis. Artillerists of the time knew that the shape of the projectile trajectory was not symmetrical, its last part being shorter than the first. Leonardo noted that this behaviour, interpreted in terms of Aristotelian physics, is the inverse of that observed in a swinging pendulum where the natural part of the motion is the first part and is longer than the violent one which is the second and shorter part.

On the basis of the dynamical assumption that the projectile reaches the culmination point of its trajectory at the end of its accidental motion, Leonardo represented shots of equal power but different elevations using lines whose highest points lie on a circle around the origin of the shot. As a matter of fact, he did so for all but one of the nine trajectories in his diagram. His dynamical account of projectile motion implied that the downward path had to be shorter and more curved than the upward path, resulting in slightly asymmetrical curves whose precise form was apparently not determined by his geometrical construction. The steeper the shot, the more symmetrical the curves; complete symmetry is reached in the case of a vertical shot. However, Leonardo's construction failed to comply with one of the most elementary experiences of practitioners, the insight that the range of a shot does not monotonously increase with the decreasing angle of elevation, but that one actually finds a nonzero elevation of maximum range. It is probably his own realization of this difficulty that induced him to violate, for the flattest shot in his diagram, the rule that the culmination points of trajectories have to lie on a circle around the origin.

A manuscript page by Thomas Harriot,[6] an English contemporary of Galileo, provides a further example of an attempt to account for practitioners' knowledge on the basis of a theoretically motivated geometrical construction (fig. 6). This attempt is one of many found among the roughly 10 000 manuscript pages covered by Harriot's research notes. Artillerists of the time knew that the range as well as the form of the projectile trajectory depends on the angle of a shot. As sketches on Harriot's manuscript page indicate, he evidently attempted to explain this dependence by a geometrical construction of those features of the trajectory that are relevant to this problem. Harriot's construction makes use of

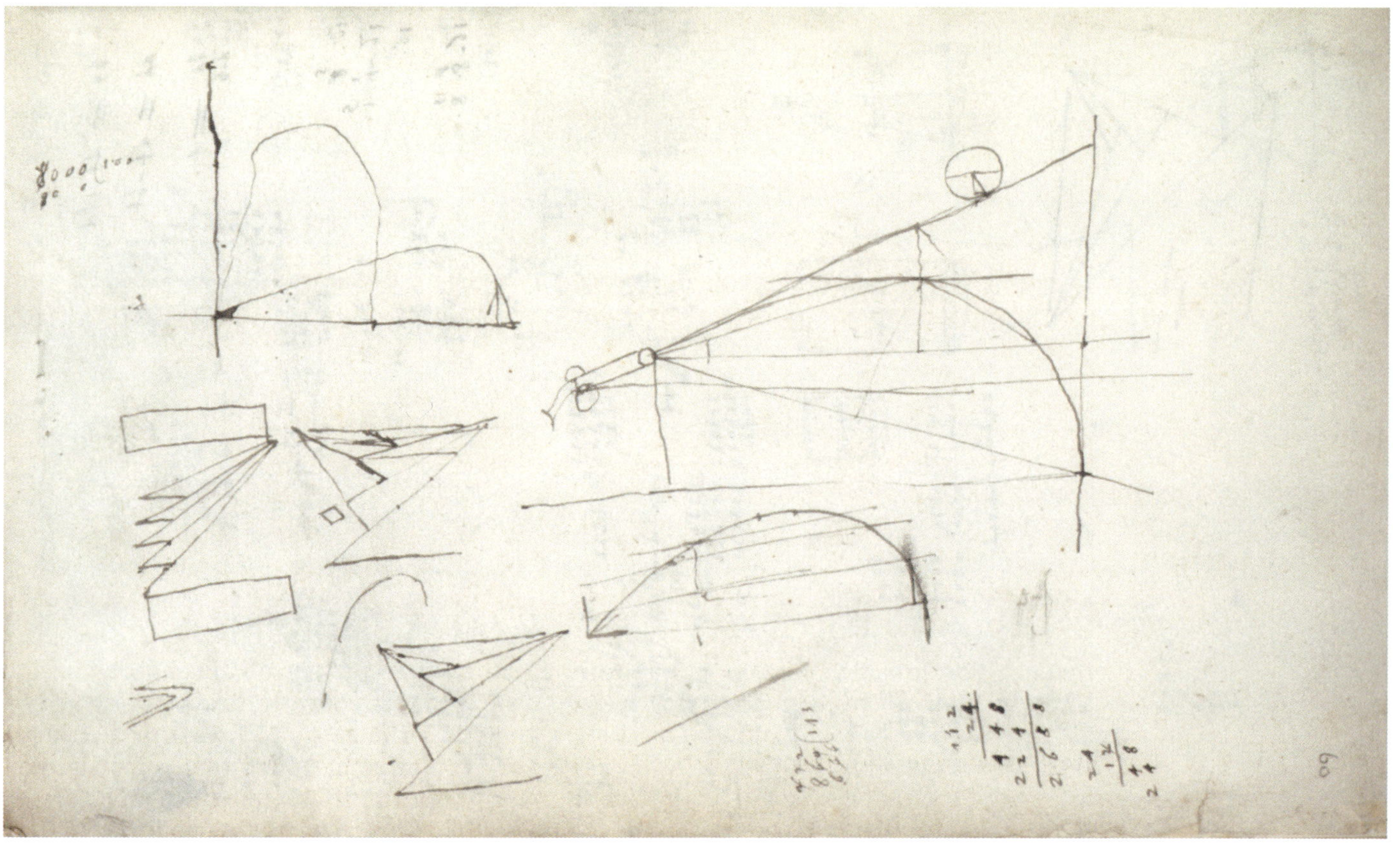

Figure 6
Attempt at explaining the mechanical effects of varying the elevation of a shot in a manuscript by Thomas Harriot, MS 6789 fol. 60r.

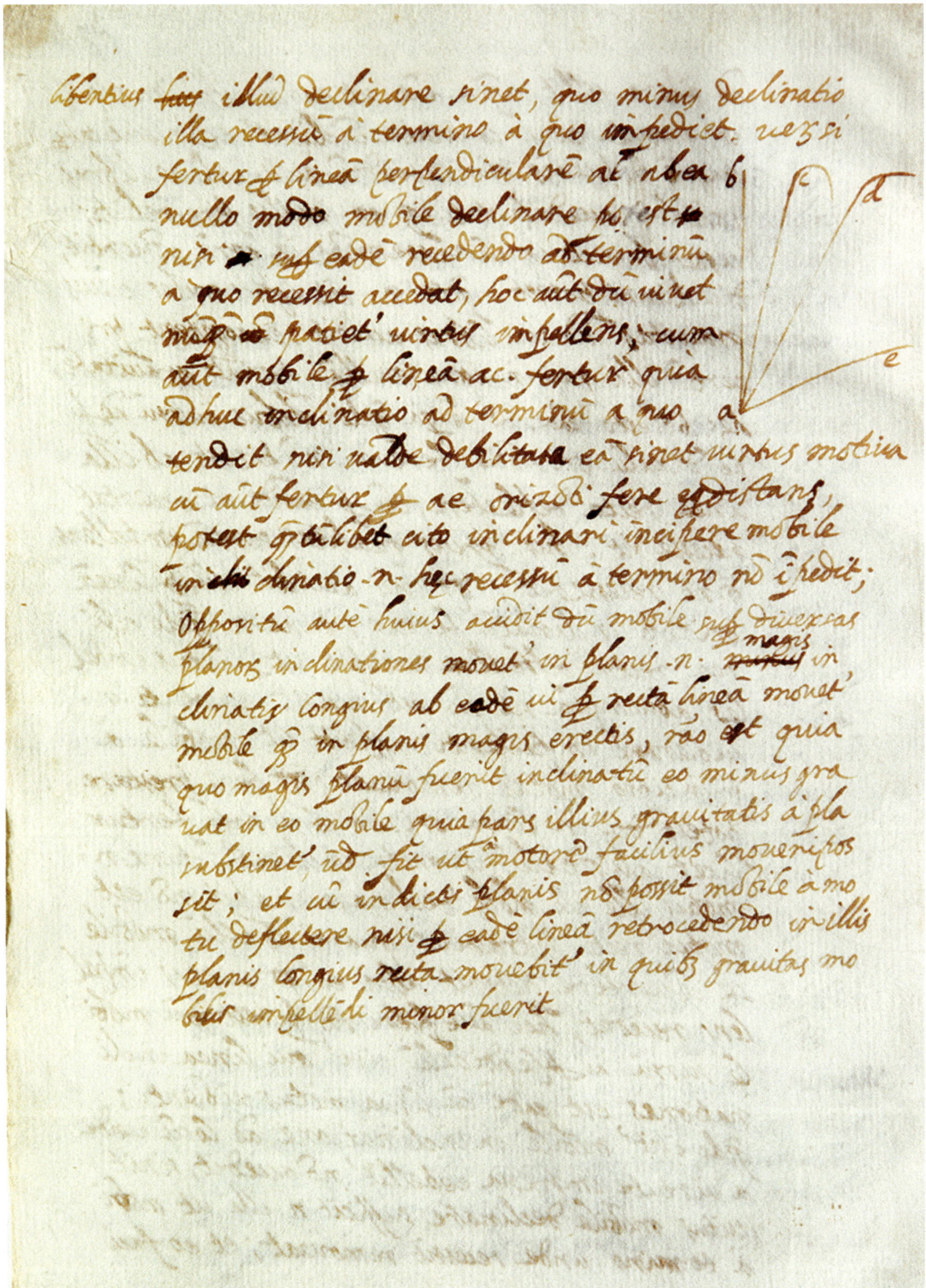

Figure 7
Sketch of trajectories at different angles in an early unpublished treatise of Galileo [after 1638].

the idea that projectile motion results from a composition of two motions, a violent one along the line of the shot and a natural one vertically downwards. In his sketch the violent component is represented by the line of the shot, i.e. a straight-line tangent to the trajectory at the origin of the shot, while the natural motion downwards is represented by a vertical straight-line tangent to the trajectory at the point where the violent motion has ceased. The asymmetric trajectory is sketched in between these two lines.

How would this construction have to be modified if a different angle of elevation were chosen? Harriot tried to solve this problem by drawing on mechanical knowledge about motions along differently inclined planes. In fact, the dependence of the violent motion on the angle of elevation can be inferred if the line of the shot is conceptualised as an inclined plane and if a law governing the motion along differently inclined planes is known. A peculiarity of Harriot's drawing of a canon ball not only confirms that he actually interpreted the line of the shot as an inclined plane but also hints at the law of motion he assumed for this plane. The ball placed on top of the line of the shot, as if on an inclined plane, is divided by construction lines according to an old proof idea concerning the behavior of a ball on an inclined plane. This idea, conceiving the ball as a lever with an unequal distribution of its weight on both sides of the point of contact between ball and plane, was handed down from ancient times and used by Pappus, Jordanus de Nemore, and, in early modern times, by Guidobaldo del Monte in his influential treatise on mechanics. However, while this idea allows us to determine the effective weight of the canon ball along an inclined plane of the same elevation as the line of shot, it gives neither a clue to the form of the trajectory nor an answer to the question of the dependency of its range on the angle.

Sketches of trajectories at different elevation angles are also found in an early unpublished treatise on motion by Galileo Galilei,[7] in a chapter on projectile motion. In this chapter he attempts to explain why steep shots behave differently than flat shots. The practical experience of artillerists suggested that, for steeper shots, the path of the projectile deviates noticeably from a straight line at a greater distance from the canon than in the case of flat shots. The trajectories drawn by Galileo have an initial virtually straight part, followed by a curved arc (fig. 7). The final part corresponding to the downward motion of the projectile is not drawn, probably because it was not relevant to Galileo's question. His text suggests that these trajectories were intended to translate the observation of the varying curvature of projectile trajectories into a simple geometrical pattern consisting of a straight part and of a curved part attached to it at different distances from the origin, depending on the elevation of the shot. The way in which the composition of straight and curved parts varies with the elevation does not follow from Galileo's construction. In his text, he accounts for this variation with an argument interpreting the interaction between violent and natural motion as being shaped by a mechanical property of the canon. He argues that, at the origin of the shot, a greater violent force is impressed upon the projectile if the line of shot is more steeply inclined due to the greater resistance with which the projectile opposes the impressed force, concluding that this stronger impressed force accounts for a longer straight part of the trajectory. Galileo explains the greater resistance incurred by a canon ball in a more elevated canon by interpreting the canon as an inclined plane on which the ball has a greater effective weight than on a less inclined plane. While Galileo succeeds in this way to account for a qualitative difference between steep and flat shots, his explanation does not yield a construction of the trajectory itself.

The rough sketches we have considered all constitute attempts to capture features of projectile motion known to practitioners by finding a geometrical representation of the trajectory and its variation with the elevation angle suitable for an interpretation in terms of Aristotelian dynamics and mechanical devices. An elaboration of such attempts towards a full account of projectile motion made, however, the inherent tensions between the knowledge components involved obvious and gave rise to mutually contradictory images of the projectile trajectory, as we shall see now.

Encountering Contradictions

The difficulty of bringing together theoretical and practical knowledge becomes strikingly apparent when we look at images of trajectories that obviously violate any intuitive knowledge about projectile motion. In a book on astronomical and geometrical problems by Daniel Santbech[8] one finds, for instance, a representation of the trajectory by a triangle, determined by the oblique line of the shot and a vertical line representing the downward motion of the projectile at the end

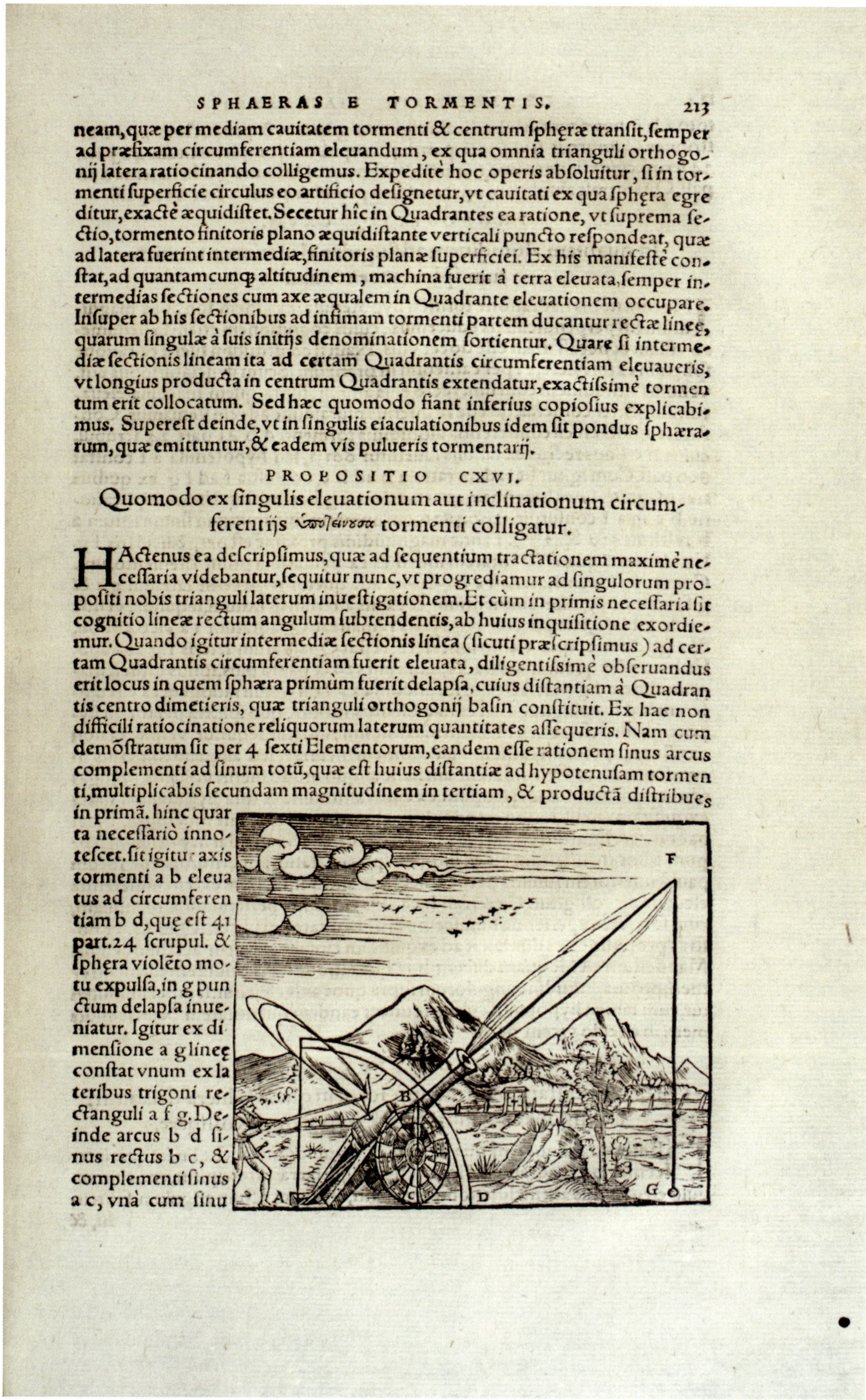

SPHAERAS E TORMENTIS. 213

neam, quæ per mediam cauitatem tormenti & centrum ſphęræ tranſit, ſemper ad præfixam circumferentiam eleuandum, ex qua omnia trianguli orthogonij latera ratiocinando colligemus. Expeditè hoc operis abſoluitur, ſi in tormenti ſuperficie circulus eo artificio deſignetur, vt cauitati ex qua ſphęra egreditur, exactè æquidiſtet. Secetur hîc in Quadrantes ea ratione, vt ſuprema ſectio, tormento finitoris plano æquidiſtante verticali puncto reſpondeat, quæ ad latera fuerint intermediæ, finitoris planæ ſuperficiei. Ex his manifeſtè conſtat, ad quantamcunq; altitudinem, machina fuerit à terra eleuata, ſemper intermedias ſectiones cum axe æqualem in Quadrante eleuationem occupare. Inſuper ab his ſectionibus ad infimam tormenti partem ducantur rectæ lineę, quarum ſingulæ à ſuis initijs denominationem ſortientur. Quare ſi intermediæ ſectionis lineam ita ad certam Quadrantis circumferentiam eleuaueris, vt longius producta in centrum Quadrantis extendatur, exactiſsimè tormentum erit collocatum. Sed hæc quomodo fiant inferius copioſius explicabimus. Supereſt deinde, vt in ſingulis eiaculationibus idem ſit pondus ſphærarum, quæ emittuntur, & eadem vis pulueris tormentarij.

PROPOSITIO CXVI.

Quomodo ex ſingulis eleuationum aut inclinationum circumferentijs ὑποτείνουσα tormenti colligatur.

HActenus ea deſcripſimus, quæ ad ſequentium tractationem maximè neceſſaria videbantur, ſequitur nunc, vt progrediamur ad ſingulorum propoſiti nobis trianguli laterum inueſtigationem. Et cùm in primis neceſſaria ſit cognitio lineæ rectum angulum ſubtendentis, ab huius inquiſitione exordiemur. Quando igitur intermediæ ſectionis linea (ſicuti præſcripſimus) ad certam Quadrantis circumferentiam fuerit eleuata, diligentiſsimè obſeruandus erit locus in quem ſphæra primùm fuerit delapſa, cuius diſtantiam à Quadrantis centro dimetieris, quæ trianguli orthogonij baſin conſtituit. Ex hac non difficili ratiocinatione reliquorum laterum quantitates aſſequeris. Nam cum demõſtratum ſit per 4 ſexti Elementorum, eandem eſſe rationem ſinus arcus complementi ad ſinum totũ, quæ eſt huius diſtantiæ ad hypotenuſam tormenti, multiplicabis ſecundam magnitudinem in tertiam, & productã diſtribues in primã. hinc quarta neceſſariò innoteſcet. ſit igitur axis tormenti a b eleuatus ad circumferentiam b d, quę eſt 41 part. 24 ſcrupul. & ſphęra violẽto motu expulſa, in g punctum delapſa inueniatur. Igitur ex dimenſione a g lineę conſtat vnum ex lateribus trigoni rectanguli a f g. Deinde arcus b d ſinus rectus b c, & complementi ſinus a c, vnà cum ſinu

Figure 8
Interpretations of the trajectory in terms of the geometry of surveying according to Daniel Santbech. Santbech [1561].

of its course (fig. 8). This representation corresponds to the strict Aristotelian distinction between violent and natural motion. As long as the violent motion conveyed to the projectile by the exploding powder of the canon prevails, the motion proceeds along a straight line in the direction of the shot. At the moment in which the violent force propelling the motion is exhausted, the natural weight of the projectile takes over, and the projectile follows its natural downward motion. Santbech's construction constitutes a simple description of the entire trajectory, making it possible to determine the ranges of shots for different elevations if a single additional physical assumption is introduced. In fact, if it is assumed that the initial straight part has the same length for any elevation if only the force of the shot is the same, the ranges of shots at arbitrary angles can be geometrically constructed once the canon is calibrated with the help of a single shot.

Constructions such as that of Santbech have been interpreted as illustrating the stubborn adherence to Aristotelian dogma, ignoring the most obvious facts of observation. For instance, in a book on the history of military sciences one reads about Santbech's representation:[9]

> *It is difficult to understand how it was possible to cling to this view: whoever observed only once the flight of an arrow or of a thrown stone or the expression of a ray of water from a fountain pipe, could not have helped convincing himself that the trajectory of any thrown body is a curve. That this did not happen demonstrates how extraordinarily great the power of preconceived opinion is even in scientific investigations, in particular when these investigations disdain the experiment.*

Do we really have to assume that a polymath like Santbech was blinded by preconceptions to such an extent that he did not even realize that the path of a projectile is always curved? Was this a discovery yet to be made by somebody bold enough to overcome the dogma solidified by a millenary tradition? A comment such as the one quoted above reveals a scheme characteristic for a history of science striving to associate any advancement of knowledge with a breakthrough accomplished by one of the heroes of science.

Even a glance at Santbech's text shows, however, that he constructed his image in this peculiar way not because he was unaware of the observational facts but because it served a specific practical function. The meaning of an illustration such as Santbech's cannot be unraveled in the framework of a history of singular discoveries but only if its function as a representation of the shared knowledge of the time is taken into account. While contemporary practitioners were well aware of the continuous effect of the bullet's weight on the trajectory as is evident also from Santbech's own text,[10] a more realistic image of the trajectory could, however, not easily be matched with clearly identifiable violent and natural parts of projectile motion as required by Aristotelian dynamics. But adhering to such a neat classification, for instance by drawing the final part of the trajectory as a straight vertical line representing the natural downward motion, also had advantages other than that of remaining faithful to Aristotelian natural philosophy. In fact, Santbech's apparently simplistic drawing actually allowed him to draw on knowledge gathered in other areas such as surveying and to apply the rich repository of surveying instruments such as the quadrant and sine tables in order to systematically determine the ranges and heights of shots at different elevations.

Where no such planning purposes were involved, as, for instance, in artistic representations of battles, there was evidently no difficulty in achieving a realistic depiction of projectile trajectories. The fact that a projectile trajectory of a real shot is continuously curved, although in some parts unnoticeably, had indeed always been taken for granted. The challenge was not to produce a realistic image of the trajectory, but rather to relate such an image to other knowledge available on projectile motion.

This challenge emerged when engineer-scientists, first of all Niccolò Tartaglia, attempted to create a comprehensive "new science" of projectile motion, aiming at explaining practitioners' knowledge on the basis of a theoretical foundation involving Euclidean geometry and Aristotelian natural philosophy. In the history of science, Tartaglia's "Nova scientia" is infamous for what appears from a modern

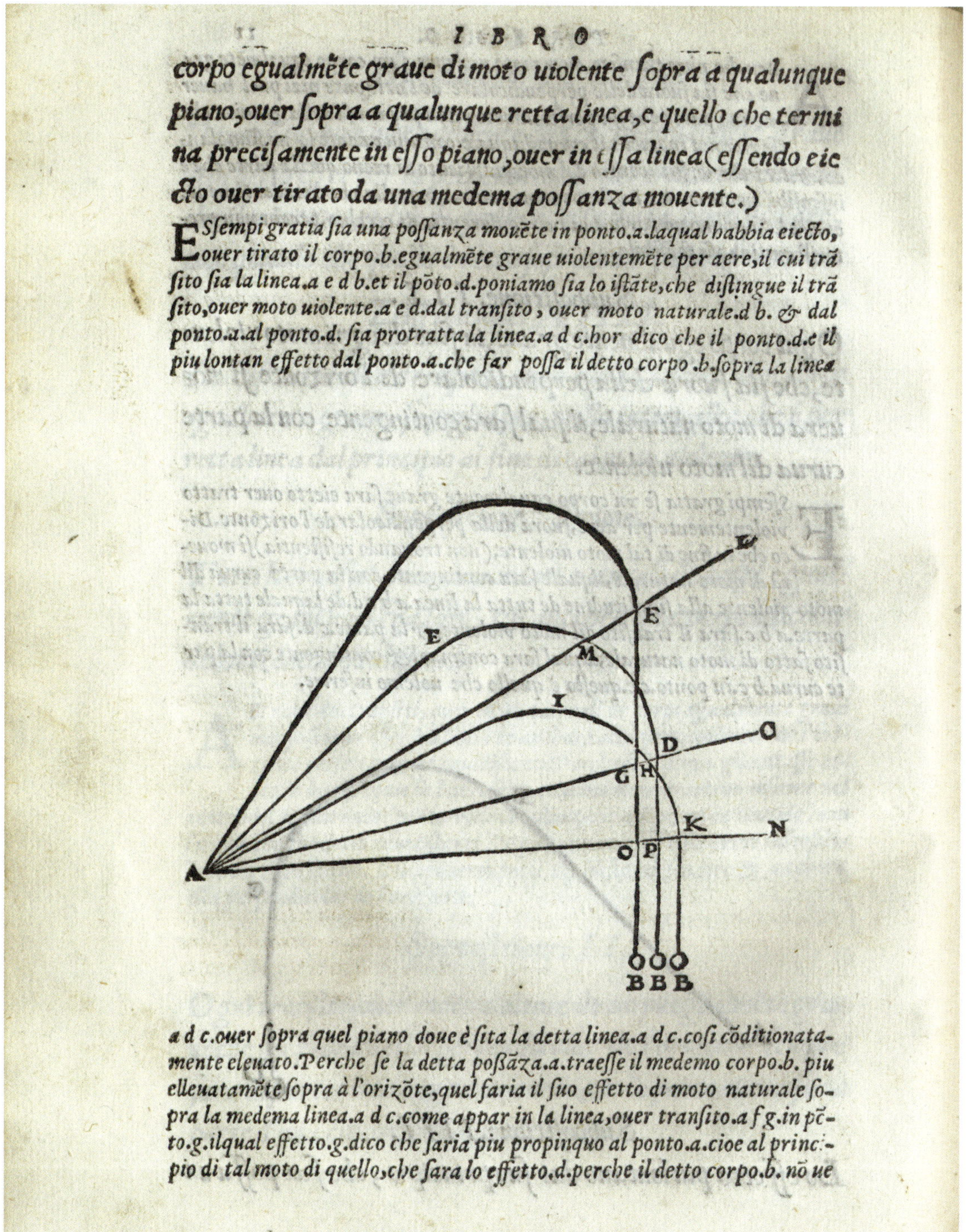

LIBRO

corpo egualmēte graue di moto uiolente ſopra a qualunque piano, ouer ſopra a qualunque retta linea, e quello che termi na preciſamente in eſſo piano, ouer in eſſa linea (eſſendo eie cto ouer tirato da una medema poſſanza mouente.)

ESſempi gratia ſia una poſſanza mouēte in ponto.a.laqual habbia eiecto, ouer tirato il corpo.b.egualmēte graue uiolentemēte per aere, il cui trā ſito ſia la linea.a e d b.et il pōto.d.poniamo ſia lo iſtāte, che diſtingue il trā ſito, ouer moto uiolente.a e d.dal tranſito, ouer moto naturale.d b. & dal ponto.a.al ponto.d. ſia protratta la linea.a d c.hor dico che il ponto.d.e il piu lontan effetto dal ponto.a.che far poſſa il detto corpo .b.ſopra la linea

a d c.ouer ſopra quel piano doue è ſita la detta linea.a d c.coſi cōditionata mente eleuato.Perche ſe la detta poſſāza.a.traeſſe il medemo corpo.b. piu elleuatamēte ſopra à l'orizōte, quel faria il ſuo effetto di moto naturale ſo pra la medema linea.a d c.come appar in la linea, ouer tranſito.a f g.in pō to.g.ilqual effetto.g.dico che ſaria piu propinquo al ponto.a.cioe al princi pio di tal moto di quello, che ſara lo effetto.d.perche il detto corpo.b. nō ue

Figure 9
Geometrically constructed trajectory in Tartaglia's *Nova scientia.*
Tartaglia [1558].

Figure 10 (*opposite*)
Realistic trajectories in the frontispiece of Tartaglia's *Nova scientia.*
Tartaglia [1558].

perspective the absurd representation of a trajectory by a geometrical patch work consisting of a straight line characterizing the beginning of the motion, a circular segment marking its culmination, and eventually a vertical line for the ending (fig. 9).[11]

However, just a glance at the title page of Tartaglia's book (fig. 10) shows that he, just as other contemporaries, was well aware of the continuous curvature of the trajectory.[12] This observation shows that a polemic separation between those who observed nature unprejudiced and those who engaged in audacious though speculative theoretical constructions makes little sense, and that the two sides could actually even be found in the work of a single individual. While modern historians debate whether progress was achieved by observation, mathematical construction, or accumulated practitioners' experience, the

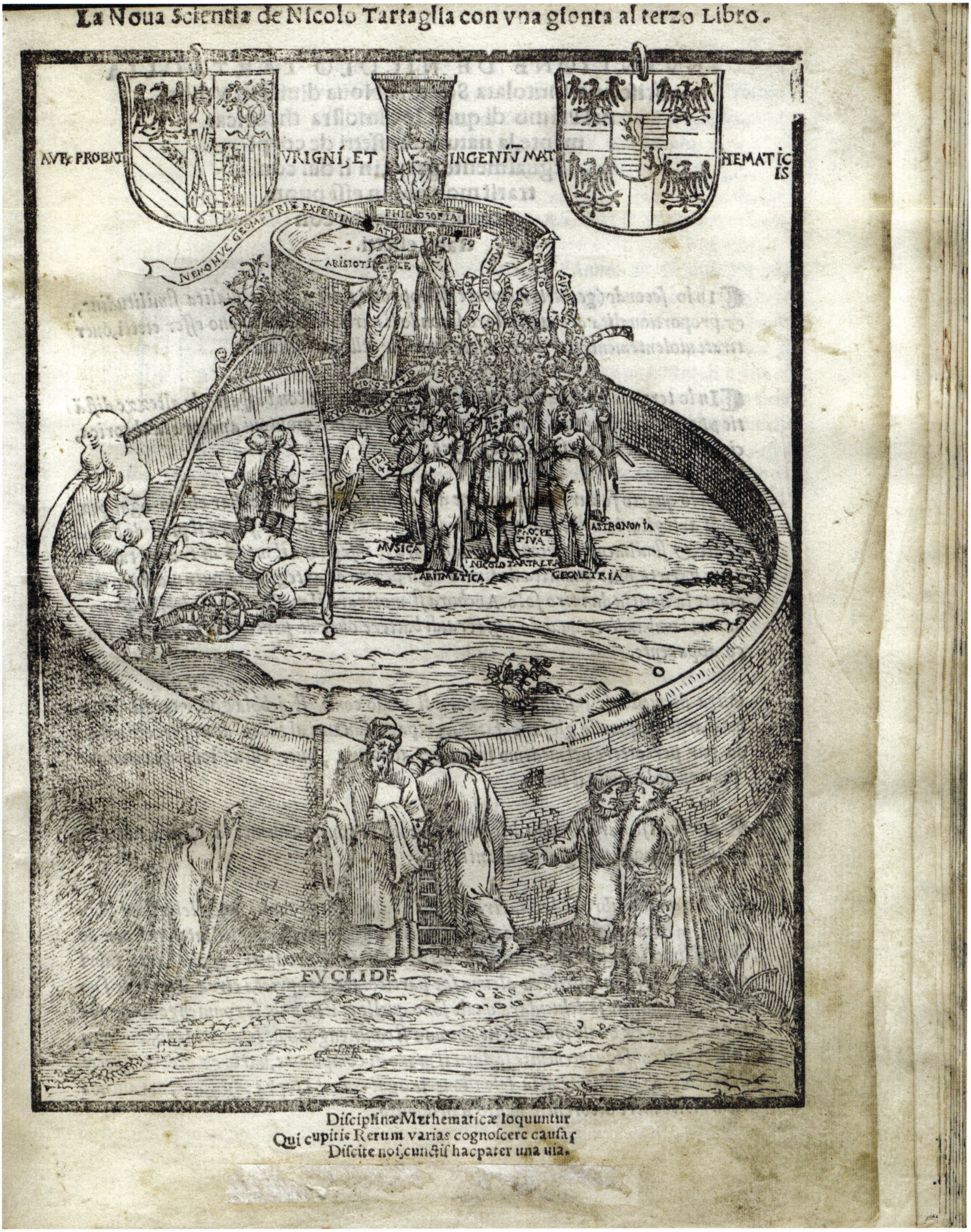
La Noua Scientia de Nicolo Tartaglia con vna gionta al terzo Libro.
PHILOSOPHIA
ARISTOTILE
PLATO
MVSICA
ARITMETICA
NICOLO TARTALEA
GEOMETRIA
ASTRONOMIA
EVCLIDE
Disciplinæ Mathematicæ loquuntur
Qui cupitis Rerum varias cognoscere causas
Discite nos, cunctis hac patet una uia.

problem for contemporary engineer-scientists such as Tartaglia was rather how to bring all this knowledge together in one coherent framework.

Tartaglia's geometrically shaped trajectory served, first of all, the calculative purposes of the practitioner, as in the case of Santbech's construction. But Tartaglia's ambitions went beyond offering quantitative rules for gunners. He aimed at nothing less than establishing a new science embracing all the wealth of knowledge accumulated by the practical tradition, for example the dependence of a shot on its elevation. Creating a new science in the early modern period implied moulding such knowledge into the scheme paradigmatically represented by Euclid's "Elements." The Aristotelian conceptualisation of motion represented another boundary condition that no attempt at a new science of projectile motion could avoid. It implied in particular the distinction between natural and violent motion which suggested to Tartaglia that the trajectory comprises straight parts that can be identified with these two types of motion. It could hence appear that a simple geometrical representation of the trajectory, as was desirable for both the purposes of practical calculation and theoretical construction according to Euclidean geometry, was also compatible with a dynamical justification of the trajectory. Of course, neither Aristotelian dynamics nor the knowledge of practitioners suggested a particular shape for the middle part of the trajectory between the two straight parts representing the initial violent and the final natural motion. But while Tartaglia's solution to represent this part by the segment of a circle may appear somewhat arbitrary, it did allow him to stay within the framework of Euclidean constructions.

Nevertheless, Tartaglia's ambitious endeavour to integrate the knowledge of the gunners within a deductive theory based on Aristotelian dynamics and Euclidean geometry failed. In his theory he was, as he had to admit, not only unable to represent even the simple fact that the trajectory is continuously curved. He also did not succeed in deriving the basic gunner rules he himself adduced, such as the existence of a maximum range. As a matter of fact, his theory did not put him in a position to solve the practitioners' problem of determining ranges for shots at different angles. None of the detailed features of his construction of the trajectory, such as its division into three parts and the relative proportion of these parts, could actually be derived from dynamical principles or matched with practical experience. Tartaglia's deductive framework could therefore hardly be more than rickety scaffolding for bringing together diverse elements of knowledge.

Adapting Constructions to the Shared Knowledge of Artillerists

Tartaglia's pioneering attempt to create a new science of ballistics did not, in spite of its failure, remain a solitary dead-end endeavour. Other contemporary engineer-scientists made similar attempts to derive rules for practitioners from constructions of the trajectory and attempted, vice versa, to adapt their constructions to the intuitive knowledge of the practitioners[13] (fig. 11). But their works were left in the shadow of Tartaglia's pioneering role and were neglected by those historians of science who focus exclusively on heroic contributions. Even if those lesser-known authors did not share Tartaglia's ambition to develop a new science they were nevertheless subject to the same framework of constraining conditions. Two examples may illustrate that the works of such authors offer an important source for reconstructing the practical knowledge of artillerists, which constitutes the shared knowledge underlying all contemporary theories however successful they may have been. These examples make it also clear to which extent any attempt to construct the trajectory was conditioned by this shared knowledge.

An example of a construction of the trajectory almost fully determined in its characteristics by rules and in accordance with practical knowledge is found in a painting (fig. 12) based on the work by Paulus Puchner,[14] an official of the Elector of Saxony serving as Master of Ordnance in charge of the armoury in Dresden, who lived a generation before Galileo. Puchner performed systematic experiments on artillery and developed an instrument for gunners based on his theoretical conception. His approach represents an advancement over that of Tartaglia with respect to the aim of solving the gunner's problem of how the range of a shot varies with its elevation. Just as in Tartaglia's case, the trajectories depicted are composed of three parts: a straight, obliquely ascending part followed by a curved part, both corresponding to forced motion, and finally a straight vertical drop corresponding to natural motion.

However, unlike Tartaglia, Puchner succeeded not only in giving a qualitative description of the trajectories but also in finding a geometrical construction that allows for quantitative statements. No accompanying text exists, but the construction on the painting together with a remark in the upper right corner: "Proportion of a square which measures 1500 Dresden cubits" suffices to reconstruct Puchner's idea. He inscribes the trajectories in a rectangle. From the intersection of the line of sight with the upper side of the rectangle a perpendicular is drawn which determines the range of the shot. Closer inspection also reveals the construction rule for the circular segments connecting the first part of the trajectory along the line of sight with the last part, the natural motion along the vertical line. The centres of the circles are all located on the diagonal running from the lower left to the upper right edge of the rectangle. No further rule seems to have been applied in constructing these curved parts of the motion.

Puchner's construction of the trajectory on the basis of his rectangle is obviously more specific than that of Tartaglia and offers a solution to the gunner's problem. However, although it provides a more realistic image of the trajectory, it resembles that of Santbech in assimilating the trajectory to a triangle. Puchner's construction thus shared a major deficiency with that of Santbech: if applied to angles below 45° it would result in shot ranges

Figure 11
Attempt to adapt the construction of projectile trajectories to the knowledge of the artillerists. Ufano, D. [1628].

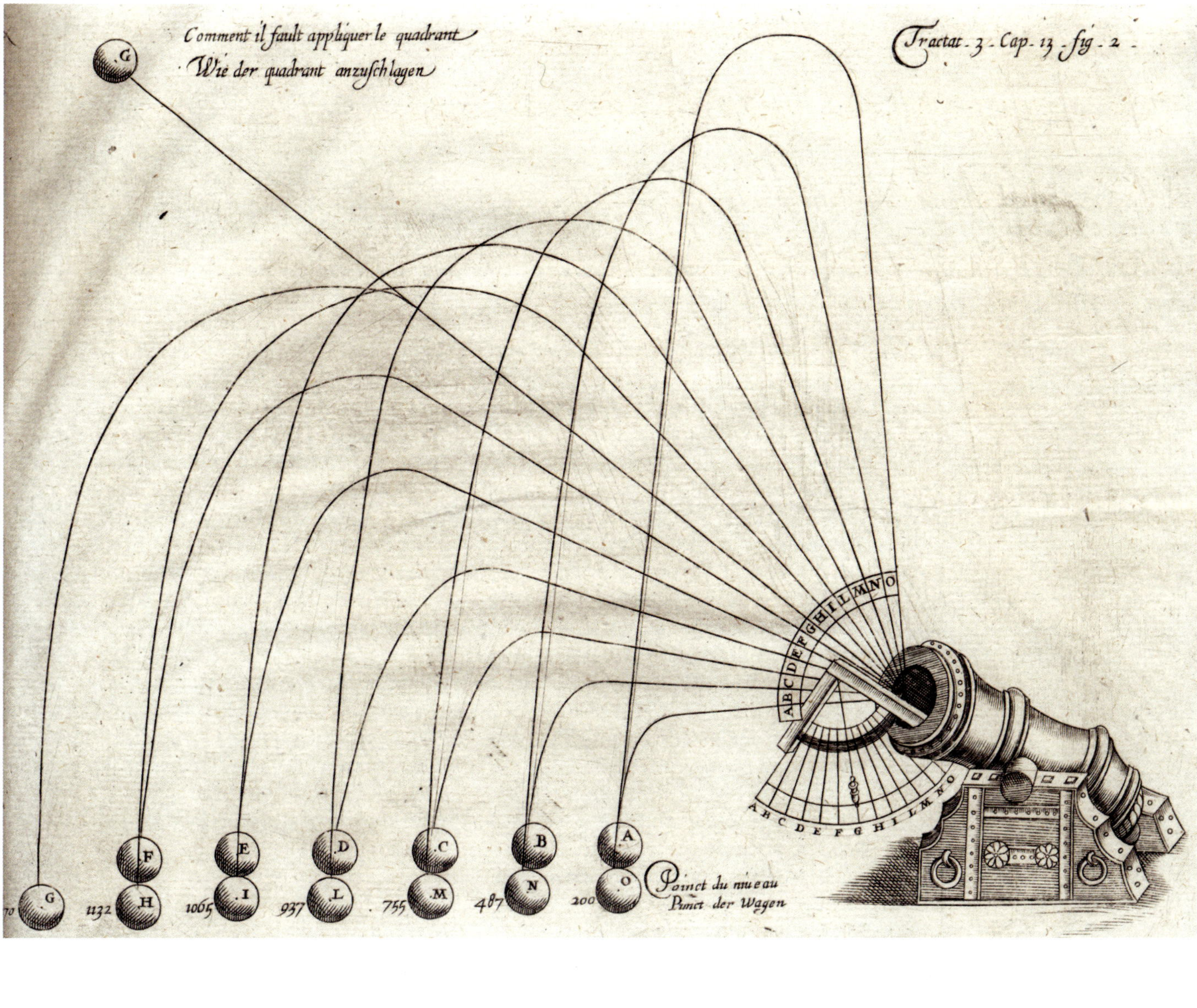

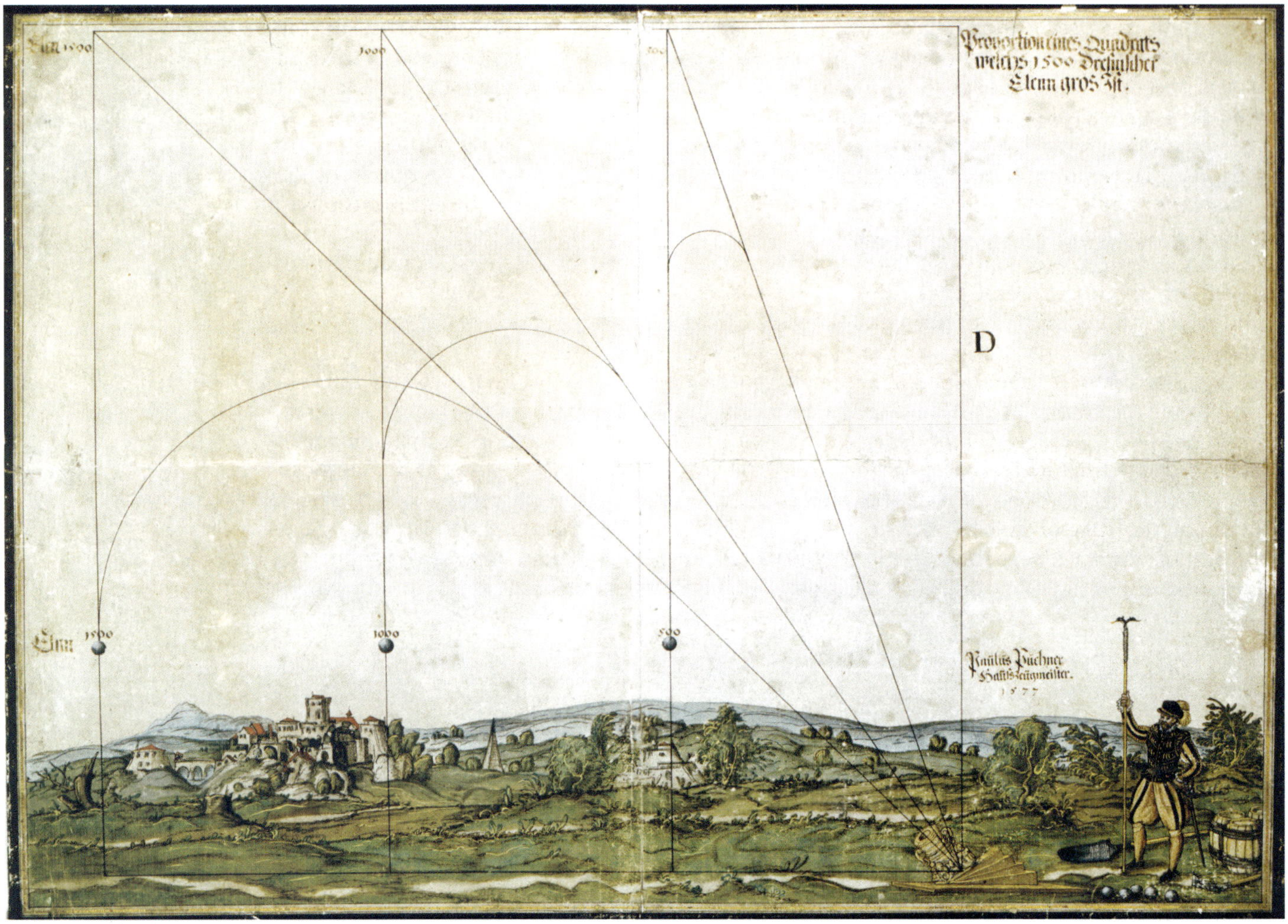

Figure 12
Painting with construction reflecting practitioners' knowledge according to Paulus Puchner. Mathematisch-Physikalischer Salon, Staatliche Kunstsammlungen Dresden. Ansicht mit Geschossbahnen von Paulus Puchner, o. Inv. Nr. Foto: Jürgen Karpinski.

monotonically increasing with decreasing angles of shot. However, in contrast to Santbech, Puchner makes use of the practitioners' knowledge of the artillerists, in particular of the insight that the maximum range of a shot is obtained for an elevation above the horizontal and of the insight that the same range may be reached with flat and steep shots. While his construction of the trajectory is not suitable for incorporating this knowledge geometrically, Puchner addresses this constraint by introducing the additional hypothesis of a symmetrical distribution of the ranges around a shot at an elevation of 45 degrees, the flattest covered by his construction. This is made evident by an aiming device designed by Puchner according to these principles (fig. 13).[15]

In summary, up to this point none of the examples considered sufficiently served both aims: to yield a complete and fully determined construction of the trajectory, and to cover the entire range of general rules suggested by practical knowledge available from artillery. Was it hence impossible to find a geometrical construction serving both aims simultaneously?

As a matter of fact, it turns out to be unnecessary to invoke fancy constructions of sophisticated modern mathematics to solve this problem. An undated Italian text of 39 pages, printed not long after 1565 and possibly authored by a certain Giusto Aquilone, makes it evident that this problem can actually be solved by a relatively simple construction complying with Euclidean geometry (fig. 14).[16] In his text, the author – let us in the following simply call him Aquilone – not only extensively discusses the implications of this construction of the projectile trajectory but also describes an instrument for gunners based on it. He furthermore adduces a report on an experiment performed in 1565 as confirmation of one of the implications of his theory.

Aquilone's construction results from his attempts to realize, with the simplest geometrical means, certain conditions compatible with the basic experiences of the artillerists. First, the trajectory is composed of a straight ascending part, a curved part, and a part representing natural vertically downward motion. Second, the initial straight motion is longest for a vertical shot and decreases with a diminishing angle of the shot. Third, the curved part is the segment of a circle and fits smoothly onto the first part. Fourth, the radius of the curved part decreases with an increasing elevation angle and equals zero for a vertically upward shot.

The construction is based on three circles, a large one called by the author "cerchio del movimento" in which two smaller circles of half the diameter are inscribed–the "cerchi della contingenze." Starting from these three

Figure 13
Puchner's aiming device. Mathematisch-Physikalischer Salon, Staatliche Kunstsammlungen Dresden. Viertelkreisinstrument, Inv. Nr. L V9. Foto: Jürgen Karpinski.

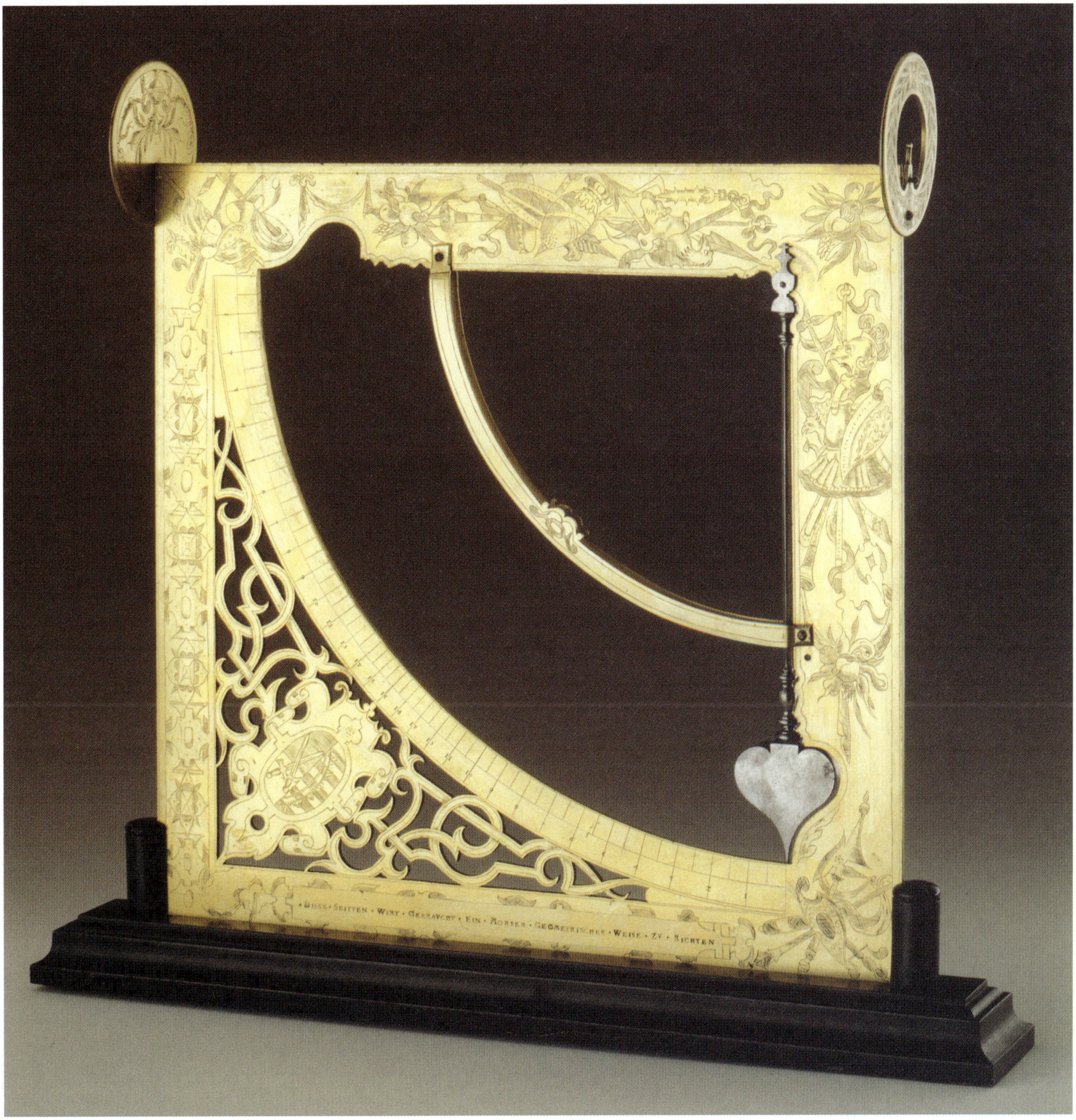

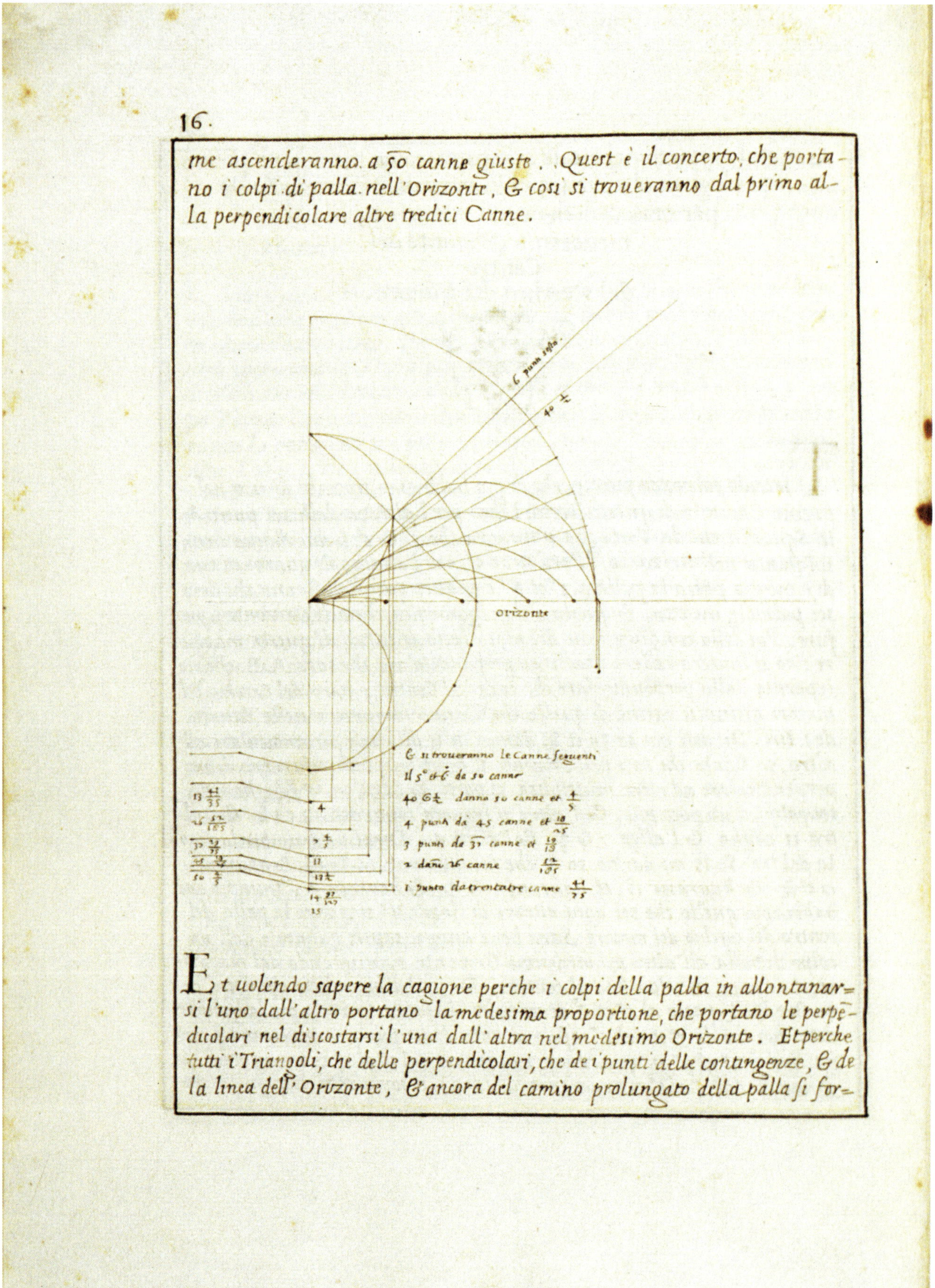

16.

me ascenderanno a 50 canne giuste. Quest è il concerto, che portano i colpi di palla nell'Orizonte, & cosi si trouerannno dal primo alla perpendicolare altre tredici Canne.

Et uolendo sapere la cagione perche i colpi della palla in allontanarsi l'uno dall'altro portano la medesima proportione, che portano le perpedicolari nel discostarsi l'una dall'altra nel medesimo Orizonte. Et perche tutti i Triangoli, che delle perpendicolari, che de i punti delle contingenze, & de la linea dell' Orizonte, & ancora del camino prolungato della palla si for-

circles the trajectories are constructed in the following manner. The first part of the motion proceeds along a straight line drawn from the centre in the direction of the shot to the intersection with the "cerchio della contingenze." The motion continues along a segment of a circle, the centre of which is determined by the intersection of a perpendicular erected at the endpoint of the straight part with the "cerchio del movimento." Where this circle intersects a horizontal line running through its centre, i.e. when the tangent is vertical, the circular movement ends and the motion continues vertically downwards.

This construction indeed implies all the characteristics of the motion of a projectile the author attributes to it. Among these implications there is also a surprising one, as Aquilone emphasizes:[17]

> *There is a dispute among the populace and it is a certain Tartaglia who has initiated it, and they say that shooting a ball with whatever piece of artillery at 45 degrees ... that it will go farther from the centre ... than in any other elevation. This is false ...*

In fact, Aquilone's construction leads, in particular, to the conclusion that the elevation angle for the maximum range deviates from that of Tartaglia. Whereas Tartaglia claimed this maximum range elevation to be at an angle of 45°, Aquilone's construction implies that it must be about 40,5°. This conclusion is, of course, incompatible with the standard theory of projectile motion in classical physics, which, for the unrealistic case of motion in a vacuum, happens to agree in this respect with Tartaglia's claim. But Aquilone's construction is actually more successful than textbook physics in capturing certain features of real trajectories and thus in matching the experience of the practitioners. While the elements of the trajectory may be more or less convincingly identified with categories of Aristotelian physics, such as the representation of the initial violent motion by a straight line, the author did not, however, even try to embed his construction in any theoretical framework based on detailed dynamic assumptions from which the trajectory can be derived.

Deriving Trajectories

The experience accumulated by practitioners and documented by the treatises and manuscripts of the engineer-scientists also became an object of study for those who strived to revise Aristotelian natural philosophy in attempts to create a new science of mechanics integrating the novel experiences of the time. Such attempts were no longer confined to treating artillery as an isolated subject, but brought it into contact with other, theoretically structured bodies of knowledge, in particular the tradition of theoretical mechanics inherited from antiquity, and the tradition of Aristotelian natural philosophy. As it turned out, geometrical constructions of the trajectory again played a key role in mediating between these different bodies of knowledge on the one hand, and the shared knowledge of the practitioners on the other. Two examples, taken again from the works of Harriot and Galileo, will be considered here. The first example is provided by the work of Thomas Harriot and demonstrates a powerful technique of deriving the trajectory from theoretical assumptions on natural and violent motions by plotting it pointwise. The second example is taken from the work of Galileo Galilei and illustrates an attempt to use the mechanical generation of a curve in order to establish a theoretical link between the trajectory and another novel object of contemporary mechanical enquiry, the hanging chain. Such attempts at integrating the practical knowledge of the artillerists with theoretical traditions by means of constructions of the trajectory eventually led, as the final section will make clear, to a revision of the underlying Aristotelian framework.

In the Aristotelian framework the trajectory must be conceived, as pointed out above, as being determined by a combination of two different tendencies, one causing a violent motion in the direction of the shot and the other causing a natural motion directed vertically downwards towards the centre of the earth. These basic characteristics of the framework do not, however, determine the actual shape of the trajectory. This makes it difficult to establish a concise link with empirical knowledge – based on the experiences of the artillerists or gained by experiments performed in order to answer the gunners' question – and thus to derive any

Figure 14 (*opposite*)
A full solution for the geometrical construction of the trajectory covering practitioners' knowledge.
National Library Florence. Prat. n. 438 mss.

consequences from empirical knowledge that might correct or modify the presupposed theoretical assumptions. In order to draw such consequences the gap between theory and experience had to be closed by precise rules prescribing how to apply the theoretical framework to constructions of the trajectory, that is by dynamical assumptions concerning the way in which the two tendencies determine its shape.

Constructions of trajectories resulting from such a specification of the dynamical assumptions may look quite similar to constructions resulting from simple geometrical translation of practitioner's knowledge as seen in the case of Puchner and Aquilone. However, contrary to such geometrical constructions, the details of a construction obtained by plotting or by mechanical generation are closely linked to specific theoretical interpretations. As to the precise characteristics of such assumptions, the generally accepted physical theory of the time left open a wide array of possibilities.

Taken by themselves, the manuscripts of Thomas Harriot, for instance, display a great variety of such specifications that result in slightly different constructions and slightly different interpretations of the same experiences of the practitioners. A very suggestive variant that can be found on one of the folio pages[18] (fig. 15) is based on the following specifications.

First, the violent upward motion of the projectile decelerates in such a way that in each unit of time one unit of space less is traversed in vertical direction; the vertical

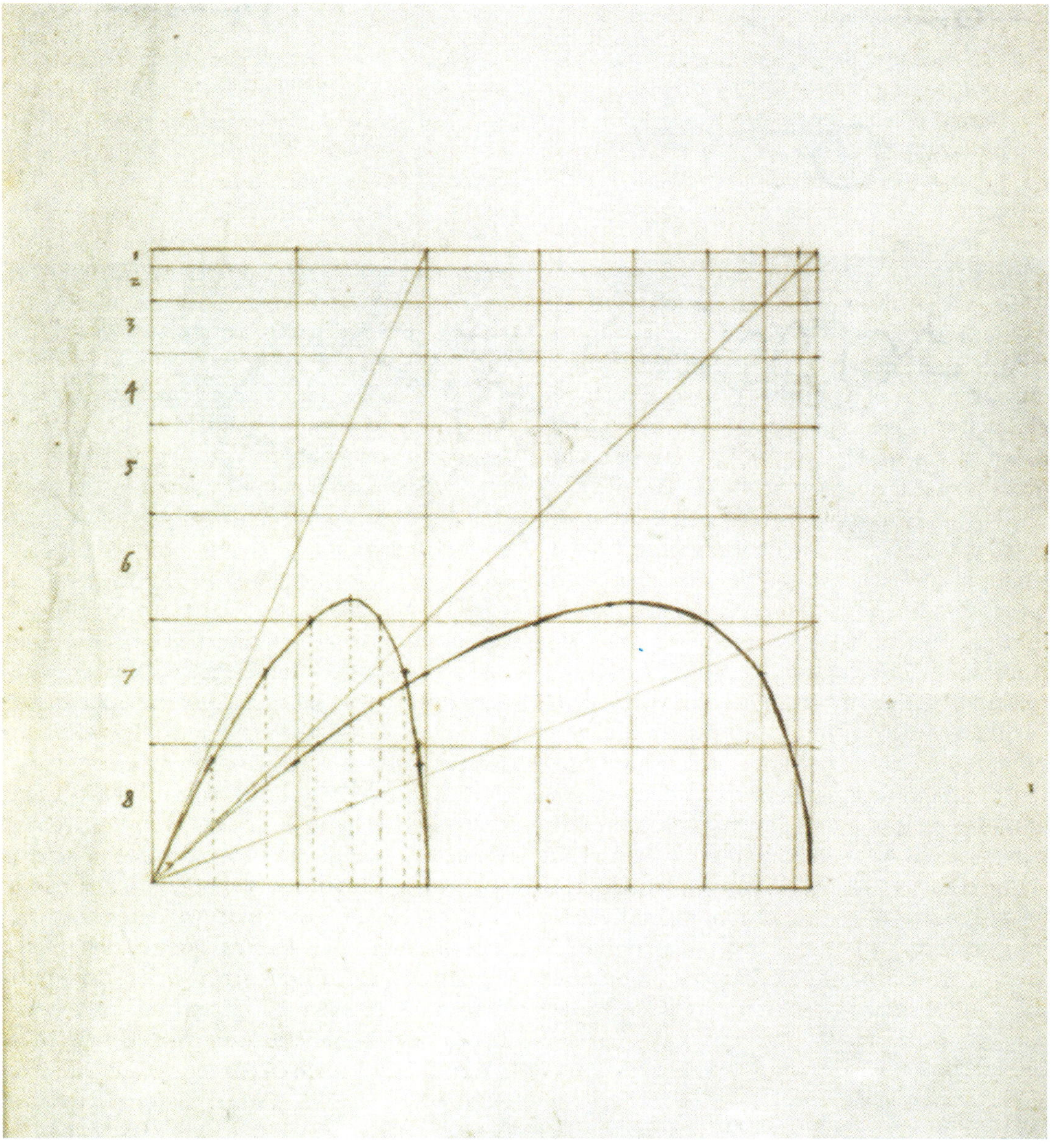

Figure 15
Plotted trajectories in a manuscript of Thomas Harriot.
British Library: Add. MS 6789 fol. 5r.
Galileo's use of a hanging chain as a model of the projectile trajectory.
Galileo Ms. Gal. 72 fol.43r.

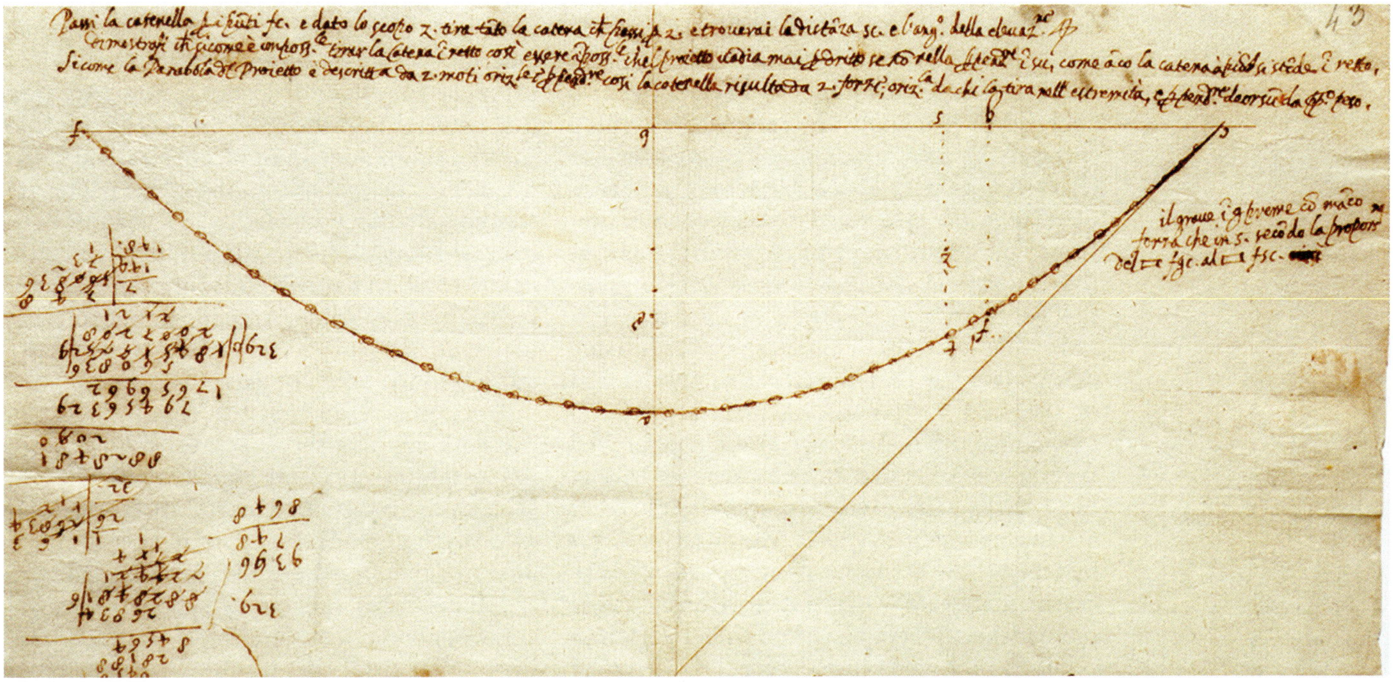

Figure 16
Galileo's use of a hanging chain as a model of the projectile trajectory. Galileo Ms. Gal. 72 fol. 43r.

distances hence decrease in an arithmetical sequence. As a consequence, the violent tendency in the direction of the shot correspondingly decreases in an arithmetical sequence. Second, the projectile moves additionally in a natural downward motion towards the centre of the earth and is accelerated also according to an arithmetical sequence. Third, all shots in whatever direction reach the same height in the same time so that they can be constructed within the same rectangular scheme using the same grid of horizontal lines. As a consequence, the range of the shot is determined by the intersection between the direction of the shot and the uppermost horizontal grid line of the rectangular scheme, below which point the projectile hits the ground.

The overall characteristics of Harriot's construction resulting from these rather arbitrary specifications of the Aristotelian theory resemble those of Puchner's construction based on a rectangle. Just as in Puchner's case, it only makes sense to construct trajectories for elevation angles larger or equal to 45 degrees because for lesser angles, the implication of the construction that flatter shots yield wider ranges becomes absurd. While in Puchner's case, however, the precise shape of the trajectory remained unexplained and could only be obtained by adding rough assumptions about its main geometrical features, as they were used in the well-known work of Tartaglia, Harriot succeeded in determining the complete curve of the trajectory by using a point-by-point construction based on rules that followed his simplistic dynamical assumptions. The fact that this construction did not allow the deduction of the main features of practitioners' knowledge, such as the non-monotonous variation of the range with the elevation of a shot, evidently motivated Harriot to explore alternative dynamic assumptions as is widely documented in his manuscripts.

Extant manuscripts by Galileo documenting his study of motion and mechanics over a period of more than forty years make it evident that he similarly struggled for a pointwise construction of the trajectory that would reproduce the knowledge of the practitioners on the basis of dynamical assumptions on the components of projectile motion. For him such dynamical assumptions were closely related to the understanding of other mechanical processes and objects such as the motion of fall, the motion along an inclined plane, the oscillation of a pendulum, and the hanging chain. In particular, he was convinced that the shape of the trajectory of projectile motion and the curve of a hanging chain must be the same because they are produced by the same constellation of violent and natural tendencies, one acting along the horizontal, the other along the vertical. From a modern point of view, both Galileo's assumption of the geometrical identity of the two curves and the supposed equivalence of their dynamic generation are evidently wrong. In fact, checks of the presumed identity of the two curves that can be found in his manuscripts made deviations between the two curves rather obvious. Nevertheless, Galileo clung to his erroneous assumption that the chain line is, just as the trajectory, a parabola. He even attempted to derive the presumed parabolic shape of the chain line from the dynamical assumptions underlying his theory, as his manuscripts also illustrate.

In one of these manuscripts,[19] probably dating from a period after the publication of the first edition of his major work on mechanics, the Discorsi, Galileo sketched several ideas related to the presumed identity between the trajectory and the curve of a hanging chain (fig. 16). He described, for instance, how a chain might be used to represent a trajectory so that artillerists could predict the angle of elevation to be chosen in order to hit a given target. He also sketched how the dynamical equivalence between a chain and trajectory may be used to demonstrate that projectile motion never proceeds in a straight line unless aimed perpendicularly. And he also jotted down the key idea of a proof aiming at deriving the presumed parabolic shape of the chain line from his theory of the rigidity of beams by conceiving the chain as an infinitely elastic beam bending under its own weight.

Even though the connection between trajectory and chain that Galileo established in his work turned out to be problematic from a later perspective, it was precisely such attempts at embedding the new experiences of artillerists in a wider framework of mechanics and natural philosophy that overturned the traditional conceptual roots of this framework. As for Galileo, his erroneous conviction that he could demonstrate the equivalence of parabolic trajectory and chain line provided him with a strong argument for claiming, against a firmly rooted and traditional doctrine, that the projectile trajectory has a symmetrical shape, a claim he could not prove otherwise.

Towards Classical Mechanics

In view of traditional historiography, it may be surprising to find Galileo's famous contributions to classical mechanics discussed in the context of the contributions of relatively unknown contemporaries such as Thomas Harriot or Giusto Aquilone. The preceding example has made it evident, however, that his conceptualisation of projectile motion showed similar deviations from classical physics, as did the understanding of his peers. He shared with them, in particular, the same conceptual roots of his understanding of motion in Aristotelian natural philosophy and he still lacked, as they did, a principle of inertia from which to derive the trajectory of projectile motion in the sense of classical physics.

Does this mean that the true founders of classical physics were yet to come? Were the essential concepts of classical physics, such as the concept of inertia or the concept of instantaneous velocity, only coined much later, for instance by Isaac Newton or Leonard Euler? But then where did these novel concepts come from and how, if at all, were they related to the traditional conceptual understanding? It turns out that diagrams played a key role as a bridge between the traditional and the novel organization of knowledge. While concepts, terms, and arguments changed, diagrams were often retained, albeit with a different interpretation. Their potential to incorporate the knowledge necessary for their construction while remaining amenable to a reinterpretation, made them a powerful medium of innovation.

This function of diagrams may be illustrated by a diagram on one of the first pages of the Fourth Day of Galileo's *Discorsi*, dedicated to projectile motion, a diagram that has been emended in Galileo's hand copy of the first edition of the book (fig. 17).[20]

The original diagram represents the parabolic trajectory of a body which is horizontally projected. The construction of the trajectory follows the specific procedure Galileo always uses for generating such trajectories. The horizontal component of the motion is assumed to be caused by the vertical fall of the body along a certain distance called "sublimity," followed by deflecting this motion into the horizontal so that it would pursue a uniform motion if supported by a horizontal plane. Letting the horizontally moved body drop down again generates the parabolic trajectory of horizontal projection that Galileo conceives of as resulting from the superposition of a uniform violent motion along the horizontal and an accelerated natural motion along the vertical.

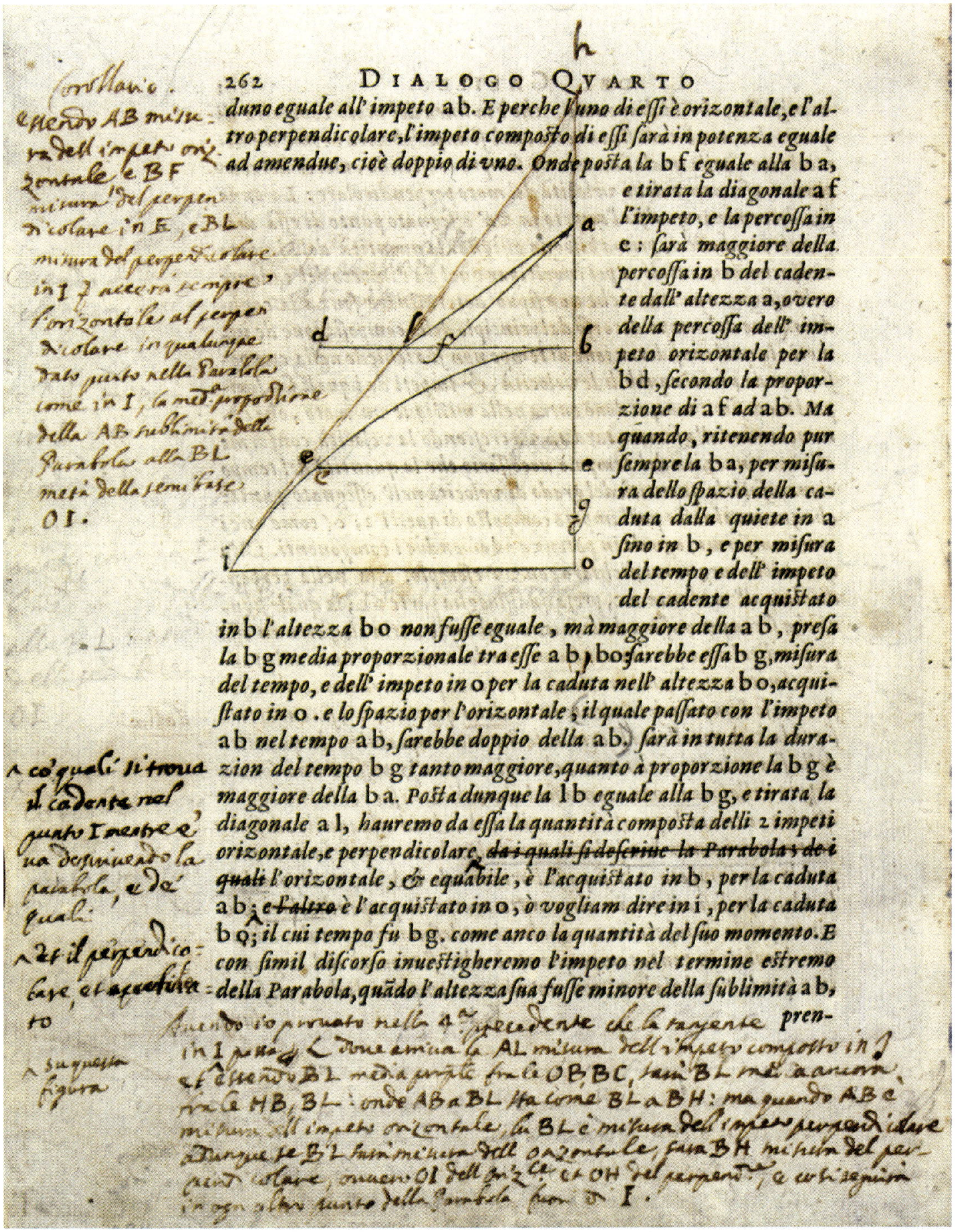
262 DIALOGO QVARTO

duno eguale all' impeto ab. E perche l'uno di essi è orizontale, e l'altro perpendicolare, l'impeto composto di essi sarà in potenza eguale ad amendue, cioè doppio di uno. Onde posta la bf eguale alla ba, e tirata la diagonale af l'impeto, e la percossa in c: sarà maggiore della percossa in b del cadente dall' altezza a, overo della percossa dell' impeto orizontale per la bd, secondo la proporzione di af ad ab. Ma quando, ritenendo pur sempre la ba, per misura dello spazio della caduta dalla quiete in a sino in b, e per misura del tempo e dell' impeto del cadente acquistato in b l'altezza bo non fusse eguale, mà maggiore della ab, presa la bg media proporzionale tra esse ab, bo sarebbe essa bg, misura del tempo, e dell' impeto in o per la caduta nell' altezza bo, acquistato in o. e lo spazio per l'orizontale, il quale passato con l'impeto ab nel tempo ab, sarebbe doppio della ab, sarà in tutta la durazion del tempo bg tanto maggiore, quanto à proporzione la bg è maggiore della ba. Posta dunque la lb eguale alla bg, e tirata la diagonale al, hauremo da essa la quantità composta delli 2 impeti orizontale, e perpendicolare, ~~da i quali si descrive la Parabola; de i quali~~ l'orizontale, & equabile, è l'acquistato in b, per la caduta ab; ~~e l'altro~~ è l'acquistato in o, ò vogliam dire in i, per la caduta bo; il cui tempo fu bg. come anco la quantità del suo momento. E con simil discorso investigheremo l'impeto nel termine estremo della Parabola, quãdo l'altezza sua fusse minore della sublimità ab, pren-

Figure 17
Emended diagram in Galileo's hand copy of his major work on mechanics, the *Discorsi*.
Galileo Ms. Gal. 79 fol. 157v; emended figure of Galileo [1638] 262.

In the context of a discussion about the first problem of the Fourth Day, that is, the problem of determining the "impetus" of the projected body in every moment, the diagram is used to find this impetus by a geometrical construction. While Galileo initially assumed that to get this impetus, the horizontal and the vertical impetuses have simply to be added, when writing his *Discorsi* he eventually compounded them by adding their squares and then extracting the square root. This rule could easily be represented geometrically by constructing right-angled triangles. Galileo thus added two such triangles representing the impetus at two different points on the trajectory at the top of the diagram, the sublimity being used hereby as a common side and as a common measure of the impetus at the two points. In the framework of Galileo's theory and in the context in which the diagram occurs in the text, the drawing is adequate and entirely correct.

However, in Galileo's hand copy of the *Discorsi* the diagram is nevertheless emended by additional lines together with a number of marginal notes written in the hand of his disciple Viviani. The first of the additional two lines is the tangent in one of the points of the trajectory for which the impetus is determined, and the second is the extension of the sublimity up to the point of intersection with the tangent. In the marginal notes describing the construction, it is explained that the new triangle constituted by the tangent, the perpendicular, and the horizontal is similar to the one used to determine the impetus so that it can also be used for the same purpose.

A further marginal note located on the next page is the most surprising element of the emendation. It contains the statement that the projected body would continue to move straight down in the direction of the tangent if it ceased to accelerate itself in the direction of the perpendicular. This is obviously a statement of the principle of inertia that does not occur in any of the published works of Galileo.

What is the story behind this emendation? Either Viviani or Galileo himself who, due to his fading sight in later years, used to have his disciples write down his ideas in the context of a reworking of the *Discorsi* for a new edition, realized the similarity of the triangle used in the *Discorsi* to determine the impetus and the triangle resulting from the emendation. This simple geometrical insight prepared the ground for a new physical interpretation of the construction that was actually realized by the author of the notes. In fact, such a simple emendation of the drawing was, in the long run, capable of triggering a cascade of discoveries with profound conceptual consequences.

It turned out, first of all, that the compounded impetus at a given point of the projectile motion can be geometrically represented by a line always lying in the direction of the tangent to the trajectory at that point. This discovery, documented in the marginal notes, made it in principle possible to devise a method for determining the tangent to any curve that can be conceived as being generated by two motions, and eventually also to conceive the impetus as the result of a derivative operation on such a curve. This insight thus opened up one developmental path along which the concept of impetus mutated to the concept of instantaneous velocity in classical physics.

But the insight that the momentary compounded impetus of the projectile motion has a direction, that of the tangent of the parabolic trajectory, had even further reaching consequences for the conceptualisation of motion. This insight made it indeed plausible to assume that, if the compounded impetus did not change, the motion would proceed along the tangent, and hence to interpret the actual projectile motion as a deviation from this uniform motion along the tangent. But since the change of the impetus was due to the motion of fall caused by the projectile's gravity, this consideration furthermore suggests that if gravity hypothetically ceased to act, the projectile would not move along the parabola but along the tangent. As speculative as this idea must have sounded to the contemporaries, it made it conceivable to construct the parabola for oblique projection as resulting from the composition of a uniform motion along the tangent and the accelerated motion of fall. This consideration thus opened up a developmental pathway along which the conceptualisation of projectile motion as being caused by a violent impetus mutated to its explanation in terms of the principle of inertia within the framework of classical physics.

Evidently the marginal notes to Galileo's *Discorsi* did not realize the full potential of the reinterpretation of the construction which essentially would have made it necessary not only to emend but also to completely rewrite his theory of motion. But the transformation of pre-classical mechanics to classical mechanics was anyway not a matter of individual discoveries. In the works of Galileo's disciples and contemporaries similar instances are displayed showing how the figurative representations of pre-classical mechanics, embodying the accumulated knowledge of the engineer-scientists of the early modern period, eventually turned into building blocks of modern physics.

Max Planck Institute for the History of Science

Notes

[1] Heath [1956] 154.

[2] Descartes [1965] 190.

[3] Guidobaldo del Monte [ca. 1587–1592] fol. 7 and fol. 8.

[4] See, for example, the artillery plate reproduced in the 1628 French edition of Diego Ufano's widespread treatise on artillery, first published 1612 in Spanish.

[5] Leonardo da Vinci [1974] fol. 147r.

[6] Harriot Add. Ms. 6789 fol. 60r.

[7] Galileo [ca. 1590] Ms. Gal. 71 fol. 124v.

[8] Santbech [1561] 213.

[9] Jähn [1889–1891] I 226; for a similar recent assessment see Arend, p. 82.

[10] Santbech [1561] 211.

[11] Tartaglia [1558] 11.

[12] Tartaglia [1558], frontispiece.

[13] Ufano [1614] Tractat 3, Cap. 13, Fig 2.

[14] See Wunderlich [1977].

[15] Schillinger [1994] 61.

[16] Aquilone [after 1565] 16.

[17] Aquilone [after 1565] 12f.

[18] Harriot Add. Ms. 6789 fol. 5r.

[19] Galileo [ca. 1602–1637] Ms. Gal. 72 fol. 43r.

[20] Galileo Ms. Gal. 79 fol. 157v; emended figure of Galileo [1638] 262.

References

Aquilone, G. ? n.d.
[after 1565] Ms. II, IV, 337 (*Della formatione de tre cerchi, ne quali si fonda tutto il transito, che fa la palla tirata per aria in qual si voglia elevatione di Artigleria*). Biblioteca Nazionale Centrale Firenze.

Arend, G.
1998 *Die Mechanik des Niccolò Tartaglia im Kontext der zeitgenössischen Erkenntnis- und Wissenschaftstheorie*. München.

Descartes, R.
1965 *Discourse on method, optics, geometry and meteorology*. Indianapolis.

Galileo.
ca. 1590 Ms. Gal. 71 (manuscript *De Motu*). Biblioteca Nazionale Centrale Firenze.

ca. 1602–1637 Ms. Gal. 72 (manuscript with notes on motion). Biblioteca Nazionale Centrale Firenze.

1638 *Discorsi e dimostrazioni matematiche: intorno à due nuoue scienze attenenti alla mecanica i movimenti locali*. Leyden.

after 1638 Ms. Gal. 79 (Galileo's hand copy of his Discorsi). Biblioteca Nazionale Centrale Firenze.

Guidobaldo del Monte
ca. 1587–1592 Lat. 10246 (*Meditantiunculae Guidi Ubaldi e marchionibus Montis Santae Mariae de rebus mathematicis*). Bibliothèque Nationale de Paris.

Harriot, T.
Add. Ms. 6789. British Library.

Heath, S. T. L.
1956 *The Thirteen Books of Euclid's Elements*. New York.

Jähn, M.
1889-1891 *Geschichte der Kriegswissenschaften*. München.

Leonardo da Vinci
1974 *Codex Madrid I: tratado de estatica y mechanica en italiano*. Frankfurt am Main

Santbech, D.
1561 *Problematum astronomicorum et geometricorum sectiones septem*. Basel.

Tartaglia, N.
1558 *La nova scientia de Nicolo Tartaglia*. Florence

Ufano, D.
1628 *Artillerie, ou vraye instruction de l'artillerie et de ses appartenances*. Berthelin.

Schillinger, K.
1994 *Kostbare Instrumente und Uhren aus dem Staatlichen Mathematisch-Physikalischen Salon Dresden*. Leipzig.

Wunderlich, H.
1977 *Kursächsische Feldmeßkunst, artilleristische Richtverfahren und Ballistik im 16. und 17. Jahrhundert*. Berlin.

Ships, Science and the Three Traditions of Early Modern Design

David McGee

Some years ago, anthropologist Jack Goody drew attention to the persistence with which we use binary and mental categories to explain large-scale historical change. Savage and civilised, traditional and modern, logical and pre-logical, these were the kind of pairings Goody had in mind, each of which has been used to assign different mentalities to those living on either side of such Great Divides as the Urban Revolution in Ancient Mesopotamia, the birth of "rationalism" in fourth century Greece, and the Scientific Revolution in Early Modern Europe. The same sort of story is told about a "scientific revolution in technology" in the nineteenth century, when technical knowledge is thought to have been uprooted from its matrix in centuries-old craft traditions, traditional techniques of cut and try empiricism were replaced by scientific methods, and craftsmen replaced by engineers as the chief agents of progress.[1] "Craft" is opposed to "science" on either side of a great historical shift. The "rational" methods of science are opposed to what, by implication, are the "irrational" habits, customs, and conservatism of craftsmen. On the basis of such terminology, the shift from a tradition-bound technics to a modern technology based on mathematics and physical science has been described as "radical transformation in the human personality," to quote Lewis Mumford's *Myth of the Machine*, as if it produced an actual change in human nature (Mumford [1966] 3). Many others have written about the advent of scientific technology in the same way.[2]

This craft-to-science formulation presents a number of historiographical problems for those interested in understanding the science-technology relationship, over and above the simplistic explanation it offers for what were clearly complex historical changes. But perhaps the most important problem is the concept of "craft" itself. On the one hand, as a sort of sub-category of mentality, craft is often regarded as set of skills or techniques literally embodied in the craftsman, a matter of tacit knowledge, and thus unknowable by definition. On the other, craft is frequently used to refer to all pre-modern technologies together, as belonging to a single category, thereby leaving us without the tools we need to tell the difference between them. Of course, we *know* there is a difference between making a pot and building a Renaissance palace. But we are unable to state what the difference is because the single category of craft fails to provide us with the necessary criteria. Without an understanding of the salient differences, I argue, it will be impossible to identify the relevant features of different contexts in which different techniques were employed, and consequently to identify those factors that either promoted or prevented the use of science in early modern technology.

Goody's approach to such explanatory problems was to search for those "technologies of the intellect" – numbers, writing, lists, tables, printing presses, computers – that the same group of people could use to make striking new achievements. Given the presence

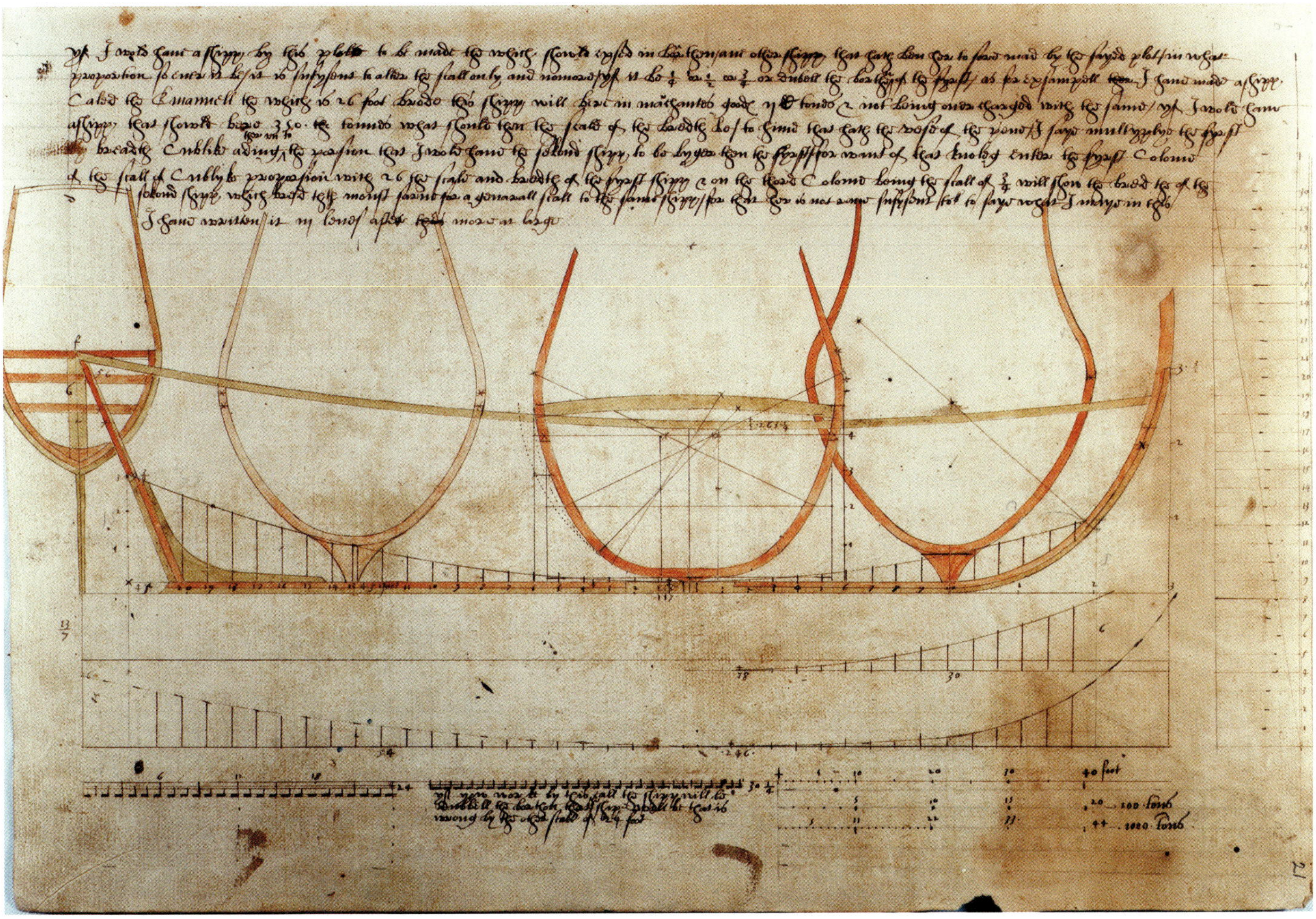

Figure 1
The first measured three-view plans used in naval architecture. From Mathew Baker's *Fragments of English Shipwrightery* of 1586, unpublished manuscript. Courtesy of the Pepysian Library, Magdelene College, Cambridge.

or absence of such technologies, Goody suggested, historians could explain apparently radical shifts in the development of civilisation, not by referring to new minds, but to new intellectual tools that allowed the old minds to accomplish a great deal more than they previously could. In this paper I adopt Goody's strategy by using criteria of graphic representation and drawing to identify three broad approaches to design in the early modern period. I refer to them as the craft, mechanical and architectural traditions. Understanding the nature of each tradition as well as the difference between them, I argue, provides a more fruitful basis for investigating early modern technology than the general category of craft. In particular, it leads to a recognition of special features of the architectural approach, where the use of measured plans allowed the application of mathematical, physical theory in design long before the modern era. It is by comparing and contrasting the three traditions through an investigation of the links between design, drawing and science that I believe we can begin to explain both the promise and the problems of science in early modern technology.

To illustrate the argument, I rely on examples from early modern shipbuilding and especially from British naval architecture prior to 1800. There are four reasons for this. One is that naval architecture was the first early modern field of technology to adopt the use of measured plan drawings after architecture itself, and the only one to do so before approximately 1750. Second, the use of measured plans in shipbuilding originated in Britain around 1580. Third, the use of plans continued to develop there over the next 200 years as the Royal Dockyards grew into the largest industrial organisation in the world.

Fourth, naval architecture was the object of a sustained attempt on the part of some of Europe's best minds to apply quantitative physical theory in design, and yet the attempt

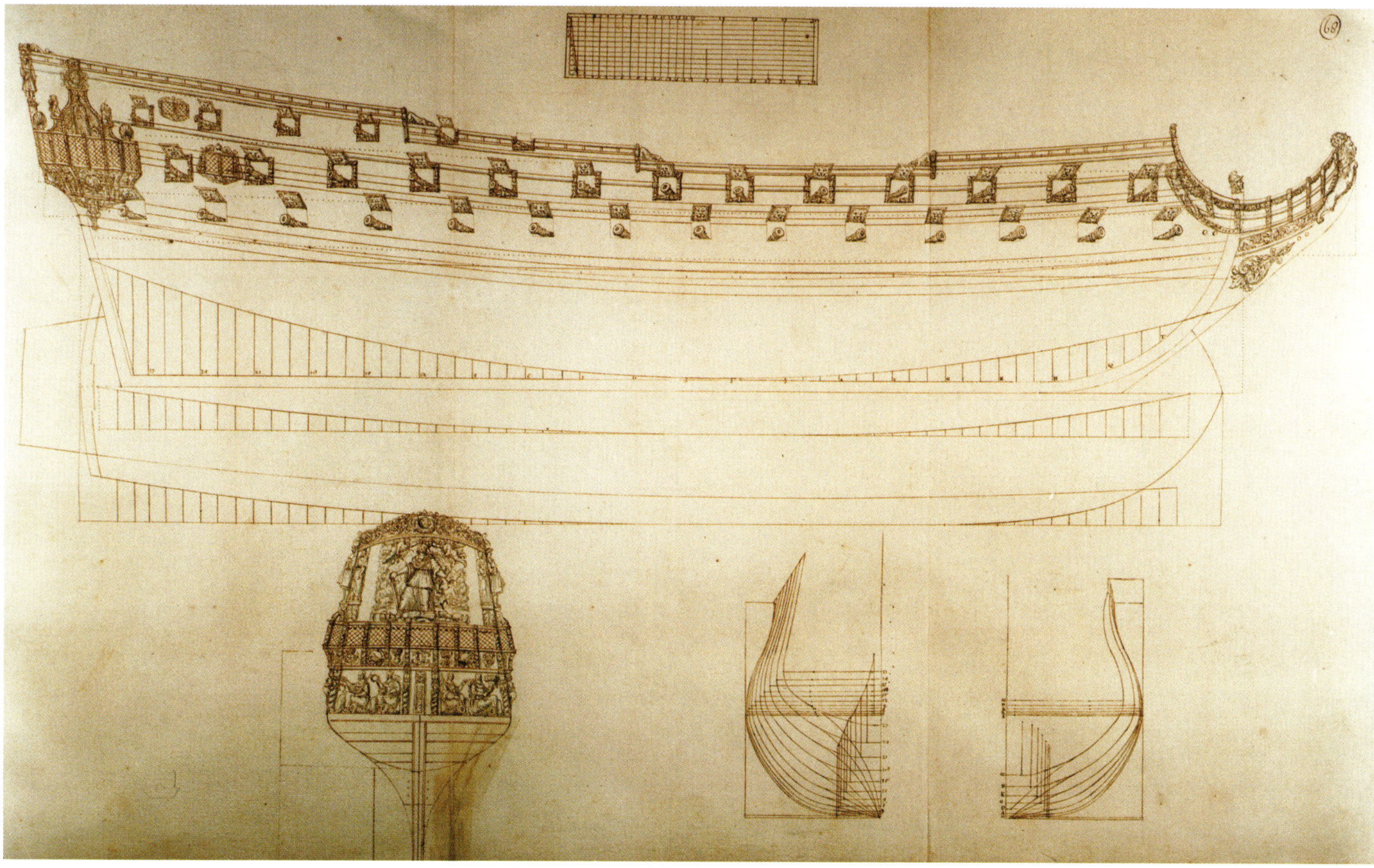

Figure 2
Sheer and body plans from Sir Anthony Deane's *Doctrine of Naval Architecture* (London 1670). Courtesy of the Pepysian Library, Magdelene College, Cambridge.

met with strikingly limited results.[3] For all these reasons, British naval architecture provides an excellent opportunity for investigating the conditions that led to the use of science in design, as well as the reasons for the limited success of the scientific approach in the early modern period.

Design and Drawing

It will be helpful to start by setting out a preliminary framework concerning the role of graphic representation in technological design. One of the best places to begin is still the account put forward by design methodologist J. Christopher Jones, for whom modern design techniques are only the continuation of a tradition of "design by drawing" that emerged in the Renaissance, superseding a "craft" approach in which no drawings were used.

According to Jones, this shift from "craftsmanship to draughtsmanship" took place because drawings offered designers advantages with respect to both construction and innovation. In construction, the use of drawings to specify dimensions in advance made it possible to *split up production work* into separate pieces made by different people. The resulting division of labour then made it possible to plan the construction of things that were *too big for a single craftsman* – like large ships and buildings – since it is only when critical dimensions are fixed in advance that parts made by different workers can be made to fit together. The division of labour also permitted an increase in the *rate* of production, since a product that would take several days for a single craftsman to make could be split up into smaller components made by several people at the same time (Jones [1970] 15-22). In short, the use of drawing made possible a *less costly* mode of production for more complicated artefacts. Interestingly, Jones illustrated his argument with a photo of a ship's plan from 1670 drawn by English naval architect Sir Anthony Deane.[4]

As for innovation, Jones argued that drawing permitted the design of *more complex* artefacts, because of the capacity of paper to store tentative decisions concerning one part, while another part was dealt with. More importantly, the use of drawings permitted a *faster pace of innovation*, since a designer working only with the geometrical aspects of artefacts on paper did not have to deal with real constraints of materials, labour, or physics, and was therefore "not prevented, either by partial knowledge, or the high cost of altering the product itself, from making fairly drastic changes in design" (Jones [1970] 22). Drawing and re-drawing in the absence of constraints allowed individual forms of greater novelty to be developed more quickly, and thus the faster evolution of new forms over time – a process Jones and others have frequently referred to as the transference of the "trial and error" process of craft onto paper.

A third aspect of Jones' thinking is extremely important, for he also observed that the use of drawings in advance of production makes *prediction* a fundamental problem. Designers, he wrote, are "forever bound to treat as real that which exists only in an *imagined* future, and to specify the ways in which the foreseen thing can be made to exist" (Jones [1970] 10f). They are forced to extrapolate from the past to the future on the basis of their knowledge of present conditions. But, Jones continued, since conditions might well change by the time an artefact is actually made, design unavoidably entails a large element of risk and uncertainty.

Jones' framework is useful for several reasons. It moves away from the psychological interpretation of design that one finds in much of design theory, identifying the determination of dimensions, not the conception of ideas, as the important question.[5] This yields a modest definition of design as *the process by which artefacts get the dimensions they actually have* and the central question becomes one of *when* final dimensions are finally determined.[6] Particularly important is the fact that Jones' model also provides criteria for distinguishing between at least two early-modern approaches to design, one which used drawings and one which did not. It further identifies cost as a potentially crucial contextual factor in determining which approach will be used in what situation. Finally, it identifies the problem of predictive uncertainty as a possible reason to adopt a scientific approach.

As we shall see, however, Jones' framework requires modification, the most important being that, using his own criteria of graphic representation, we can identify not two traditions of early modern design, but three.

The Craft Tradition

The first tradition of design to be identified on the basis of graphic representation is one in which no graphic representations are used at all, at least in its purest form. This we can refer to as "craft" proper, the method by which the vast majority of artefacts were made prior to 1800.

Perhaps the most crucial feature of "craft" is that craftsmen work immediately with their materials while actually making an artefact. But it is important to recognise that craftsmen don't work like machines. Rather, they compensate for errors, inaccuracies, and inconsistencies in the material as they work. They change their mind as they go along. They consider what they have already done and respond to it, maintaining a sort of feedback loop with the object they make.[7] This is the process usually referred to as "cutting and fitting," and its centrality allows us to identify basic characteristics of the craft approach.

The defining characteristic of craft is that *final* dimensions are determined only as the materials are actually worked and the artefact actually made. As a result, there is *no separation of designing from making*. The two activities take place at the same time. There is *no separation of designer from maker*. They are the same person. Craft methods may also be said to be *inherently wasteful of materials*, in the sense that each part must start out oversize, then be reduced. Since there is always the possibility that this "wasting" process will be carried too far, and the materials will have to be thrown away, great care must be taken in gradually adjusting one piece to another. Craft methods may therefore be said to be inherently both *time consuming* and *skilled-labour intensive*.

Given these constraints, a hallmark of the craft tradition has been a lengthy apprenticeship in which apprentices are taught how to arrive at the desired form with the fewest mistakes and least waste, at the lowest cost. Some crafts use representations in the form of patterns or lasts to help achieve these goals. Some even make use of *partial* graphic representations. Early modern merchant shipbuilders, for example, made full-scale drawings of three or four of the major transverse frames of their vessels.[8] Templates were then made to the drawings, frames cut to the templates, and frames set up in the yard. Long, thin wooden battens were then bent around the frames from bow to stern, and dimensions for the other frames taken from the battens. Despite the initial use of partial graphic representations, however, early merchant shipbuilding remained a craft as defined above. The materials themselves were used to determine final dimensions during construction. Designing was not fully separated from making.

Now, given that such large and complicated artefacts as wooden ships were made by craft methods right into the twentieth century, we can confidently reject Jones' suggestion that the craft approach must be limited to artefacts of small size and limited complexity, involving no division of labour. But it will nevertheless be true that the larger the scale and the more complex the object, the more problematic constraints of time, labour and materials will become. This has a direct bearing on the discussion of innovation in the craft tradition, where there is a long history of explaining a supposedly slow pace of change in terms of ingrained customs and conservatism of the craftsman, and many attempts have been made to eliminate the role of conscious intention altogether.[9]

The value of approaching the issues from the point of view of graphic representation is that we can describe the problems of innovation in the crafts without having to delve into problematic accounts of mentality. In the absence of drawings, there is no symbolic means of predicting whether a new kind of artefact will work or serve its intended purpose. The only way to find out is to make the artefact in question, but the craftsman is thereby exposed to the potential waste of all of the time, money and materials involved. Craft innovation thus unavoidably involves relatively high risks because the only way forward is by "trial and error," which is to say by actually constructing artefacts and subjecting them to the conditions of use. From all of this we can conclude that the typical context for a craft approach will be one where labour and materials are relatively cheap and time is not a pressing concern. Where time, labour and materials are absolutely or relatively cheap, we might actually expect a great deal of innovation (as the seemingly infinite variety of ancient pottery types might indicate).[10] But where time, labour and materials are expensive, we would expect to find either a slow pace of innovation or a change in design methods.

The Mechanical Tradition

According to Jones, the use of drawings permits a much faster pace of innovation in design at the same time as it offers huge advantages in construction. However, Jones did not specifically acknowledge that *different kinds* of drawings are used to achieve these different purposes.[11] Modern designers still use rough, back-of-the-envelope sketches or perspective drawings to work out new ideas. But they use measured plans to control construction. One of the fascinating features of early modern technology is that there appears to have been a distinct separation of the design traditions that used these different kinds of drawing prior to approximately 1750. Until then, while mechanical designers made great use of rough sketches and perspective drawings, they do not appear to have used measured three-view plans at all (Baynes and Pugh [1981] 34f). For this reason the second tradition of early modern design may be called the "mechanical" tradition.

The emergence of the mechanical tradition can be traced back at least as far as the notebooks of Villard de Honnecourt (ca. 1328), who recorded a number of the machines in use in the fourteenth century. Thereafter, as seen in the works of Renaissance engineers like Guido da Vigevano, Mariano Taccola, and Francesco di Giorgio Martini, it was recognised that drawings could be used to invent as well as record. The invention of the linear perspective played a further role, allowing the development of drawing into the kind of rigorous means of investigation and innovation seen in the work of Leonardo da Vinci, as well as the kind of orderly presentation found in Agostino Ramelli's famous theatre of machines published in 1588.[12] But to get at the characteristics of the mechanical tradition of design, I present two later drawings. One was part of a proposal submitted

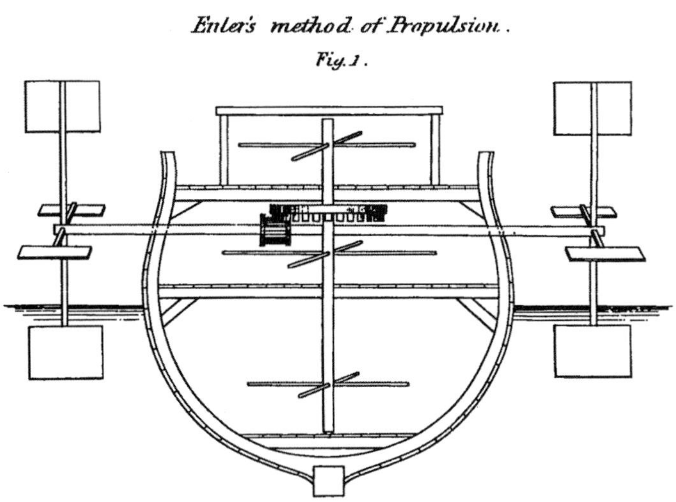

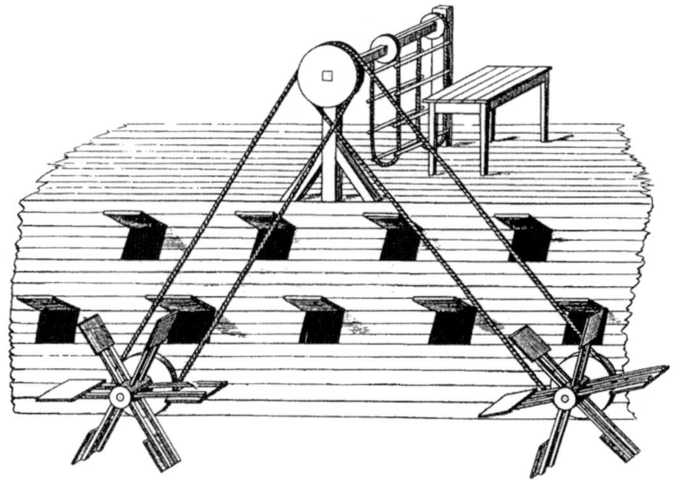

Figure 3 (*left*)
Leonard Euler's proposed method of ship propulsion of 1753. From Fincham: *A History of Naval Architecture* (London 1851) 280. Courtesy of the Hart Nautical Collection, Massachusetts Institute of Technology.

Figure 4 (*right*)
Mathon de la Cour's proposed method of ship propulsion of 1753. From Fincham: *A History of Naval Architecture* (London 1851) 280. Courtesy of the Hart Nautical Collection, Massachusetts Institute of Technology.

by Leonhard Euler to a French Royal Academy competition in 1753. The second is a proposal submitted by Euler's fellow competitor Mathon de la Cour.[13]

What strikes the eye is how quickly both drawings declare their purpose – propelling ships in the absence of wind. This they do by clearly illustrating the working principles involved. In Euler's case, handspikes are used to turn a vertical shaft with a toothed wheel. The wheel turns a lantern gear attached to a horizontal shaft, which rotates the paddles. In de la Cour's device, the paddles are turned by endless cords attached to pulleys, which are powered by a rope treadmill. In both cases, visual clarity is achieved by adopting the principle of "one object, one drawing," using the conventions of perspective to show the object from a single viewpoint. Euler has complicated matters by trying to place the viewer in a position to see a perfect cross-section. This created a difficulty in the depiction of the handspikes, which would have appeared as horizontal lines whose purpose was difficult to determine. A different viewpoint was therefore chosen from which to depict the handspikes only. An "apparent" shape was substituted for a "real" shape, resulting in something of a visual distortion.[14] In de la Cour's drawing, we see the top, side, and front views simultaneously and, although it is more "realistic," it is actually de la Cour's drawing that contains the greatest amount of "distortion," since the use of perspective has resulted in the substitution of "apparent" for "real" shapes throughout. Note that neither drawing contains any measurements.[15]

Now we may simply agree with Jones that this type of drawing provides an intuitive medium for the rapid development of new forms, because it allows the designer to work solely with the geometrical aspects of a machine in the absence of any actual constraints of time, labour, materials, and physics. But it will be important to recognise that the creation of such drawings does *not* represent the transference of craft "trial-and-error" onto paper. In the crafts, the trial took place when an artifact was *actually* built and subjected to *actual* conditions of use. As long as a design remains on paper, no such trial takes place. Nor is there a mental equivalent because, as we have just seen, the drawings were created in the absence of the constraints that would constitute such a trial. What *has* been transferred to paper, as the designer draws a line here and erases one there, is the "cutting and fitting" aspect of a craft – but with a major difference. The result of cutting and fitting in the crafts was the exact determination of final dimensions. The ersatz cutting and fitting involved in early machine drawing produced only a *general* idea of form and arrangement.

What this means will become clear if we consider building either of the machines in our drawings. The first problem is that no measurements are supplied. That is, no dimensions are actually determined. Each will therefore have to be guessed at, a task made more difficult by the substitution of apparent for real shapes. The second problem is that not all of the parts of the arrangement have been depicted. For example, Euler's picture doesn't show how the horizontal shaft passes through the side of the hull, or how the vertical shaft is to be seated in the hold or fixed at the top. It does not tell us the diame-

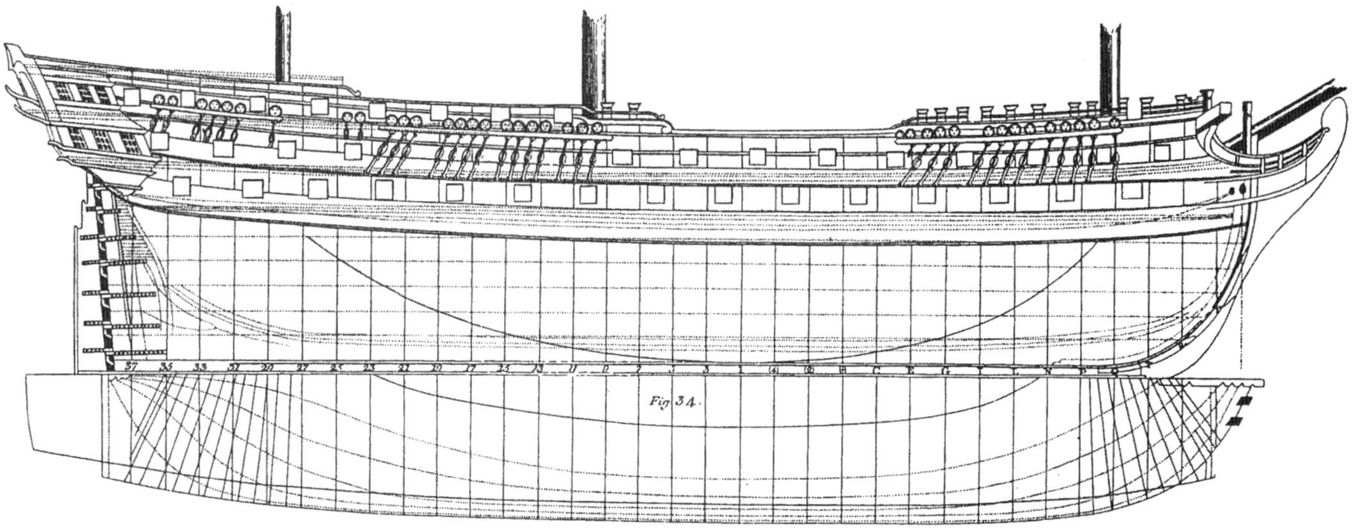

Figure 5
Side view or "sheer plan" from the article on "Shipbuilding" in *Encyclopaedia Britannica* (2nd ed., Edinburgh 1787). Courtesy of the Burndy Library, Dibner Institute for the History of Science and Technology, Massachusetts Institute of Technology.

ter of the toothed wheel, nor how many teeth it has. Similarly, de la Cour's design fails to specify either the manner in which the paddle arrangements are joined to the hull, or the way in which the treadmill frame is secured to the deck.

The result of these omissions is that the mechanical tradition of design *ultimately must rely on craft methods of production*. Each part will have to be cut, fitted and shaped until all of the parts fit together. Missing parts will have to be invented on the spot. Thus there can only be a *partial separation of designing from making*, because there is only a *partial determination of dimensions* in advance of construction. There can only be a partial separation of *designers* from *makers*, since either the inventor must be present in the yard to explain what is needed, or else the workers must take on the designer's role by determining dimensions and adding arrangements for themselves. Of course, all this cutting and fitting implies at least the same waste of time, labour and materials associated with the crafts. But, in fact, building a device from an unmeasured perspective drawing will likely be *more expensive* than building a craft artefact, because no craft tradition can possibly exist for making a hitherto unknown machine at the least cost with the fewest mistakes in the shortest time.

Given potentially high costs, we would probably like to know if the device would actually work before going to the trouble of building it. For example, can enough men be put to work at Euler's handspikes to provide the force necessary to move the ship? Will the wooden gear-teeth stand the stress? Can enough men be put on de la Cour's treadmill to move the ship? Or would the ropes break first? Unfortunately, even if we possess a relevant body of mathematical theory, it will not help us because there are no measurements. So there is nothing to quantify. True, we might guess at dimensions in order to make some calculations, but this would actually be rather pointless since the cutting, fitting and adding of parts during construction means that the finished device will *not* be the same as the one we calculated.[16] To this we may add the further problem of trying to determine whether the device in question would really be useful. For example, when the ship is under sail, Euler's paddles will increase resistance while the long handspikes will apparently flail about on their own as the ship moves forward. Can a ship packed with hundreds of men be operated under these conditions? Not without a number of cracked heads!

To sum up, the drawings used in the mechanical tradition do not provide accurate predictions as to dimensions, arrangement, cost, whether a device will work, or whether it will really be useful. To find out, we will have to build the machine, knowing for certain only that it will be expensive. In fact, the more innovative the object, and the larger the scale, the more expensive it will be. It thus appears that the context must be one where money is no object, or where the need for a new solution is so pressing that cost is largely irrelevant. In either case, the realisation of a paper invention will likely require a patron with deep pockets. In the absence of wealthy patrons, high costs will act as a heavy constraint on the development of actual machines. Thus it is no accident that mechanical designers like Leonardo worked for wealthy princes and states, and that most of the Renaissance machines invented on paper were never built.

The Architectural Tradition

Having looked at the craft and mechanical traditions in terms of graphic representation, we now come to the third tradition of early modern design. This is the "architectural" tradition, as defined by the use of measured multi-view plans. It is illustrated here with a discussion of "naval architecture," an approach to ship design whose origins can be traced back to Elizabethan master shipwright Mathew Baker, who appears to have been the first to adapt measured architectural-style plans for use in ship design.[17] Only after the work of French artillerists, but more importantly after Boulton and Watt began to manufacture steam engines in the second half of the eighteenth century, did the use of measured three-view plans spread to fields other than architecture and naval architecture (Alder [1997]; Baynes and Pugh [1981] 34–37). Interestingly, the use of plans did not become the norm in merchant shipbuilding until far into the nineteenth century. Until then, plans were used almost solely for the construction of warships.

Figure 5 shows the kind of drawings used in eighteenth century naval architecture as illustrated in the *Encyclopaedia Britannica* of 1787.

On top is the side elevation or "sheer plan." Below the line of the keel is the top or "half-breadth" plan which, since ships are symmetrical about their axis, shows only half the vessel from above. Figure 6 is the "body plan," showing multiple cross-sections of the hull at regular intervals along its length.

The lines to the left and right of the centre represent cross-sections tapering towards the bow and stern. Each line also represents the outside shape of an actual frame whose location along the keel is shown by the straight, perpendicular station lines in the other two drawings.[18] In contrast to the one-object, one-drawing, one-view approach taken in the mechanical tradition, we have three views of the ship at right angles to each other, in which each drawing and each line is intended to show true rather than apparent shapes, any perspective "realism" being limited to the inconsequential details. The result, however, is a set of drawings that are relatively incomprehensible to the layman, in the sense that it is extremely difficult to understand what many of the lines are for, or to visualise what the shape of the hull is really going to be. On the other hand, we need to note that we have

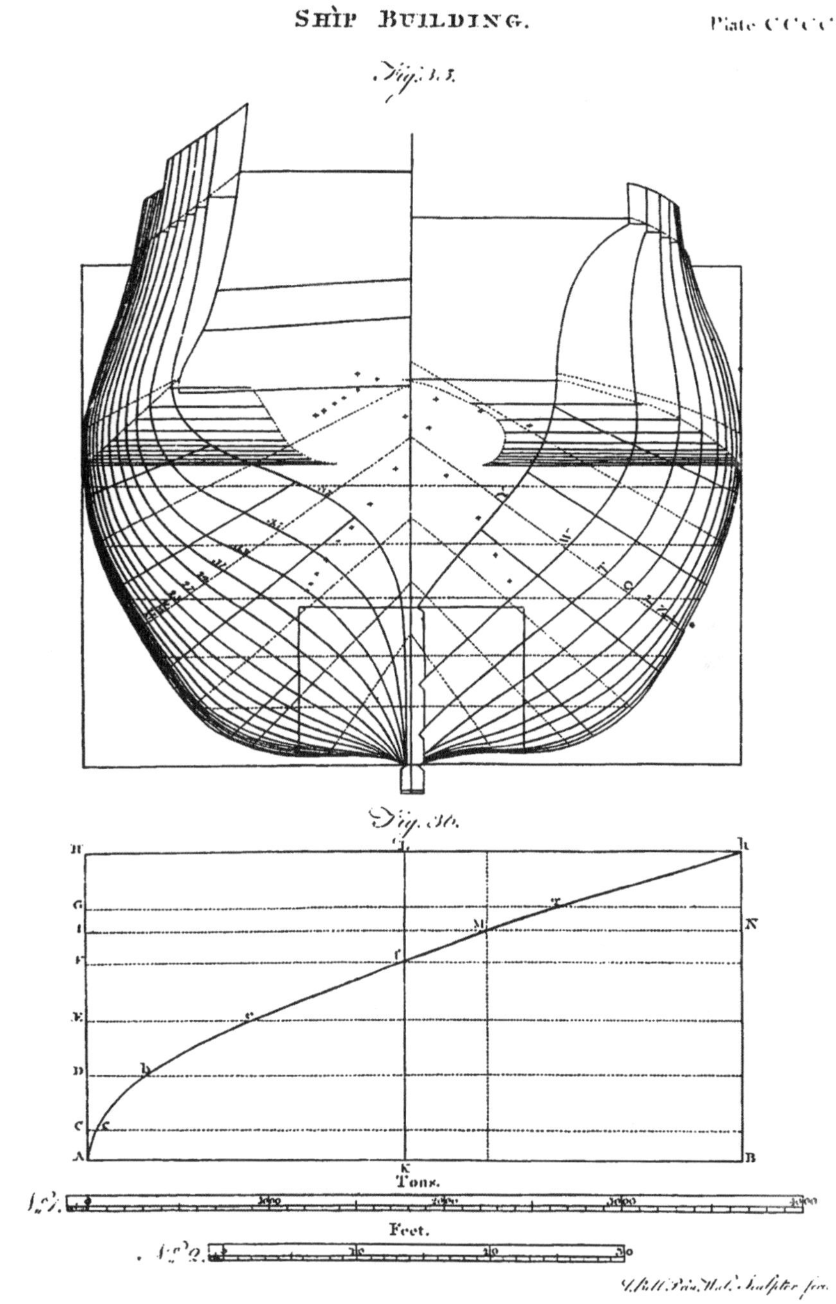

Figure 6
Body plan from the article on "Shipbuilding" in *Encyclopaedia Britannica* (2nd ed., Edinburgh 1787). Courtesy of the Burndy Library, Dibner Institute for the History of Science and Technology, Massachusetts Institute of Technology.

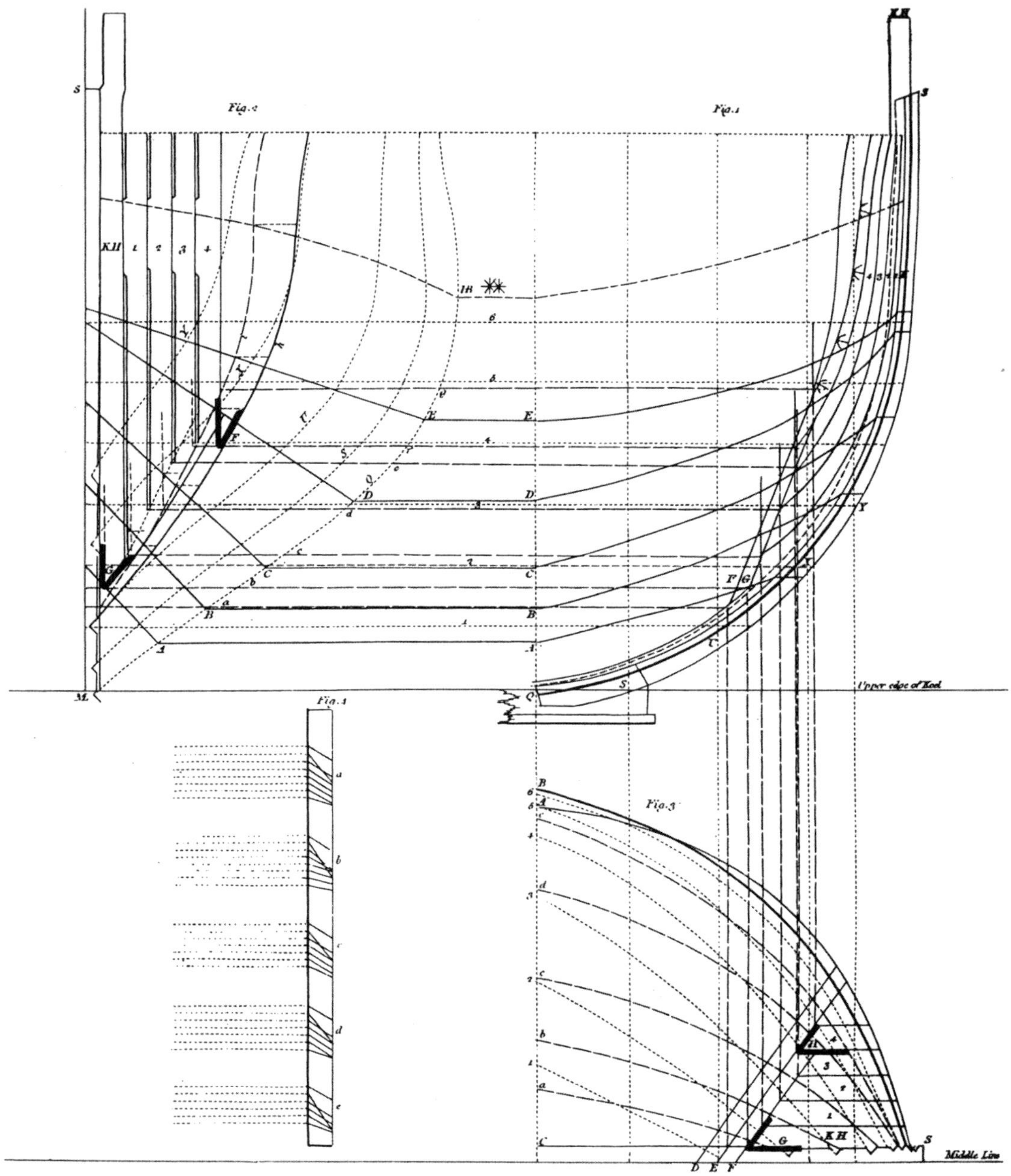

Figure 7
Geometrical manipulation of the plans in laying-off on the mould-loft floor. From Rees: *Shipbuilding* (London 1819). Courtesy of the Burndy Library, Dibner Institute for the History of Science and Technology, Massachusetts Institute of Technology.

here *measured plans*, a fact indicated by the scale found at the bottom of each drawing.

According to Jones, these are the kind of drawings in which the final dimensions of an artefact are determined in advance of construction. It may be objected, however, that nowhere near all the parts of the hull are represented in Figures 5 and 6.[19] For example, not all of the frames of the hull are depicted in the body plan. Again, therefore, it is necessary to point to the importance of "moulding" carried out in the dockyard, where the lines of the plans were reproduced full-size on the "mould loft floor," then manipulated geometrically to derive additional lines and dimensions as required. As in merchant practice, templates were made to the resulting drawings and given to sawyers and shipwrights who cut the timbers to match. However, the process of moulding was carried much further in warship design and came to include the determination of the dimensions of *every* part of a hull. Figure 7 illustrates the extremely complicated procedures needed to determine the dimensions of timbers at the bow and stern, where the curves of a ship were most complex. Determining lengths, location and bevels of the hull planking also required surprisingly sophisticated procedures.[20]

In other words, the plans used in early modern shipbuilding *did* contain the dimensions of every part of the hull, in what might

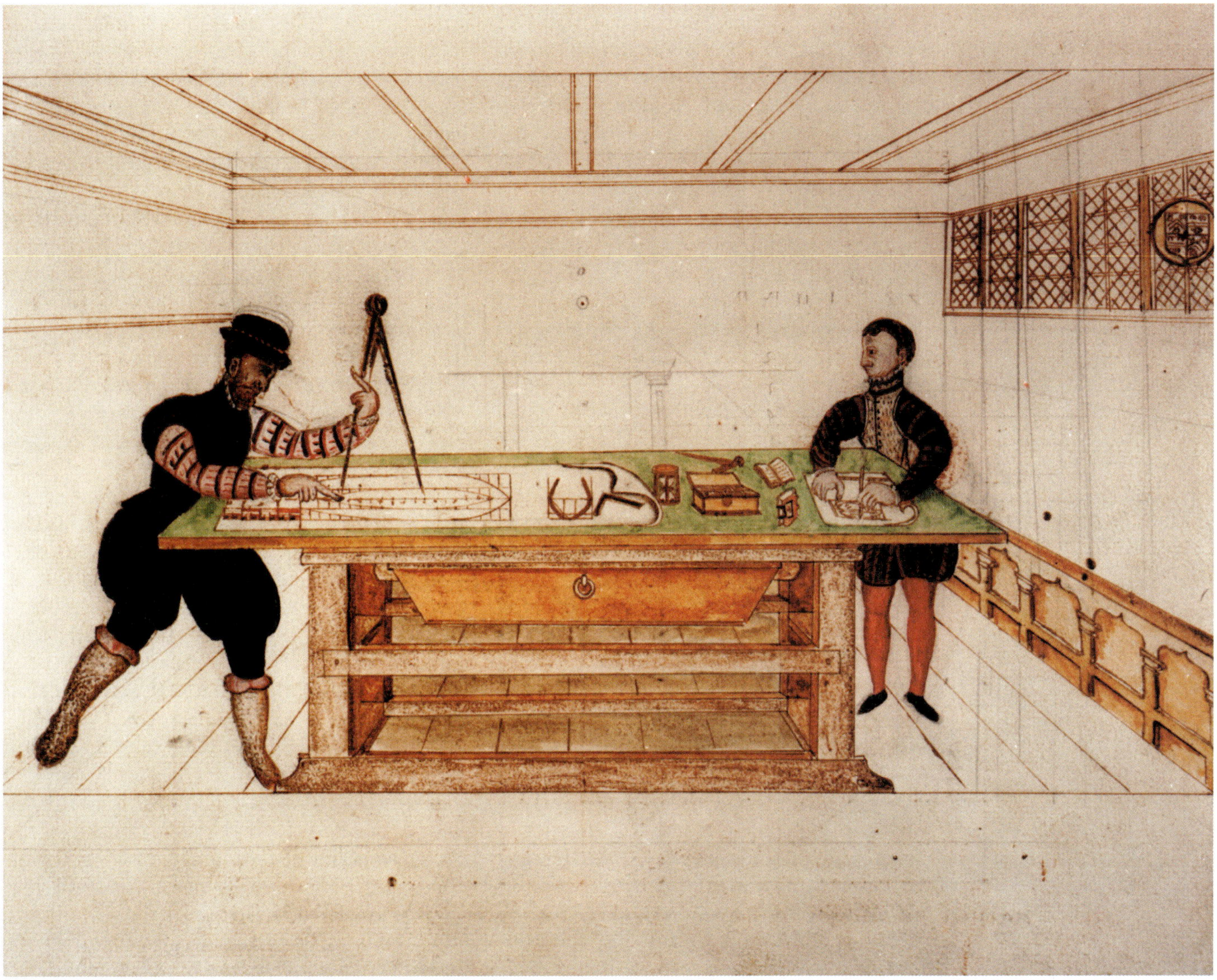

Figure 8
Mathew Baker's illustration of the naval architect at work in the office, not the shipyard. From Mathew Baker's *Fragments of English Shipwrightery* of 1586, unpublished manuscript. Courtesy of the Pepysian Library, Magdelene College, Cambridge.

be called geometrical principle. A central feature of British naval architecture was therefore the *separation of designing from making*. These activities took place at different times. There was also *a separation of designers from makers*. For those who made the plans and those who made the parts were no longer the same people. In fact, making plans became the prerogative of the Surveyor of the Navy, whose offices were in London, miles away from any dockyard. The result of all these separations is that British naval architecture was formally regarded as being divided into two parts: the "plan or form of the construction of Ships, comprehending all that falls within the Science or Theory of Naval Architecture, being the first part;" and the "execution of the plans formed, or the Art of Shipbuilding, being the second" (House of Commons [1806] 193).

The separation of the designer from the shipyard was already presaged in Baker's manuscript of 1586, which shows the naval architect at work on a table in an office rather than in the mould loft.[21]

Clearly, a naval architect working on paper in an office miles away from the dockyards is not to be regarded as a "craftsman." But it will be important to recognise that after the introduction of measured plans dockyard workers were no longer craftsmen either. In the crafts, final dimensions were determined only when the materials were actually worked. In naval architecture, dimensions were determined in advance. In the crafts, the determination of dimensions entailed a process of cutting and fitting, which required skill, creativity, and an active contribution from the worker. Following the introduction of measured plans in shipbuilding, cutting and fitting was to be eliminated. Creativity was *not* wanted. All the worker was supposed to do was execute the shapes determined by somebody else, and any "creative" contribution could only be a *mistake*, in the sense that a part altered by one worker would no longer fit together with those made by another. To put the point more strongly, it may be said that the use of measured plans in naval architecture was deliberately *anti-craft*, for the purpose of determining dimensions in advance of construction was precisely to eradicate the waste of time, materials, and labour inherent in craft methods.

Stressing the "anti-craft" nature of naval architecture is important because it points to the context in which plan drawings were introduced, where the determining factor appears to have been the high cost of warships to the state.[22] As John Brewer has estimated, the fixed capital invested in early eighteenth-century factories rarely exceeded £10,000. A first-rate warship of the time already cost up to £40,000, and the cost rose to £80,000 by the end of the century. Brewer further estimates that the *total* fixed capital invested in the 243 mills of the large West Riding woolen industry in 1800 was £402,651. The replacement value of the ships of the navy in that year was £2.25 *million* (Brewer [1989] 34f). In other words, the cost of building a navy was enormous, and so much larger than anything tackled in the craft or mechanical traditions that economy of production became a crucial issue. This is particularly true of Britain, where the expense of a permanent navy was a constant problem for governments from the time of Elizabeth I.

Innovation and Risk in Naval Architecture

Measured plans may have provided a cheaper and more precise means of controlling construction than the cutting and fitting practiced in both the craft and mechanical traditions, but it did not provide a medium for rapid innovation in shipbuilding.

One reason for this was the laborious method used to create the plans. Instructions were generally of the kind "connect A to B, make AC 1/3 of AB, raise a perpendicular, draw the line," and so on. Hundreds of such steps followed each other in deliberate sequence as each line was carefully drawn in one view, then transferred to the other two views. Each step had to be followed exactly, otherwise subsequent lines would not meet and there could be errors in construction. As a result, it took upwards of two months to create a set of ship plans. Moreover, because of the complex geometrical relationship between them, altering a line in one view required a time-consuming reconstruction of the other two views. In short, the measured plans used in naval architecture did not allow the kind of fast and intuitive drawing and re-drawing that led to the rapid development of new forms in machine design.[23]

However, a second and more important reason for the slow pace of innovation in naval architecture had to do with the fact that changing lines in drawings could have extremely unpredictable effects on the behaviour of ships at sea and innovation thus entailed huge risks. In order to understand why, we need to step back and consider ship design from a broader perspective.

The place to begin is with the recognition that ship designers in the age of wood had a consistent list of the qualities a good ship should have. A typical list is found in the *Encyclopaedia Britannica* article on shipbuilding for 1787, which called for ships that would pitch moderately, sail well, sail into the wind, make little leeway, have enough stability to carry a good spread of sail, and enough displacement to carry the guns high out of the water (Anonymous [1787] 376f). Naval architect John Knowles gave essentially the same list in his influential *Elements and Practice of Naval Architecture* of 1822, calling for ships that steered well, worked well, pitched and rolled easily, had enough stability, and carried their guns high enough out of the water that they could be used in heavy weather (Knowles [1822] 123).

Ship designers were also well aware of the hull shapes needed to obtain these qualities. They knew that ships with fine lines aft were easier to steer, and that narrow ships with sharp bows sailed the fastest. They knew that greater breadth gave greater stability, while deeper hulls gave less leeway, fuller bows reduced pitching and so on. However – and this is the crucial point – they also knew that the qualities desired in a ship *conflict* with each other in an extremely complicated way. Speed requires a narrow ship, but narrow vessels have less stability and less capacity to carry large amounts of sail. Full bows result in moderate pitching, but ships that drift to leeward. Fine bows reduce leeway, but make pitching worse. Carrying the guns requires a full body, but speed requires a narrow underwater body, and so on.[24]

The result of all this was that naval architects knew it was not possible to combine *all* the desirable qualities in a single ship. As the *Encyclopaedia Britannica* explained, to obtain all the desirable qualities "very opposite rules must be followed; and hence it appears to be impossible to construct a ship so as to be possessed of them all" (Anonymous [1787] 377). As Knowles put it:

> *The properties which every ship ought to possess are, in a manner, subversive of, or in opposition to, each other. One figure is required for extraordinary swiftness, another for extraordinary strength or capacity; and all are regulated, more or less, by peculiar and local circumstances. The great art, however, in all places and under all circumstances, consists, in so forming the body, that none of the desired qualities shall be entirely wanting (Knowles [1822] 121).*

But as Knowles emphatically concluded: "to unite, in one ship, all these desirable qualities, some of which are subversive of others, is impossible" (Knowles [1822] 127).

How was the designer to decide which qualities to emphasise, and which not? To this question naval architects also had a uniform answer, namely that the qualities to emphasise had to be chosen in light of the *purpose* of the vessel. As seventeenth-century Secretary of the Navy Samuel Pepys noted, "no one shape can be in general said to be the best; for every distinct use requires a different shape, and the skill lies in building best for the particular purpose designed."[25] According to the *Encyclopaedia Britannica* in 1787, "a ship ought to be constructed so as to answer the particular purpose for which she is intended," which meant that the hull "must be so formed, that as many of these properties may be retained as possible, always observing to give the preference to those which are most required" (Anonymous [1787] 376). According to Knowles, "the first object must be, to consider the various purposes that [the ship] is intended for, and the various impediments that it may meet with." Only thus could the ship be given the "arrangement of its parts and combination of its principles; especially where it will be necessary that contradictory powers shall be blended together, and wherein one principle shall not predominate too much over another."

Since it was impossible to unite all the desirable qualities, Knowles wrote, designers had to be satisfied to "gain that one in an eminent degree which formed the main point of design" (Knowles [1822] 122). But because the dimensions governing the sailing qualities interacted with each other in such a complicated fashion, it was impossible to predict precisely how a new blend of qualities would behave. Secretary of the Navy Pepys was struck by this feature of ship design, writing that it was "worthy of note how small things are sometimes found to mar or mend a ship's quality." Two centuries later, George Atwood was still stressing how seemingly trivial departures from established rules could "wholly change the qualities of a ship from bad to good, or the reverse."[26] Tales were frequently told of sister ships, built to the same plan, which had radically different sailing properties – one good, the other bad – driving home the point that even *small* variations in ships could lead to inexplicably large changes in sailing properties (Pepys [1926] 14).

It was this tremendous uncertainty that held back the pace of innovation in naval architecture. After all, ships were extraordinarily expensive, carried as many as 800 men, and their failure might have dire consequences even for the state. In a context like this, naval architects simply could not afford to forget about physics, changing lines in drawings at will the way mechanical designers did. On the contrary, the most sensible strategy was to stick closely to previous designs, known to perform well, making as few changes as possible, and while the pace of change may have been slow according to our standards, we ought to realise that this was not the result of some inbred "craft conservatism," but the result of the inevitable uncertainties arising out of the need to blend conflicting qualities. That is to say, the difficulty arose out of the complex nature of the design problem, not the conservative human nature of the designers.

However, it is for just this issue of uncertainty that the use of plan drawings in naval architecture presented an intriguing possibility, namely the possibility of applying quantitative mathematical theory to overcome problems of prediction in ship design.

Science and Design in Naval Architecture

In discussing of the relationship between science and design in early modern naval architecture we are already in position to lay down two fundamental principles. One may be stated as: *drawings first, then science.* For without the use of measured plans there could be nothing to quantify, and no possible application of mathematical physical theory to design whatsoever. The second is that without the use of measured plans to achieve accuracy in construction, there could be *no point* in applying scientific theory, since there could be no point in calculating the behaviour of a "paper" ship if the real ship was going to be built (and therefore behave) differently. But how did the application of science work in practice? To answer this question we can turn to the use of Archimedean theory to calculate displacement, a calculation that was made for British warships from at least 1670 onward, and a calculation that represented the only really useful application of physical theory to ship design prior to 1800.

The first step in determining whether a ship possessed adequate displacement was to calculate the weight of everything to be carried on board including iron, lead, masts, sails, rigging, men, provisions, and so on. Since warships were classified by "rates" for which such things as provisions and compliments were set down in tables, this was relatively easy. Next, the weight of the hull had to be found. This meant calculating the weight of every individual timber which, as the *Encyclopaedia Britannica* noted in 1787, was "indeed a very laborious task, upon account of the several pieces of timber, &c. being of so many different figures, and the specific gravity of some of the timber entering the construction not being precisely determined" (Anonymous [1787] 411). Nevertheless, by using the plans to determine the volume and weight of each timber, the total weight of the hull could be found. Next, the designer had to calculate the cubic capacity of the hull below the waterline. This too was a laborious task, given the irregular shape of a ship's hull, but the basic idea was to use the lines of the plans to divide the hull into rectangles whose volume could be easily calculated. The volume of the irregular remainder could then be found by methods of exhaustion. Once the volume of the underwater hull was known, it could be multiplied by the weight of a cubic foot of water to determine the displacement of the hull at the intended load line. The designer could then compare the weight of the displaced water to the total weight of the hull and its contents to see whether or not the ship would float at the intended waterline.

Determining the displacement of wooden warships was important since it could be rather dangerous to open the ports of ships that floated too low. It was also important to divide the volume of the hull amidships to see if each half of the ship was equally buoyant, since an unevenly supported hull would begin to "hog" or bend. Such a ship might also tend to pitch heavily, working and weakening the entire structure as one end plunged more deeply into the waves than the other. Without the use of measured plans, difficulties like these could not be detected and corrected before a ship was built, thereby saving lives, as well as the money and materials that would have been wasted on a failure. Here was the *promise* of science in ship design. But to get at the *problems* we may follow the question of displacement a little farther, for according to the *Encyclopaedia Britannica*'s calculation of the ship represented in Figure 5 and 6, the rear portion of the hull was too small by 14,166 cubic feet. The aft portion therefore had to be "filled out" by 7083 cubic feet and the fore part "drawn in" by the same amount (Anonymous [1787] 415).

The fundamental thing to note here is that such seemingly precise quantification says almost nothing about *which* lines in the plans should be changed, and almost nothing about *how much* they should be changed to correct the problem. For example, to fill out the stern the designer could either expand the breadth at the waterline or the lower part of the hull near the keel. To draw in the bow the designer could narrow the bow itself or the mid-body further back, either at the waterline or below it. The choice was the designer's, acting by "eye" in light of the purpose of the vessel. In principle, the only way the designer could be sure that such changes actually solved the problem was to redraw the plans, recalculate the weight of the ship, and recalculate its cubic capacity. This would be laborious enough. But worse still, supposing the problem *wasn't* solved, the designer was then back where he started, faced with the unpleasant prospect of re-drawing and re-calculating all over again.

This brief discussion allows us to make four additional points about science in early naval architecture. First, scientific calculations actually gave very limited guidance to the designer on how to correct the all-important lines of the plans when problems were identified. Second, as a direct result, the use of quantitative theory entailed the use of new feedback loops of drawing, calculating and redrawing into the design process. Third, if initial calculations were tedious and time consuming, subsequent loops only added to the tedium and delay. The fourth is that even if the designer completed several tedious feedback loops, this did little to achieve the overall goal of reducing uncertainty – not just because the designer eventually had to stop calculating and trust his judgement, but because the designer adjusting lines to approximate desired properties (in this case, the required displacement) was simultaneously *altering the dimensions that governed all the other sailing qualities*. Thus, the attempt to reduce uncertainty about displacement actually *introduced* uncertainty about the ship's other properties.

In this connection it is worth noting that the only other sailing property that could be calculated in any sensible fashion prior to 1800 was stability. But if it took two months to make a ship's plans, and another two months to calculate displacement, it could take two *years* to determine the stability – and even then only for a single small angle of heel (Lavery [1984] 25). As with displacement, the results did not tell the designer precisely how to change his plans to correct an anticipated problem. Nor could he be certain about the effect of the changes made without calculating everything all over again, taking even more time. The fact is that many wooden warships could be built in less than two years – that is, they could be built faster than they could be calculated.

These points are important to keep in mind because very early on the possibility of using physical theory in conjunction with ships' plans to solve problems for the state seems to have attracted the attention of many prominent natural philosophers. Harriot, Briggs and others were involved in the improvement of naval architecture circa 1600. Hooke, Boyle, Wren, Petty, and even Newton made contributions.[27] The French took the lead in the eighteenth century as a result of the efforts of Bouguer, Euler, Bossut, d'Alembert and others.[28] What is of interest here is that many of these natural philosophers articulated the view that predictive problems in shipbuilding could be *eliminated* by taking a deductive approach. That is, they believed that if you knew the laws of nature you could *deduce* the proper shape of a ship and eliminate uncertainty altogether. Thus Father Paul Hoste, referring to the Colbert's campaign to improve French shipbuilding in the 1680s, wrote that:

> *It is true that for some years past, the French shipbuilders have laboured at improving their art. Some have begun to make plans, in which they have determined all the frames or models of the bow and stern; but as they have not necessary principles, they labour with little certainty. Their ships are not better than those which were built without the knowledge of either reading or writing; they do not sail better – often they do not carry sail so well – they rather hog – they are less durable; in a word, the constructors of the present day agree with the ancients, that it is not yet known what the sea requires.* [29]

This is certainly a powerful rhetoric. We would all grant an ontological priority to the laws of nature. We would all agree that a ship behaves at sea as a result of these laws. So it might seem plausible to assume that if you could determine the laws you could deduce the proper shape of a ship and thus know precisely how it would behave in the future.

By paying attention to the role of graphic representation in design, however, we can see why this view is mistaken. The basic problem facing ship designers was the conflict between the dimensions governing the desirable sailing qualities. A *choice* therefore had to be made about which qualities to secure and which qualities to sacrifice. The only way to make this choice was to decide on the purpose of the proposed vessel. The resulting compromise could then be embodied in measured drawings – but only then could theoretical calculations be made. Which is to say, ships *do not* get their shapes by "what the sea requires." They get their shape by what the *purpose* requires. To think that you could *begin* designing with the laws of nature is simply to mistake ontological priorities for the real temporal priorities of the design process.

Conclusion

In this article I have followed the lead of Jack Goody and J. Christopher Jones, examining early modern technics in terms of drawing and design. This has provided criteria for the identification of three different traditions of design in the early modern period. One was craft proper, where no complete drawings were used, final dimensions were determined only as materials were actually worked, and there was no separation of designing from making. The second was the mechanical tradition, where unmeasured perspective drawings and sketches were employed to evolve new designs on paper, but an ultimate reliance on craft cutting and fitting in production meant there could only be a partial separation of design from construction. The third is the architectural tradition, where the use of measured plans entailed the elimination of craft methods of construction and a full separation of designing from making. The identification of these three traditions, I believe, provides a more fruitful basis for investigating the technology of the past than the generalised notion of craft. In particular, it provides a comparative basis for investigating the conditions under which a scientific approach to design can emerge.

As we saw in the case of naval architecture, one of the most important conditions was the use of *measured* drawings to provide a symbolic medium for calculation; otherwise no application of quantitative science can take place. A second was the use of measured drawings to control construction, so that calculations of the behaviour of the ship as drawn would apply to the ship as built. A third condition was the high cost of labour and materials. A fourth was the high degree of uncertainty about the results of innovation, and thus the risk of wasting all the money, labour and materials used.

Recognition of these points helps to explain the lack of science we perceive in early modern technology. To begin with, neither the craft nor the mechanical tradition employed the kind of representations that allowed calculation, or ensured that completed artefacts were the same as those drawn on paper. That is to say, in the vast majority of technologies of the early modern period, there was no way and little point in making use of science in design. Even in naval architecture, where these conditions were met, there were still major problems. Calculations were extremely long and tedious. The results provided little guidance on how to change lines and correct problems. Alterations had to be made by eye according to the judgement of the designer, necessitating feedback loops of redrawing and recalculating that made the process even longer and more tedious for only marginal gains in certainty. As a result, scientific calculations of anything other than displacement did not become common, at least in British shipbuilding, until the 1860s when the balance between the risk, the burden of calculation and the utility of the results began to change (McGee [1998]). Until then it was far more sensible – "rational" in fact, rather than "conservative" – to rely on known models that were known to work well rather than engage in radical innovations.

Finally, taking all these points together, I would like to suggest that approaching early modern technology from the point of view of graphic representation offers the possibility of bridging the divide between "craft" and "science" with which I began. Rather than a gap, the principle of "drawings first, then science" points to the conceptualisation of two broad historical stages. The first would entail the widespread adoption of measured drawings in the early nineteenth century, when capitalists began to act like states, and heavy investments led to the use of measured drawings as a way to reduce the costs of production (Baynes and Pugh [1981] 11; Booker [1963]

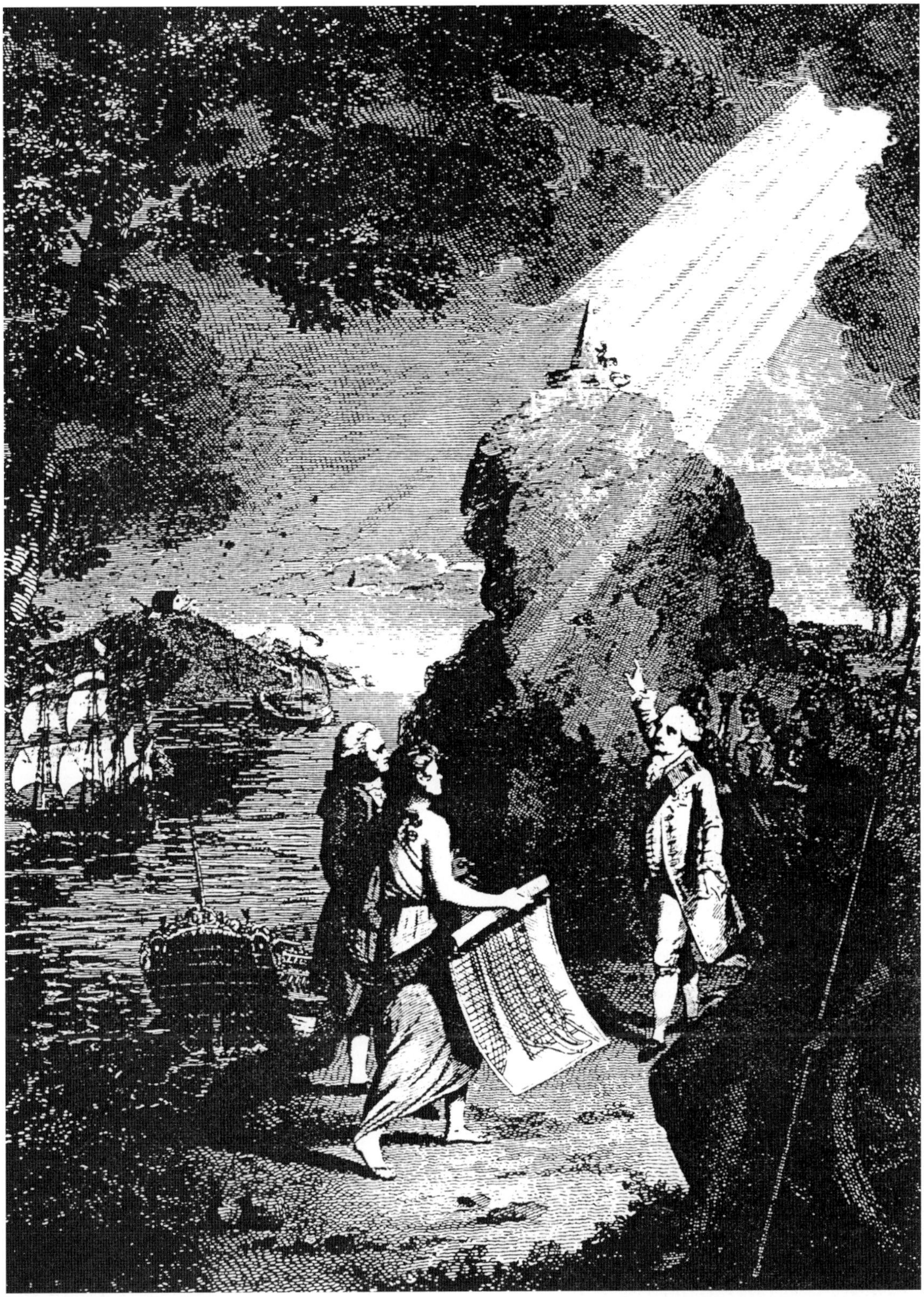

Figure 9
The frontispiece to Marmaduke Stalkartt's *Naval Architecture: or the Rudiments and Rules of Shipbuilding Exemplified in a Series of Draughts and Plans* (London 1781). Courtesy of the Hart Nautical Collection, Massachusetts Institute of Technology.

128-139; Brett [1995]). The widespread use of drawings then represented the essential foundation for the second stage in which, as costs continued to rise, it was recognised that measured drawings could be used to predict the future behaviour of machines and artefacts and thus reduce costs still further. According to this scenario, however, it is worth noting that by the time industries adopted the use of measured plans and found the costs and risks high enough to call for predictive calculations, *they were no longer crafts*. Indeed, according to the criteria set out in this paper no craft industry could *ever* turn directly into a scientific technology, for the simple reason that by the time they were scientised they were by definition no longer crafts at all.

Alas, the complete story of draughting and drawing in the nineteenth century still awaits its historian.[30] I only argue that if we want to know how we got from craft to science we shall have to take the advice offered in the frontispiece to Marmaduke Stalkartt's *Naval Architecture* of 1786, following the path of drawings to enlightenment.

Max Planck Institute for the History of Science, Berlin

Notes

1 I paraphrase Layton [1971]; and Layton [1979]. A list of others who have presented the emergence of modern technology this way can be found in Fores [1988].

2 I deal with a number of these figures and the logical problems of their position in McGee [1995].

3 British naval architecture has been subject to a particularly violent strain of the craft/science historiography. Writing in the 1920s, for example, naval historian Frederick Robertson argued that the British neglect of naval science was the "great blunder of two centuries,“ leaving shipbuilders to grope along the path of trial and error, guided by "blind lore“ and erroneous principles with no "rational“ foundations, held back by "hereditary principles and dogma“ and "cautious timidity.“ More recently, Stanley Sandler has written of the craft "rule-of-thumb system that had served the Royal Navy so badly.“ A. Rupert Hall has argued that French investigations into ship stability and resistance were ignored by British shipwrights because of the "anti-theoretical conservatism of British shipyards,“ a subset of the "anti-theoretical conservatism“ of British science. See Robertson [1921] 1f., 34, 56f; Sandler [1979] 33; Hall [1978-80].

4 The same point has been noticed by Lubar [1995] 71.

5 Classic emphasis on the importance of conception and the psychology of design in history of technology can be found in Ferguson [1977] and [1992]. Among design theorists see, for example: Lawson [1980]; and Rowe [1987].

6 It should be noted, however, that my definition is *not* the same as Jones‘, who defined design in rather large but vague terms as "initiation of change in man made things.“

7 Newport [1978]. Jones makes the same point, Jones [1970] 15-20.

8 For an illustrated account see Greenhill [1989]. Partial drawings were also used in the Ventian Arsenal. See Lane [1934].

9 For a classic account that minimizes the role of conscious intent in the crafts, see Sturtt [1923]. See also Alexander [1964]; Basalla [1988]. Similar points are made in Jones [1970] 20; and Lawson [1980] 14. Many of these accounts are couched in evolutionary terms. A good critique of biological notions as applied to design is found in Steadman [1979].

10 Similarly, Adrian Forty observes that handicraft chair makers were able to produce many more types of Winsor chairs than mechanized producers in Forty [1986] 88f.

11 Jones [1970] 14f. Ken Baynes and Francis Pugh offer a more complete typology in Baynes and Pugh [1981] 14f.

12 For a survey of the early development of mechanical design drawings, see Gille [1966]; and Ferguson [1992] 75-130.

13 These illustrations are taken from Fincham [1851] 280f.

14 Interestingly, the "distortions“ in Fincham's illustration of Euler's machine are actually an attempt to correct the oddities of the depiction of the handspikes in the slightly different original illustration, which may be seen in Euler [1974] 200.

15 Much of this analysis follows Booker [1963] 16-22.

16 Euler's treatise does contain all sorts of algebraic formulas concerning propulsion. I argue they were of little use in determining the actual dimensions of the device Euler himself proposed.

17 Baker's *Fragments of English Shipwrightery* is dated to 1586 and held in Pepsyian Library, Cambridge. For a detailed analysis of the whole manuscript, which contains more than just shipbuilding, see Johnston [1994] 107-165. For a closer analysis of the manuscript as it relates to naval architecture, see Barker [1988]. For an admirable summation with a bibliography of the recent literature on early methods of ship design see Loewen [1998].

18 The scale employed was typically 1 to 48, so that in reality the drawing found in Figure 5 would be some four feet long.

19 The reader may also object that none of the various rooms and cabins are shown, and it should therefore be noted that warships were often built and placed "in ordinary," lying about in rivers and harbours until they were needed, sometimes many years later. It was important *not* to install the rooms and cabins of ships lying in ordinary. This would impede air flow and thereby promote decay. For this reason, the "design" of a ship was traditionally considered to be complete once the shape of the hull had been determined. See the testimony of Sir William Rule (a surveyor of the Navy from 1793 to 1813), House of Commons, "Third Report of Commission for Revising and Digesting the Civil Affairs of the Navy." *Sessional Papers* [1806] V/312: 291.

20 Sir William Rule testified that it would take six apprentices and two joiners about six or seven weeks to make the moulds for a first-rate warship. See House of Commons [1806] 291. A detailed technical account of laying-off can be found in Rees [1819].

[21] This is in dramatic contrast to any number of medieval illustrations in which the Master Shipwright (often the biblical Noah) is shown giving direction to the workers in the shipyard. See the illustrations in Unger [1991], especially Figures 59, 62, 63, 66.

[22] Complexity was an important issue, but since wooden vessels of considerable size and complexity were built by craft methods right through the nineteenth century, complexity alone would appear not to have been the *crucial* factor.

[23] According to Sir William Rule it took a draughtsman six weeks to make the draught of a first rate warship, not counting the attention given to the design by the surveyors themselves. Moreover, he noted that, "it frequently happens that when ships are ordered to be built by the Draughts of Ships already in the service, with certain alterations therein, the trouble of making those alterations is as great as making new Draughts." See House of Commons [1806] 293.

[24] Jean Boudriot has an excellent discussion of these complicated oppositions in Boudriot [1986] 18-22.

[25] The point is repeated several times by Pepys. See Pepys [1926] 9f., 15, 201, 312.

[26] Pepys [1926] 217; Atwood [1798]. See also Atwood [1796]. Both of Atwood's papers contain lengthy discussion of the relative merits of "theory" and "practice."

[27] Evidence concerning the interest of British natural philosophers in shipbuilding is scattered throughout the diaries and correspondence of John Evelyn and Samuel Pepys. See the entries for July 16 and Aug 22, 1662 in Evelyn [1955]. See also Pepys [1925] vol. I: 66, vol. 2: 115; and Pepys [1926] 127, 201f. Newton suggested in Book II of the *Principia* that his conclusions concerning the solid of least resistance "may be of use in the building of ships." See, the Scholium to Proposition XXXIV, Theorem XXVIII, Section VII, Book II, in Newton [1729, 1960] 333f.

[28] A brief but good history of early naval science, focusing on French developments in particular, can be found in the introduction to Fincham [1851].

[29] Quoted in Fincham [1851] xvi.

[30] However, an excellent beginning, with many references to the available literature may be found in Brown [2000].

References

Alder, K.
1997 *Engineering the Revolution: Arms and Enlightenment in France, 1763-1815.* Princeton.

Alexander, C.
1964 *Notes on the Synthesis of Form.* Cambridge, Mass.

Anonymous
1787 "Shipbuilding." In *Encyclopaedia Britannica*, second edition. Edinburgh.

Atwood, G.
1796 "The Construction and Analysis of Geometrical Propositions ... Determining the Stability of Ships, and other Floating Bodies." *Philosophical Transactions of the Royal Society* 86: 46-130.

1798 "A Disquisition on the Stability of Ships." *Philosophical Transactions of the Royal Society* 88: 127.

Barker, R.
1988 "Fragments From the Pepsyian Library." *Revista da Universidade de Coimbra* 34: 161-178.

Basalla, G.
1988 *The Evolution of Technology.* Cambridge.

Baynes, K. and F. Pugh
1981 *The Art of the Engineer.* Guildford, England.

Booker, P. J.
1963 *A History of Engineering Drawing.* London.

Boudriot, J.
1986 *The Seventy-Four Gun Ship; A Practical Treatise on the Art of Naval Architecture*, vol. I, trans. David H. Roberts. Annapolis.

Brett, D.
1995 "Drawing and the Ideology of Industrialization." In *Design History: An Anthology*, ed. D. P. Doordan, 3-16. Cambridge, Mass.

Brown, J. K.
2000 "Design Plans, Working Drawings, National Styles: Engineering Practice in Great Britain and the United States, 1775-1945." *Technology and Culture* 41: 195-237.

Euler, L.
1974 "De Promotione Navium Sine Vi Venti." *Opera Omnia*, Ser. 2, vol. 20. Bern.

Evelyn, J.
1955 *The Diaries of John Evelyn*, ed. E. S. de Beer. Oxford.

Ferguson, E.
1977 "The Mind's Eye: Non-verbal Thought in Technology." *Science* 197: 827-836.

1992 *Engineering and the Mind's Eye.* Cambridge, Mass.

Fincham, J.
1851 *A History of Naval Architecture.* London.

Fores, M.
1988 "Transformations and the Myth of 'Engineering Science': Magic in a White Coat." *Technology and Culture* 29: 77.

Forty, A.
1986 *Objects of Desire, Design and Society Since 1750.* London.

Gille, B.
1966 *The Renaissance Engineers.* London.

Greenhill, B.
1989 *The Evolution of the Wooden Ship.* New York.

Hall, A. R.
1978-80 "Scientia Navalis." *Transactions of the Newcomen Society* 51: 158.

House of Commons
1806 "Third Report of Commission for Revising and Digesting the Civil Affairs of the Navy." *Sessional Papers* V/312.

Johnston, St.
1994 *Making Mathematical Practice: Gentlemen, Practitioners and Artisans in Elizabethan England* (Unpublished PhD thesis). Cambridge.

Jones, J. C.
1970 *Design Methods: Seeds of Human Futures.* London.

Knowles, J.
1822 *Steel's Elements and Practice of Naval Architecture*, 3rd. ed. London.

Lane, F.
1934 *Venetian Ships and Shipbuilders of the Renaissance*. Baltimore.

Lavery, B.
1984 *The Ship of the Line, vol. 2: Design, Construction and Fittings*. London.

Lawson, B.
1980 *How Designers Think*. London.

Layton, E. T.
1971 "Mirror-Image Twins: The Communities of Science and Technology in Nineteenth-Century America." *Technology and Culture* 12: 562.

1979 "Scientific Technology, 1845-1900: The Hydraulic Turbine and the Origins of American Industrial Research." *Technology and Culture* 20: 64.

Loewen, B.
1998 "Ships and Shipbuilding: Recent Advances in Ship History and Archaeology, 1450-1650: Hull Design, Regional Typologies and Wood Studies." *Material History Review* 48: 45-55.

Lubar, S.
1995 "Representation and Power." *Technology and Culture* (Supplement).

McGee, D.
1995 "Making Up Mind: The Early Sociology of Invention." *Technology and Culture* 36: 773-801.

1998 "The Amsler Integrator and the Burden of Calculation." *Material History Review* 48: 57-74.

Mumford, M.
1966 *The Myth of the Machine: Technics and Human Development,* vol. 1. New York.

Newport, N.
1978 "Design History: Process or Product?" In *Design History: Fad or Function?*, ed. Terry Bishop, 89-97. London.

Newton, I.
1729, 1960 *Mathematical Principles of Natural Philosophy and his System of the World,* ed. F. Cajori. Berkeley, Calif.

Pepys, S.
1925 *The Private Correspondence of Samuel Pepys*, ed. J. R. Tanner, vol. 1-2. New York.

1926 *Naval Minutes*, ed. J. R. Tanner. London.

Rees, A.
1819 *Shipbuilding*. London.

Robertson, F.
1921 *The Evolution of Naval Armament*. London.

Rowe, P.
1987 *Design Thinking*. Cambridge, Mass.

Sandler, S.
1979 *The Emergence of the Modern Capital Ship*. Newark.

Stalkartt, M.
1781 *Naval Architecture: or the Rudiments and Rules of Ship-building Exemplified in a Series of Draughts and Plans*. London.

Steadman, P.
1979 *The Evolution of Designs: Biological Analogy in Architecture and the Applied Arts*. Cambridge.

Sturtt, G.
1923 *The Wheelwright's Shop*. Cambridge.

Unger, R.
1991 *The Art of Medieval Technology: Images of Noah the Shipbuilder*. New Brunswick, N.J.

Art and Artifice in the Depiction of Renaissance Machines*

Paolo Galluzzi

During the course of the Quattrocento, emancipating itself from the occasional and marginal roles formerly played, drawing developed into a fully autonomous language whose unrivalled expressive power was soon universally acknowledged.

The evolution of this new language has been studied very carefully in many of its various fields of application: especially in relation to the depiction of man, landscape, and architecture. On the contrary, almost no attention has been devoted to the depiction of technical subjects. And yet, in this field, in the course of a few decades, a revolutionary change occurred which would lead to a radical reform of the traditional conception of machines.

Ancient architects and engineers certainly produced visual demonstrations of machines. It is almost established, for example, that Vitruvius's *De architectura* included images of the machines described in his text. None of these figurative archetypes is extant. Thus the machine portraiture of the Greeks and the Romans is documented only by indirect sources, such as the images in the bas-relief of the Lateran Museums that illustrates a Vitruvian crane activated by a treadmill (fig. 1).

The documentation must not have been any more plentiful in the Quattrocento. One need only remember that the news of the arrival from the East of a Greek codex of the eleventh century, bearing numerous colour sketches of war machines, induced Lorenzo Ghiberti to travel to Rome especially to see these graphic documents, which were evidently considered exceptional.

There remains scarce iconographic documentation of technical subjects after the classic age. One might quote only the chariots armed with scythes, military bridges and

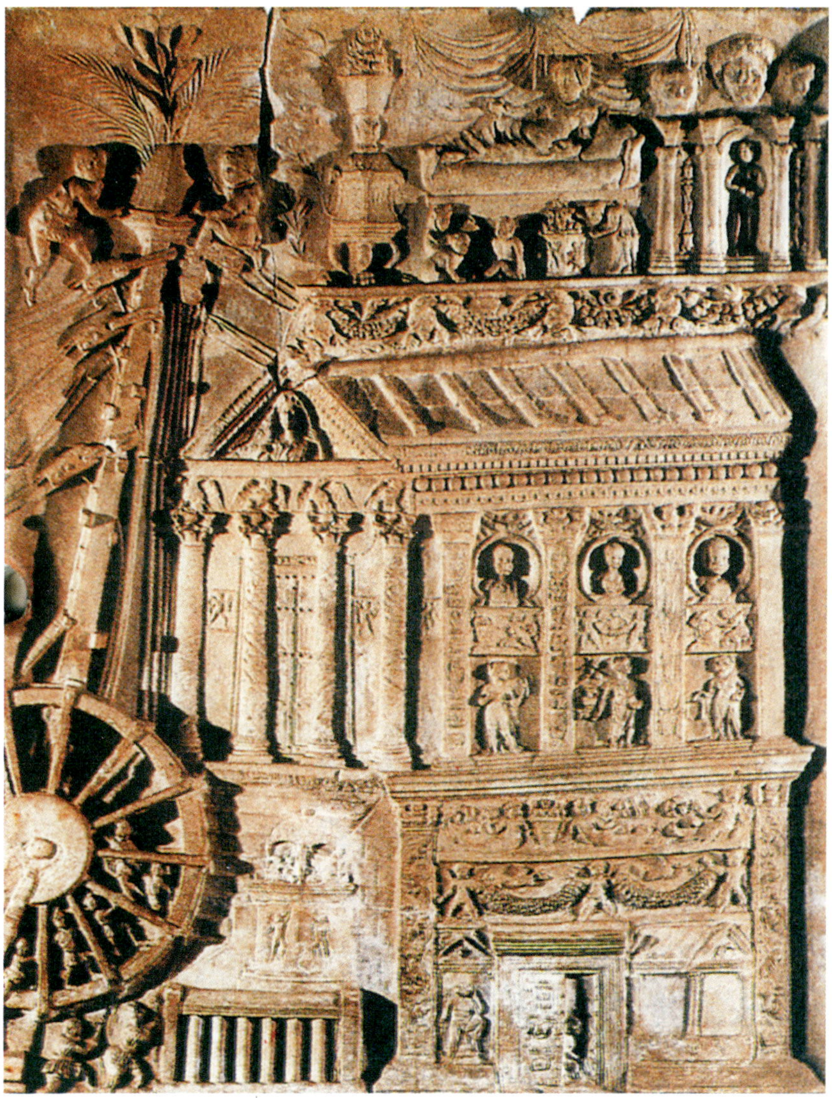

Figure 1
Vitruvian crane from the bas-relief on the Hateri tomb (Lateran Museum, Rome).

* Lord Acton Lecture, New York University, December 18, 1997.

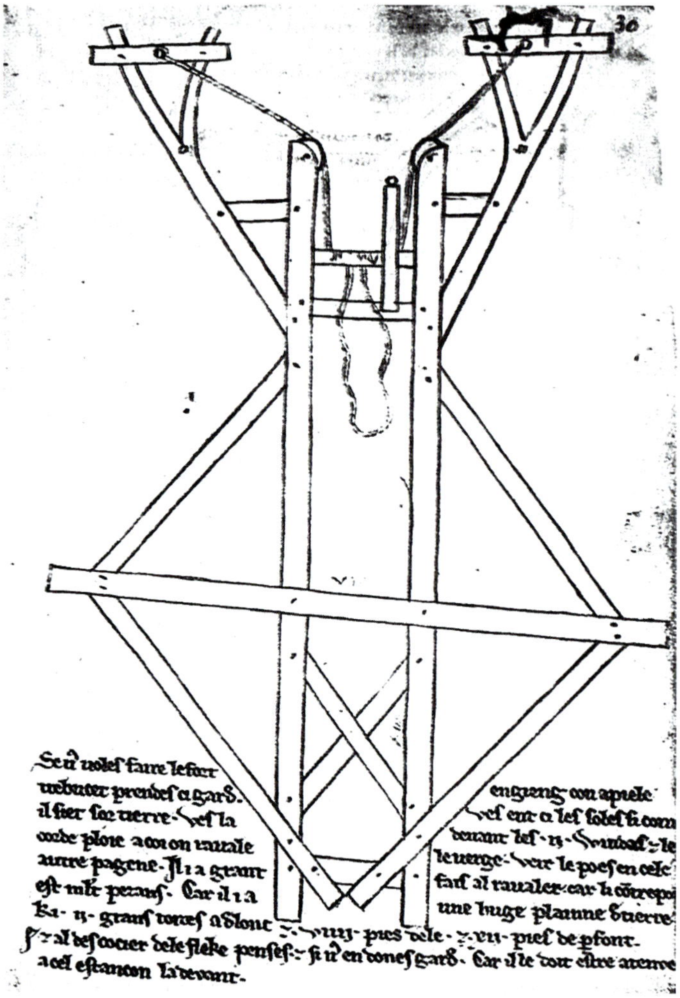

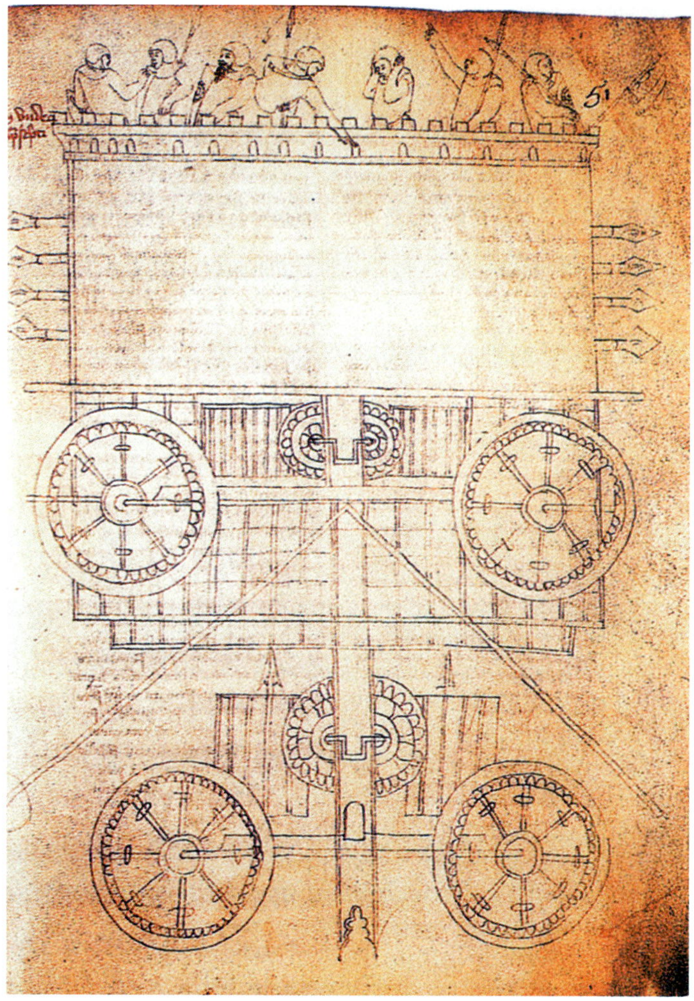

Figure 2 (*left*)
Trebuchet. Drawing by Villard de Honnecourt (Ms. Franc. 19093, Bibl. Nat. Paris, c. 30r).

Figure 3 (*right*)
Mobile assault tower. Drawing by Guido da Vigevano (Ms. Lat. 11015, Bibliothèque Nationale, Paris, c. 51r).

crossbows in the *De rebus bellicis*, a late-classic treatise on military arts, or the few drawings of mechanical devices in the *Notebook* of Villard de Honnecourt (thirteenth century) (fig. 2) and the illustrations of the *Texaurus* of Doctor Guido da Vigevano from Pavia, written in France in 1348 (fig 3). At the end of the Trecento, the technical activity of the German world had found its visual expression in the *Bellifortis* of Conrad Kyeser.

For the most part these early drawings rely upon such rudimental visualisation techniques that our eye, educated to perspective, finds it difficult to decipher the machines: usually the mechanisms are 'squashed' in planimetric view, or 'compressed' in section, with a particularly unhappy effect on the legibility of the spatial structure of their parts and for the understanding of their functions.

This primitive technique is common to Greek codices, Arab technical manuscripts, Villard de Honnecourt, and Guido da Vigevano, while in Kyeser, and later in Valturio and Giovanni Fontana, one discerns the attempt to propose a perspectival view of machines; which still, however, appear "warped."

This was the modest state of the art between the end of the Trecento and the beginning of the Quattrocento, when a process of specialisation in drawing of machines began. In only a few decades this process would bring a definition of criteria and techniques capable of conferring on this sector of artistic activity an unprecedented relevance.

One of the most effective motors behind this process was the need to restore clarity to the technical knowledge of the ancients. Rendering comprehensible largely corrupted texts, like the *De architectura* of Vitruvius, meant above all reinventing the drawings which the ravages of time had cancelled. In fact, a technical text without images is substantially "mute." Thus, the restoration of clarity could not be achieved by the philol-

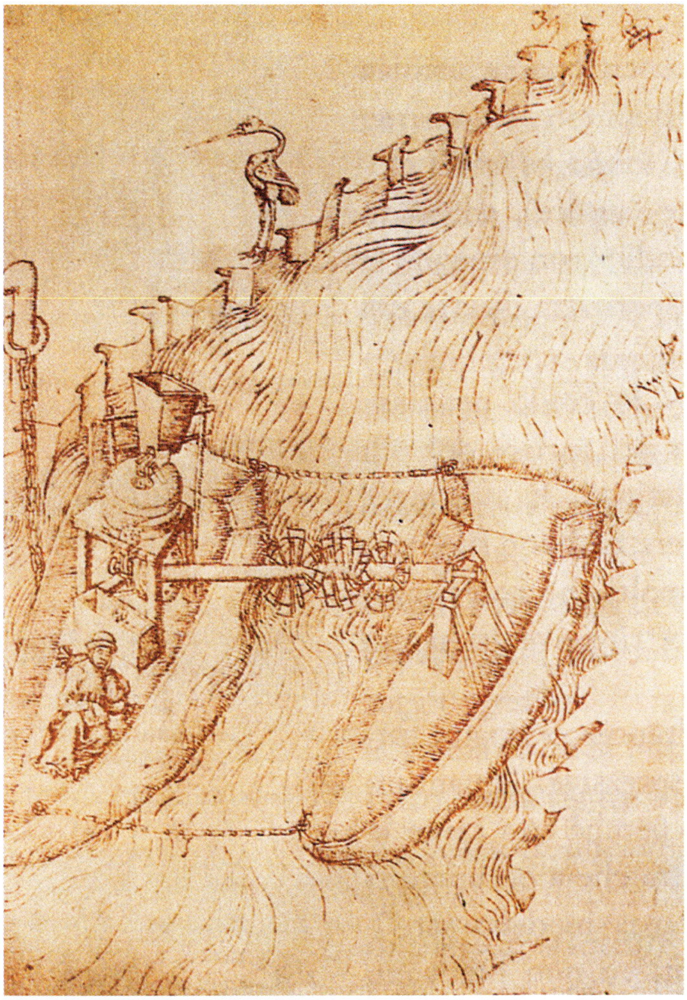

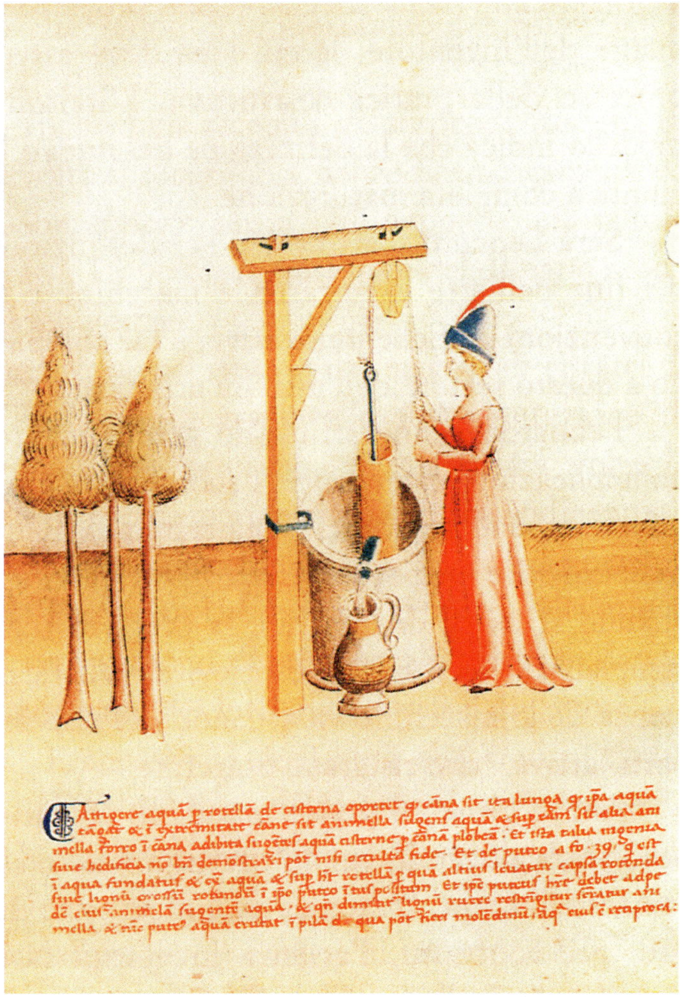

Figure 4 (*left*)
Floating mill with sleeping miller. Drawing from Taccola: *De ingeneis III-IV* (Ms. Palat. 766, Bibl. Naz Florence, c. 39r).

Figure 5 (*right*)
Female figure operating a piston pump. Drawing from Paolo Santini (after Taccola): *De machinis* (Ms. Lat. 7239, Bibl. Nat. Paris, c. 42v).

ogist alone. The collaboration of the artist, expert in machines and able to visualise them through drawing, was also necessary.

One perceives this tension in Mariano di Iacopo, called Taccola. A contemporary of Brunelleschi, Taccola was a Sienese artist who, in more than forty years of activity, produced an iconographic corpus of technical subjects which, for the variety of themes: hydraulics, military and construction techniques, etc., is in strident contrast with the rare visual documents of the ancient and medieval tradition. For Taccola, who draws in perspective, but still with some uncertainty, the machine is not yet a subject worthy as such of portrayal. Beyond the frequent insertion of machines into the landscape, which appears almost always functionally redundant, we observe in Taccola's drawings of technical devices the recurring presence of the human figure, to which the machine remains subordinate: man serves to express intolerance for the heavier mechanical operations, or exults in his liberation from labour thanks to mechanical devices, as in the scene of the miller sleeping on the floating mill (fig. 4). The gracious technology of Taccola's world even allows women to play an active role in the mechanical universe (fig 5).

Taccola, who also nourishes humanist ambitions, strives to restore visibility to the technical knowledge of the Greeks and Romans. This is shown, among many other examples, by his graphic divination of the inverted siphon of the Roman aqueduct at Toledo (fig 6). Moreover, he conceives of daring projects aimed at recovering the submerged documents of the ancient civilizations, in an exercise of what we would today call submarine archaeology (fig. 7).

This trend was accelerated thanks to another Sienese artist, Francesco di Giorgio. Active in Siena, Urbino, and Naples in the second half of the Quattrocento, Francesco di Giorgio introduces a series of solutions that allow him to reach an unprecedented pre-

Figure 6 (*right*)
Imaginary depiction of the inverted siphon of the Roman aqueduct at Toledo. Drawing from Taccola: *De ingeneis I-II* (Bayer. Staatsbibl. Munich, c. 72v).

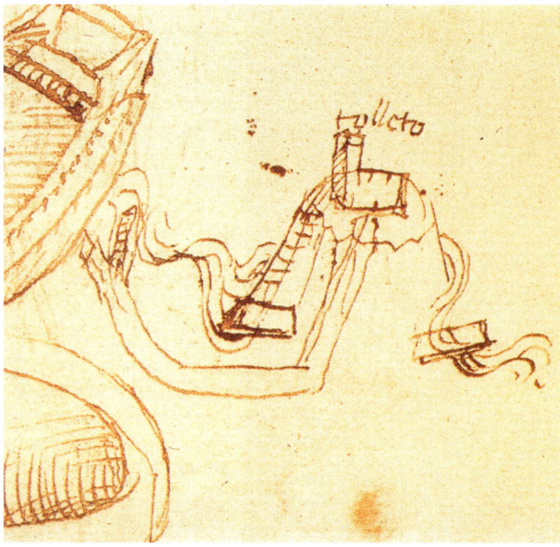

Figure 7 (*below*)
Diver and pontoon for recovering a column. Drawing from Taccola: *De ingeneis III-IV* (Ms. Palat. 766, Bibl. Naz Florence, c. 18r).

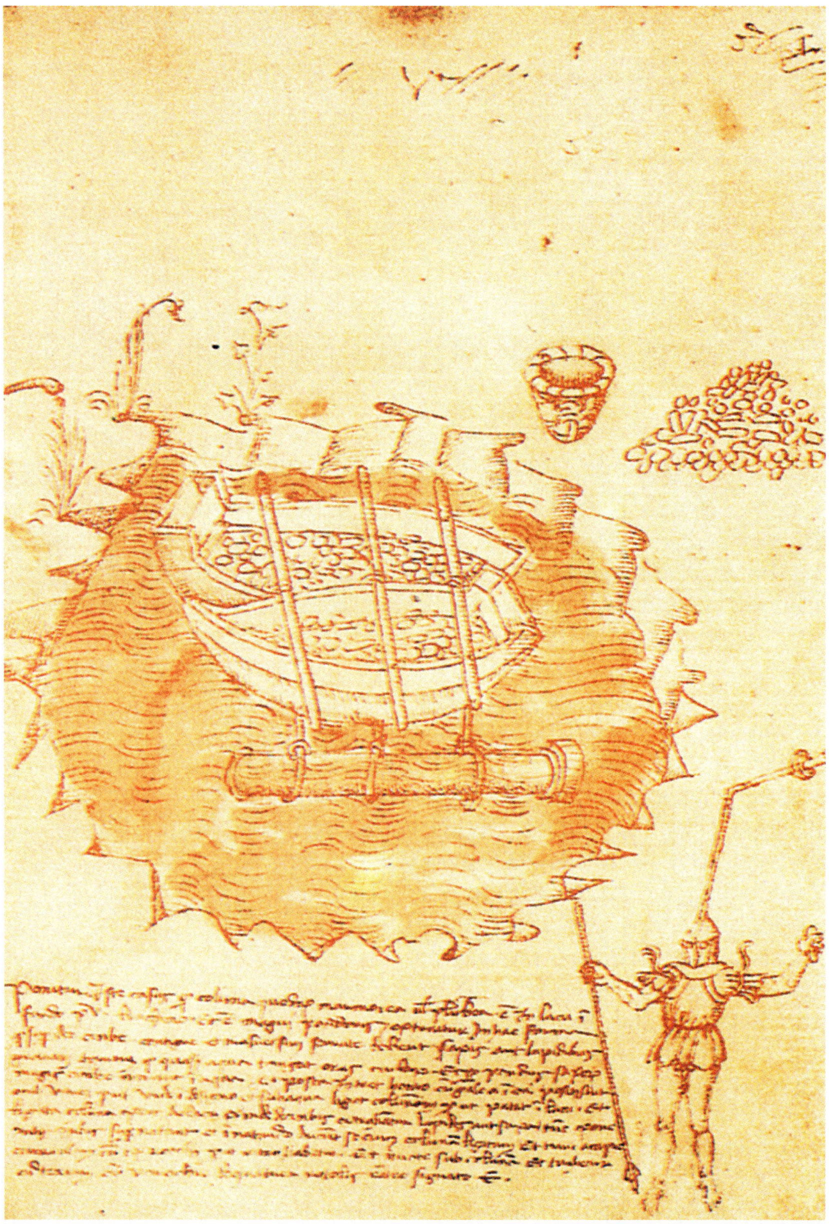

cision in portraying machines. For the first time, the machine begins to free itself of landscape and, above all, of the human figure.

Francesco displays great ability in suggesting the spatial structure of mechanisms of remarkable complexity. He lays them out systematically in open box structures (fig. 8). This allows him not only to show the distribution in depth of the various parts, but also to supply a visual scheme which renders the gears and outlines clear and tangible. We are also indebted to Francesco for a collection of stupendous drawings of technical equipment and devices which appear in his parchment codex dedicated to Federico da Montefeltro, for the first time without a single word of explanation. Full confidence in the expressive power of the drawing has by now rendered verbal description superfluous.

Thanks to the quality represented by the masterly portraits of mechanical devices by Francesco and thanks also to his reputation as a qualified engineer, machines for the first time are selected as appropriate elements for monumental decorations, as in the extraordinary bas-reliefs of mechanical subjects (fig. 9) which adorned the stone bench outside the Palazzo Ducale in Urbino. This was a clear sign of the new value which Renaissance princes were beginning to attribute to antique and modern machines, as well as to their inventors and draughtsmen.

Later in his career Francesco dedicated himself to the compilation of a *Trattato*, a fairly systematic text with many learned quotations, in which, following the Vitruvian model, architecture and machines appear juxtaposed. Here the interplay between the essential descriptive text and the thick texture of excellent portraits of machines is exemplary and new (fig. 10). In its elegant and functional impagination, the *Trattato* appears as a prophetic anticipation of the modern illustrated technical encyclopaedia.

This mature work of Francesco di Giorgio eloquently reveals his strong motivation to become an author, entering thus into competition with the humanists: the artist proudly claims his privileged position, depending on direct experience of the technical processes as well as on his draughtsman's ability. It is not by chance that Francesco will sneer at "those" – I believe he was thinking of Leon Battista Alberti – "who illude themselves to restore clarity to the texts of the ancient architects and engineers, simply through Greek and Latin grammar." Elsewhere, he will assert peremptorily that one can not be a good philologist nor a capable architect if one lacks draughtsmanship.

Even though he continually exalts drawing, Francesco still shows some reservations about his expressive potential. He repeats that drawing and descriptive text can not provide full expression of the project as conceived in the mind of the inventor, given that the original idea must be adapted and defined through the practical process of construction. The unresolved tension in Francesco between drawing and model indicates that the development of the new language had not yet reached complete maturity.

It will be Leonardo who, through a new definition of the function and ends of the art of portraying machines, and, thanks to the introduction of innovative graphic conventions, will make a qualitative leap for this branch of artistic activity.

Leonardo gradually refined this art by extenuating application. His youthful drawings (fig. 11) in fact show imposing technical structures immersed in landscape and architectural elements which are still of a medieval style. The perspective here seems uncertain, especially when he attempts to show the meshing of gears or threadings which are not complanate. The effort to render realistically the power and the efficiency of the machines plays against respect for the proportions of the "active" parts, which often appear exaggerated.

Also the studies of flying machines (fig. 12) of Paris Ms. B, which are, again, early studies, show this emphatic attitude towards power and the motive parts. The utopian magniloquence of these drawings reveals by contrast the absence, in this phase, of mature theoretical investigation of force and resistance and of the nature of the fluid air.

The less dramatic drawings of the later *Codex on the flight of birds* (1505-1506) are in reality much more innovative (fig. 13). These fix the theoretical results of Leonardo's penetrating observation of birds' flight; introduce the consciousness of the determining influence of air, wind, and currents; and show the functioning of the bird machine and its directional instruments through refined schematisations, which suggest daring analogies: "The bird is an instrument that operates

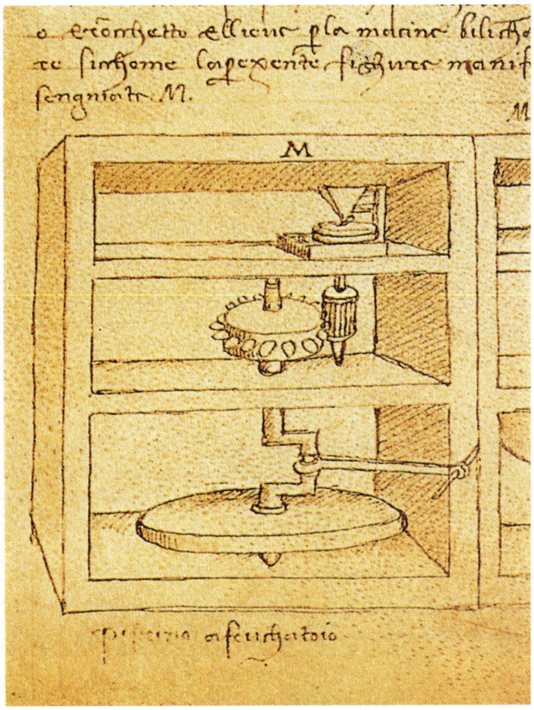

Figure 8 (*left*)
Mill mechanism. Drawing from Francesco di Giorgio: *Trattato di architetura e macchine* (Ms. Ashb. 361, Bibl. Med. Laurenz. Florence, c. 34r - detail).

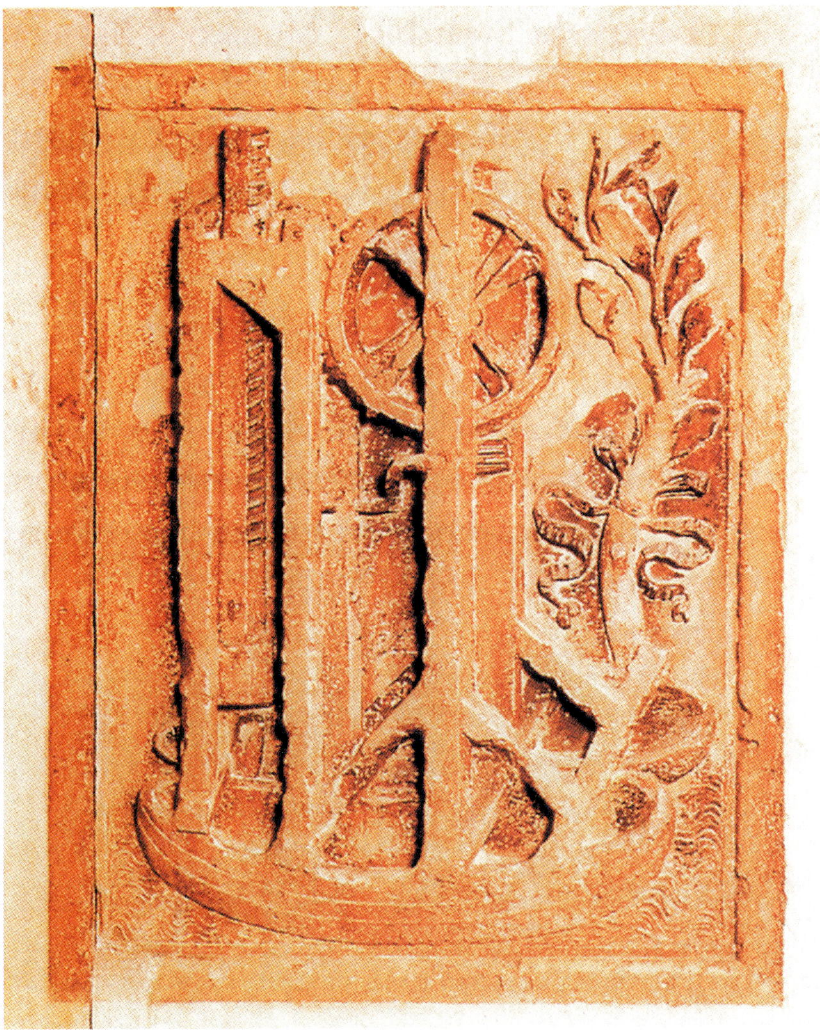

Figure 9 (*below*)
Floating pile driver (after Francesco di Giorgio). Stone relief on the Palazzo Ducale at Urbino.

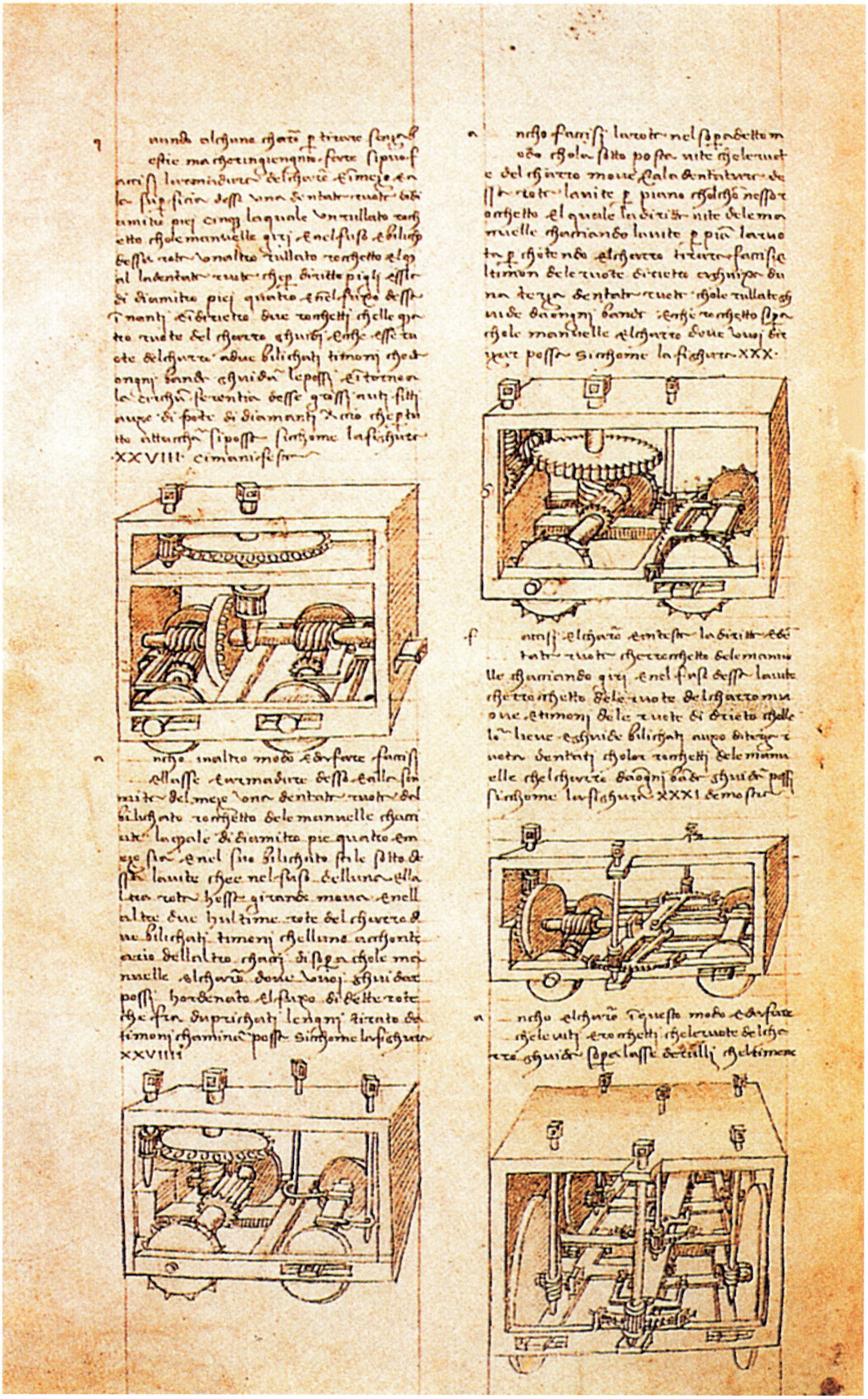

Figure 10
Various types of cars. Page from Francesco di Giorgio: *Trattato di architettura e macchine* (Ms. Ashb. 361, Bibl. Med. Laurenz. Florence, c. 40r).

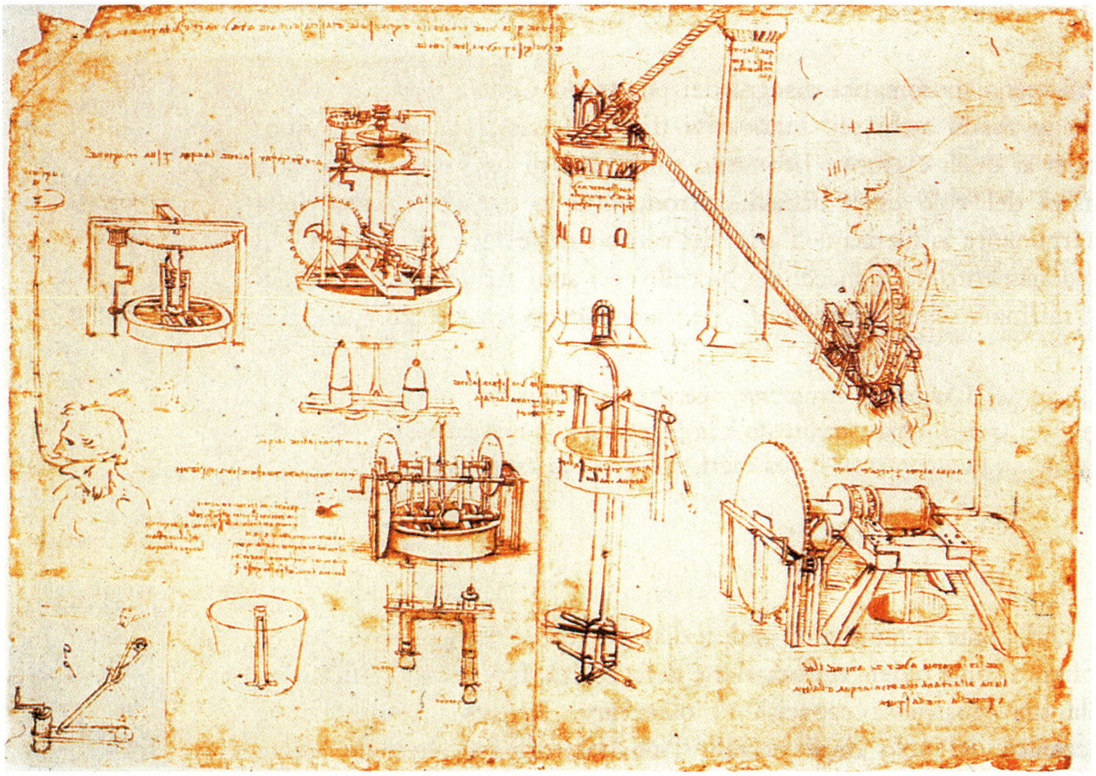

Figure 11
Hydraulic devices. Drawing by Leonardo da Vinci (Cod. Atlant., Bibl. Ambros. Milan, c. 1069r).

according to mathematical laws, an instrument that it is in man's power to reproduce, but not with its natural force."

Here Leonardo's technical drawing must not be intended – as has often been suggested – as a faithful reproduction of reality, fruit of his enormous capacity for observation. Rather, his strictly mechanical interpretation of bird flight (the wing as wedge, the tail as rudder, the hydrostatic laws as a constant guideline) does in fact urge him to transfigure reality through graphic registrations.

This is the first symptom of the new function assigned to technical drawing; that of giving visual demonstration of theoretical hypotheses or of purely mental experiments.

The invasion of theory into the sphere of representation is thus for Leonardo not circumscribed to his activity as a painter. His performance in this field, as is known, became an ever more challenging task, as a consequence of his ambition of including in his paintings the results of his efforts to decipher the mechanisms and the schemes of living nature and of the inorganic world: the body of man, plants and animals, the structure of the earth and the eternal battle, rigorously governed by geometrical laws, between light and shadow.

This essential tension will culminate in the Madrid Ms. I, from the last decade of the Quattrocento, in which the most advanced elaborations of his treatise on "elementi macchinali" (the elements, or foundation, of mechanics) are conserved. This is the title under which Leonardo intended to collect all of his thoughts on both practical and theoretical mechanics.

At the beginning of his career Leonardo is one of the many talented but "unlettered" artisans active in Renaissance artists' workshops. He does, however, achieve professional qualification through great application. He attempts to assimilate classical mechanics and to learn the "elements" of Euclidean geometry. This long and intense self-training program will produce new awareness and a new method of working. But he will never forget the practical dimension of the artists' workshop. The studies of his maturity in fact aim at establishing a mechanical theory conceived as a tool to deal with real machines, machines that operate with a physical structure, create friction, and that are made of materials with limited resistance.

The initial section of Madrid Ms. I, in which he analyses a series of mechanical devices, is particularly noteworthy. He clearly declares that he intends to produce a real "anatomy of

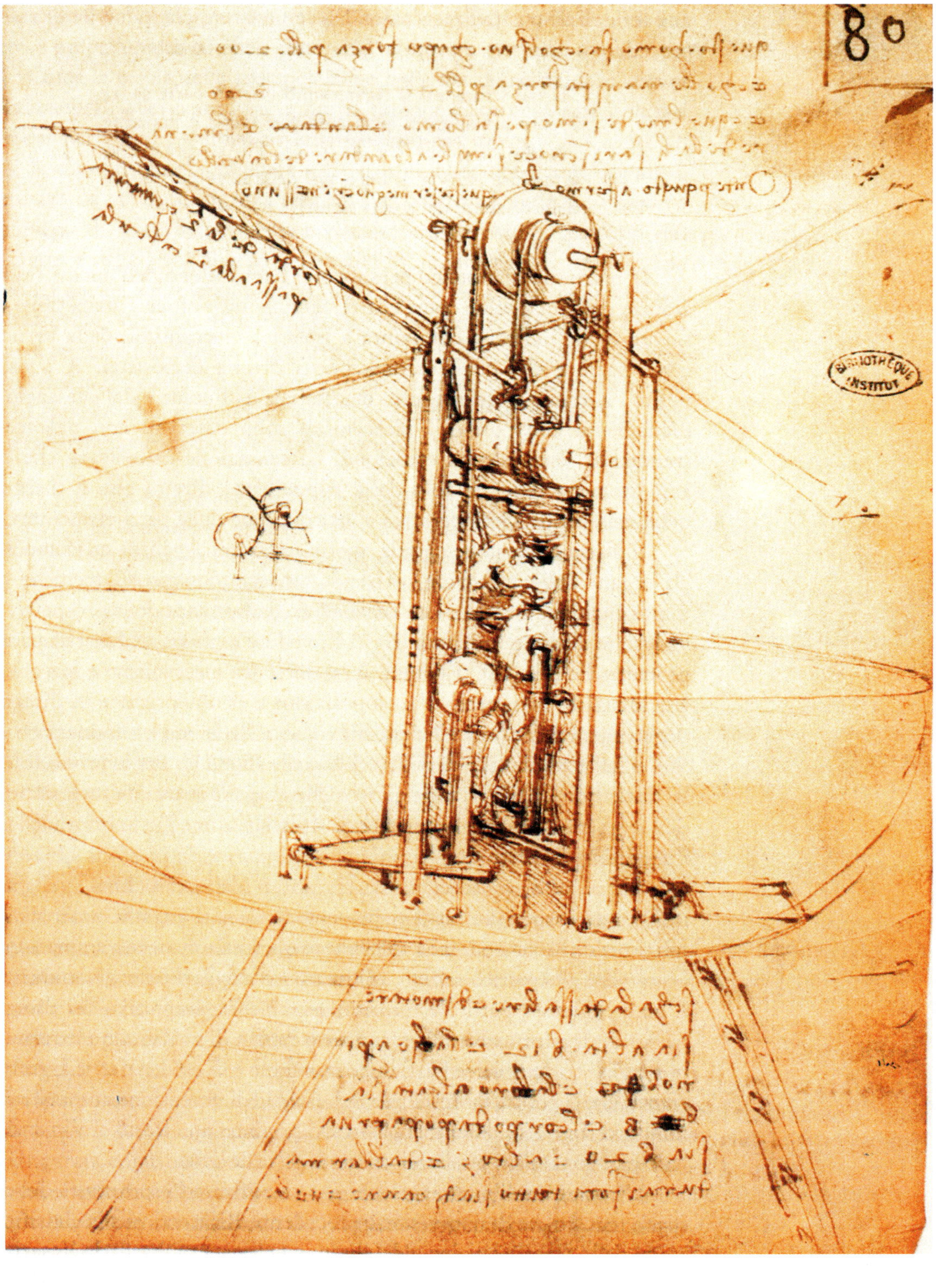

Figure 12
Man operating a flying machine. Drawing by Leonardo da Vinci (Ms. B, Inst. de France Paris, c. 80r).

machines": "And such instruments will for the most part be presented without their armatures or other structures that might hinder the view of those who study them. Then these same armatures shall be described by means of lines, after which we shall describe the levers by themselves, then the straight of the supports."

This anatomical approach presupposes leaving behind the traditional analysis according to which each machine was an indivisible and unrepeatable entity, in favour of an examination tending to isolate the elementary mechanisms whose combinations give rise to a variety of machines. At this point the autonomy of the mechanical subject from the human figure and landscape is fully achieved. The mechanisms are now observed with cold objectivity. The sensation of beauty that the drawing of these devices emanate depends solely on the direct response between functionality and the articulation of the elementary parts.

Drawing instructs and guides this inquiry. Leonardo dissects and recomposes the elementary components of the machines, thanks to a series of illustrative techniques which are based on an exceptional mastery of perspective: views in elevation and in plane (fig. 14), exploded views, and geometrical schematisations (fig. 15). At times, the rapport between mechanisms and machine is underlined by the deconstruction and reconstruction in drawing the elementary mechanical elements, as in the screw jack with ball and roller bearing (fig. 16), or the machine for milling iron staves for cannons (fig. 17). For the first time, he distinguishes the various parts of the machines by lettering and numbering, which permits a precise dialogue between the information in the text and that conveyed by visual demonstration.

At other times, the basic mechanisms are extracted from the machine and portrayed suspended in the air, in order to emphasise their abstract functional potential. Using the same techniques as in modern animation, cinematic chains are effectively suggested (fig. 18). The shading loses the purely decorative function of his youthful technical drawings and is transformed into an instrument capable of visualising the spatial relations between the various parts of the machines and highlighting their reliefs and outlines. The drawing of an automatic flintlock for a firearm (fig. 19) is exemplary of the three-dimensional effect produced by the *chiaroscuro*. Here the eloquence of the drawing is comparable to that of a physical model.

The use of shading to increase the expressiveness of technical drawing was without precedent. Making reference to the classification of diverse categories of drawing which Giorgio Vasari later proposed, one could say that in Leonardo's mechanical illustrations we see the passage from "disegno delineato" al "disegno finito" (from the simply 'outlined' to the 'finished' drawing); that is, to the true 'portrait.'

Just as the theory of dissection of the machines into their basic organs had pro-

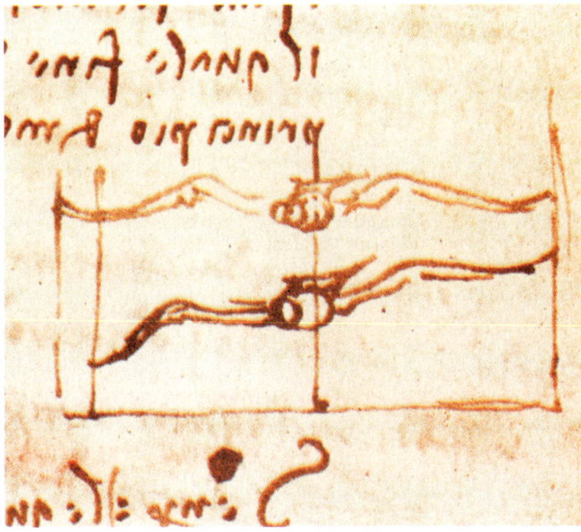

Figure 13 (*left*)
Study of the flight of birds. Drawing by Leonardo da Vinci (Cod. sul volo degli uccelli, Bibl. Reale Turin, c. 14r).

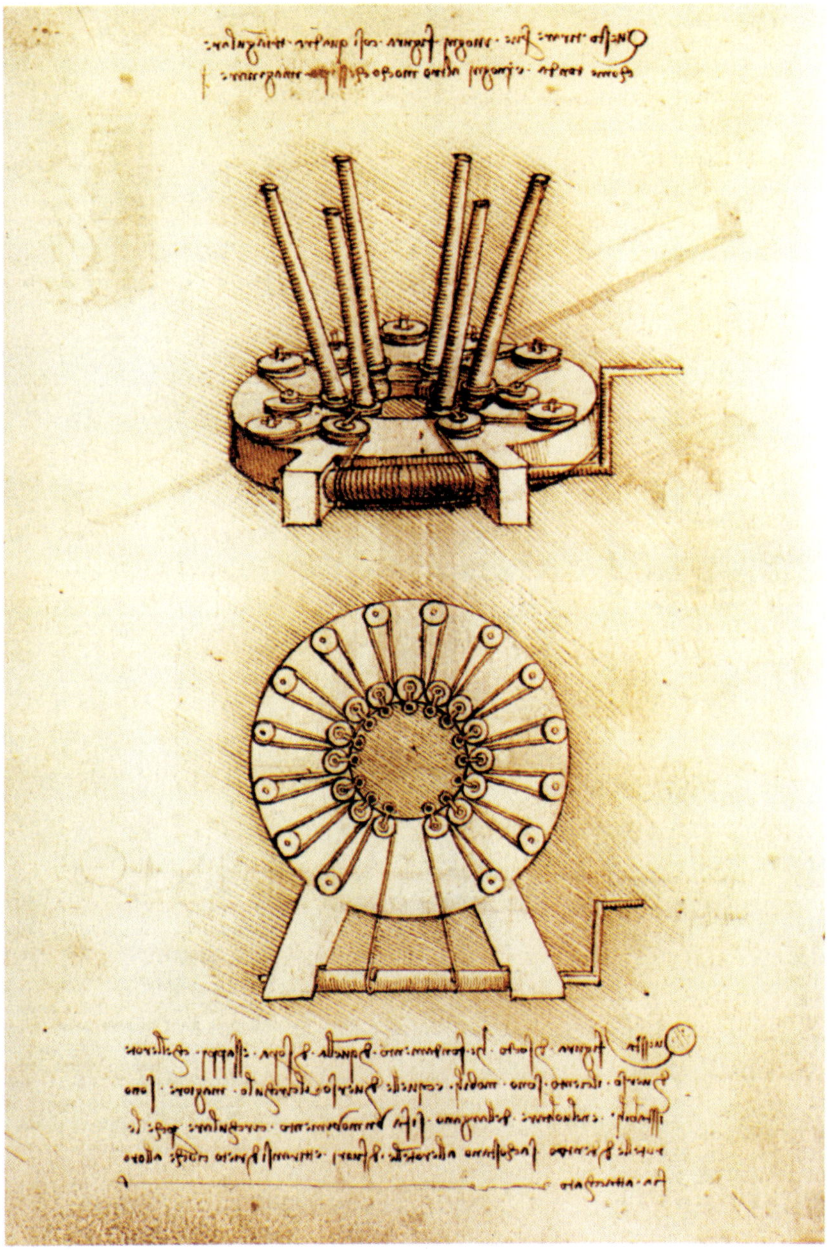

Figure 14 (*below*)
Stretching device. Represented as elevation and plan by Leonardo da Vinci (Cod. Madrid I, Bibl. Nac. Madrid, c. 44v).

Figure 15 (*right*)
Equivalence of screw and inclined plane. Drawing by Leonardo da Vinci (Cod. Madrid I, Bibl. Nac. Madrid, c. 86v - detail).

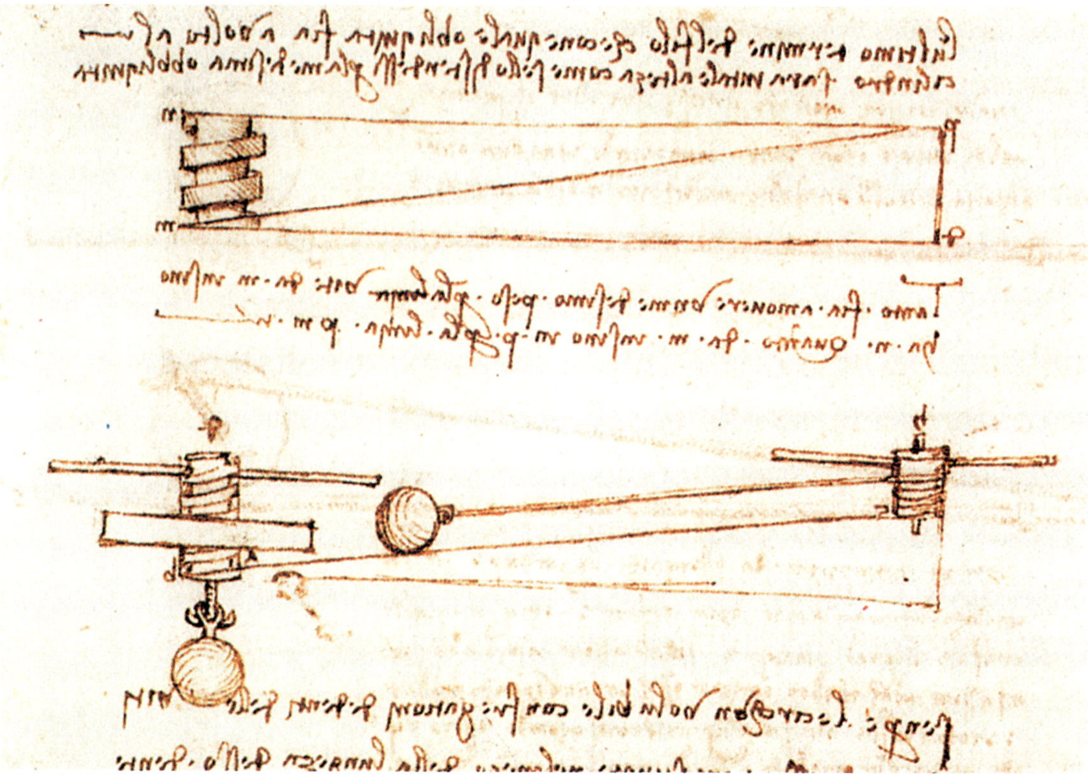

Figure 16 (*below left*)
Screw jack with bearings. Drawing by Leonardo da Vinci (Cod. Madrid I, Bibl. Nac. Madrid, c. 26r).

Figure 17 (*below right*)
Machine for milling cannon barrels. Drawing by Leonardo da Vinci (Cod. Atlant., Bibl. Ambros. Milan, c. 10r).

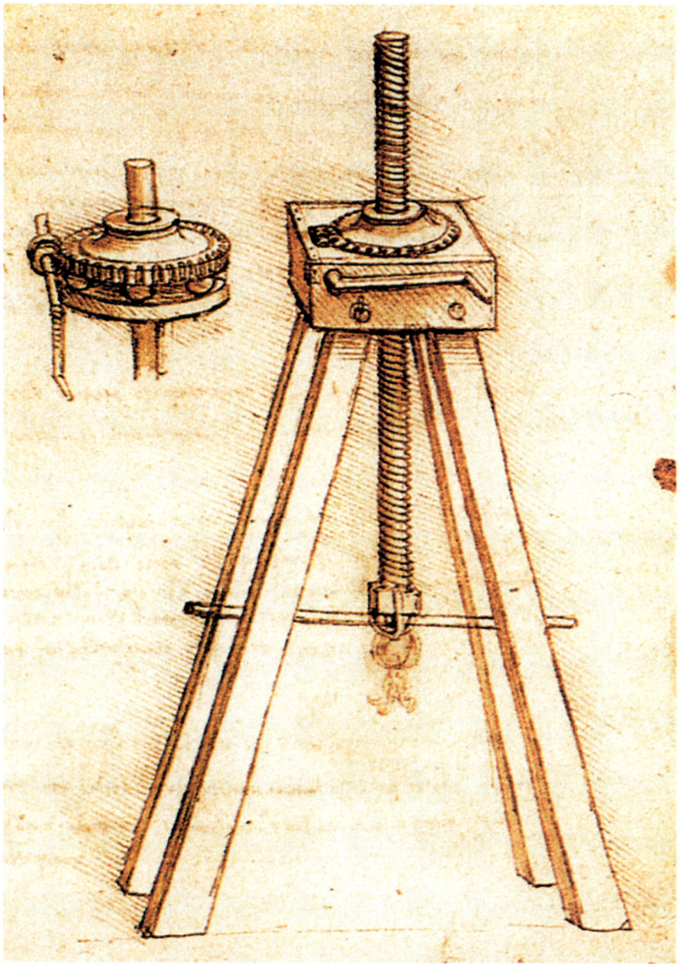

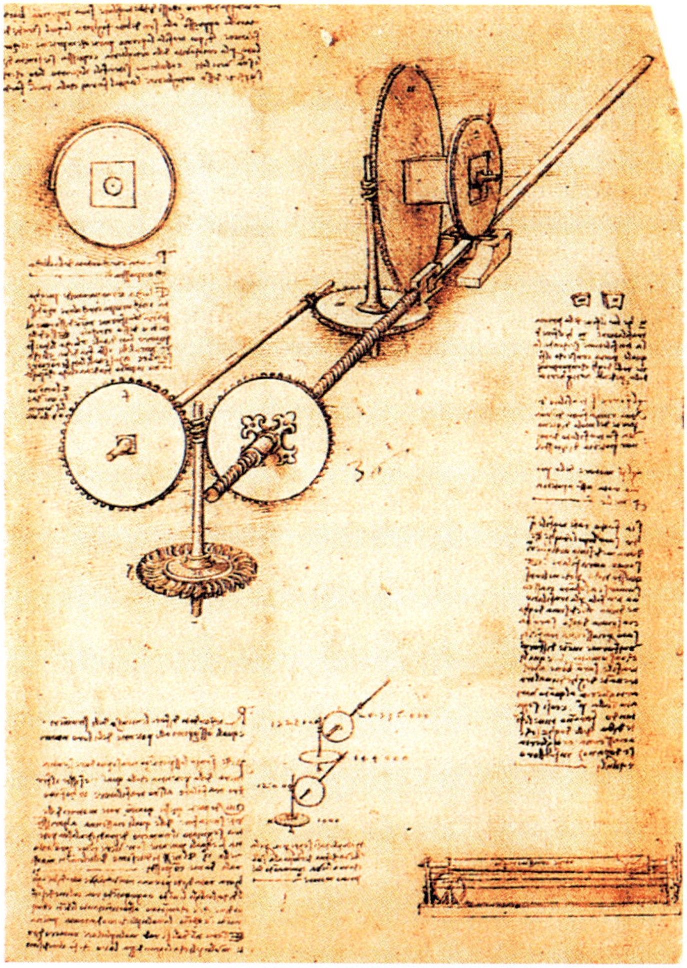

duced the new drawings of isolated mechanisms suspended in the air – a singular anticipation of Calder's mobiles – so also the care taken in modelling the devices with shading derived from another theoretical necessity. In the Madrid Ms. I, in fact, one encounters a shading by means of curved lines to show the erosion produced by the contact and rubbing of the moving parts of several devices. Leonardo studied the friction of dragging and rolling theoretically as well, and proposed to show the effect of attrition by this new shading technique. In this manner he succeeds also in suggesting the dynamics of contact and rotation. Leonardo's full awareness of this new graphic convention emerges in the text appended to a drawing of a "universal joint" consumed by friction (fig. 20): "I found that all axles, no matter what kind of metal they are made of [...] are worn in such a way that that part of their length which is nearer the centre, becomes thinner and, conversely, that part which is farther from the centre, remains thicker. In this way the axle which was originally made up of straight lines, becomes constituted of curved lines."

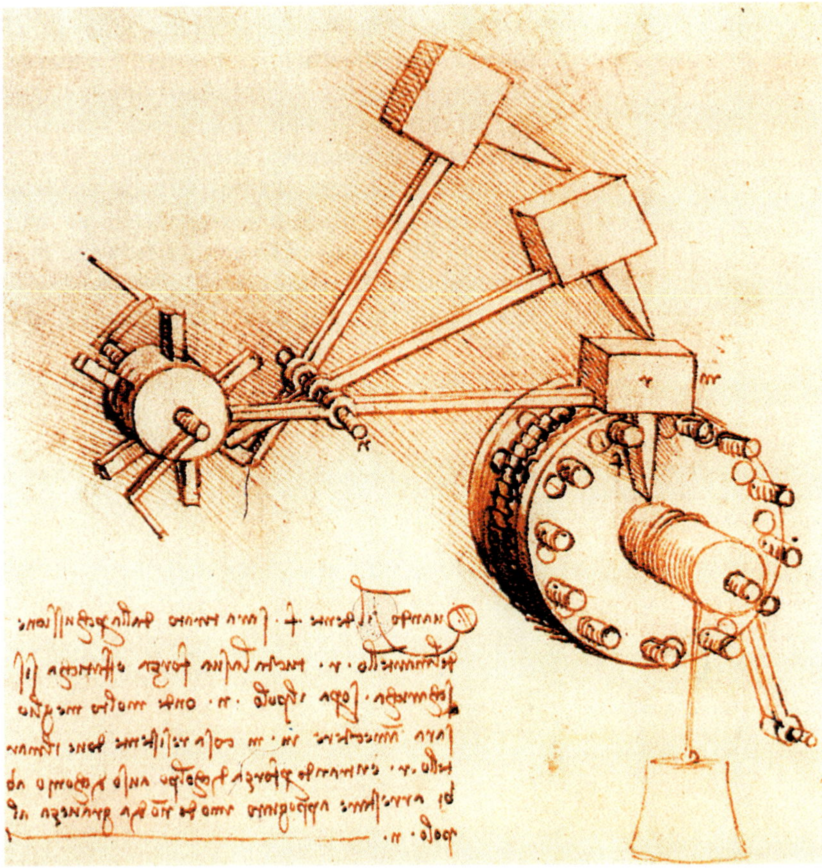

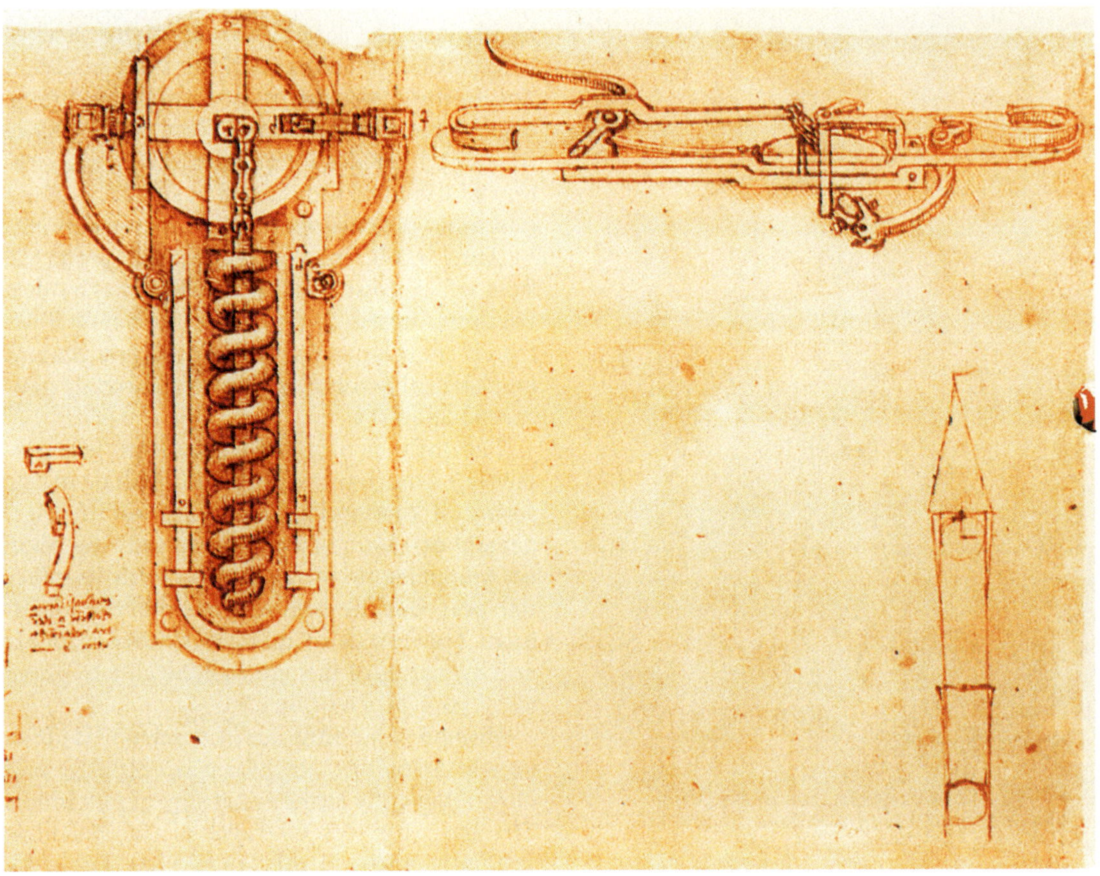

Figure 18 (*above*)
Trip hammers operated by a camshaft. Drawing by Leonardo da Vinci (Cod. Madrid I, Bibl. Nac. Madrid, c. 92v - detail).

Figure 19 (*left*)
Automatic wheel-lock of a firearm. Drawing by Leonardo da Vinci (Cod. Atlant., Bibl. Ambros. Milan, c. 158r).

This is eloquent proof of the density of information stratified in a single Leonardo drawing and of his continual striving towards the refining of graphic techniques, impelled as he was by theoretical motivations.

Shading by curved lines will be used by Leonardo in other fields as well, as can be seen in studies of trees and flowers, ballistics and fortifications (to suggest the deviation of bullets produced by their impact on the rounded surface of the bastions) and in those on birds' flight. These masterly drawings show another aspect of theory at work: the perennial dynamic contrast between the subject's action and the reaction of the medium: "as much force is exerted by an object against the air, as by the air against the object."

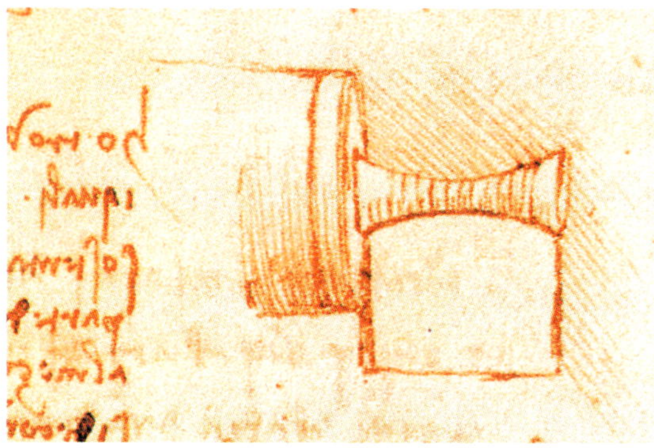

Figure 20 (*right*)
Wear caused on a pole by it's support. Drawing by Leonardo da Vinci (Cod. Madrid I, Bibl. Nac. Madrid, c. 119r - detail).

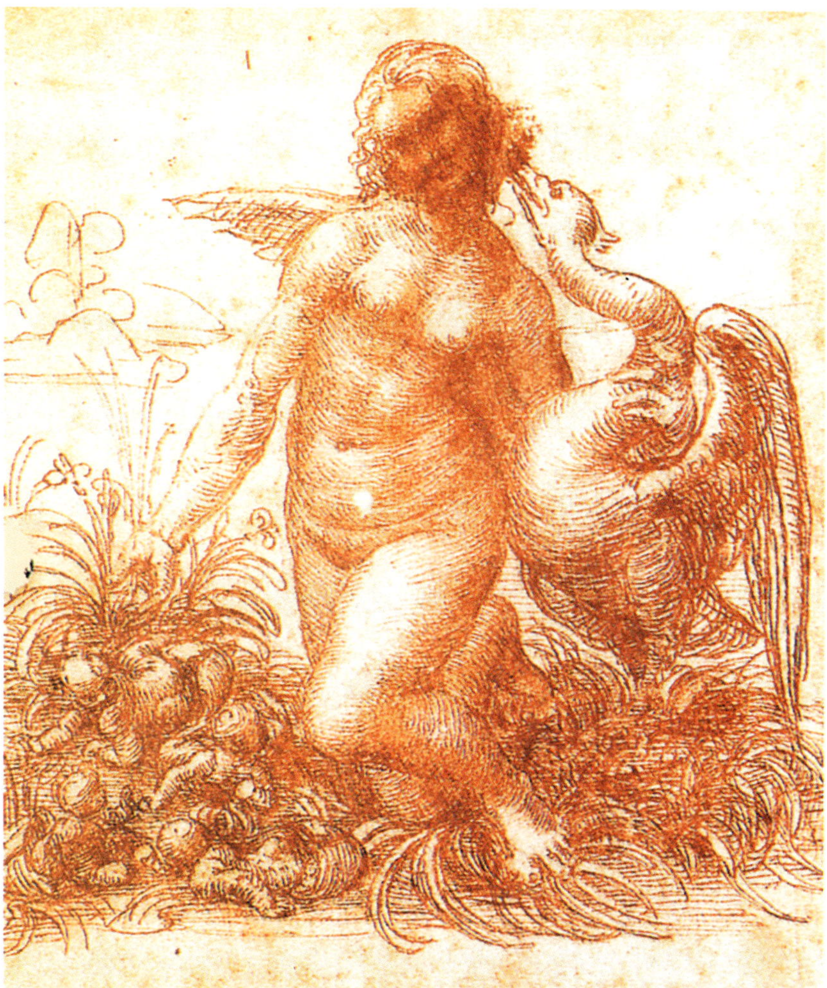

Figure 21 (*below*)
Leda kneeling with swan. Drawing by Leonardo da Vinci (Chatsworth, Devonshire Collection).

This technique seems even more evident in the study of the kneeling Leda (fig. 21), a classical subject which Leonardo reproposes as a symbolic point of intersection between the irrepressible forces of life and the iron laws of mechanics that govern nature. Leda is pictured as a spring (fig. 22) which releases accumulated energy. Her rising in a spiralling motion is emphasised by the heavy curvilinear hatching. The shading sensually suggests the friction between the female body and the air. Here, most appropriately, Sir Kenneth Clark has spoken of the "curves of life." The function of drawing as a hybrid melting-point between the organic and the artificial, that is, between the production of man and the production of nature, appears striking here.

The 'dynamic' anatomy of machines was a model which Leonardo attempted to render universal after 1500. This method finds application in architectural studies where the laws of the "elements of mechanics" are introduced to explain the lateral thrust of the arches in quantitative terms. Buildings are analysed not as static structures, but rather as 'living' organisms in dynamic equilibrium. Ever more frequently the drawing appears as a registration of the ideal moment of suspension of the perennial process of cosmic transformation.

After 1500, man becomes the preferred field of exercise for Leonardo's mechanics. He draws up his own program for a new illustrated encyclopaedia, based on a few general mechanical principles and on rigorously geometric analyses.

The influence of his treatise on "the elements of mechanics" appears evident in the series of anatomical drawings of the first decade of 1500: "Arrange it so that the book on the elements of mechanics with its practice precedes the demonstration of the movement and forces of man and other animals and by means of this you will be able to prove all of your propositions."

Human anatomy for Leonardo is thus largely tributary to his conceptually and chronologically earlier anatomy of machines. The basis for this courageous transferral is explicitly stated in the Madrid Ms. I: "Every

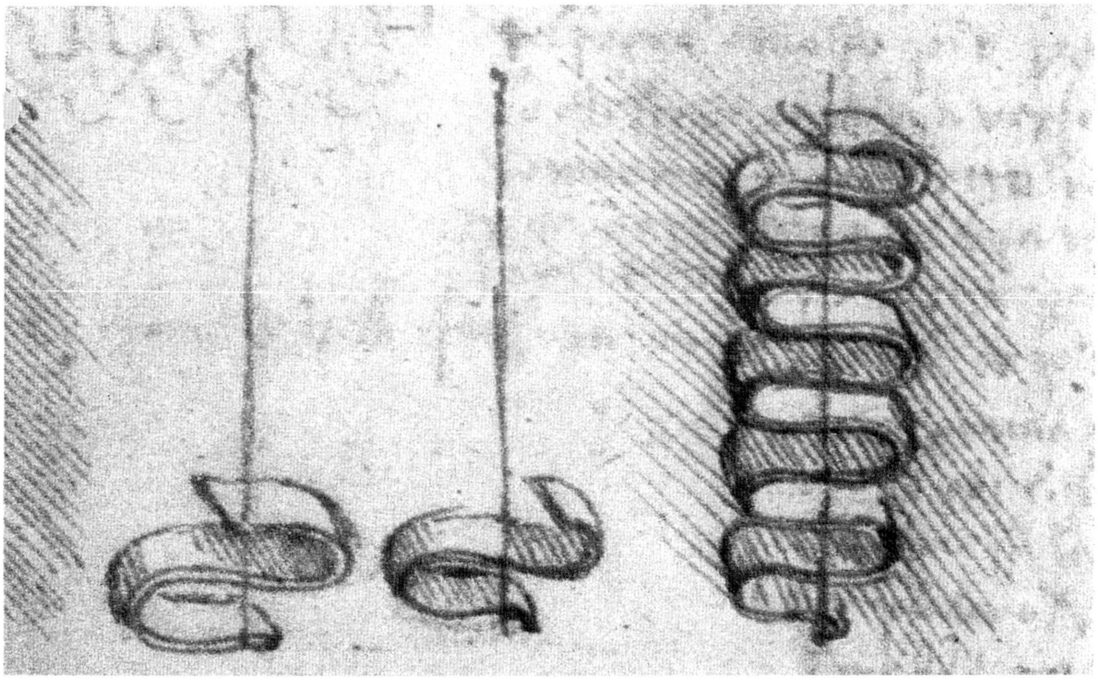

Figure 22
Spring with detail of its coils. Drawing by Leonardo da Vinci (Cod. Madrid I, Bibl. Nac. Madrid, c. 85r - detail).

body requires its members and every art its instruments. And the moment that the whole is created, its parts are also created."

This means that the parts come before the whole in the productions of man as in those of nature. Nature is never proposed by Leonardo as "creator," but rather as "composer": it mixes in various ways the fundamental elements in accord with universal mechanical laws. It follows that reality can not be revealed at the level of surface appearances, where one sees only the composite already formed, but in the basic elements and in the processes of composition. Drawing is thus necessary in order to grasp the intimate essence of things. Drawing is the sharpest analytic instrument that man possesses; guided by geometry and mechanics, it opens the way to the understanding of the compositive process.

The extension of the principles of the anatomy of machines to the inspection of organic structures is based on the conviction that mechanical laws reign over nature's productions as well: "Nature cannot give movement to animals without mechanical instruments, as I demonstrate in this book." And again: "Instrumental or mechanical science is most noble and useful beyond all others since by means of it all animate bodies that have motion perform all their operations."

Besides the general principles, also the technical terminology and, above all, the illustrative techniques are derived from his earlier mechanical investigation.

Leonardo's anatomical studies present the human body as an extraordinary complex of devices: "It does not seem to me that coarse men with lewd habits and little reasoning power deserve so beautiful an instrument or so many varieties of mechanical arrangements."

The multiple articulations of the body are analysed as ball and socket joints. Leonardo continually stresses the similarity between the articulation of the shoulder and the universal joint. Moreover, he defines tendons as "mechanical instruments [...] which carry out as much work as is assigned to them" (*Corpus of the Anatomical Studie*s (K. D. Keele and C. Pedretti) 1772).

Another mechanical term, "fulcrum," is used to indicate the ankle joint: "The fulcrum is where a man balances his weight through the tendons *mn* and *op* which are to the shank of the leg above the said fulcrum as the shrouds are to the mast of ships." Here is introduced that mechanical analogy with the mast of the ship and its shrouds that Leonardo will use also in his studies of the structure that sustains the head (fig. 23).

The analogy between articulations and universal-joints induces the search for antifriction mechanisms in the human body as well, which Leonardo will detect under the big toe: "Nature has placed the glandular

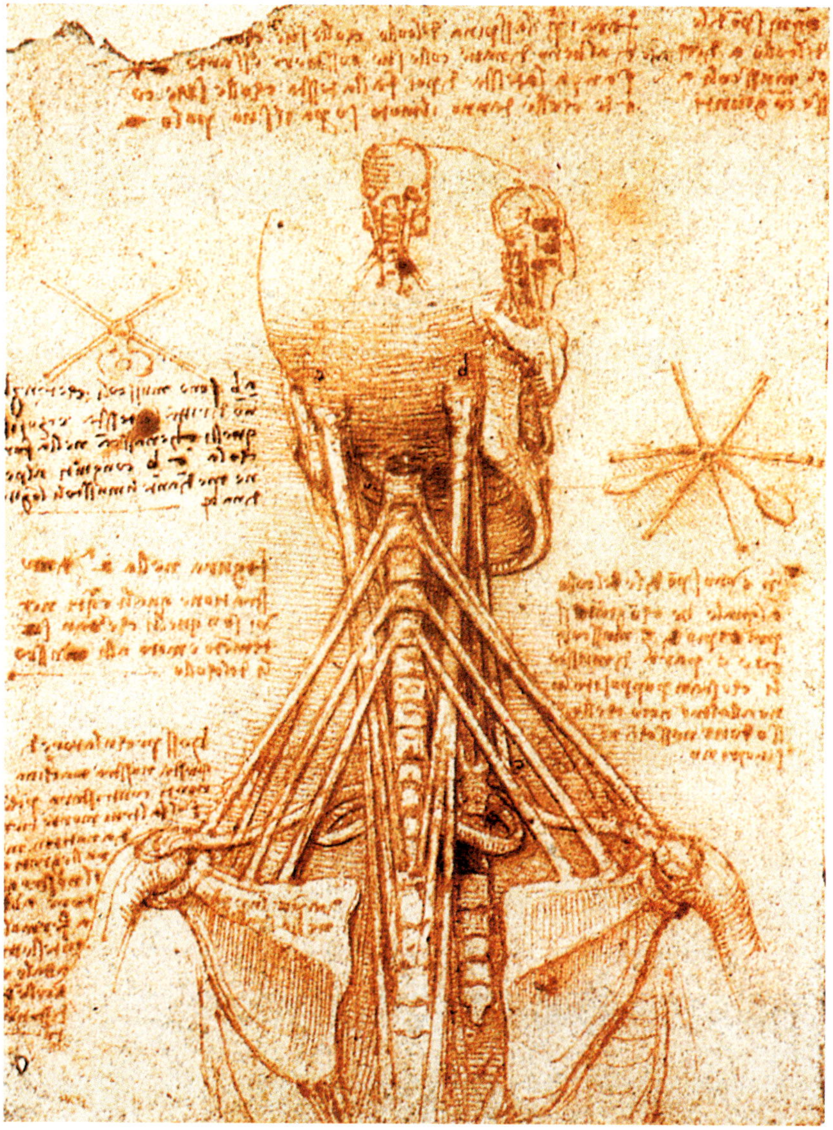

Figure 23
Structure supporting neck and head. Drawing by Leonardo da Vinci (Corpus of the anatomical studies, Royal Library Windsor, K/P 179v).

bone under the joint of the great toe of the foot, because if the tendon to which this sesamoid bone is joyned were without the sesamoid, it would receive damage from the friction made under so great a weight."

Elsewhere, the action of the muscles is compared to the "wedge," while the terms "lever" and "counter-lever" are frequently used to explain the motion of the upper and lower limbs.

Other characteristic features show that the model of the "elements of mechanics" is at work in the illustration of the organic machine. Leonardo in fact individually examines many of the devices of the human body. One typical example is the drawings of single teeth (fig. 24), from whose mechanical action he derives the respective forms and functions; another is in the graphic catalogue of a group of contrivances of the organic machine: the nerve, sinew, artery, muscle, each of which is associated with a precise function.

And lastly, he frequently refers to the geometrical schematisation of the organs to underline the mechanical laws that preside over their functioning, as in the demonstration of the action of the intercostal muscles in breathing or in the subjection of the motion of the jaw to the laws of the lever, where the related text displays the characteristic conciseness of the proportions of mechanics: "that tooth which is more distant from the center of the movement has less power in its bite."

It must be emphasised that we are dealing once again with unprecedented methods of analysis and manipulation: no one had ever before conceived such a daring project of decomposition and recomposition of the human figure.

The complexity of the organic machine pushes Leonardo to constantly refine his drawing technique. The number of views in continuous rotation (fig. 25) increases, the exploded views (fig. 26) are frequently used, as are the penetrations of the structures in transparency, showing the surface with only a few essential lines and exploiting – as one can see in the extraordinary drawing of the lungs in transparency (fig. 27) – his capacity to convey outline and depth by straight and curved hatching.

Leonardo also faces the problem of global visualisation of the information contained in the anatomical data. He at first considers that three views are sufficient to show the limbs. Being unsatisfied with the results obtained, he later uses four views. At this juncture, he discovers that eight views are necessary to fully portray each member. These are accurate indicators of his growing consciousness of the extreme complexity of a project to illustrate the organic machine, as it theoretically presents infinite points of view.

In addition, the human body also has significant variants at the level of structural strata, from the skeletal support out to the skin. This imposes the production of sets of drawings of successive strata of the dissected organs starting from the skin and ending with the bone, or, vice versa, putting the strata back together from the bare bone up to the skin. We have a stupendous example of this technique in the series of drawings of the hand (fig. 28) and the torso.

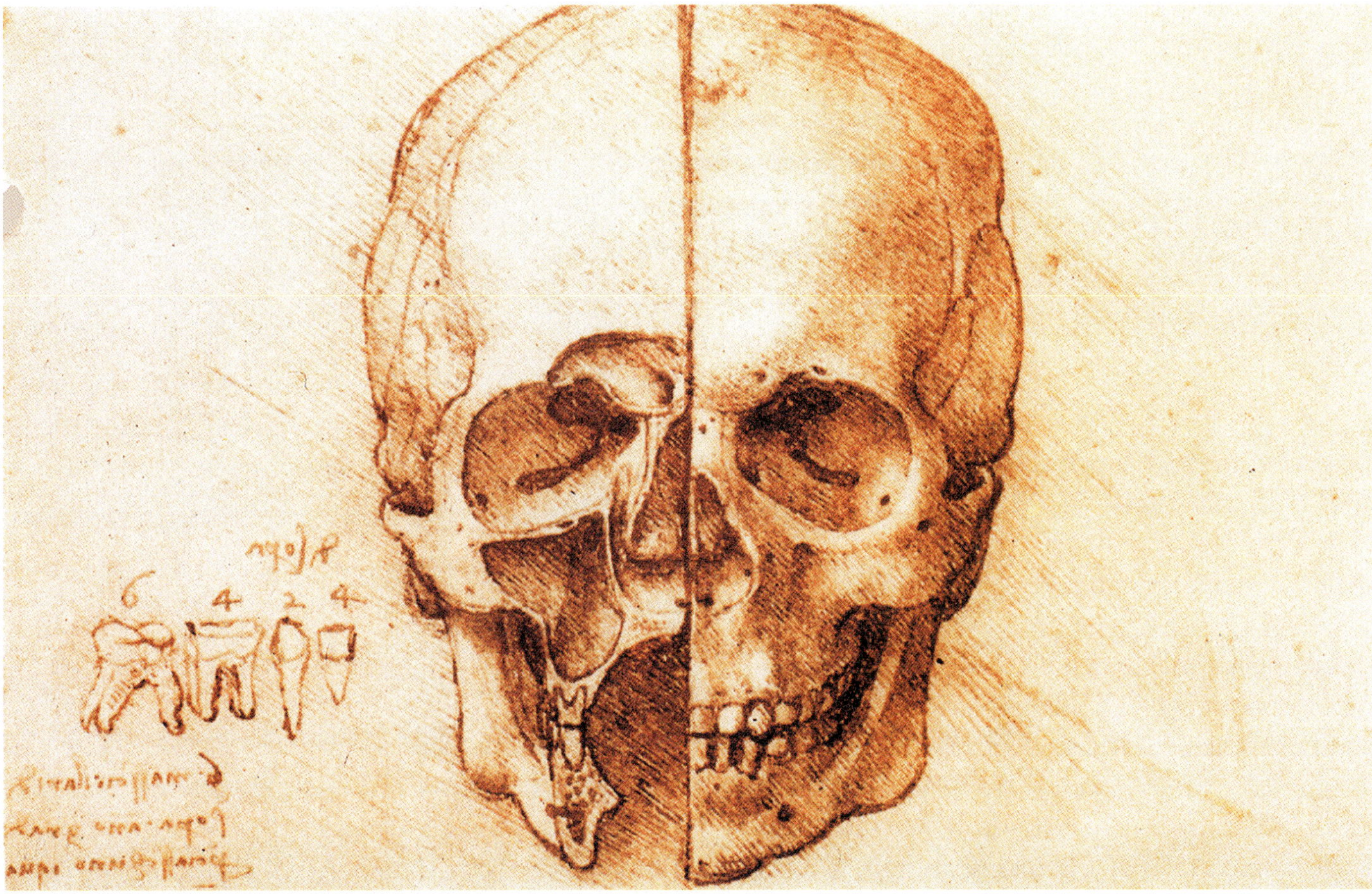

Figure 24
Front view of the skull with details of the teeth. Drawing by Leonardo da Vinci (Corpus of the anatomical studies, Royal Library Windsor, K/P 42v).

In order to underline the analogies between organic and artificial machines Leonardo adopts graphic conventions that distance him from realistic representation. "If you want to demonstrate clearly the thin muscles you have to use rows of threads. Thus you will be able to represent them one upon the other as Nature has placed them [...]. And having given such knowledge you will draw alongside this the true size, shape and position of each muscle. But remember to make the threads that denote the muscles in the very same position as the central lines of each muscle."

This means recognition not so much of the impossibility of faithful illustration of the fine muscular structure, but rather the inadequacy of such an effort when proposing to show the functional relations of the muscles and tendons or the mechanics of the articulations. In order to achieve this goal, the drawing must turn to simplification and schematisation, working from the presupposed analogy between organic and artificial machines. Thus the muscular masses have to be replaced by threads which indicate the lines of force of their operation, as in mechanical models (like that of the lower limb (fig. 29), which Leonardo will actually build, using copper threads to outline the mechanical action of the muscles).

We are faced with a graphic solution which is extraordinarily innovative. Once again, it does not derive from a dilation of observation, but appears rather as a transfiguration of reality through drawing, in order to visualise the direct and necessary connection between observed evidence and intrinsic causes.

It is this ambiguous level of reality that Leonardo once suggestively defined as "infra l'anatomia e il vivo" ("between anatomy and life"). That is, an intermediary situation between the cadaver (which may be dissected, but is dead) and a "live" organic being (in which it is not possible to examine the transformations produced at the level of fine structure by various motions). Leonardo contrasts the impotence of the naked eye to the capacity of drawing to construct that which we would today call a "virtual" reality, in which the information culled from dissection as well as from simulation of functional mod-

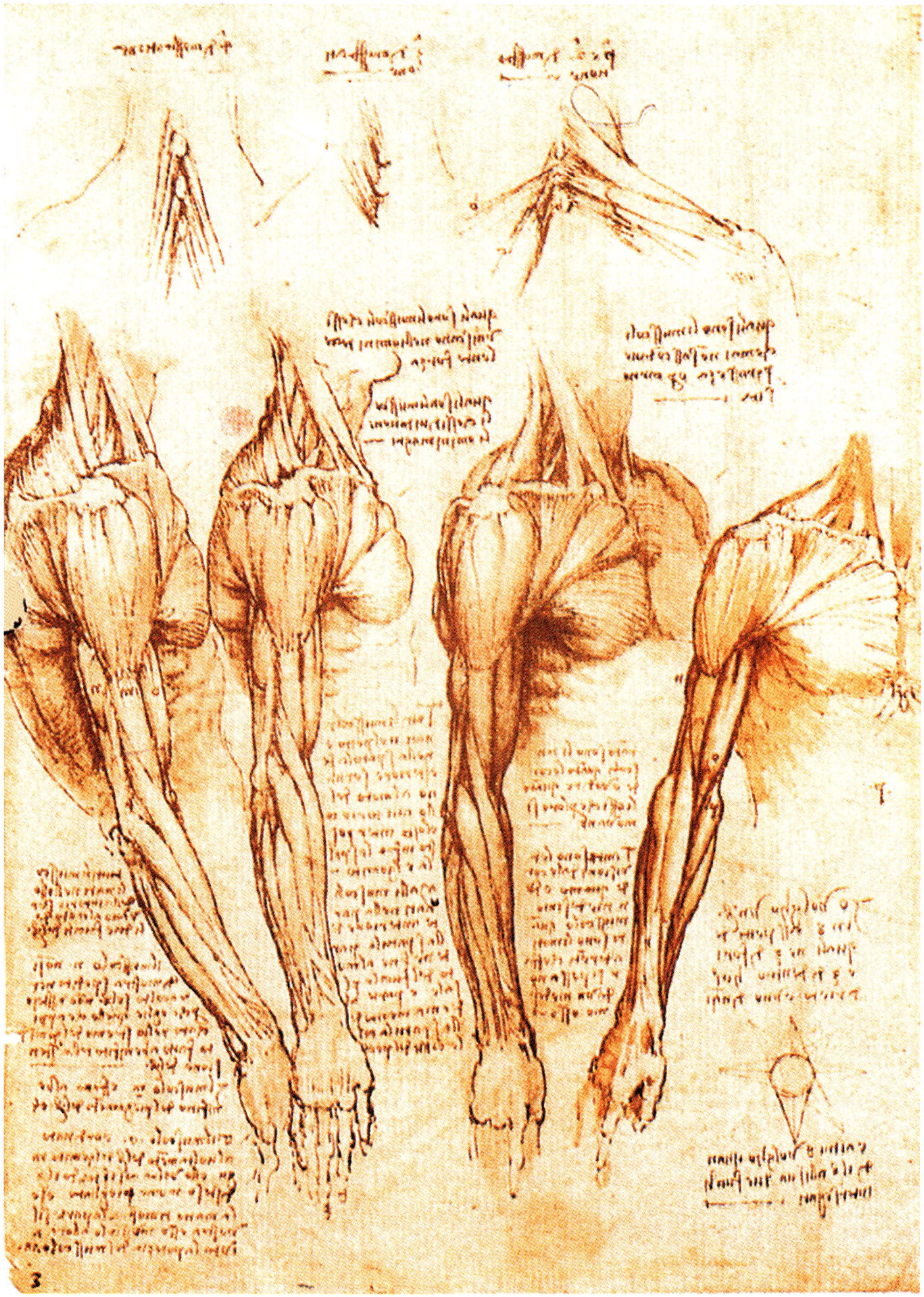

Figure 25
Four views of the shoulder and the arm in continuing rotation. Drawing by Leonardo da Vinci (Corpus of the anatomical studies, Royal Library Windsor, K/P 140v).

ifications of the living body is channelled together. The drawing analyses and recomposes in a unified view the level of the fine structures and that of their various functioning which remain otherwise, irremediably separate.

It is not inappropriate here to speak of "virtual reality." In fact, Leonardo's ambitious plan developed between 1495 and 1497, to realise an anthropoid capable of simulating human movements, is directly connected to the theme of graphic simulation as a concep-

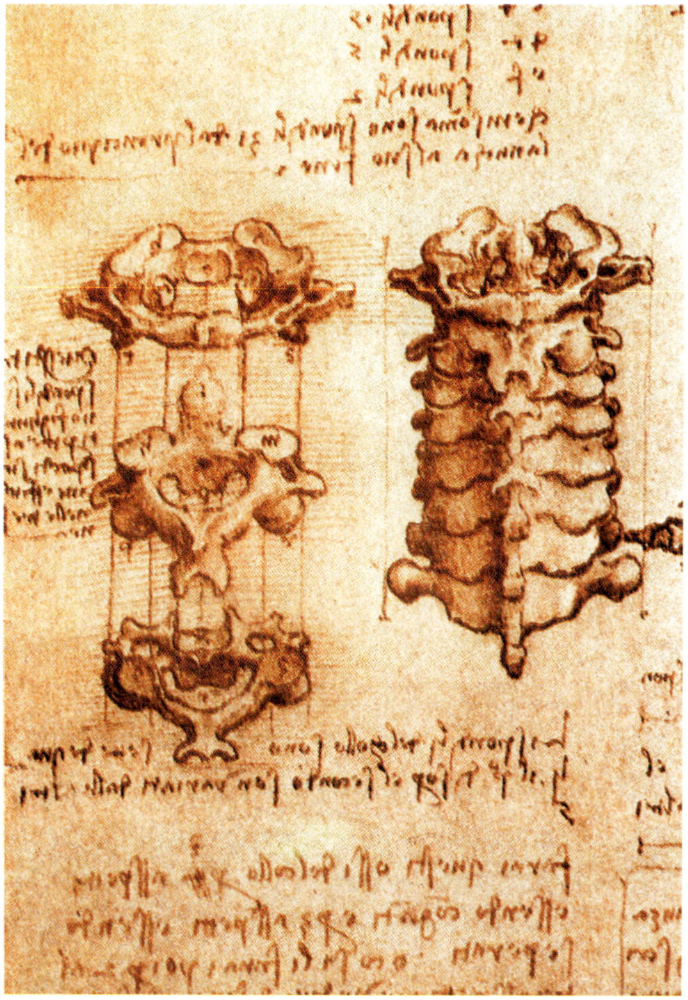

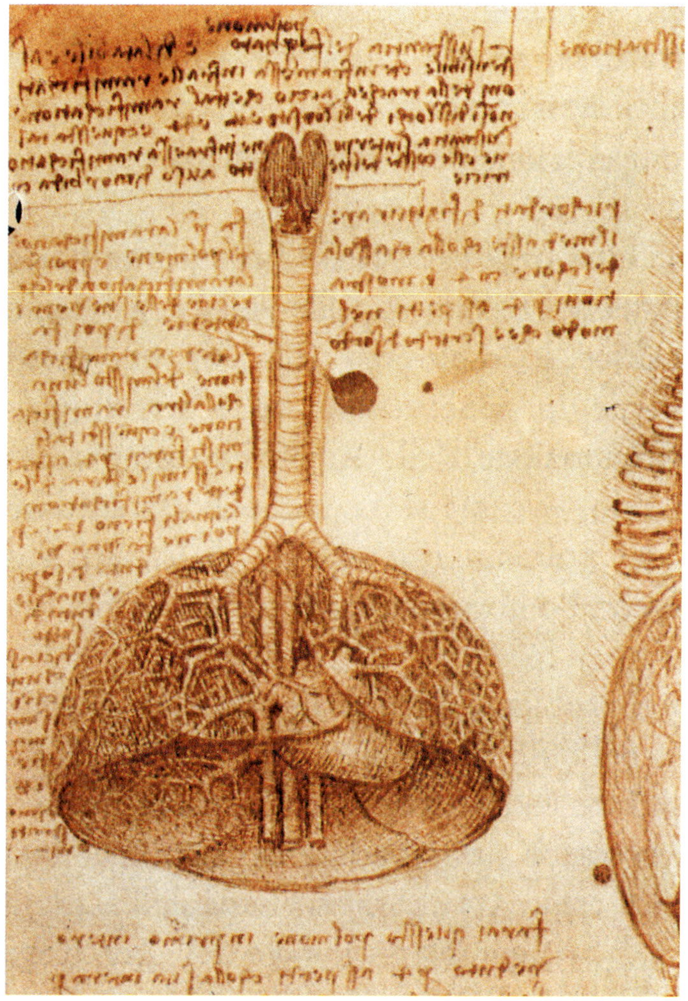

tual basis for the construction of mechanical simulators. Leonardo's lost robot has recently recovered a more precise physiognomy. As illustrated in fragmentary drawings, Leonardo's robot had the appearance of a soldier covered by a robust armature. The robot consisted of two independent systems: three-degree-of-freedom legs, ankles, knees, and hips; and four-degree-of-freedom arms with actuated shoulders, elbows, wrists, and hands. The visor, neck, jaw and possibly the spine may also have been actuated. The orientation of the arms indicates it was designed for whole-arm grasping, which means that all the joints moved in unison. A mechanical, analog-programmable controller within the chest provided power and control for the arms. To drive the arms, the controller had a cylindrical, grooved cam that triggered high-torque worm gears attached to a central pulley. A central shaft, perhaps splined, provided power while still permitting the robot to stand and sit. The legs were powered by an external crank arrangement driving the cable, which was connected to key locations in the ankle, knee, and hip.

This project indicates the direction in which Leonardo was pushed by his own pioneering research and the ideas which inspired it: to cancel the borders between natural and artificial anatomy, exalting the creative power of man, not simply imitator of nature but almost competitor with it.

All of Leonardo's graphic work, not just in the anatomy of machines and of the organic world, is indeed characterised by the effort to utilise drawing as an instrument to illustrate the creations of nature and man, not according to their sensory appearances but as a necessary expression of the workings of universal laws. Leonardo liberated drawing from the restrictions that bound it tightly to the figurative arts, transforming it into a powerful tool of investigation and demonstration. Vasari's perfect understanding of the sense and the scope of the new function assigned

Figure 26 *(left)*
Exploded view of the spinal column. Drawing by Leonardo da Vinci (Corpus of the anatomical studies, Royal Library Windsor, K/P 149r - detail).

Figure 27 *(right)*
Transparent rendering of the lungs. Drawing by Leonardo da Vinci (Corpus of the anatomical studies, Royal Library Windsor, K/P 53v - detail).

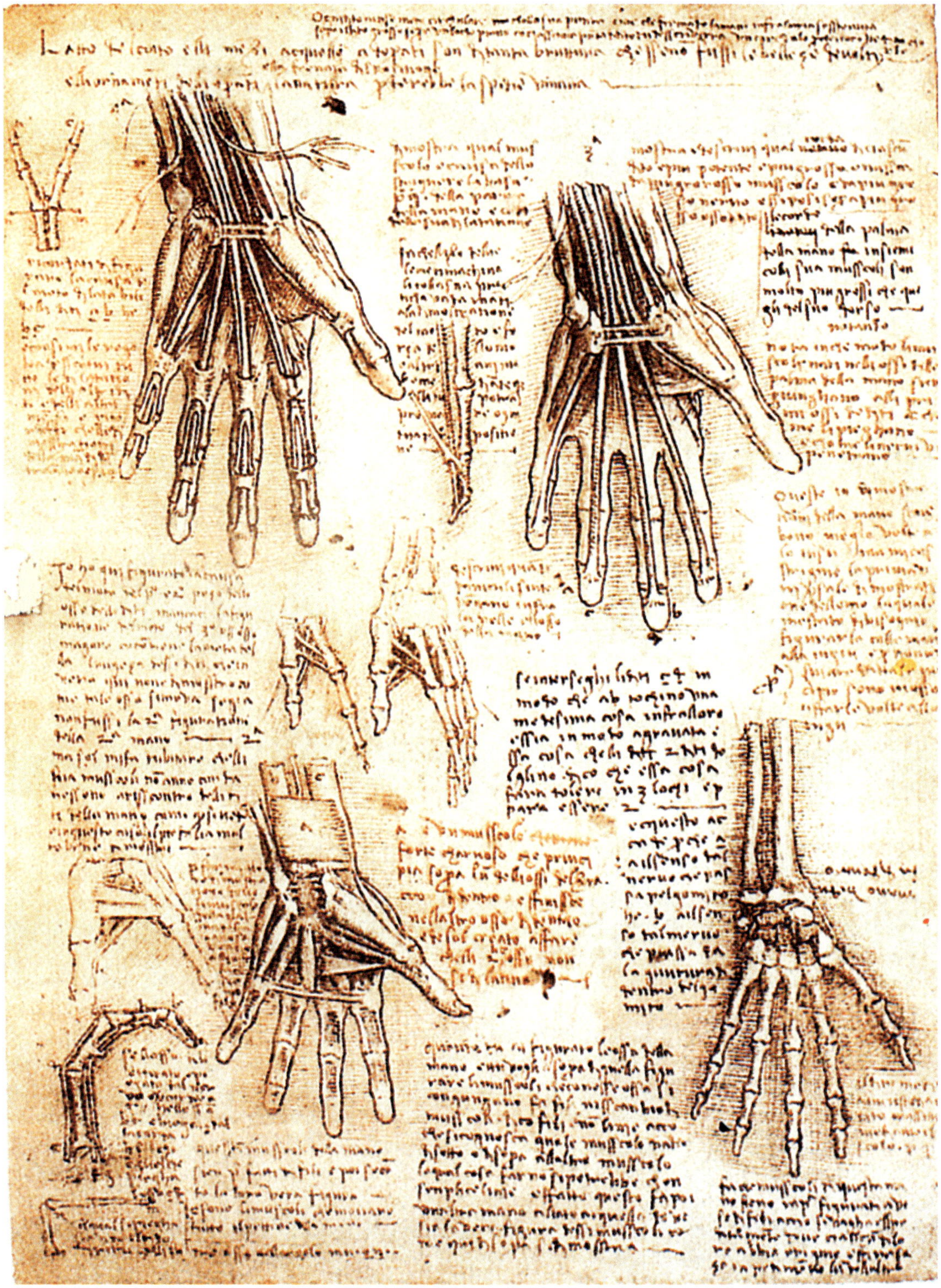

Figure 28
Sudies of the hand. Drawing by Leonardo da Vinci (Corpus of the anatomical studies, Royal Library Windsor, K/P 143r).

by Leonardo to drawing is mirrored in the definition he proposed of drawing: "similar to a form, or idea, of all the things of nature [...]," and again, "visual expression and declaration of the concept that one has in one's soul." This definition echoes fundamental Platonic motifs which are also clearly expressed in the Leonardian definition of painting as "cosa mentale" (a product of the mind).

As a confirmation of this reading, it is worth stressing that many of Leonardo's studies on the vortical motion of water, (often quoted as evidence of his exceptional observational capacity), are simply bold visuali-

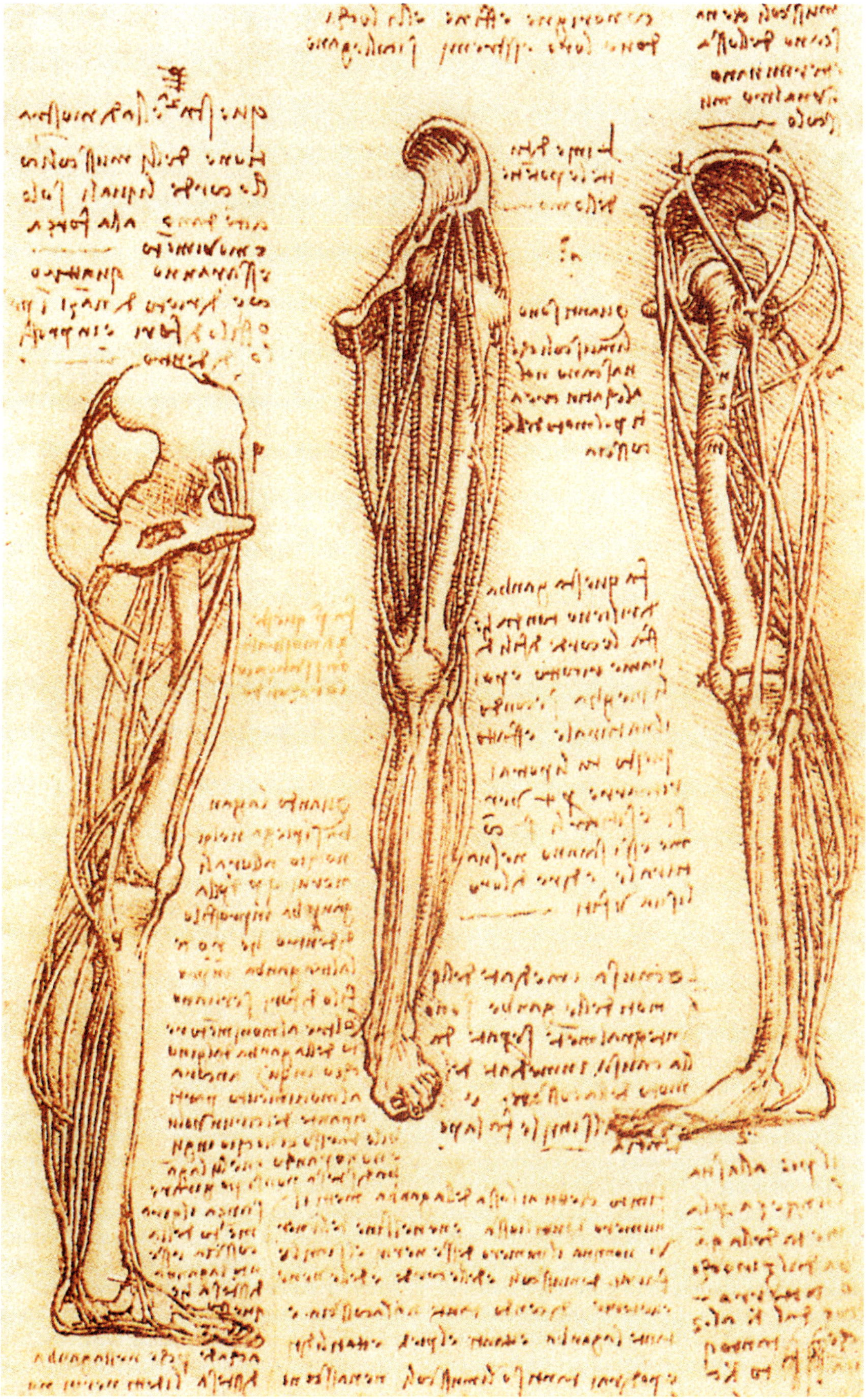

Figure 29
Model of the lower limb. Drawing by Leonardo da Vinci (Corpus of the anatomical studies, Royal Library Windsor, K/P 152r).

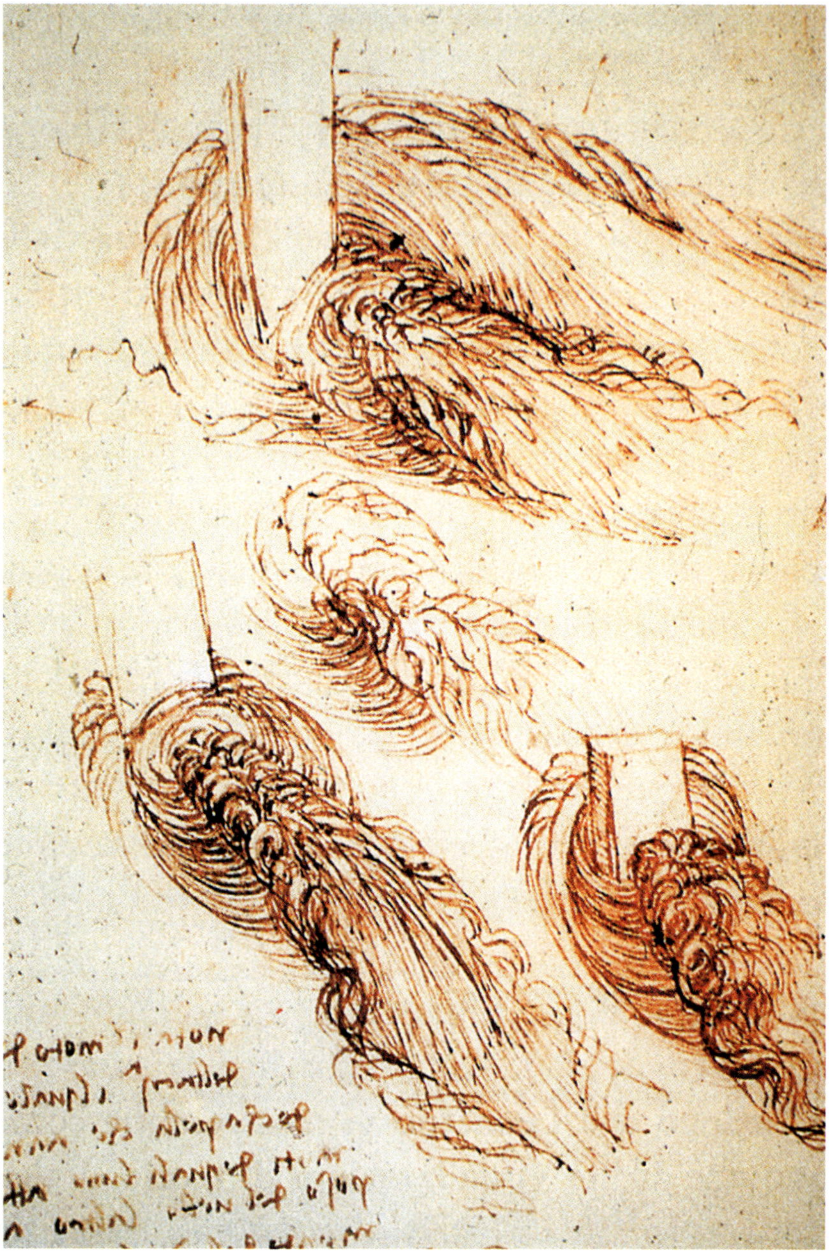

Figure 30
Water vortices. Drawing by Leonardo da Vinci (Royal Library Windsor, 12579r).

sations of theories. Let us consider, as a telling example, the comparison which Leonardo establishes, in many wonderful drawings, between the vortices of water (fig. 30) and wavy masses of curly hair (fig. 31). The question here is not about Leonardo's extraordinary capacity to perfectly reproduce reality thanks to accurate observation. One must in fact admit that between curly hair and vortices of water there are no appreciable analogies on the plane of pure appearances. Leonardo suggested them on the basis of the schematisation of the motion of water with lines of force (figs. 32-33), a schematisation founded on the mechanical analogy which inspires his drawings of machines and of human anatomy, and which constitutes the theoretical presupposition for a purely "mental" perception. "Observe the movement of the plane surface of water, which is similar to that of the hair. This last has two movements: one depends on its weight and the other derives from its vaulted shape."

The drawing, then, is better able than the model, which is too rigidly objective, to give evidence of the universal essentiality and constancy of the laws of nature, through its rigorously controlled and codified transfigurations. Art and nature thus come closer; in the sense that art seems much more natural, while the products of nature appear ever more as marvellous artifices. In an extraordinary statement Leonardo explains that this convergence is made possible by the demiurgic nature of drawing: "necessity compels the painter to transmute his mind into nature, to become the interpreter between nature and art."

Thus, no instrument is more effective than drawing in exploring the extreme complexity of the human figure. After having outlined a program of illustration of the human body which would have required thousands of drawings, he must have noted the paradox in a project which was not only impossible to complete, but that many would have judged unable to compete with direct observation of cadavers. Reacting in a lively manner to this potential objection, Leonardo reaffirmed the superiority of a language in which experience and theory melded, and because of this give an integrated vision of effects and latent causes:

> *And you who say that it is better to see an anatomy performed than to see these drawings would be right if it were possible to see all these things which are demonstrated in these drawings in a single figure. In this with all your ability you will not see nor obtain knowledge of anything but some few vessels. In order to obtain a true and full knowledge of which I have dissected more than ten human bodies, destroying all other organs and taking away in its minutest particles all the flesh which was to be found around the vessels without causing them to bleed except for the imperceptible bleeding of the capillary vessels. And one single body*

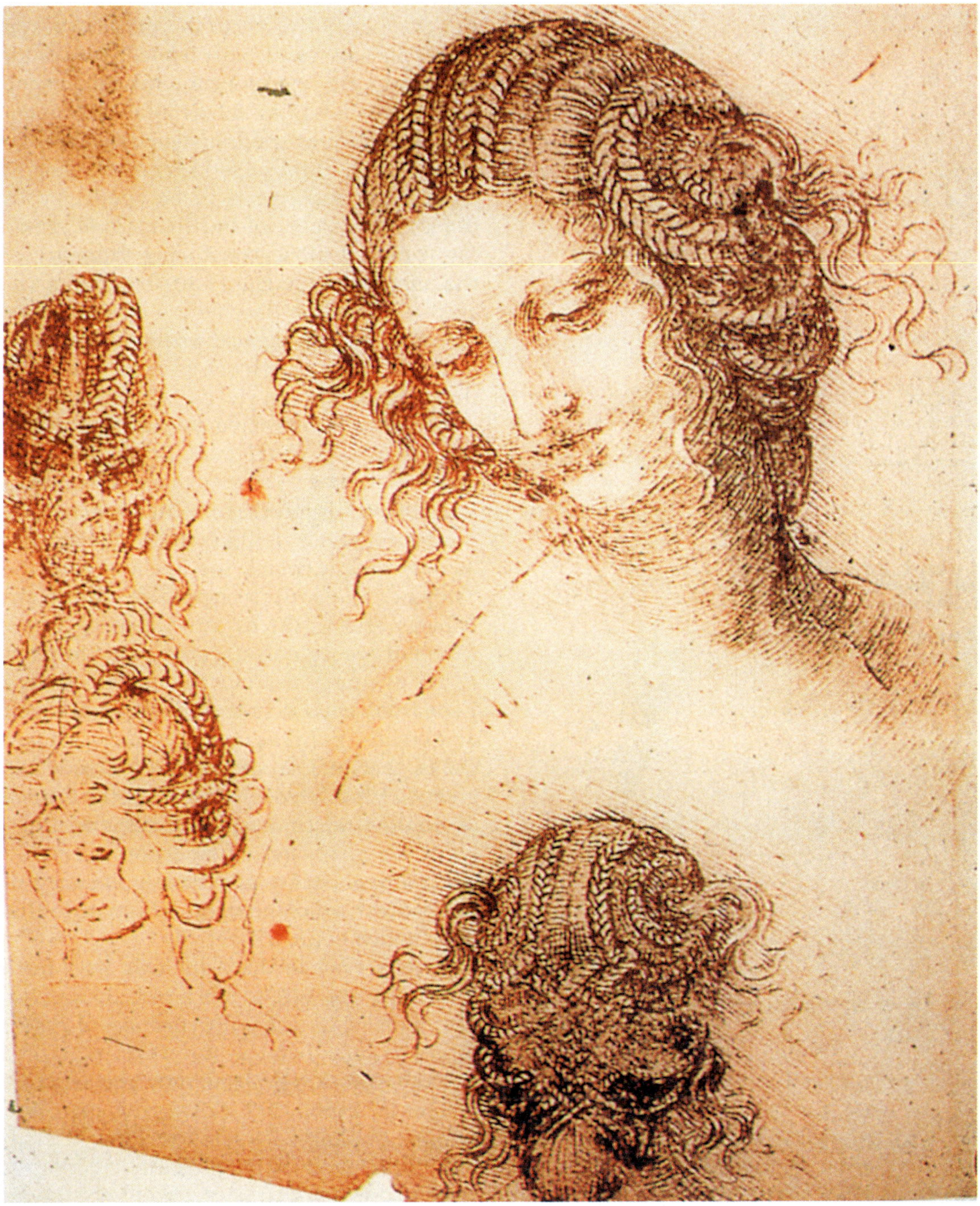

Figure 31
Study for the head of Leda. Drawing by Leonardo da Vinci (Royal Library Windsor, 12516r).

> *was not sufficient for enough time, so that it was necessary to proceed little by little with as many bodies as would render a complete knowledge. This I repeated twice in order to observe the differences.*

It is precisely the "complete knowledge" – that is, the combination of observation of superficial appearances and reflection as to causal processes – that renders drawing the most appropriate instrument for realising a perfect imitation of nature. Thus it is not surprising that for Leonardo the comparison between verbal expression and drawing is unequal:

> *O writer, with what letters will you describe with such perfection all that is depicted here in drawing? Because of your lack of information, you write confusingly and give little notion of the true shape of things, with which, deceiving yourself, you believe you can fully satisfy*

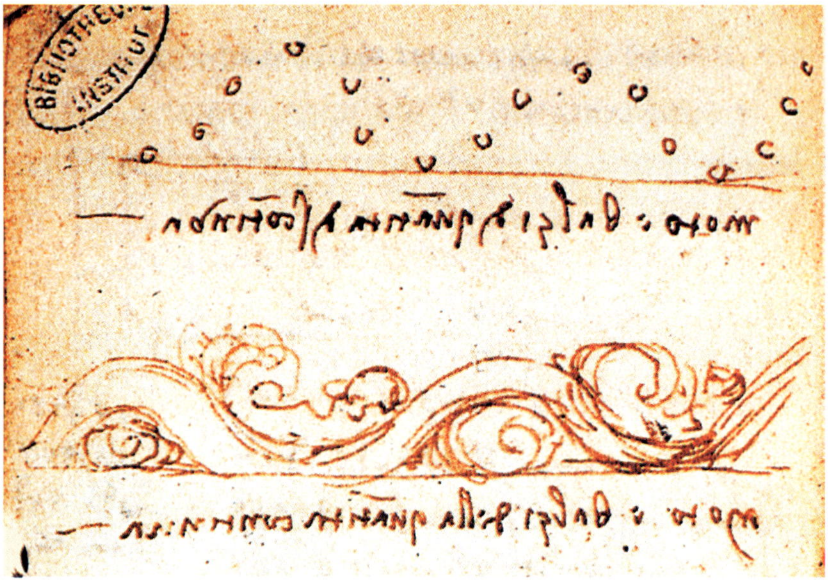

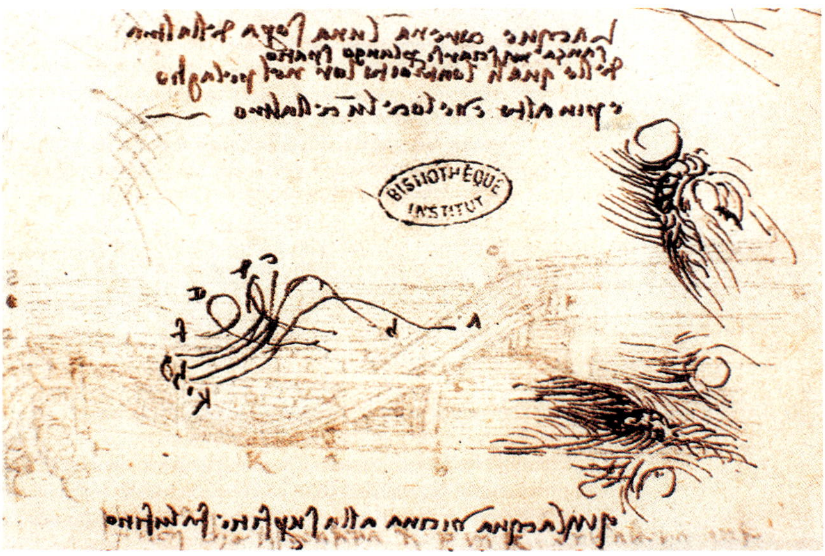

Figure 32 (*above*)
Schematic representation of wave motion. Drawing by Leonardo da Vinci (Ms. I, Inst. de France Paris, c. 67v).

Figure 33 (*below*)
Schematisation of a water vortex using lines of force. Drawing by Leonardo da Vinci (Ms. F, Inst. de France Paris, c. 20r - detail).

your listener [...]. But I want to remind you that you shouldn't get entangled in words – unless you have to talk to the blind – or, if you insist on making your point with words for men's ears rather than for men's eyes, then talk about things, substances, and natures; don't try to convey to the ears those things that pertain to the eyes, because the painter will be vastly better at it than you.

Few excerpts could more eloquently confirm the thesis suggested by Lucien Febvre that the most characteristic trait of the passage from mediaeval culture – based on the model of commentary – to Renaissance knowledge – marked by the visual processes of the artistic world – is the transition from the primacy of the ear to the primacy of the eye. This last must be intended, for Leonardo; not as the eye of the forehead but as the "eye of the mind," an intellectual instrument which penetrates and transfigures reality.

The claim to the superiority of drawing in anatomical studies has to be connected to the many passages of the *Treatise on painting*, which exhibits convincing examples of the import that Leonardo attributed to a scientific foundation (mechanics, optics, theory of vision, natural science, anatomical knowledge) in the formation of a painter "worthy of the name."

In works constructed by man, as in the great machines of the body or of the world, drawing captures and visualises the schemes and operative processes of proliferate nature, illustrating them not as static effects, but as processes in which the live action of the "motors" is clearly recognisable. This contributes also to the establishment of a new canon of beauty, now understood as the resonance of energy and movement. Underlining Leonardo's success in embodying "movement and breath" in his figures, Vasari once again shows that he has understood perfectly the great da Vinci lesson. In the era of the *Corpus of Anatomical Drawings* and numeric drawing it is perhaps not unimportant to emphasise the complexity and the significance of the process which brought definition to this new branch of artistic activity.

If, as a gloss to that which we have been discussing, we were to add a paradoxical, but not inadequate, general definition of Leonardo's graphical endeavour, we could affirm that he was always and only an artist of machines, even when portraying man and nature. Forcing the paradox even further, and intentionally using anachronistic terminology, one could even define him – again, not inappropriately – as the first multimedia author; an artist who expanded, refined, and above all strove to integrate fully all the analytic and expressive media, driven by an utopian design: to show the latent meaning of all things, through portraying the theatre of the world governed by necessary and universal laws incessantly pervaded by movement.

Istituto e Museo di Storia della Scienza, Florence

The Limits of Pictures

Cognitive Functions of Images in Practical Mechanics – 1400 to 1600

Wolfgang Lefèvre

Introduction

The images that are subject of this essay pertain to the realm of engineering in the fifteenth and sixteenth centuries. That is the time of Leonardo da Vinci and Albrecht Dürer, the time of technical treatises like Georg Agricola's *De re metallica* with its telling illustrations, and the time of the famous *Theatri machinarum*. These last works in particular, with their rich collections of splendid and suggestive images, tempt one to focus on their rhetorical resourcefulness and power. This essay, however, will neither inquire into this aspect of these images nor examine them as documents of the technology and techniques of that age. Rather it is concerned with the cognitive functions which these graphical representations may have had for engineers and technicians in the fifteenth and sixteenth centuries.

As is generally known, the sixteenth century and the first decades of the seventeenth century made up the period in which persons like Tartaglia, Cardano, Benedetti, Guidobaldo del Monte, Stevin and Galileo developed the decisive requisite conditions which enabled the generation of Huyghens, Newton, and Leibniz to bring forth the classical mechanics of modern times. As is also well known and acknowledged, at least in principle, the theoretical efforts of the above-mentioned preclassical figures were undertaken in close contact with practical mechanics. The relationship between practical and theoretical mechanics in this period, however, is still not entirely understood (see Lefèvre [2000]). One reason lies in the fact that we know only very little about how the mechanicians themselves understood the technical devices they created. What kind of mechanics shaped the thinking of the mechanical practitioners?

Apart from a few exceptions,[1] the mechanicians didn't write down their notions, ideas, understandings, or "theories." There are a variety of reasons for this: A great part, if not the bulk of their knowledge, was "tacit knowledge" and in addition much of it was kept secret. Even the few documents we have shouldn't be taken at face value – think, for instance, of Leonardo's famous letter to the court of Milan –; and last but not least, the mechanicians of this period acquired their knowledge not in schools but in the traditional way, that is as apprentices. There was consequently no need for textbooks. (Later on, I will point out the fact that in those days the first important change in the system of teaching technicians took place.)

Against this background, it seems to be of interest to find out whether the technical images of that age can reveal something about the mechanicians' thinking. Is it possible to reconstruct some aspects of their thinking by reconstructing the cognitive functions of these images? The answer depends of course on which cognitive functions these images actually had. What will turn out, however, is that in the period under scrutiny, graphical representations had only very limited cognitive functions for mechanicians. This phenomenon can be reconstructed by analysing the type of graphical representations drawn or used by technicians. In order to do this, the essay has to start with some very general considerations regarding the art of producing images at that time.

Printed Images and Techniques of Projection

As is generally known, the Renaissance did not invent the use of images in connection with cognitive processes. This use of sketches, plans, maps, pictures, and constructions can be traced back to the dawn of human culture. And, regardless of which intellectual occupation of Antiquity one considers truly scientific – including those of Ancient Mesopotamia and Ancient Egypt, or reserving the title for those of Greek Antiquity, among the documents which testify such scientific activities, one always finds images as well.

Nevertheless, the Renaissance did introduce an important innovation in this context. Until then, the only way to reproduce images had been to copy them by hand. During the fifteenth century, techniques were introduced to reproduce them by print: woodcuts, engravings, etchings, and, of course, the printing press. The significance and consequence of these inventions for the scientific image cannot be overestimated. As William Ivins has shown with images in botany and zoology, without techniques which allow the exact reproduction of images, the scientific image remains a mere idea.[2]

For the purpose of this essay, perhaps even more important than this invention was another discovery made during the Renaissance, a discovery which was at first in a way a rediscovery (see Edgerton [1975]) but which at once had its own characteristic features and immediately took its own course of development – the discovery and development of techniques of projection which allowed the construction of unambiguous graphical representations of three-dimensional space and solids on a plane surface.[3] Regarding the development of the techniques and theories of linear perspective during the Renaissance, it may suffice here just to mention the contributions of Filippo Brunelleschi, Piero della Francesca, and Albrecht Dürer. By merely mentioning the name Gerhard Mercator, all knowledge of the different kinds of projection used in mathematical geography are brought to mind, a great part of which had already been contrived in the sixteenth century. Less known perhaps may be the fact that the modern techniques of projection applied by engineers and architects for drawing different kinds of plans can also be traced back to that period, and especially to the work of Dürer.

Pictures and Plans

With respect to these different techniques of projection, I would like to introduce a distinction which will turn out to be of significance for the cognitive functions graphical representations can have. In this essay, I will distinguish between "pictures" and "plans." To start with a black-and-white definition, "pictures" endeavour to represent objects as they appear to the eye whereas "plans" try to represent objects as they are.

Pictures[4] try to show how an object, regardless of whether real or fictive, would be seen by an observer looking (with one eye and motionless) from a certain point towards the object. The different kinds of linear perspective share the aim of constructing in an unambiguous way how objects or an assemblage of objects would be perceived from a certain observation point. In order to achieve this goal, the graphic elements of the picture and the spatial relations among them have to be interrelated with the spatial structure of the object in a definite way which is specific to linear perspective. The illusion particular to this kind of representation can only be produced by representing systematically distorted certain elements of the object – certain angles, certain (scaled) distances, etc. (fig. 1 above). The rules regulating which elements are to be represented correctly and which distorted in a fixed way can be called the "syntax" of linear perspective.[5] It is this "syntax" which determines the limits within which pictures are useful in the realm of technology.

A different type of "syntax" for graphical representation is characteristic of *plans*. Plans try exactly to avoid distortions whilst representing the spatial elements of the object.

A ground plan of a building, for example, is only useful if it depicts true angles and correctly-scaled distances. The same is true with respect to elevations and cross-sections and other simple instances of plans. Furthermore, whereas a single picture is sufficient for representing the appearance of a three-dimensional object, at least two and often more interrelated plans are necessary for representing the thing as a three-dimensional object. There are different special kinds of "syntax" for plans, depending upon the projection technique chosen. With respect to the plans which play a role in the context of this essay, only the simple "syntax" of orthographic projections (fig. 1 below) and the ways in which they were interrelated have to be considered.

Intellectual Requirements

Regarding the knowledge necessary for constructing either pictures or plans by techniques of projection, one can call such images "learned images" for two reasons. First, their production required the application of techniques which could only be developed on the basis of scientific geometry regardless of whether the men who applied these techniques were familiar with this geometry or not. Secondly, the competence needed to produce images by applying those techniques could only be acquired by specialised professional training. It was just these problems of teaching and learning the art of producing such drawings which affected the traditional apprenticeship and with which the first elements of a school education of technicians emerged.

Two events may point to the change which then took place in this respect: First, at the end of the fifteenth century, Matthaeus Roriczer's *Büchlein von der Fialen Gerechtigkeit* appeared which can be characterised as one of the first textbooks written to pass on knowledge necessary for practitioners – namely for the "stone-cutters", i.e., the stonemasons who had to engrave geometrical constructions on stones (fig. 2). Secondly, in 1563 the famous *Accademia del disegno* was founded in Florence, offering instruction for practitioners of "those arts based upon the principles of 'disegno,' i.e., the arts of painting, sculpture, and architecture" (Ward [1972] 5), including instruction in anatomy and applied mathematics (see Olschki [1926] and Kay [1976]).

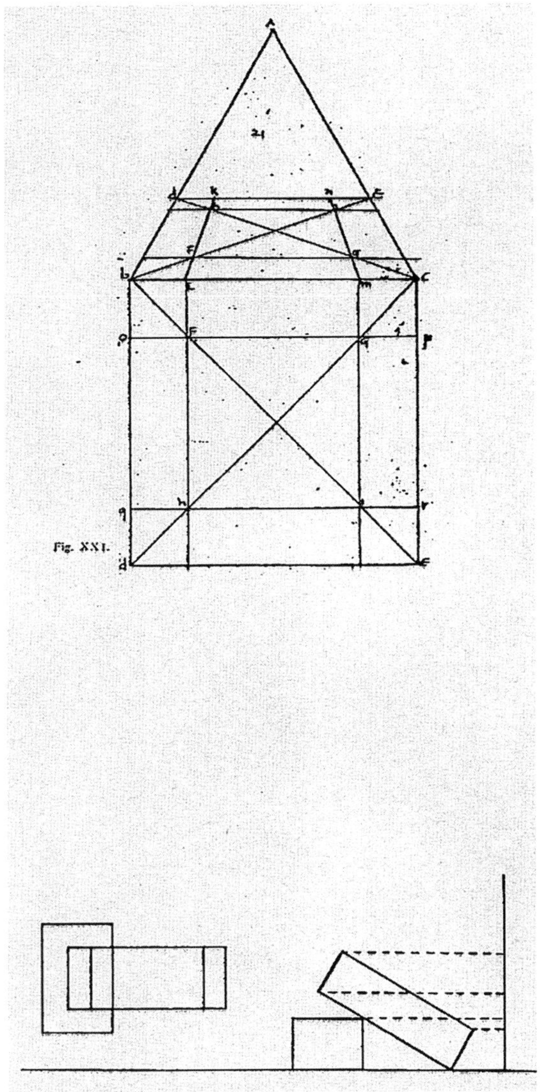

Figure 1
Above: Projection according to linear perspective. The quadrangle bcde at the base, appearing as square when seen from a bird's-eye view, is represented immediately above according to linear perspective as a quadrilateral with only one pair of its opposite sides parallel and with shortened distances among the still parallel horizontal lines. (Figure from Francesca [1984] table IV.)
Below: Orthographic projection. The same two cubes are seen on the left from above (ground plan) and on the right from the side (elevation). The sight-lines (broken lines) are always parallel to each other and at right angles to the plane of projection. (Figures from Samuel Marolois, depicted in Booker [1963] 41.)

The significance of this development is not diminished when one adds that the instruction which practitioners received from those textbooks and institutions probably had the principle character of passing on recipes for solving given problems (see Sellenriek [1987] 153) and not that of teaching the principles of these techniques. At least one has to state that the art of technical drawing in the sixteenth century did not keep pace with the theoretical foundations which Albrecht Dürer had developed for constructing interrelated plans at the beginning of the century and which were not continued and developed further earlier than in the seventeenth century by, for example, Samuel Maralois (see Booker [1963] 39f.) and Gerard Desargues (see ibid. 61f. and Sellenriek [1987] 153f.).

The knowledge required to construct a picture in accordance with the principles of

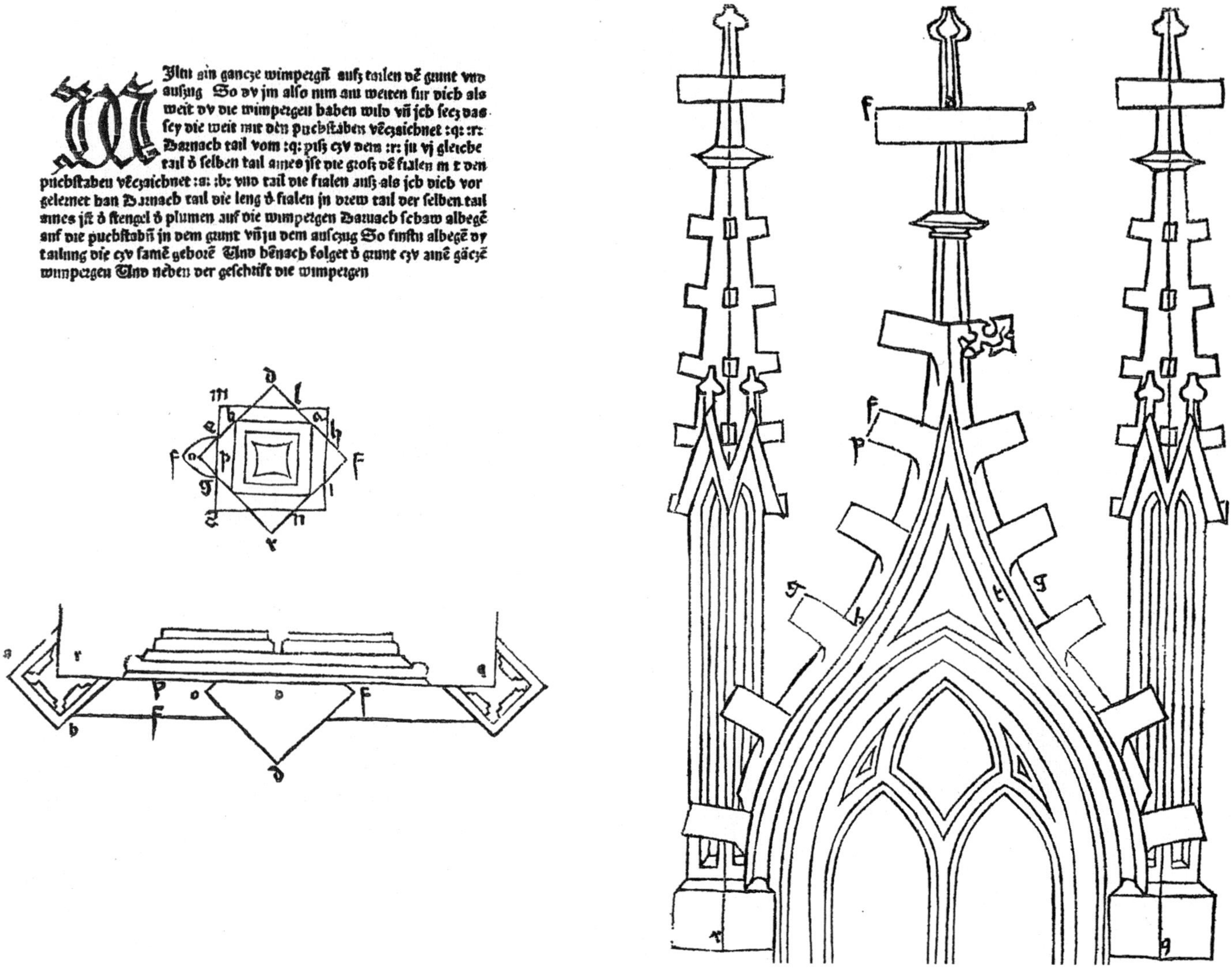
Wiltu ain gancze wimpergē aufz tailen võ grunt vnd aufzug So dv jm also nim ain weiten für dich als weit dv die wimpergen haben wild vñ ich secz das sey die weit mit dem puchstaben verzaichnet :q: :r: Darnach tail vom :q: pisz czv dem :r: in vj gleiche tail d selben tail aines ist die grosz võ fialen m t den puchstaben verzaichnet :a: :b: vnd tail die fialen ausz als ich dich vor gelernet han Darnach tail die leng d fialen in drew tail der selben tail aines ist d stengel d plumen auf die wimpergen Darnach schaw albegē auf die puchstabñ in dem grunt vñ in dem aufczug So findest albegē dy tailung die czv samē geborē Vnd dēnach folget d grunt czv ainē gäczē wimpergen Vnd neben der geschrift die wimpergen

Figure 2
Page from Matthäus Roriczer: *Das Büchlein von der Fialen Gerechtigkeit* (Regensburg 1486) folio 5v.

linear perspective on the one hand, and that required for constructing interrelated plans on the other is, all in all, the same. But the knowledge required for understanding pictures differs considerably from that required for understanding plans. What characterises a well executed perspective drawing or painting of an object is simply that it is immediately understandable for everybody; regardless of his or her education, he or she will recognise the depicted object so far as it is recognisable just by viewing it (fig. 3).[6] On the other hand, only very simple plans are readable for non-professionals whereas more sophisticated ones are understandable only for experts (fig. 4).

Concrete Pictures – Abstract Plans?

The differences between pictures and plans discussed so far already give some clues regarding the different roles pictures and plans can play in the realm of technology.

Pictures are suitable for communicating information about objects. In addition, when correctly constructed, pictures have the special virtue of communicating information to everybody without discriminating against non-professionals. They can, however, communicate only that information which can be gathered from the given appearance of the object. Contrary to the intuition one may have when thinking of the appeal pictures have for our senses, the seemingly concrete pictures are a much poorer means of conveying information than the seemingly abstract plans. Plans are not restricted to information provided by the appearance of the object for an observer. In principle, one can represent the structures of an object completely by combining interrelated plans (fig. 5). On the

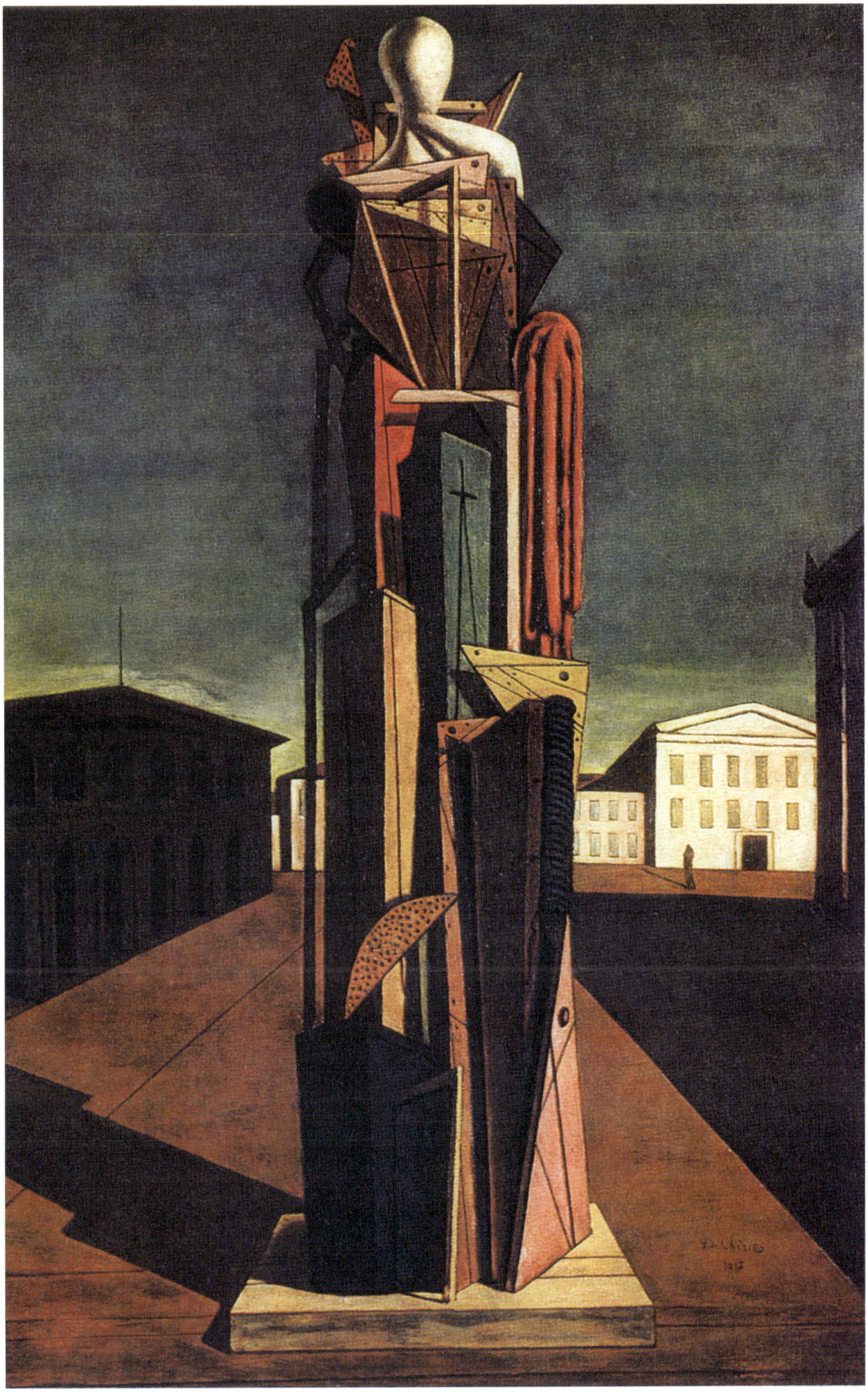

Figure 3
An understandable picture – whatever its meaning may be. Painting by Giorgio de Chirico from 1917 (from Argan [1977] plate XXXVII).

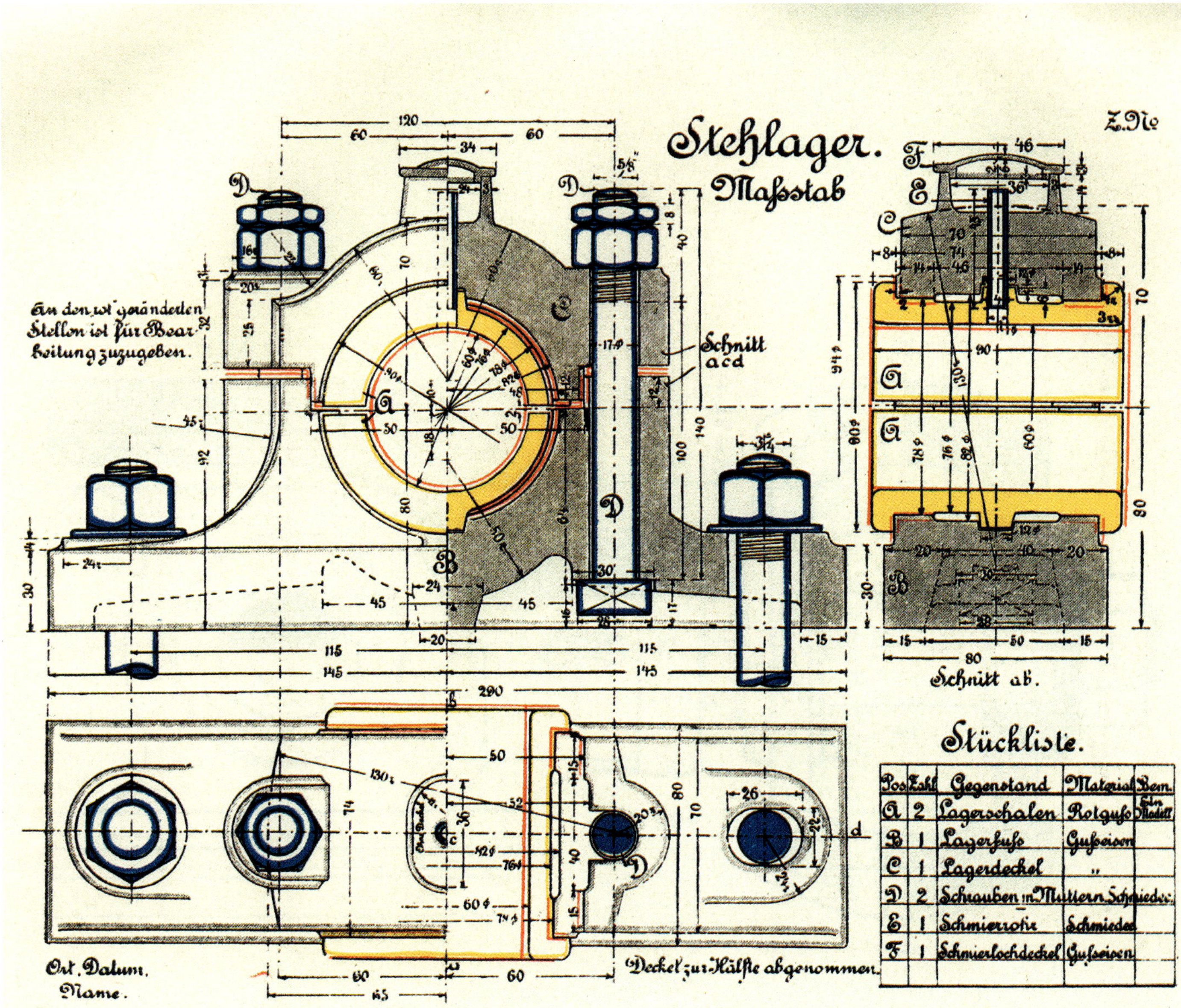

Figure 4
Constructional plans – understandable for professionals. Plan from O. Schulz: *Konstruktionszeichnen* (1908) (from Nedoluha [1960] 93).

other hand, plans are only suitable for communicating information to a broad audience when they are kept very simple; for at a certain degree of sophistication they are only meaningful to professionals.

Construction of Drawings – Construction by Drawings

The capacity of plans to represent the structures of an object completely, at least in principle, leads us to a further difference between pictures and plans which deserves particular attention in connection with engineering. So far, we have considered pictures and plans as resulting from construction since the Renaissance. Not only since the Renaissance, however, were plans used by architects and engineers as a means of construction. With respect to this function as a means of construction in architecture and engineering, pictures and plans reveal their most important difference. One can of course record ideas of a planned building or a planned technical device by sketches drawn with the "syntax" of pictures. Architects and engineers usually outline their first ideas in this way. But even carefully drawn pictures will very soon show their limited capacity of planning anything in detail. With plans, on the other hand, it is possible to design a building or a machine in detail in all its structure and, in the case of a machine, also to devise the kinetic relations amongst its parts. It is of course not by accident that there is an etymological connection between "plan" meaning a drawing and "plan" meaning the activity of designing, and furthermore, "plane" meaning a level surface.[7] Plans serve as a means of producing intricate artefacts, i.e., as a means in the processes of designing and manufacturing them. In contrast to pictures, plans are not only a means of communication but also – and primarily – a means of designing and manufacturing used by engineers and architects.

Images in the Realm of Engineering in the Fifteenth and Sixteenth Centuries

Turning now to graphical documents from the realm of engineering in the fifteenth and sixteenth centuries, two points must be stated in advance. First, for various reasons, this essay will not consider architectural sketches and plans and will deal primarily with graphical representations of machines. Secondly, since this essay is not concerned with the history of technology, these graphical representations will not be discussed here as documents of techniques and technology of the age. Rather, the main point of interest is the cognitive functions which these representations may have had for the mechanicians of the age. The essay will therefore focus especially on those graphical documents which had actual or possible significance for the practitioners; in other words, those documents which were either drawn and/or used by the historical practitioners themselves or can be derived from those drawings.[8]

Although not truly immense, the corpus of graphical documents matching these constraining requirements is quite sizeable. Thus, in order to get a first survey, it may be useful to divide this corpus into several subdivisions and name the most important documents pertaining to each one. It would be appropriate to chose such criteria of division which bear more or less directly on the possible cognitive functions of these documents.

With respect to a most obvious cognitive function of graphical representations, namely conveying information, one can divide the corpus into published and unpublished documents (fig. 6).

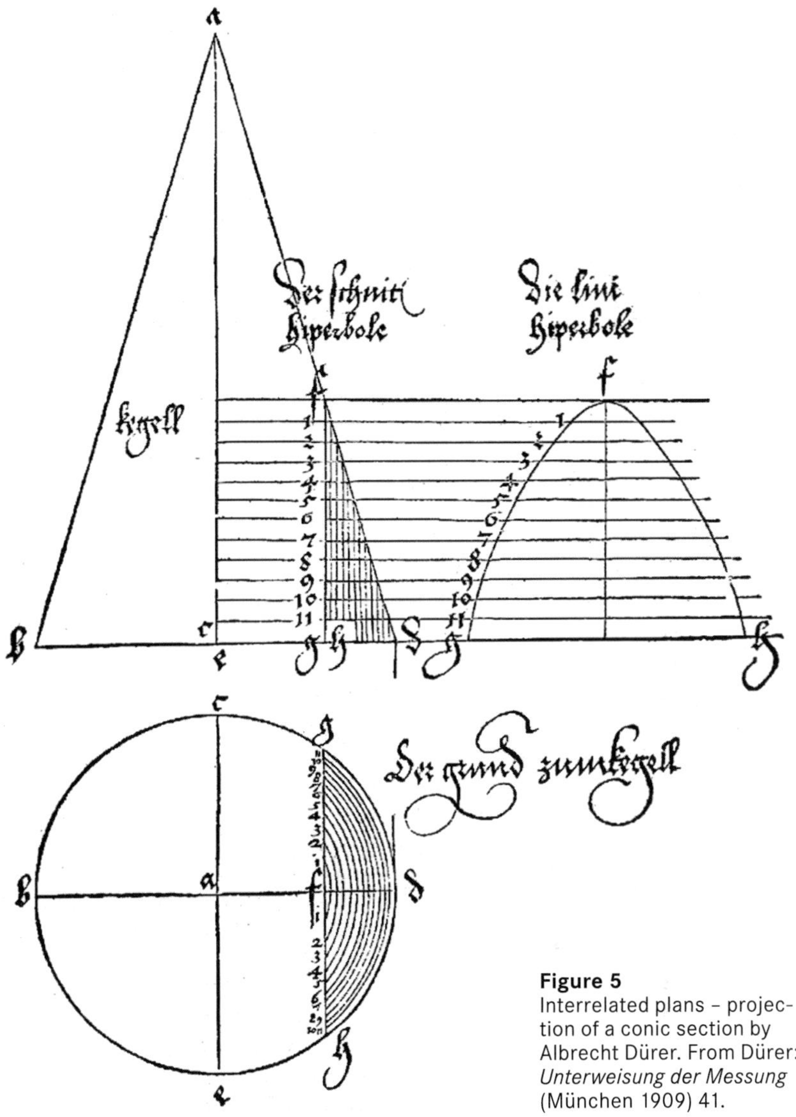

Figure 5
Interrelated plans – projection of a conic section by Albrecht Dürer. From Dürer: *Unterweisung der Messung* (München 1909) 41.

Unpublished Documents

The category of unpublished documents comprises mainly technical drawings found among the private papers of practitioners – the most prominent example is probably Villard de Honnecourt's collection of patterns *(Musterbuch)* from the thirteenth century (Honnecourt [1972]) – or in private notebooks like the famous ones of Leonardo da Vinci. It seems probable that documents of this kind exist undetected among various archival materials and can be expected to be discovered in the future.

Published Documents

As to the published documents, it seems suitable for the purposes of this essay to classify them with reference to the relationship between image and text and to distinguish

Unpublished Documents	**Published Documents**		
Drawings in Personal Collections on Notebooks	**Collections of Drawings with Explaining Texts**	**Illustrations in Technological Treaties**	**Illustrations in Theoretical Treaties on Mechanics**
A) Collections	A) Before Printing	A) Before Printing	
"Musterbücher" From Villard de Honnecourt (13th century) to the engineers of the 16th century	"Bilderhandschriften" c. 1400 Kyeser *Bellifortis* c. 1450 Taccola *De Machinis*	c. 1480 Giogio Martini *Trattati di architettura ingeniaria*	
B) Notebooks	B) Printed	B) Printed	
Leonardo da Vinci	Theatri Machinorum 1569 Besson 1584 Errard 1588 Ramelli 1607 Zonca 1613 Zeising 1615 Veranzino 1615 de Caus 1617 de Strada	1521 Vittruvius (Cesariano) 1525 Dürer *Etliche underricht* 1540 Biringuccio *Pirotechnia* 1556 Agricola *De re metallica* 1596 Lorini *Delle fortificatione*	1537 Tartaglia *La nova scientia* 1550 Cardano *De subtilitate* 1577 Guidobaldo *Mechanicorum liber* 1585 Benedetti *Diversarum speculationum* 1586 Stevin *Beghinsel der Weegkonst* 1638 Galileo *Discorsi*

Figure 6
Images in the realm of engineering in the fifteenth and sixteenth centuries.

accordingly between (i) "Collections of drawings with explanatory texts" and (ii) "Illustrations in technological treatises." "Illustrations in theoretical treatises on mechanics" should form a particular subdivision (iii) among the published documents.

Collections of Drawings with Explanatory Texts

Published collections of drawings with explanatory texts did not first emerge with the invention of woodcuts, engravings or the printing press. Rather, picture manuscripts *(Bilderhandschriften)* preceded printed books and pictures and did not vanish immediately with those inventions. The most prominent examples come from the first half of the fifteenth century – Conrad Kyeser's *Bellifortis* (ca. 1405) (Kyeser [1967]) and Mariano Taccola's *De machinis* (1449) (Taccola [1971]).

Printed collections of drawings with explanatory texts were not produced and published earlier than the last third of the sixteenth century. Once released, the whole series of the famous *theatri machinorum* appeared within just 50 years: Besson's in 1569 (Besson [1578]), Errard's in 1584 (Errard [1979]), Ramelli's in 1588 (Ramelli [1991]), Zonca's in 1607 (Zonca [1985]), Zeising's in 1613 (Zeising [1708]), Veranzino's in 1615 (Veranzino [1615]), de Caus' in 1615 (Caus [1977]), and de Strada's in 1617 (Strada [1617]).

Illustrations in Technological Treatises

This subdivision again comprises manuscripts as well as printed books. Perhaps the most important manuscript is Giorgio Martini's *Trattati di architettura ingeniaria* (ca. 1480) (Giorgio Martini [1967]). A list of the most eminent printed books of this kind must include Dürer's *Etliche underricht* (1525) (Dürer [1980]), Biringuccio's *Pirotechnia* (1540) (Biringuccio [1977]), Agricola's *De re metallica* (1556) (Agricola [1556]), and Lorini's *Delle fortificatione* (1596) (Lorini [1596]). The illustrations in Cesariano's edition of Vitruvius likewise deserve special attention in this context (Cesariano [1969]).

Illustrations of Theoretical Treatises on Mechanics

Almost all theoretical treatises on mechanics of the time include illustrations. Thus the whole series of famous Renaissance books on mechanics – Tartaglia's *Nuova scientia* (Tartaglia [1537]), Cardano's *De subtilitate* (Cardano [1550]), Guidobaldo del Monte's *Mechanicorum liber* (Monte [1577]; in Italian: Idem [1581]), Benedetti's *Diversarum speculationum* (Benedetti [1585]), Stevin's *De beghinselen der weeghconst* (Stevin [1586]), and Galileo's *Discorsi* (Galileo Galilei [1638]) – can be listed here.

Although this general survey of graphical documents in the field of engineering in the fifteenth and sixteenth centuries is confined to the most prominent and important documents, one has to add that it encompasses the bulk of material currently known. As already mentioned, new material which might affect our current understanding can be expected to emerge only from the subdivision of unpublished documents.

Designing and Manufacturing Machines Without Plans

Turning now to the cognitive functions these images may have had for the technicians of that age, one can start by investigating whether such functions can be ascribed to the images within the processes of designing and manufacturing machines and other technical devices of this kind. In this respect the distinction established between pictures and plans may prove to be a helpful heuristic means.

Browsing through these documents, one will find that plans occur almost exclusively on those drawings which depict buildings – for instance in Francesco di Giorgio Martini's *Trattati*, and of course in Cesariano's edition of Vitruvius, or in treatises on fortification like those of Dürer or Lorini. The same applies to the drawings in private collections or notebooks. In Leonardo's notebooks, for instance, one will encounter plans only on sheets dedicated to architectonic problems. The graphical representations of machines and related technical devices of the fifteenth and sixteenth centuries, on the other hand, are apparently always drawn as pictures but almost never as plans.

Even in the rare cases where one finds drawings which show different aspects of the object in a seemingly co-ordinated way, for instance, one drawing depicts the object as seen from the side and the other as seen from above, – the drawings cannot be regarded as an elevation and a ground plan but remain essentially pictures (fig. 7).[9] Plans such as we can find in a picture manuscript of Berthold Holtzschuher (1558) are absolute exceptions (fig. 8).

The drawing and use of plans in the realm of architecture was of course not new. The apparently oldest document which testifies to the use of plans in architecture is dated no later than about 2100 BC.[10] Remnants of such plans stemming from Ancient Egypt, Greek and Roman Antiquity, and from the Middle Ages have been discovered and preserved. The changes which took place during the Renaissance in comparison with the Middle Ages consist in the theoretical foundations of the techniques of constructing plans which were newly appropriated and evolved. Just as the monumental buildings of the Roman

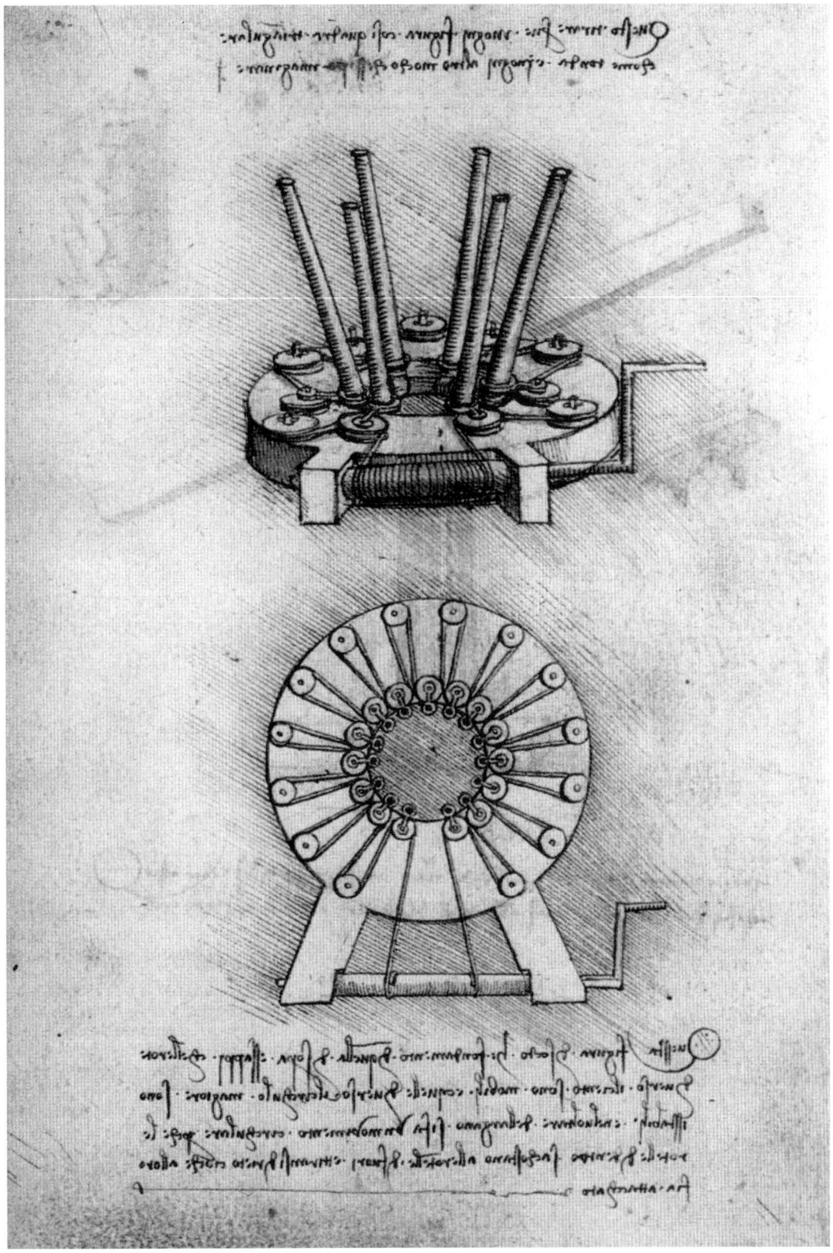

Figure 7 *(above)*
Stretching device – interrelated pictures by Leonardo da Vinci (Cod. Madrid I, folio 44v; from Galluzzi [1996] 205).

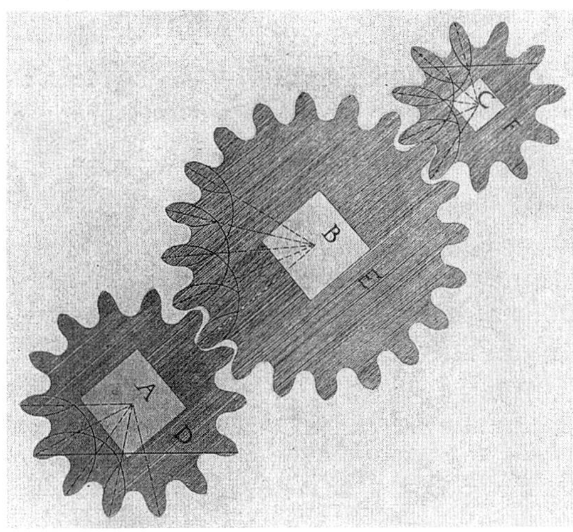

Figure 8 *(left)*
A plan of cogwheels by Berthold Holtzschuher (1558) (from Feldhaus [1995^{2}] 36).

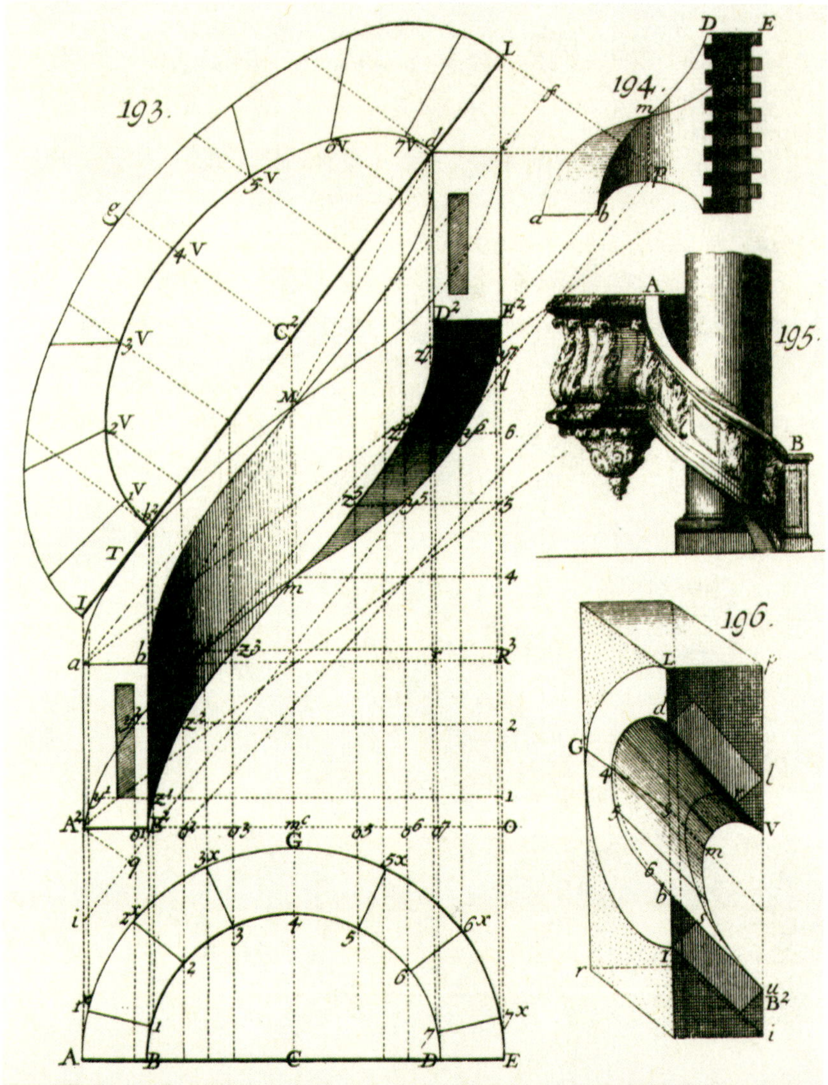

Figure 9
Ground plan and elevation by A. F. Frézier (1737) (from Sellenriek [1987] 162).

emperors were inconceivable without sophisticated methods of constructing plans, known through the writings of Ptolemy, so would be the baroque architecture without the men of the early modern period who developed those methods anew (fig. 9).

Mills were the first machines to be designed by plans from about the turn of the sixteenth to the seventeenth century. But it is telling that this happened only because mills are machines which are integrated in buildings. Their being represented by plans was thus at first only a side effect of the architectural practice of employing plans to design buildings.

The first mills were designed by non-architectural plans no earlier than at the beginning of the eighteenth century (fig. 10). The first field of engineering independent of architecture, in which plans were normally used for designing and manufacturing complex technical devices, was apparently that of shipbuilding (fig. 11).

Thus it must be stated: Apart from a few exceptions, plans of machines do not appear until the end of the sixteenth century. There is no reason to assume missing links in the source material found to date. One has rather to conclude from this fact that during this period, machines were actually manufactured without plans. Apparently drawings did not pertain to the means by which they were contrived and built. This conclusion raises many open questions. To mention just a few: which role could models have played in the process of designing and manufacturing machines? The machines designed by Brunelleschi in connection with the building of the cupola of Florence Cathedral, for example, represent the highest level of machine engineering achieved in the fifteenth century. No engineering drawings of Brunelleschi are known, however, and it seems that this is not due to missing sources. Rather, there are sufficient clues to indicate that Brunelleschi made broad use of models of all kinds (see Degenhardt et al. [1982] 117). To take another example, what about the warnings made by Lorini about trusting too much in conclusions deduced from experiences made with models? (See Lorini [1596] 175.) Closer to the topic at hand, did relationships between models and drawings exist, and if so, what kind? However interesting these questions may be, what must be pursued with priority here is whether the fact that plans were not actually used for producing machines in this period had consequences for other cognitive functions of graphical representations.

Technical Images as a Means of Recording and Communicating

That graphical representations were not used as a means of designing and manufacturing technical devices does not of course mean that these representations were of no use at all to the mechanicians of that time. Technical devices belong to the objects which reveal language's limited ability to represent things. It is almost impossible to depict an unknown device with words. It is of course best to present the device itself, or at least its model. In the many cases where that is impossible, graphical representations are an indispensable means of recording the shape and structure of an object either for one's own memory or for the purpose of communication. It is exactly for these purposes that almost all of the graphical representations were drawn and used by technicians. They registered devices they had seen or ideas for devices, – as in the case of drawings for private collections or notebooks – for their own personal use, or – as in the case of the published drawings – for others.

Regarding published drawings, we can distinguish between those serving the exchange of information between professionals and those which are also or even primarily intended for interested non-professionals. In the given surveying scheme (see fig. 6 above), the latter category is found mainly under "Collections of Drawings with Explanatory Texts." The precious picture-manuscripts of Kyeser and of Taccola, for instance, were compiled and drawn for princely persons who were interested in technology, especially for the purposes of war. More or less the same can be said of the printed *Theaters of Machines* from the last third of the sixteenth and the first decades of the seventeenth century. Splendid picture-books as they are, they were so expensive that practically no technician could afford them.

The fact that they were intended in the first place for non-professionals is probably one of the reasons why so many of these pictures depict the use, application, and advantages of the machines and devices in detail but only outline sketchily how they worked or might be manufactured (fig. 12).

A lot of these pictures therefore convey very little to an expert, and one wonders if and how many of them had a truly formative impact upon the imagination, understanding and thoughts of experienced technicians. One should not forget, however, that these collections contributed without doubt to the dissemination of ideas and knowledge from centres like Venice, Florence, Paris or Nuremberg to remote places.

The technological treatises, on the other hand, were addressed explicitly to professionals, not excluding curious laymen. With works like Agricola's *De re metallica* or Biringuccio's *De la pirotechnia* a new age dawned. They constitute the beginnings of the modern Western culture of public discourses on technological subject matter. These works strove to convey a precise understanding of the most advanced technologies used in the first half

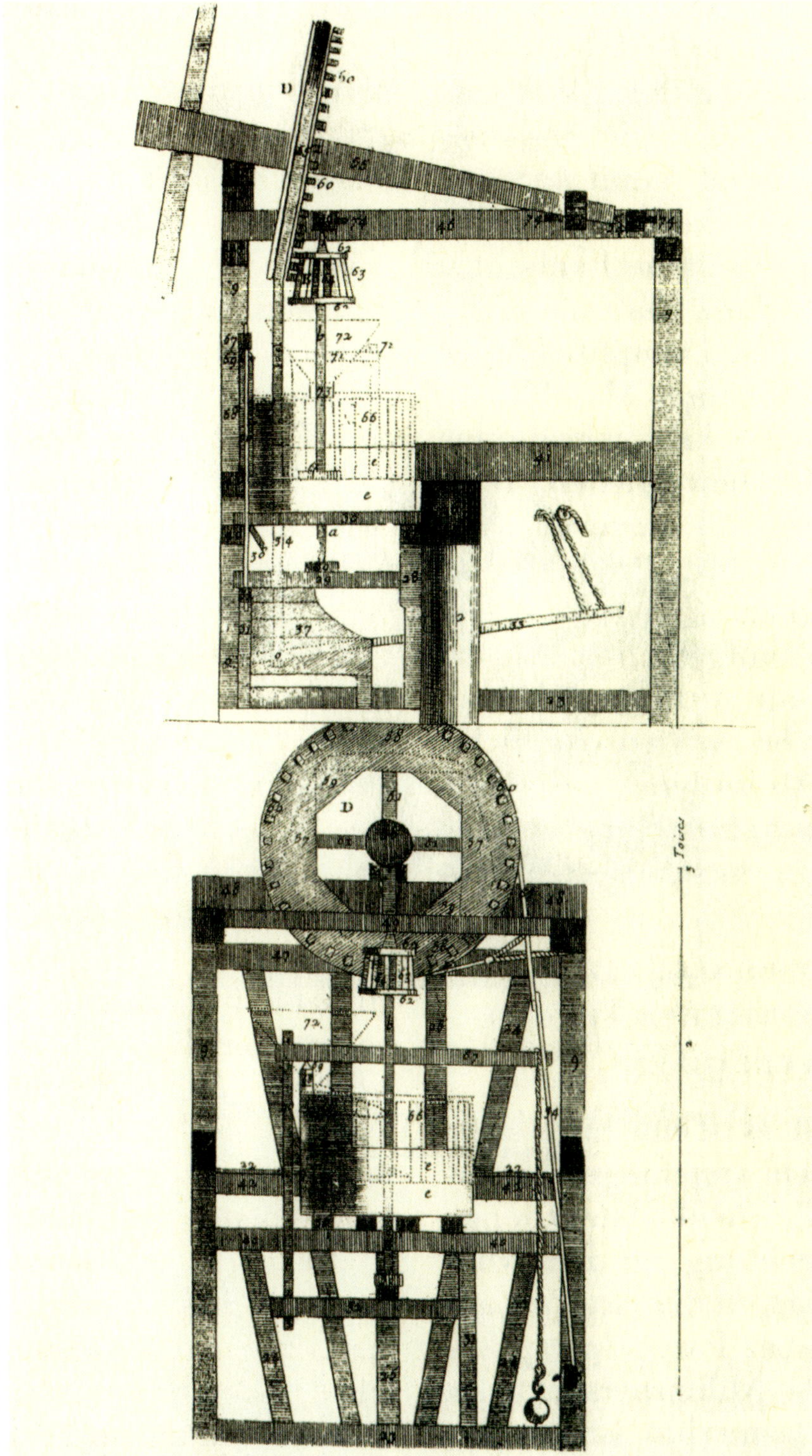

Figure 10
Cross sections of a mill by P. de La Hire (1702) (from Wißner [1948]).

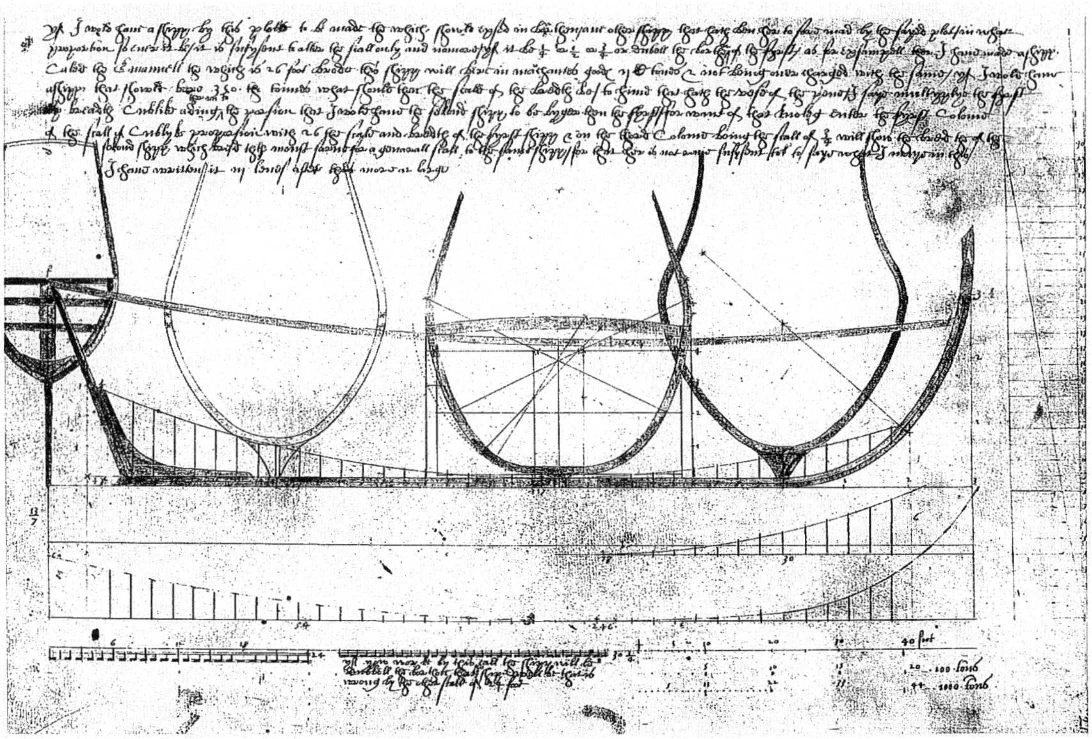

Figure 11 (*right*) Elevation and sections of a ship (ca. 1586) (from Baynes et al. [1981] 74).

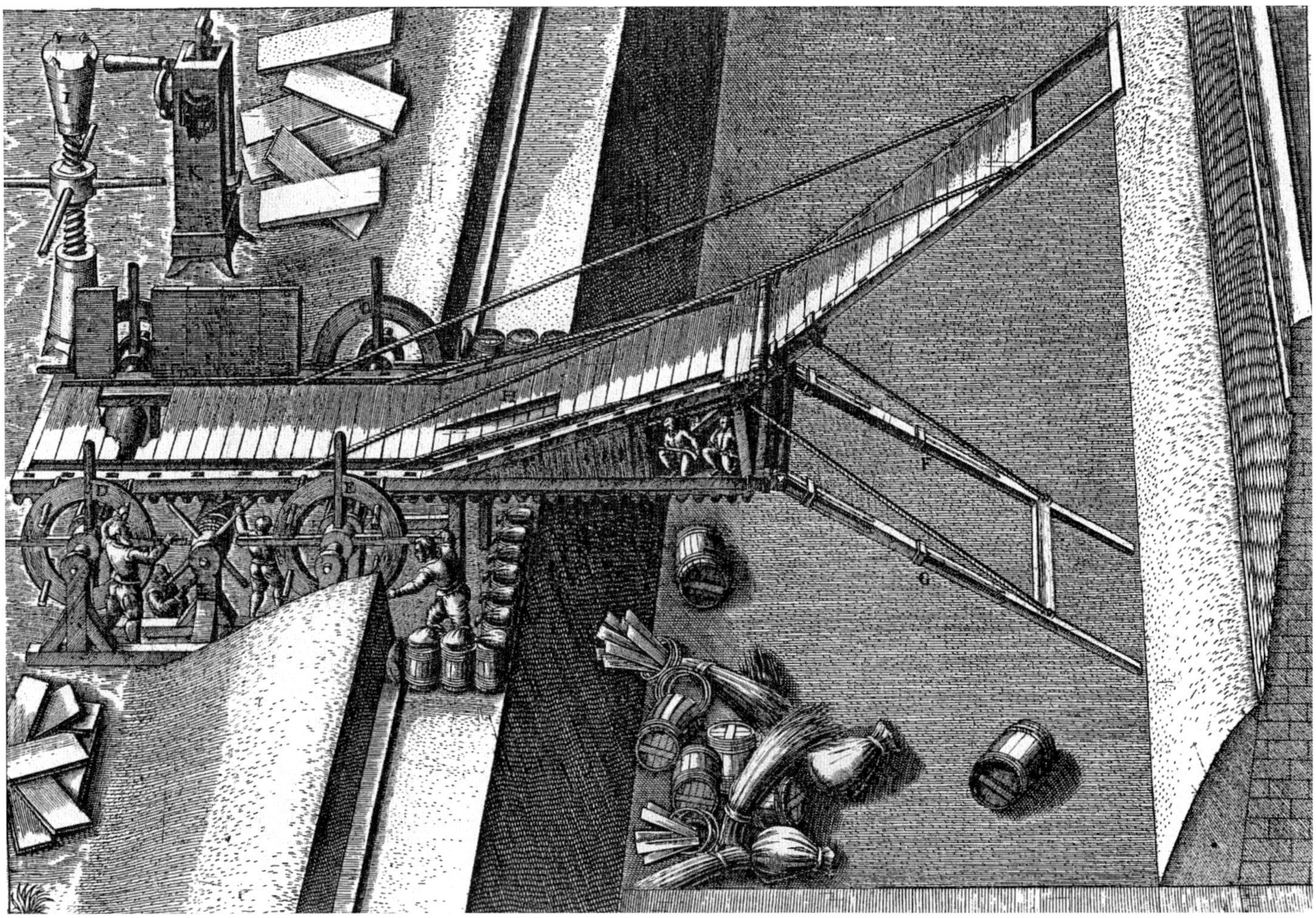

Figure 12 (*below*) Stressing the benefit of a device. Plate 140 of Ramelli: *Le diverse et artificiose machine* (Milan 1991).

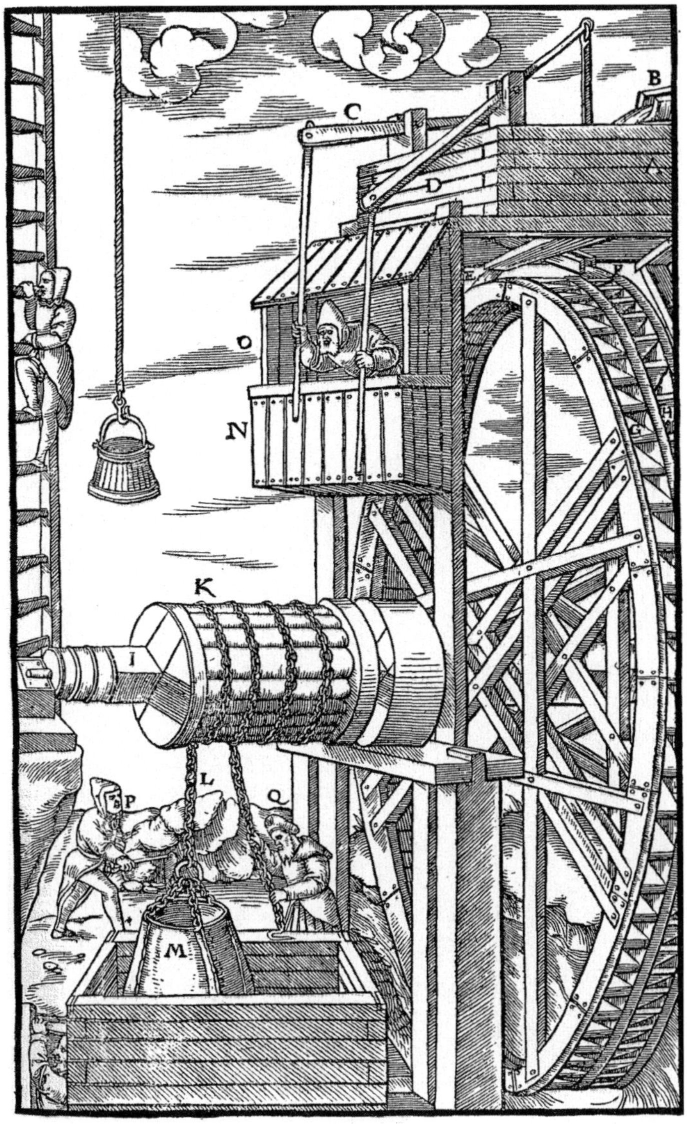

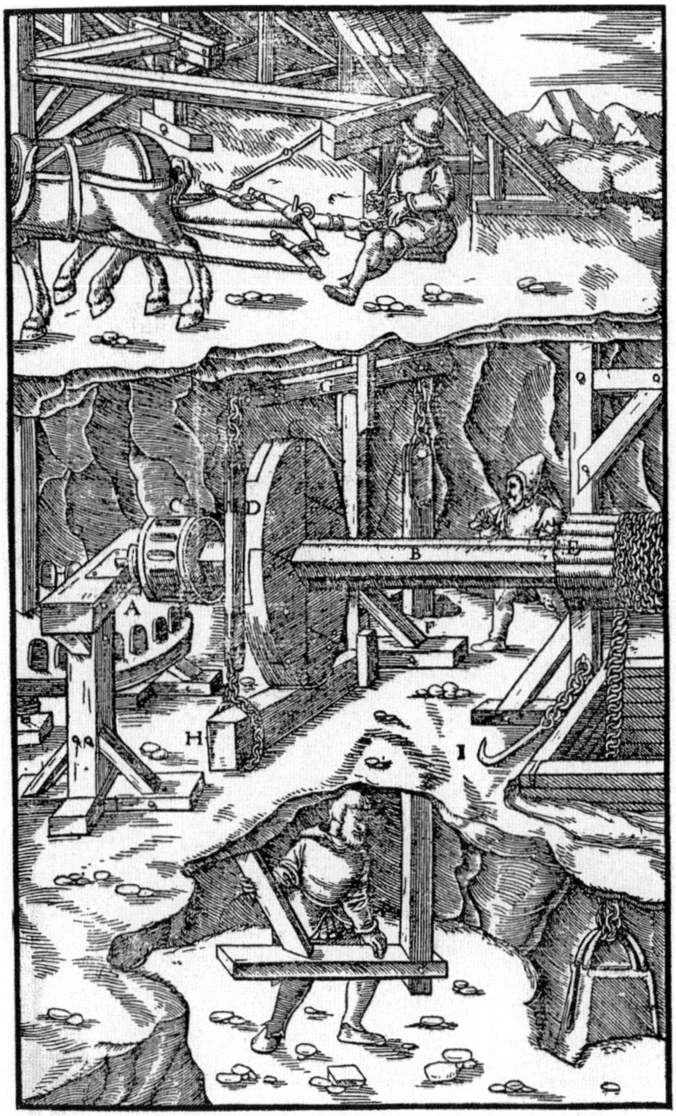

Figure 13 (*left*)
Two-directional hoist.
Woodcut in Agricola: *De re metallica* (Basel 1556) 158.

Figure 14 (*right*)
Horse operated windlass.
Woodcut in Agricola: *De re metallica* (Basel 1556) 125.

of the sixteenth century. Therefore, the pictures in these books are justly regarded as being among the best graphical representations of technology of the time. They are reliable, depict devices that actually existed, and try to convey how these devices worked (fig. 13).

No doubt, these pictures genuinely strive to fulfil the task of communicating useful and telling information concerning their topics. Because of that, they reveal the limits peculiar to pictures as a medium of information. This becomes especially noticeable where these pictures show astonishing resourcefulness in displaying much more than only the appearance of their objects.

Sometimes, these pictures try to represent the whole structure of a device although, in reality, one can only ever see one part of it at a certain time. Figure 14, for instance, applies a cut-away technique for displaying a machine at different levels above and beneath the earth.

Sometimes, these pictures depict an integral machine and, at the same time, in addition, its several parts as disconnected pieces next to it (fig. 15).

Sometimes, these pictures render transparent what is, in reality, opaque. Figure 16 again displays a machine at different levels in the underground as well as in its entirety and with its separate parts. What is most remarkable here is that we can look into a closed wooden box (upper right side).

Combined with text, these pictures are the most telling documents about technology

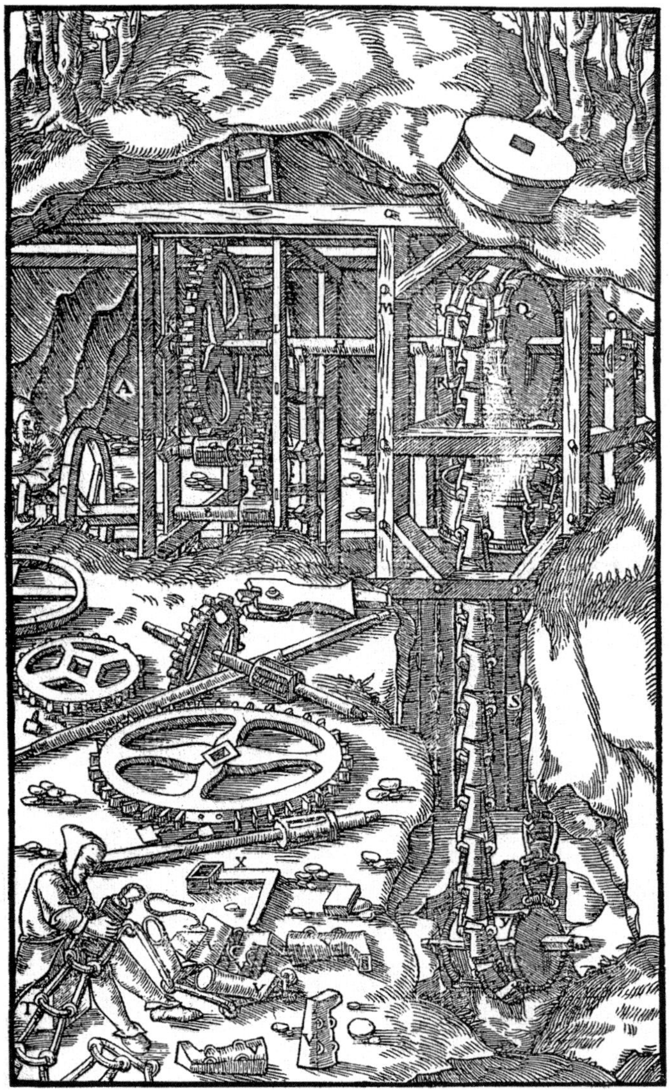

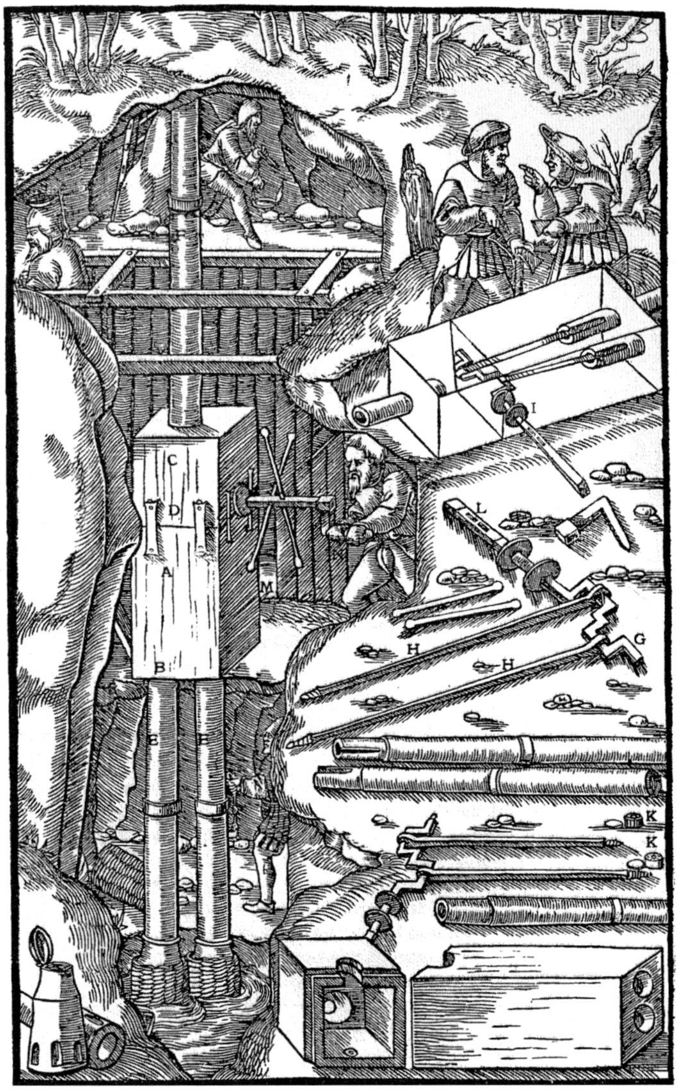

Figure 15 (*left*)
Mine pump. Woodcut in Agricola: *De re metallica* (Basel 1556) 131.

Figure 16 (*right*)
Duplex suction pump. Woodcut in Agricola: *De re metallica* (Basel 1556) 139.

existent from the sixteenth century. And yet, how many details remain implicit in these pictures, partly since they were no doubt obvious to professionals – for instance, what is to be provided so that an axis will revolve smoothly? etc., partly since they display only a general type of a certain machine, whereas its accurate details – for instance, the sizes and measures of the parts – remain to be determined by engineers according to the circumstances of a concrete situation. Nevertheless, one has to understand that even if a historical actor had wanted the pictures to deliver those details it wouldn't have been possible due to the limits inherent in the "syntax" of pictures. An image published just fifteen years later depicting a comparatively simple technical device may demonstrate this point. By depicting (at the bottom) the different parts of the bridge's stand in the "syntax" of plans, figure 17 clarifies all constructional questions. Not surprisingly, the figure comes from a treatise on architecture.

Technical Images as Means of Reflection

Besides their function of communicating information, technical images had yet another cognitive function which also appears to be dependent on the "syntax" chosen. Much more than physical models of technical devices, images are free to select what they choose to display and what not. When drawing, one can depict only certain features of an object, one can emphasise certain sides and

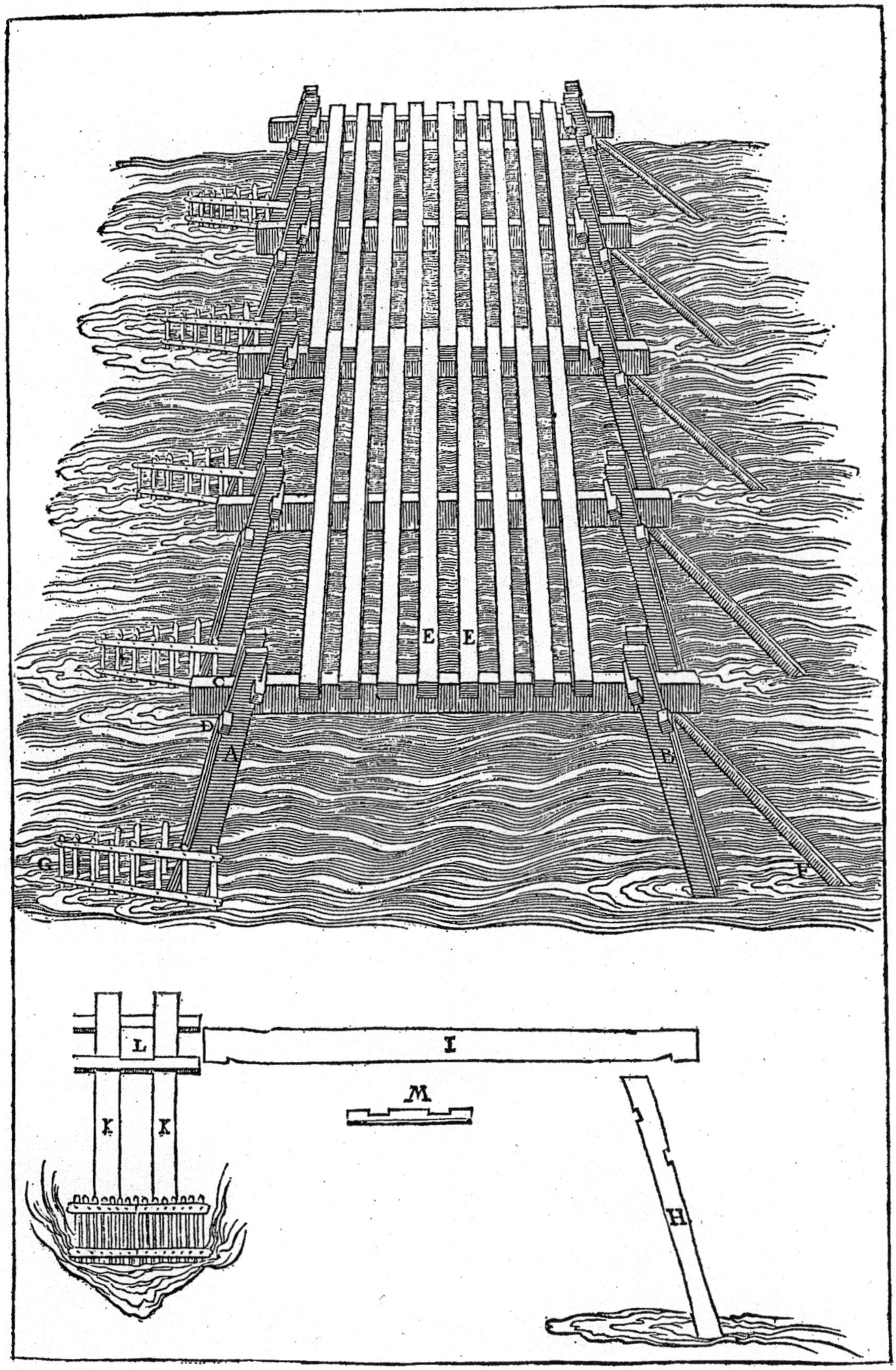

Figure 17
Wooden bridge (from Palladio [1570] book III chap. 6).

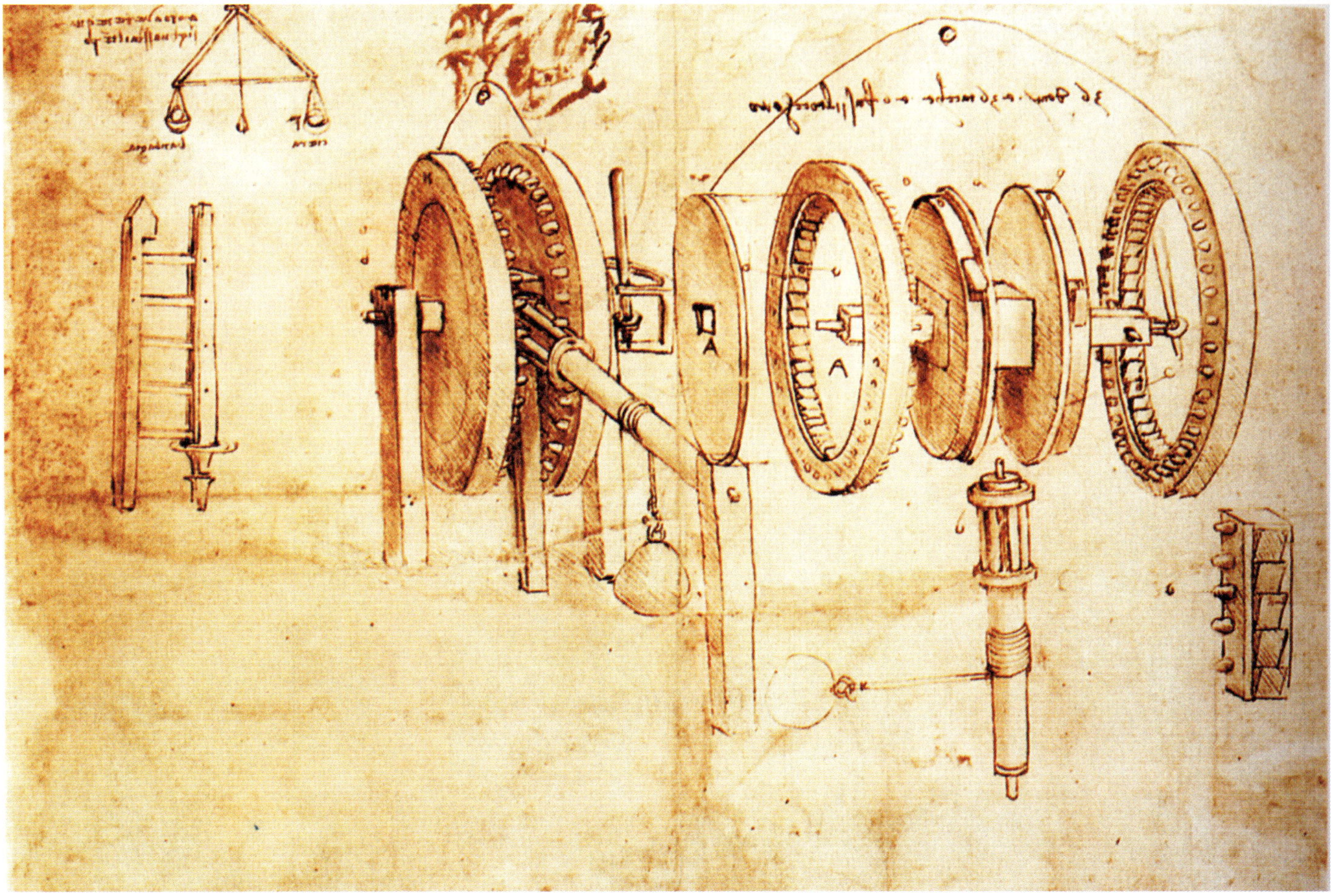

Figure 18
Reflecting by drawing. Drawing by Leonardo da Vinci (Cod. Atlant. (BAM), folio 30v; from Galluzzi [1996] 202).

omit others, or separate parts which in reality are inseparable. While drawing, technicians thus had a rich arsenal of analytical means at their disposal. Images can therefore be a suitable means of reflection. A proper discussion of this capacity of images would require (and deserve) an independent essay.[11] For the realm of engineering in early modern times, however, it may be sufficient to remember the well known technical drawings of Leonardo which impressively show how images were actually used as a means of reflection on technological questions.

The drawings of Leonardo are without doubt the most excellent instances in this respect. They display in an unparalleled manner how technological problems can be considered and studied by drawing. Because of their masterly quality, they are at the same time the best example with which to prove that this way of using drawings for reflection leads to a dead end. Firstly, Leonardo's way of using drawings for reflection depends on his skills as a draftsman. Only very few people, if any at all, can therefore follow his example. In order for images to become a normal medium for technicians, normal technicians must be capable of producing them. That is to say, in order to pertain to the standard competence of technicians, the techniques by which such suitable images are drawn must be teachable and learnable. As is well known, what actually became the teachable and learnable art of engineering drawing, starting just in the sixteenth century, was the art of constructing plans, and not the art of drawing pictures in Leonardo's way.

However, the main reason why drawings like Leonardo's will appear as an insufficient means of reflection lies once again in the fact that they are drawings of pictures and not of plans. As soon as one has to reflect minutely on how the parts of a technical device can be arranged in the most effective way, the graphical representation sufficiently reliable must represent the device rather than the appearance of the device. This is particularly true in the case of machines. In order to anticipate what exactly will happen when a cog-wheel or a piston-rod moves this way or that, one needs graphical representations with true scaled distances, correct angles, etc.; that is to say, one needs plans.

Up to now, it has become apparent that the reasons why pictures are restricted as a means of reflection do not differ from the reasons why mechanicians can make only limited use of pictures for planning. What was discussed earlier, with regard to their use in designing machines also applies to their function as a means of reflection. With respect to reflection, a further difference between pictures and plans depending on their "syntax" becomes visible, a difference of no little consequence: Whereas it is possible to interrelate plans with graphical representations in the more theoretical treatises on mechanics of that time, it is not possible to link pictures with them. That is to say, the technicians needed plans, not pictures, as a means of reflection on technological questions to draw the results of theoretical mechanics into their reflections (fig. 19).

As is well known, at least until the middle of the eighteenth century, techniques of geometrical demonstration were widely applied for reasoning within theoretical mechanics and connected with geometrical constructions. For the purposes of geometrical proofs, it is of no real significance whether or not these constructions were absolutely correct, but they are supposed to be; for instance, that certain angles should be exact right angles if that is required for the proof. What is thus assumed is that they are constructed according to a "syntax" peculiar to plans.

Regarding only the geometrical proof, these constructions represent nothing other than ideal geometrical entities. However, when one regards the physical meaning attributed to them in mechanics, certain parts or elements represent structures of physical processes or physical objects – for instance, the trajectory of a projectile, an inclined plane, the arms of a lever, etc. Sometimes one even finds constructions which can be easily traced back to graphical representations of technical devices (fig. 20). In these cases, we can always observe that the transformation into a construction suitable for reasoning includes the translation into a drawing with the "syntax" of plans. These transformations prove from another angle that the "syntax" of plans is essential for images as means of reflection in mechanics, particularly if such reflections try to apply theorems or doctrines of theoretical mechanics. Plans thus appear as an important means of intermediation between practical and theoretical mechanics – a point which without doubt deserves further investigation.

For the sixteenth century, however, we must establish the fact that such geometrical representations of technical devices are found within treatises dealing mainly with theoretical questions, whereas plans were not yet a means of designing and manufacturing

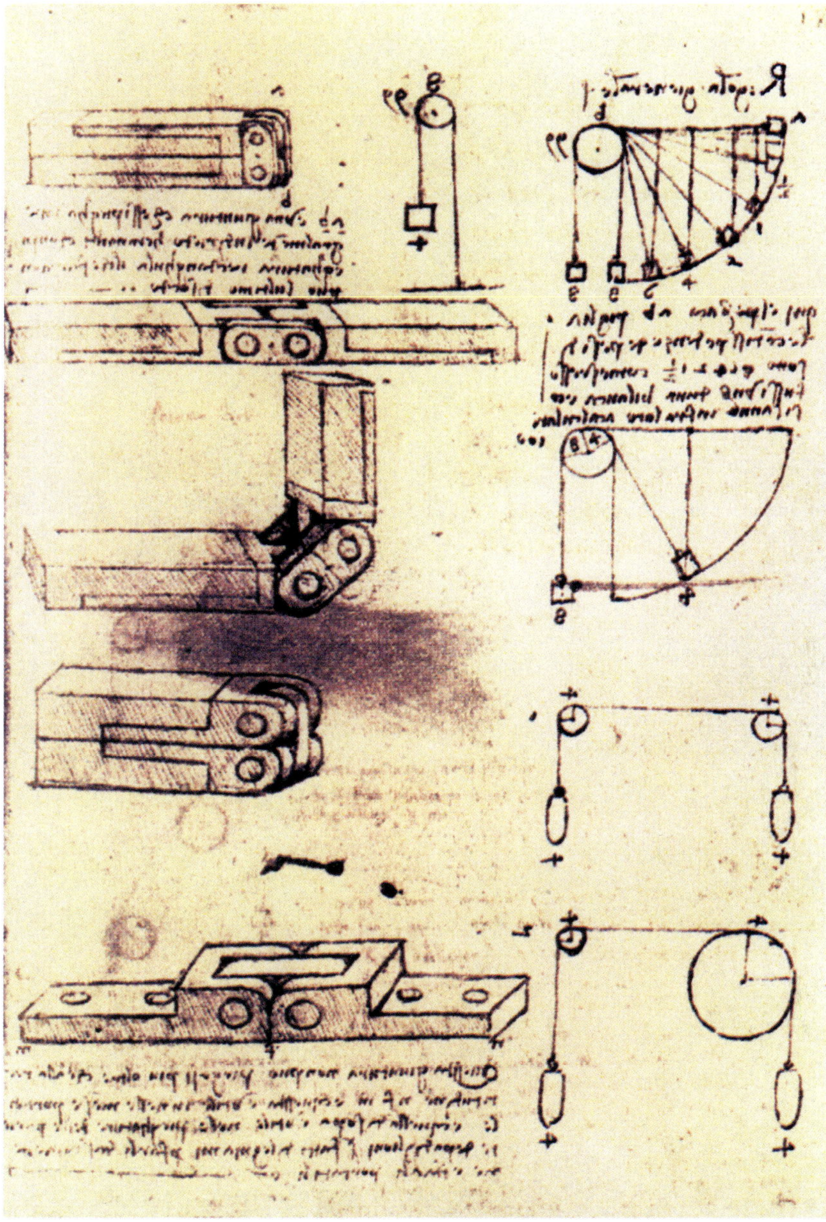

Figure 19
Picture and geometrical construction. The pictures on the left side serve as a medium of reflection on two kinds of joints, whereas the drawings on the upper right side serve as a means for theoretical conclusions on forces. Drawings by Leonardo da Vinci (Cod. Madrid MS. I, folio 172r; from Galluzzi [1987] 143).

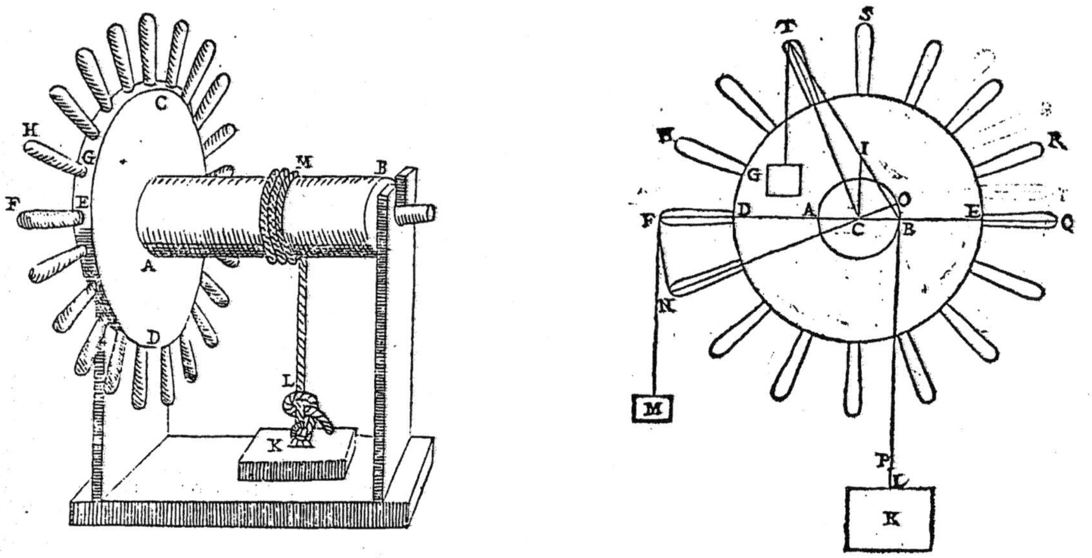

Figure 20
Transformation of a picture into a geometrical construction. From Guidobaldo da Monte: *Le Mechaniche.* (Venice 1581) 102f.

machines for the practitioners. Whereas practical mechanics had already become subject matter of theories, theoretical mechanics was apparently still of no significance to the mechanicians. The fact that the graphical representations of technical devices consisted almost exclusively of pictures rather than plans seems to be symptomatic of this situation.

Max Planck Institute for the History of Science, Berlin

Notes

[1] The most prominent treatises of actual practising engineers of that age are probably: Vannoccio Biringuccio's *Pirotechnia* (1540), Giuseppe Ceredi's *Discorsi* (1567), Buonaiuto Lorini's *Delle fortificatione* (1596), and Vittorio Zonca's *Teatro di machine* (1607). Looking, however, more closely at Ceredi's and Zonca's treatises, for instance where doctrines of contemporary theoretical mechanics are obviously applied, one may wonder whether they mirror the thoughts of common engineers, let alone common mechanicians.

[2] See Ivins [1969], especially the discussion in connection with the quotation from Pliny (p. 14).

[3] Not only the geometrical basis of these techniques of projection is a achievement of Antiquity but such techniques themselves. Nevertheless, regarding linear perspective, no coherent unified system of representation is known from Antiquity which could be compared with those of Brunelleschi and his followers. See still Panofsky [1927], particularly part II. Christopher S. Wood translated this essay into English: Panofsky [1997].

[4] The following statements are not about pictures in general. The ways in which, for instance, pictures of the Middle Ages depict their objects are not topics of this essay. Rather, it is only concerned with those pictures which are either the result of projection or imitate constructed ones. Since at least the sixteenth century, linear perspective has ruled our vision so powerfully that even pictures produced without construction strive to look like constructed ones.

[5] Though the particular semiotic meaning of "syntax," viz. the rules of constructing signs (see, for instance, Goodman [1976] 130ff.), may first come to mind in connection with images, the term is used here in the common linguistic sense, i.e. for a system of rules governing the combination of signs into larger units which are meaningful.

[6] This statement doesn't mean that a linear perspective depiction is more "natural" than other kinds of graphical representation. Only if it is about the depiction of the *appearance* of the object and if the picture on the retina is the standard for judging naturalness can one call linear perspective in a way "natural." (If, additionally, the degree of correspondence with the true geometrical properties of the object is taken into account, one could even say that a linear perspective picture is more "natural" than the picture on the retina because the latter comprises more distortions than the former due to its being a projection – not on a plane, but on a spherical surface. What linear perspective actually tries to construct is the (inverted) appearance of objects in plane mirrors or comparable devices – see Wartofsky [1979] 217f.) But to depict an object by just depicting its appearance to the eye (or in a mirror) is by no means natural. Rather, pictorial representation through representation of the appearance emerged as result of particular historical and cultural developments and can be regarded as an expression (with Panofsky: as "symbolic form") of a peculiar culture. That is why in early modern times linear perspective pictures

were immediately understandable only for natives of this culture. For the cognitive problems involved in the perception of these pictorial representations see Hagen [1980].

7 This applies not only to English but to all languages which use words derived from the Latin "planum."

8 Excluded therefore are pictorial representations of technical devices or processes as they occur on paintings or engravings, etc. created by artists.

9 One has to admit, however, that in figure 7 the drawing below seems to be on the border-line between picture and plan. See also Booker [1963] 46.

10 As an attribute of a statue from Lagash, now in the Musee National du Louvre, Paris. See for example Sellenriek [1987] 31.

11 An elaborate discussion should comprise an investigation of the conception of "images of thought" introduced by *Gestalt* psychologists. See still Arnheim [1969].

References

Agricola, G.
1556 *De re metallica*. Basel.
1950 *De re metallica*, eds. H. C. and L. H. Hoover. New York.

Argan, G. C. (ed.)
1977 *Die Kunst des 20. Jahrhunderts 1880–1940* (= Propyläen Kunstgeschichte Bd. 12). Berlin.

Arnheim, R.
1969 *Visual Thinking*. Berkeley.

Baynes, K. and F. Pugh
1981 *The Art of Engineer.* Guildford.

Benedetti, G. B.
1585 *Diversarum speculationum mathematicarum et physicarum liber.* Turin.

Besson, J.
1578 *Theatrum instrumentorum et machinarum.* Lyon.

Biringuccio, V.
1977 *De la pirotechnia*, ed. A. Carugo. Milan.

Booker, P. J.
1963 *A History of Engineer Drawing.* London.

Cardano, G.
1550 *Hieronymi Cardani Medici Mediolanensis De svbtilitate.* Nuremberg.

Caus, S. de
1977 *Von gewaltsamen Bewegungen: Beschreibung etlicher, so wol nützlichen alss lustigen Machiner.* Hanover.

Cesariano, C. (ed.)
1969 *Vitruvius: De architectura*, ed. C. H. Krinsky. München.

Degenhardt, B. and A. Schmitt
1982 *Corpus der italienischen Zeichnungen 1300–1450*, vol. II-4. Berlin.

Dürer, A.
1909 *Unterweisung der Messung*, ed. H. Thoma (= Süddeutsche Monatshefte). München.
1980 *Etliche underricht zu befestigung der Stett, Schloß und flecken.* Unterschneidheim.

Edgerton Jr., S. Y.
1975 *The Renaissance Rediscovery of Linear Perspective.* New York.

Errard, J.
1979 *Le premier livre des instruments mathématiques méchaniques.* Paris.

Feldhaus, F. M.
1995[2] *Geschichte des technischen Zeichnens.* Wilhelmshaven.

Francesca, P. della
1984 *De Prospectiva Pingendi*, ed. G. Nicco-Fasola. Florence.

Galilei, G.
1638 *Discorsi e dimostrazioni matematiche : intorno à due nuoue scienze attenenti alla mecanica i movimenti locali.* Leyden.

Galluzzi, P.
1987 (ed.): *Leonardo da Vinci – Engineer and Architect.* Montreal.
1996 *Renaissance Engineers. From Brunelleschi to Leonardo da Vinci.* Florence.

Giorgio Martini, F. di
1967 *Trattati di architettura ingegneria e arte militare*, ed. L. M. Degrassi. Milan.

Goodman, N.
1976 *Languages of Art.* Indianapolis.

Hagen, M. A. (ed.)
1980 *The Perception of Pictures.* New York.

Honnecourt, V. de
1972 *Kritische Gesamtausgabe des Bauhüttenbuches*, ed. H. R. Hahnloser. Graz.

Ivins Jr., W. M.
1969 *Prints and Visual Communication.* Cambridge MA.

Kay [Ward], M. A.
1976 "The Accademia del Disegno in Late Renaissance Florence." *Sixteenth Century Journal* II/2: 3–20.

Kyeser, C.
1967 *Bellafortis*, ed. G. Quarg. Düsseldorf.

Lefèvre, W.
2000 "Galileo Engineer – Art and Modern Science." In *Science in Context*, vol. XIII/3-4.

Lorini, B.
1596 *Delle fortificatione.* Venice.

Monte, G. del
1577 *Guidiubaldi e Marchionibus Montis mechanicorum liber.* Pesaro.
1581 *Le Mechaniche.* Venice.

Nedoluha, A.
1960 *Kulturgeschichte des technischen Zeichnens.* Vienna.

Olschki, L.
1926 "L'Accademia Fiorentina del Disegno nella Storia delle Arti e delle Scienze." *Nuova Antologia*: 470–477.

Palladio, A.
1570 *I quattro libri dell'architettura*. Venice.

Panofsky, E.
1927 "Die Perspektive als 'symbolische Form.'" In *Vorträge der Bibliothek Warburg 1924–1925*, 258–330. Leipzig.
1997 *Perspective as Symbolic Form*. New York.

Ramelli, A.
1991 *Le diverse et artificiose machine*, ed. G. Scaglia. Milan.

Roriczer, M.
1486 *Das Büchlein von der Fialen Gerechtigkeit*. Regensburg.

Sellenriek, J.
1987 *Zirkel und Lineal*. Munich.

Stevin, S.
1586 *De beghinselen der weeghconst*. Leyden.

Strada, J. de
1617 *Künstlicher Abriss allerhand Wasser-, Wind-, Ross- und Handt-Mühlen*. Frankfurt.

Taccola, M.
1971 *De machinis*, ed. G. Scaglia. Wiesbaden.

Tartaglia, N.
1537 *Nova scientia inventa da Nicolo Tartalea*. Venice.

Veranzino, F.
1615 *Machinae novae*. Florence.

Ward, M. A.
1972 *The Accademia del Disegno in sixteenth-century Florence*. Ann Arbor (U.M.I. Dissertation Information Service).

Wartofsky, M. W.
1979 "Rules and Representation: The Virtues of Constancy and Fidelity Put in Perspective." In idem, *Models – Representation and the Scientific Understanding*, 211–230. Dordrecht.

Wißner, A.
1948 *Die Entwicklung der zeichnerischen Darstellung von Maschinen*. Diss. Munich.

Zeising, H.
1708 *Theatrum machinarum*. Leipzig.

Zonca, V.
1985 *Novo teatro di machine et edificii*, ed. C. Poni. Milan.

Reframing the Language of Inventions: The First Theatre of Machines

*Luisa M. Dolza**

> … we prefer sight to practically all the other senses. The reason for this is that of all the senses sight best helps us to know things, and reveals many distinctions.
>
> (Aristotle, *Metaphysics* I, 980a.)

Introdution

The supremacy of the sense of sight with respect to cognition is an essential part of philosophical tradition.[1] "As if they were alive," the illustrations have an enormous potential to elicit emotions, to amuse, and to delight. They use a universal, attractive and immediate language that requires no translation yet permits individual interpretation. During the Renaissance, when painters and then engravers learned to use *chiaroscuro* – the artistic technique of rendering light and shade – and such perspective conventions as the vanishing point, illustrations assumed a strategic cultural role.[2] As a result, illustrated books became increasingly popular, and publishers began discovering new and improved methods for reproducing illustrations alongside texts. The illustrated book, which was ever more attractive and affordable, had become, in the words of Montaigne, "la meilleure munition que j'aye trouvé à cet humain voyage," a cornerstone of religious and cultural life of the period (Montaigne, III, 2).

Since the 1930s, scholars from different backgrounds have stressed the importance of better understanding these images in their complex entirety.[3] Among others, Lucien Febvre called for more in-depth study of what he referred to as "les supports sensibles de la pensée aux diverses époques."[4] While the relationship between literature and illustration has been studied carefully, the use of illustrations for scientific and technical purposes, as well as their role in the development of science and technology, has received insufficient attention from historians of science, notwithstanding the significant affirmations of Erwin Panofsky and George Sarton.[5] The ideas advanced by these two well-known scholars on the importance of illustrations in the scientific literature of the Renaissance have been debated over the years by historians of philosophy, culture, and the arts.

Over the last few years, historians of science and technology, a considerable number of which had largely doubted the importance of illustrations, and have generally considered the study of these images to be somewhat irrelevant and outside their field of interest, have felt the need to re-examine and redefine the role of illustrations using new methods and approaches.[6] While strongly felt by historians of science, this need is of utmost importance for historians of technology, as the issues related to technical illustrations are possibly even more complex than those related to scientific illustrations, and require supplementary analysis.[7] First, due to the limitations and ambiguity of language at the time, and to the lack of precise terms for naming and describing the intricacies of mechanical structures, it is rare to find drawings as a simple complement to the text. In most cases, the so-called "technical" illustration is considerably independent from the text. Indeed,

* I am grateful to Andrea Goldstein, Penni Oren and Hélène Verin for their invaluable help.

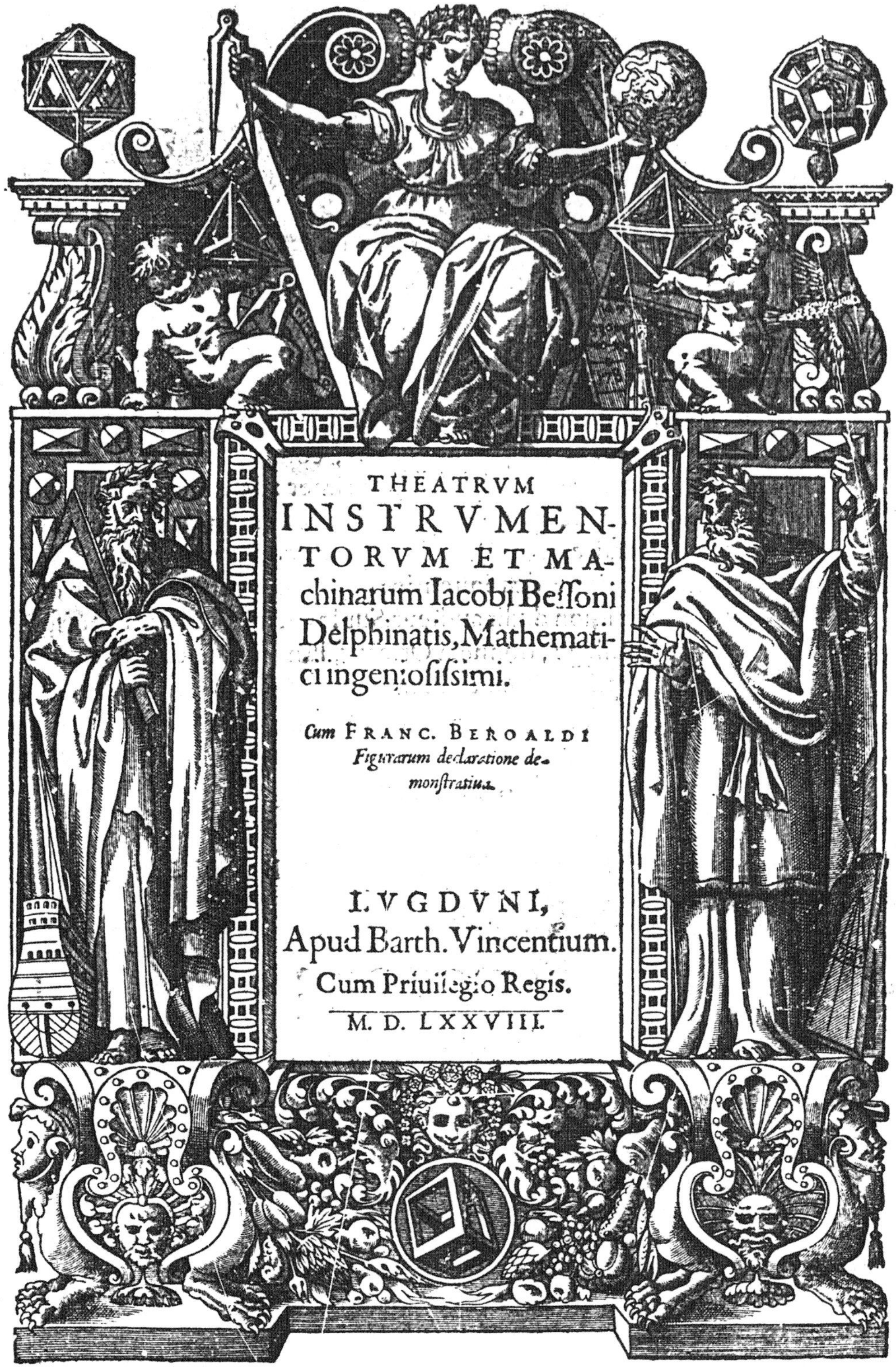
THEATRVM
INSTRVMEN-
TORVM ET MA-
chinarum Iacobi Beſſoni
Delphinatis, Mathemati-
ci ingenioſiſsimi.

Cum FRANC. BEROALDI
Figurarum declaratione de-
monſtratiua.

LVGDVNI,
Apud Barth. Vincentium.
Cum Priuilegio Regis.
M. D. LXXVIII.

Figure 1
Title page of Besson: *Theatrum instrumentorum et machinarum* (Lyon 1578).

in some cases, such as the case of the *theatres of machines* examined here, illustrations are virtually self-explanatory, presented with only a few brief introductory words.

For historians today, these theatres, as well as the manuscripts of the fifteenth century, represent one of the few available sources for improving our knowledge and understanding of sixteenth-century technology.[8] Despite the differences noted by scholars, in general all of these theatres show various types of machines in motion, displaying both how

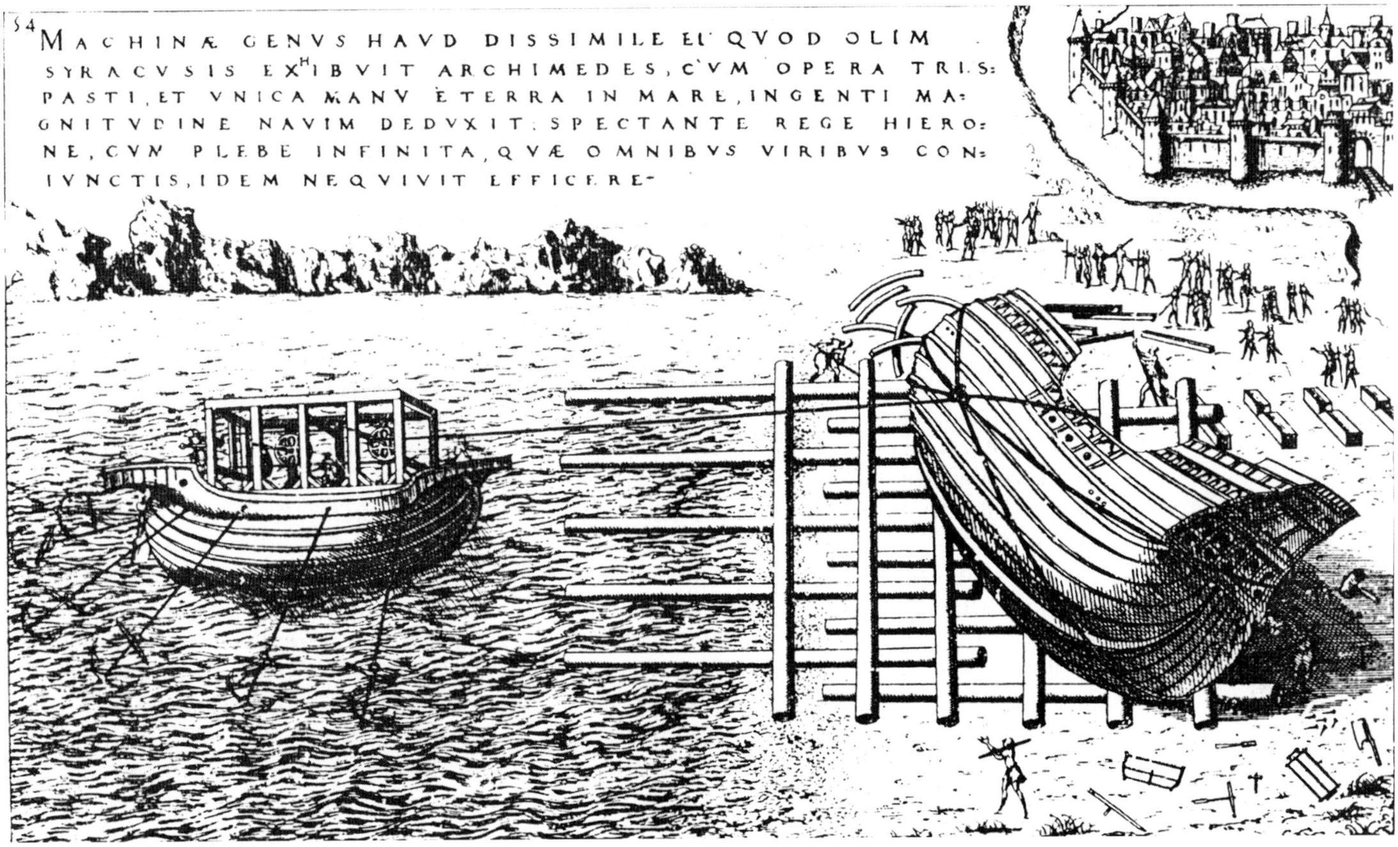

Figure 2
"Tripaston" or "Charistion", a hauling device for lifting ships attributed to Archimedes. Besson reconstructed this device in a way that, following his description, a system of screws allowed to lift a ship only by one hand.
(Besson: *Theatrum* (Lyon 1578) plate 54).

they worked and how they were assembled. They also suggest ways in which new technical combinations might be devised to carry out different, but related, tasks. Among the mechanisms commonly depicted were mills, pumps, fountains, and cranes, as well as those for lifting water and waging war. More or less all theatres follow a precise layout: a dedication to the sovereign, or some other high-ranking official, followed by a short preface. An introduction in praise of mathematics sometimes leads to the heart of the book, the illustrations of the machines. The role of these introductions, and more generally the relationship between technology and mathematics in the sixteenth century, are important issues that merit further study. However, in these theatres, drawings are the predominant element; while there is always a description of the machines, it is generally no more than two pages long.

Due to their structure, and to the emphasis which is placed almost exclusively on the illustrations, it was generally assumed that the principal objective of the theatres was the transmission of technical knowledge. It was also assumed that the authors of these theatres were technicians and inventors who were generally unknown, and who used these books as a vehicle to make their skills and inventions known to potential patrons and the public. Therefore, historians of technology have focused their attention almost exclusively on the machines illustrated, attempting to assess if they were new, to what extent the images communicate information about how the machine works, and whether or not a viewer could construct a working model using solely the illustrations. However, while relevant, this exclusive interest in the illustrations of the machines up to date has given little insight into the true meaning of the theatres. In order to fully understand the significance of these books, and to appreciate their cultural and historical importance, more profound study is underway.[9] Thus, the first challenge and the goal of this contribution is to demonstrate the complex nature of the theatres, and how they were profoundly linked to the historical period in which they were produced. Contrary to what has generally been assumed, the first theatre ever published, Besson's *Théâtre des instruments mathématiques et mechaniques,* is a complex work by a noted author, rooted in an environment profoundly influenced by the emblematic culture of the period in which the representation of an object was not necessarily an accurate depiction of visual reality, but often had a deeper significance.[10]

Painting is Silent Poetry: Emblems[11]

Books of emblems, which are small collections of images in which the imagination plays a fundamental role, were extremely popular among the educated public in sixteenth- and seventeenth-century Europe.[12] They also had an important influence on the first theatres of machines, not least on their graphic layout.[13] In the beginning, these emblems were an attempt by the Humanists to create a modern equivalent of ancient pictorial scripts. For the ancient Egyptians, Greeks, and Romans, hieroglyphics were allegorical, with sacred, religious, and philosophical meanings. *Hieroglyphica,* an ancient hieroglyphic text attributed to the mythical Egyptian priest Orapollo (or Horappollo) by the Greek translator Filippe, was discovered in 1419 by Cristoforo di Buondelmonte in Andros. It was first published in Venice in 1505 by Aldo Manunzio, and was enthusiastically received by scholars eager to access the ancient wisdom contained in its symbols. Numerous subsequent editions followed, as well as commentaries. From its first Latin translation by Filippo Fasanini, to a commentary written by Piero Valeriano in 1567, as well as those of Bernardo Trebazio and Jean Martin, *Hieroglyphica* was presented in the form of a dictionary, offering those interested in the subject a unique and fascinating "alphabet" leading to archaic knowledge which could be deciphered only by the initiated.[14]

One of the most beautiful illustrated books published by Aldo in Venice is *Hypnerotomachia Poliphili* which was printed in 1499, and contained one hundred and seventy engravings, which some have attributed to Andrea Mantegna and Jacopo Bellini.[15] *Hypnerotomachia* – which was published anonymously, but contained a reference to the Dominican friar Francesco Colonna, identified by some as Prince Colonna – is in the form of a novel written in the first person, by *Polifilo,* who recounts in great detail his allegorical voyage in which very little action occurs.[16] Using also the illustrations, the author recounts his travels, not to far-off lands, but to places only the mind can go, using cryptic references and suggestions, to create an intellectual pathway that leads readers through a maze of diverse monuments: fountains, temples, gardens, cemeteries, and bridges.[17] For the first time, the engravings contain pictorial symbols and figures which are the precursors of emblems: a symbol along with a descriptive epigram (Pozzi [1982]). In his dedication to Guidobaldo, the duke of Urbino, Leonardo Crasso, the Veronese jurist who financed the Venetian edition of *Hypnerotomachia,* states:

> *che essa contiene non solo tanta scientia, ma tanta ricchezza che, leggendolo ti parrà non pure d'aver letto tutti i libri degli antichi ma di avere penetrato ne secreti della stessa natura.*

In response to the growing interest in illustrations, more engravings were added to the original one hundred and seventy plates in subsequent editions of the book, such as those published by Kerver in 1546, and Jean Martin in 1554 (see Pozzi [1982]). Kerver and Martin, who was an active translator of Italian literary and architectural texts, worked together as editor and translator on the 1543 French edition of *Hieroglyphica.*[18] No longer a simple complement to the text, illustrations had become a necessary part of it; furthermore, as elegantly expressed by Leonardo in his *Trattato,* they needed no translation:

> *essa ha il suo fine comunicabile a tutte le generazioni dell'universo [...] non ha bisogno di interpreti in diverse lingue come hanno le lettere, e subito ha satisfato all'umana specie (Leonardo da Vinci [1890] c. 3).*

While there had been a long tradition of emblematic literature, it was only with the successful publication of *Emblemata liber,* by the Milanese jurist, poet, and humanist Andrea Alciati, that the popularity of books of emblems began to spread.[19] So great was its popularity, that hundreds of editions were published in a number of different languages. In a letter dated May 26, 1528, Andrea Alciati writes to Boniface Amerbach on the subject of a new edition of Pliny's *Natural History:*

> *Je voudrais que chapitre par chapitre soient ajoutés eikonikos depicta, sous forme d'images peintes, animaux, paysages, poissons, herbes, [...] car on trouve ici (Avignone) des graveurs sur bois capable de confectionner avec la plus grande facilité les plaques exigés par de tels sujets: or ce serait lá non plus seulement enseigner les choses de la Nature, ce que fait Pline, mais les mettre veritablement sous les yeaux meme pour les ignorants.*[20]

In 1534, in Paris, where he had been called to teach, Alciati published his book of emblems entitled *Emblematur liber* (see Alciati [1997]). As he had been unsatisfied with the earlier Augsburg edition of his book published by Heinrich Steyner (who may have acted on his own initiative without the author's permission), Alciati chose Chrestien Wechel, one of the most respected Parisian editors of illustrated texts, to edit a new version of his book. Little is known about Wechel, in spite of the fact that he was responsible for the publication of many original works in French as well as many translations of scientific and technological texts. For reasons of personal safety, Wechel, a declared Protestant, transferred his editorial activities to Frankfurt in 1570. There, his heirs later decided to merge his publishing house with that of the famous editor and engraver Theodore de Bry, author of numerous illustrated books, including an anatomical theatre. Together, they made significant historical contributions to early sixteenth-century culture.[21]

Wechel markedly improved Alciati's work, not only in the quality of the engravings, but the overall layout as well, placing each emblem on a page by itself, with a few lines of *Inscriptio* and a short explanatory poem, or *Subscriptio.* The result was an elegant work, a perfectly-balanced fusion of text and images.[22] While the first edition published in Augsburg contained one hundred and four engravings, Wechel's edition added nine more, which were included in all subsequent editions. Over time, even more plates were added, with the 1547 edition containing one hundred and ninety, and culminating with the 1551 Lyonese edition by Mathias Bonhomme, which contained two hundred and two extremely fine engravings.[23]

Scholars of Alciati's emblems consider this book to be an erudite work of humanistic scholarship intended for a well-educated audience.[24] The fact that it was written in Latin, but even more so the level of knowledge required to appreciate the epigrams, are evidence that this book was destined for the cultured elite. At the same time, the possibility of selling to a wider audience was extremely appealing to the editors, who, conscious of the growing interest in illustrations, added engravings and made changes to the layout of Alciati's book.

Evidence that the impetus for the illustrations came from the editors of these books is found in the preface to the Augsburg edition of Alciati's work written by editor Heinrich Steyner, where he claims personal responsibility for the decision to add engravings to this text for the first time. In this well-known passage, Steyner admits that while the art of woodcutting had not yet reached the level of refinement necessary to accurately reproduce the desired images, the illustrations by Jorg Breu were an important tool for better understanding the subtleties of the text:[25]

> *haud merito, candide lector, nostram desiderabis diligentiam in his tabellis quae hiuc operi adjectae sunt, elegantiores namque picturas et authoris gravissimi auctoritas et libelli dignitas merebantur [...] utilisssimum [...] nobis iusum est, si notulis quibusdam rudioribus gravissimi authoris intentionem significaremus quod docti haec per se colligent (Alciati [1531], Prefatio).*

Four years later, Wechel, in his new edition of Alciati's work, declares in the preface:

> *facile ab eo impetravi ut ad lineam revocaret et foetum illum immaturum informemque ursi instar lambendo conformaret. Mendas itaque quibus scatebat undique sustulit, plurima etiam retractavit et correxit, addidit item non oauce [...] quod ad me adtinet pro viribus contendi me in formandis iconibus.*

Fifteen years later, another preface to the same work opens with some thoughts on the

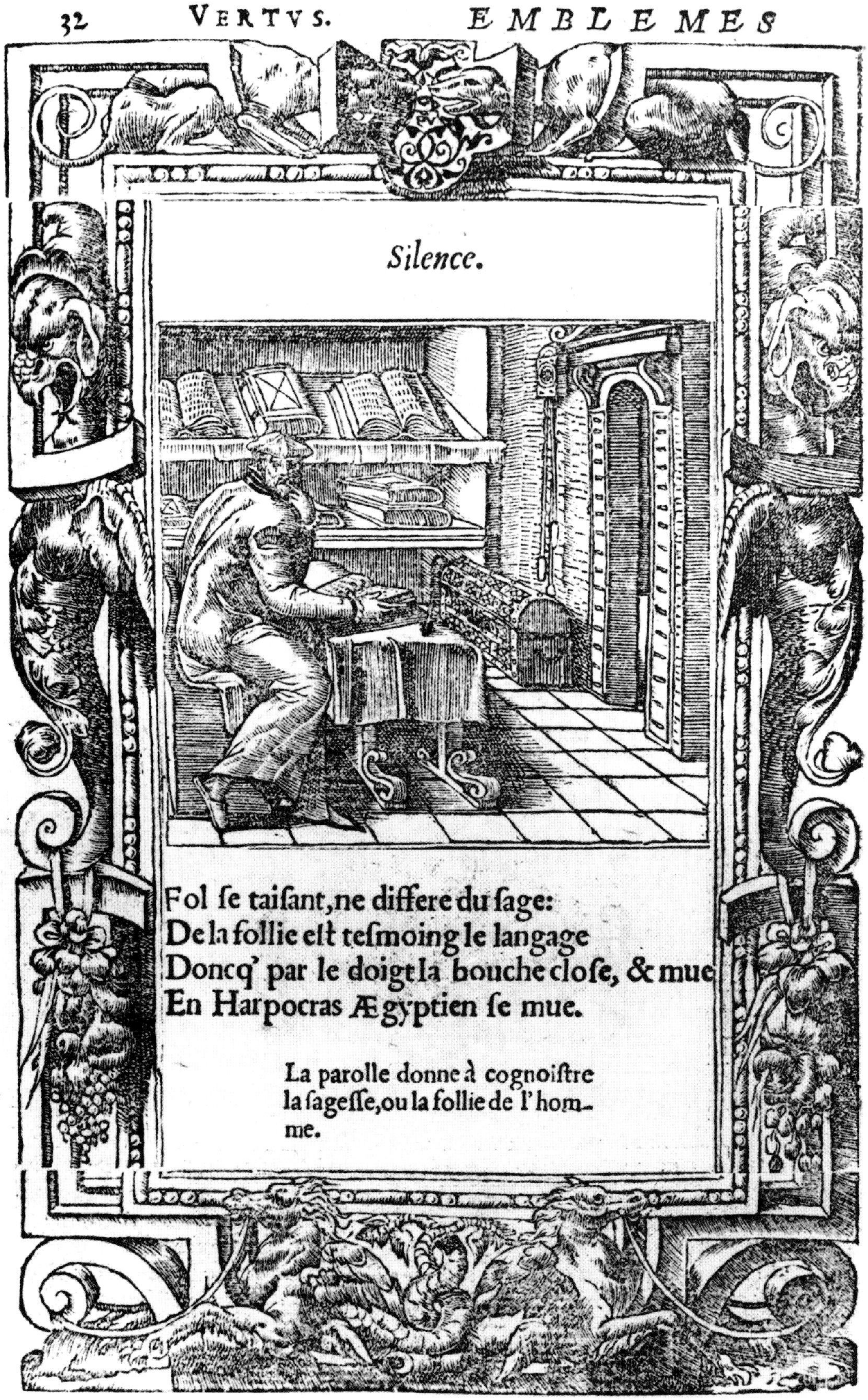

Figure 3
The emblem "Silence." Engraving from Alciati: *Les Emblemes* (Lyon 1549) 32. By kind permission of the Mazarine Library, Paris.

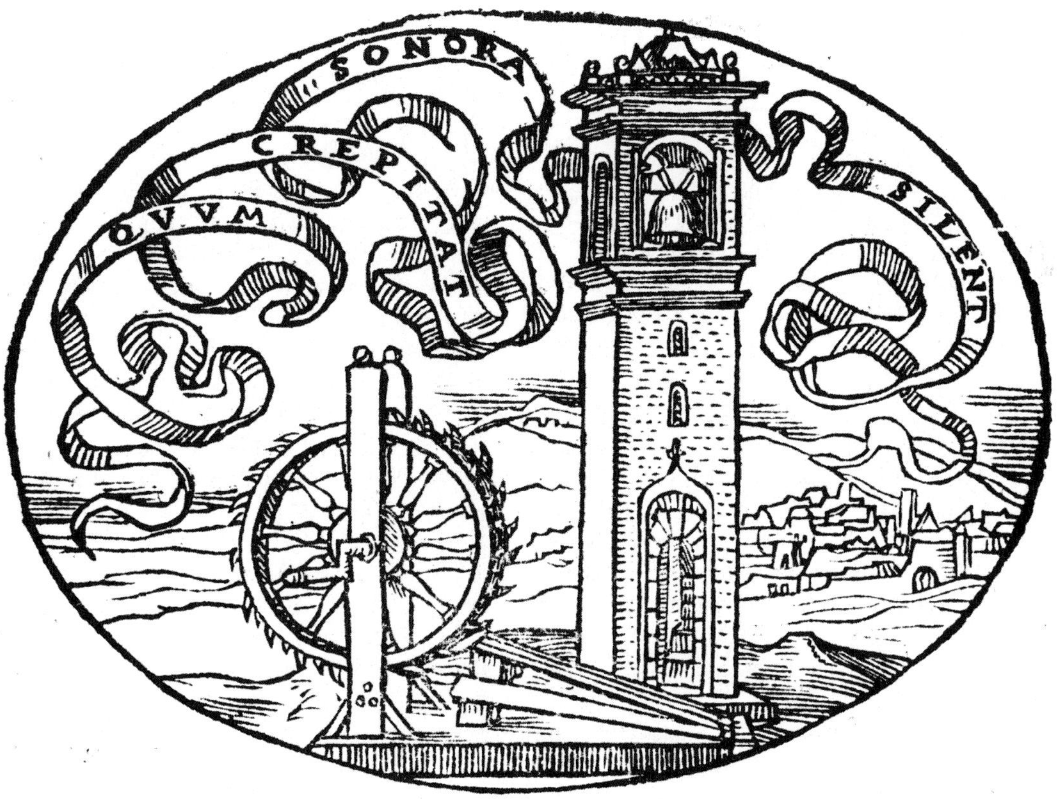

Figure 4
A Raccola symbolizing those who talk too much about irrelevant issues, but remain silent on important matters (Giovio: *Dialogo dell'imprese militari* (Lyon 1574) 124. By kind permission of the Mazarine Library, Paris.

supporting role of images to words, adding another concept which by this time was receiving more scholarly attention, that of the visual pleasure of images:

> *or l'usage des Emblemes ou Entregectz oultre la grace & plaisir de la ioyeuse nouveaulté (qui allege l'ennuy) la briesve trenche des sentences (qui poingt l'esprit) la doulceur delectable des vers (qui adulcit les oreilles) la picture non vianw des images (qui repaist les yeulx) oultre tout cela, encore tel est l'usaige e l'utilitéé: que toutes & quantessoys que aulcun vouldra attribuer, ou pour le moins par fiction applicquer aux choses vuydes accomplissement, aux nues aornement, aux muetres parolle, aux brutes raison, il aura en ce petit livre, comme en un cabinet tresbien garny, tout ce qu'il pourra & vouldra incripre, ou pindre (Aneau [1549] 6f.).*[26]

Various elements demonstrated readers' appreciation of illustrated works: the popularity of illustrated fables, bestiaries, Boccacio's illustrated comedies, and religious works; the growing demand for texts such as the *Ars moriendi*; and the success among the educated classes of the various editions and translations of Colonna's book as well as of books of emblems. These editorial successes emphasised to publishers the importance of including images alongside text. In 1548, Mathias Bonhomme published what is considered, because of the high quality of the engravings by Bernard Salomon, to be the finest edition of Alciati's *Emblematur Liber.*[27] The artist, who also illustrated editions of the Bible and Ovid's *Metamorphosis* published by Bonhomme, and who, like the editor, was a Huguenot, is referred to in the text as a painter "autant excellent qu'il n'y en a point en nostre hemisphere." Salomon completely transformed the layout of the book, giving absolute priority to the pictures by framing each image and its corresponding motto with a richly decorated border.

This style would eventually be applied to a variety of illustrated works, including books of emblems, mottos, and theatres of machines.[28] It consists of one page (usually on the right side) showing the emblem – in the form of an object, an animal, or a machine, often in a natural setting – and its motto. On the opposite page is a short explanation, offering further insight into the significance of the emblem.

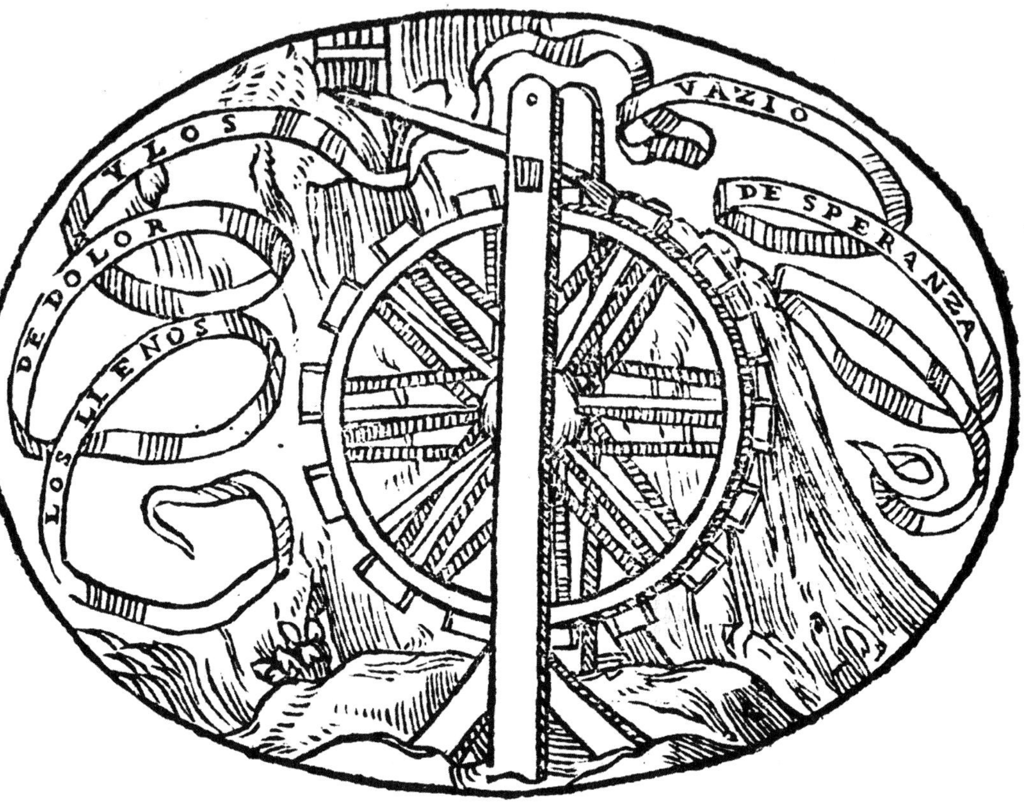

Figure 5
The emblem represents the *Noria*, a type of water wheel, whose buckets symbolise the profoundness of great sorrow (Giovio: *Dialogo* (Lyon 1559) 32). By kind permission of the Mazarine Library, Paris.

In 1559 in Lyon, Guglielmo Reviglio published a collection of emblems by Paolo Giovio.[29] In the preface of this book, dedicated to Alessandro Farnese, prince of Parma and Piacenza, the editor, after apologising for a delay in publication, discusses five essential elements that are key to a successful book of emblems:

> *quantunque io non dubiti ch'el veder fuori à quest'ora il dialogo dell'imprese di mons. Giovio vi porga una subitana meraviglia, tutta volta tengo per certo che intesa la cagione, cesserete di meravigliarvi. [...] Ma prima che io venga a questi particolari, è necessario, ch'io vi dica le condizioni universali che si ricercano a fare una perfetta impresa, il che forse è la più difficile, che possa essere ben colta da un ingegno perspicace e ricco d'invenzioni, la quale nasce dalla notizia delle cose scritte da gli antichi. Sappiate adunque Lodovico mio, che l'inventione o vero impresa, s'ella debba avere del buono, bisogna ch'abbia cinque condizioni: Prima giusta proportione d'anima e corpo, seconda, ch'ella non sia oscura, di sorte, c'habbia mistero della sibilla per interprete a volerla intendere, ne tanto chiara ch'ogni plebeo l'intenda. Terza, che sopra tutto abbia bella vista, la qual si fa riuscire molto allegra, entrandovi stelle, soli, lune, acqua, arbori verdeggianti, instrumenti meccanici, animali bizzarri, e uccelli fantastici. Quarta non ricerca alcuna forma umana. Quinta richiede il motto, che è l'anima del corpo (Giovio [1559] 9).*

In this passage, Reviglio explains perfectly the significance that images had acquired in the second half of the sixteenth century: "not overly obscure, but at the same time not so clear that everyone can easily decipher their meaning, born of ancient wisdom," and that, in particular, "they are attractive to look at."[30] It is particularly significant to note that in this book, Giovio, following in the tradition of Leonardo, includes mechanical devices among his emblems[31] (fig. 5). Included among these devices are typical workshop tools, such as hammers, a small compass with the motto "aspicit unam," and an instrument for shoeing horses with the motto "pour domter follie," as well as a hydraulic wheel and a *raccola*. Already known from a sketch by Villard de Honnecourt,[32] the *raccola*, is described by Giovio as "a mechanical instrument con-

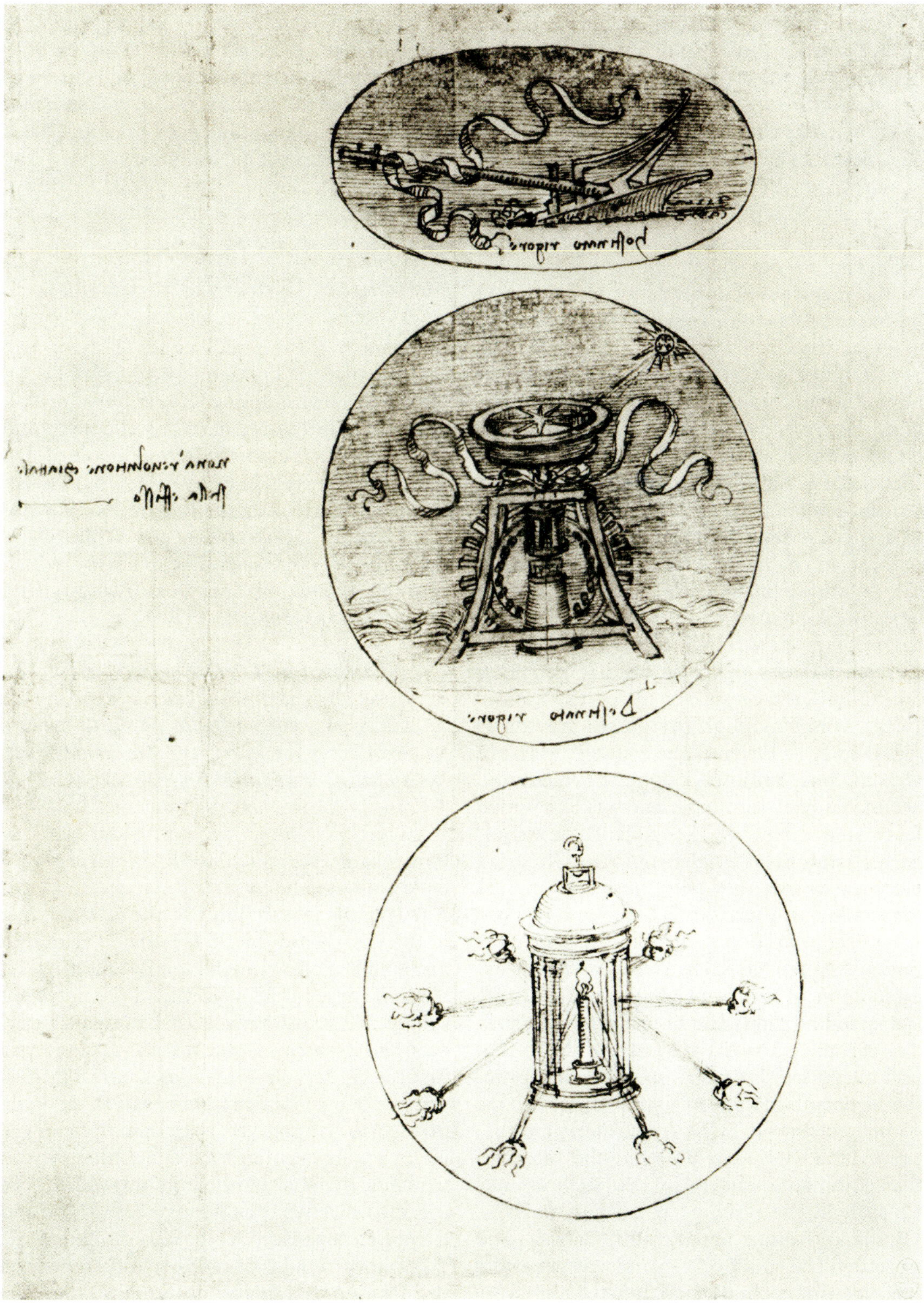

Figure 6
The drawing by Leonardo (Windsor 12701) is the first one, to our knowledge, that uses the representation of a mechanical device as an emblem.

sisting of a wheel containing many hammers which makes a great noise, and which is placed on bell towers at sunrise on holy days." Placed on top of bell towers during Lent, this instrument is used by Giovio to symbolise those who talked too much about irrelevant issues, but remained silent on important matters, thus the motto "cum crepitat sonora silent." Another example of a machine used by the author as a symbol is the *noria*, a type of water wheel, whose buckets represent the profoundness of great sorrow:

Quei vasi che levano l'acqua e la gittano fuora e perché di punto in bianco quasi la metà di essi si truova piena pigliando l'acqua e l'altra vota per gitarla fuora, nasceva da quei vasi un motto in questa guisa: llos llenos de dolor y los vazios de speranza (Giovio [1559] 54).

As is evident, machines had left the workshop and become part of the language of symbols: their function was no longer solely mechanical. Machines used in this way were of great interest to another famous author of a commentary on Giovio's *Discorsi*, Girolamo Ruscelli, translator of Ptolemy, but certainly better known by the pseudonym Alexis Pedemontanus under which he wrote a highly successful book of secrets.[33] Using language which was still highly influenced by Aristotle, Ruscelli described illustrations as "potentially natural in mankind," while "representing things and intellectual activity using only words is not natural." [34] Without questioning the primary importance of verbal language, Ruscelli stresses that humans are generally naturally interested in visual representation. In his works, as in all the others previously cited, images become the vehicle – if not the only one, definitely the most evident and accessible one – of the arcane wisdom which fascinated so many in the sixteenth century.[35] In his analysis of emblems, Carlo Ginzburg emphasises that they were based on images and could easily cross linguistic borders, even when they were not written in an international language such as Latin. The wide circulation of books of emblems in Europe surpassed religious and national boundaries.

Near the end of the century, illustrated and allegorical literature underwent a process of decodification in which the language of images moved in several different directions. In some cases, all symbolic meaning was eliminated, and in others, such as may be seen in the writings of authors such as Giordano Bruno, images still "moved and unsettled the soul." [36] While it was generally believed that technological illustrations were devoid of any symbolic significance, in reality the first theatre of machines was strongly rooted in the emblematic literature of the time. The layout of this theatre is very similar to that of a book of emblems. Each machine is depicted in a natural setting and is accompanied by a short explanation in Latin, more like a motto than a technical explanation. Furthermore, it is significant that Jacques Besson, the author of this book, had close ties with the French translators of *Hyperomachia Poliphili*, a text in which each machine has a precise emblematic meaning.

The *Livre de la fontaine perilleuse*: Gohory and Besson

François Béroald and Jacques Gohory had three things in common. Both worked on the French translation of *Hyperomachia Poliphili*, were interested in reform theology, and knew Jacques Besson.[37] Béroald, son of one of the closest friends of Besson, would eventually write the commentary for a posthumous edition of Besson's theatre. Gohory was the founder of a scientific academy which counted among its members Besson, who proudly declared:

combien qu'il y ait environ douze ou treze ans, mes benevoles e bien aymez auditeurs, que je lisois en ceste celebre Academie de Paris, les Mathematiques en latin (Besson [1567] A3).

Jacques Gohory, known to scholars of Paracelsus and Machiavelli, is surprisingly little known to historians of science, notwithstanding the fact that in 1571 he founded one of the first, if not the first, Parisian scientific academy.[38] Following the family tradition, he began a career as a lawyer and diplomat, even though he soon realised that his true interests lay elsewhere: "gousté en jeunesse par institution de voie forcée les Cours tant des princes que de Justice." He began to dedicate himself to "de son gré à la contemplation de Nature pour negotier avec elle seule hors des troubles, vices et confusions du monde," in order to discover "de beaux secrets à grand travail et dépense pour l'usage de l'homme." [39] Gohory would never formally renounce his position as a lawyer at the Parliamentary Court in Paris, but in reality he devoted himself completely to the study of literature, music, the occult, philosophy, and medicine. In keeping with the motto which would accompany all his works, *Envie, d'envie, en vie,* he dedicated himself to the translation of numerous texts: *Hyperomachia Poliphili*, as well as Amadis de Gaule, and writings by Macchiavelli, Titus Livius and Paracelsus.[40] Estienne du Tournet greeted Gohory's translation of the third volume of Amadis in the following manner:

> *car, ayant repoli tant de nobles sciences, cherché tant de secrets, fait tant d'experiences, tu-raflaichis ton âme en ce dous passetemps.*[41]

Fed by strong religious sentiment, the growing influence of Paracelsus' natural philosophy, founded in part on Neoplatonic thought, stimulated interest in studying the world, or nature's book, as one of the few human means of achieving a deeper understanding of God. In this environment, influenced by Neoplatonism, and the parallel revaluation of the importance of the sense of sight, the use of symbols, hieroglyphics, images, and numerals provides a means of interacting with a deeper and more secret level of reality. In an attempt to define the role played by symbols, and numbers in particular, in deciphering a hidden reality, Gohory uses the following analogy:

> *sono come il Sileno di Alcibiade, una specie di coperta ridicola posta sopra cose meravigliose, o come un vestito da poco che copra una bellissima ragazza. Ma entrate, qui stanno gli Dei, diceva Eraclito.*[42]

In 1567, after years of study, Gohory published his *Compendium,* dedicated to the philosophy and medical theories of Paracelsus. In this book, he expressed his unconditional approval of the analogy of the micro- and macrocosm, and more significantly, he stressed the importance within Paracelsian philosophy of magic, the cabala, and numbers (Gohory [1567]).[43] In 1571 he decided to found an academy, the *Lycium philosophal,* for the study of alchemy, botany, medicine, mechanics and the art of divination. While the members of the Baff academy, who were mainly interested in poetry and music, met regularly at the Louvre, the members of the *Lycium philosophal* met periodically at the home of Gohory in Foubourg Saint Marceau, which had a specially-designed space for conducting scientific experiments. Very little archival evidence remains of the meetings held in Gohory's home, with the exception of a few references in his books, and an engraving which shows the surgeon Ambroise Paré in a park dominated by a hill on which stands a windmill.[44] The *Lycium philosophal* takes its name from another of Gohory's works, *Instruction sur l'herbe petum,* which is the richest source of autobiographical information about the author. In this book, published in 1572 by the Parisian editor of one of Besson's works and enriched by several woodcuts, Gohory describes his garden, the rare plants he cultivated, and the climatic conditions in which they grew, his sundial, the fountains, the labyrinths, and pavilions where chemical experiments were carried out, and finally the friends and visitors who shared his interests, among whom was the "mathematician," Jacques Besson (Gohory [1572a]). In yet another book, *Livre de la fontaine perilleuse,* on the back of page 35, Gohory again mentions Besson, referring to his theatre:

> *Et notre Besson en son livre d'invention machinales met un artifices: Ut motus levis preter naturam ratione navis ita contruatur ut tranquillo mari ipsam ratione navis ita contruatur ut tranquillo mari ipsam promoveat et remisso vento cursum acceleret aut nimio temperet (Gohory [1572b]).*

In the same book, Gohory, who collaborated on the French translation of the *Poliphili,* demonstrated his appreciation for the symbolic role of the images, defining them as *"fiction contenant haute sapience."*[45] The enchanting power of the symbols contained in *Poliphili* and that of nature's arcane mysteries eventually captivated François Béroald as well. A family friend, and the first to write a commentary on Besson's theatre, Béroald published in 1611 a translation of *Polifilo,* significantly entitled *Tableau des riches inventions couvertes du voile* (Verville [1611]; Mauri [1992]). Besson's strong ties to Gohory's academy do not necessarily mean that his illustrations had a symbolic meaning. At the same time, it would be mistaken to overlook the cultural environment from which Besson's works emerged, a world which was fascinated by symbols, seeking to discover and decipher nature's secrets, and to expose them through the art of illustration.

Centre Alexandre Koyré, Paris

Notes

[1] This is indeed the conclusion reached by several philosophers, including Aristotle and Augustine. On the relationship between sight and science see Lindberg and Steneck [1972]; and Kemp [1990].

[2] On the relationship between illustrations and literature and culture during the Renaissance, see Ossola [1971]; Gordon [1975]; Fumaroli [1994]; and Pozzi [1993]. On Renaissance image-building, see Baxandall [1972].

[3] The need for historians, not only of art, to analyse iconography has been acknowledged in the wake of Abi Warburg's research on the use of drawings as sources of history. See Gombrich [1972] and [1982]. On Abi Warburg, the Warburg Institute until Gombrich, and the debate, also outside of the Institute, on the use of drawings in historical scholarship, see Gombrich [1970]. See also Ginzburg [1986].

[4] In 1942, Lucien Febvre, in his *Rabelais*, invited scholars to pay greater attention to what he defined as "les supports sensibles de la pensée aux diverses époques." It is well known that the French historian saw the sixteenth century as one characterised by a "retard de la vue," in contrast to the seventeenth century with an unprecedented "promotion" of visual perception. See Febvre [1962] 471. The motives for Febvre's lack of acknowledgement of the importance of visual arts in the 1500s are examined in Jay [1993]. Febvre's hypothesis, by now completely refuted, is also analysed by Chartier [1983]. A more recent study is Havelange [1998]. This book, while with certain limitations (there is no introduction nor name index, secondary sources are almost all French, no mention is made of the original date of publication of various Latin texts, for which quotes are taken from the French translations only), offers some useful analytical cues.

[5] As it is well known, the former held that the technical innovations of Renaissance art, in particular linear perspective and *chiaroscuro*, laid the essential foundations for the Scientific Revolution of the seventeenth century. The latter sustained that scientific illustrations had been as path-breaking as the discovery of printing and that for historians of science the analysis of graphic materials should be a key part in the study of written texts.

[6] The reception of Sarton's thesis (Sarton [1955]) among historians of science has varied, some being in full agreement while others have put the role of drawings in scientific development in perspective. For a survey of this debate, see Eisenstein [1979]; see also Shirley and Hoeniger [1985], especially Ackerman [1985], Edgerton Jr. [1985], and Mahoney [1985].

[7] On the *problematique* of scientific drawings, see Baroncini [1996]. On the importance of non-verbal communication in the history of science, see Mazzolini [1993]; and Gingerich [1986]. A more recent study of techno-scientific illustrations is Beretta [1998]. See also Hackmann [1994].

[8] For the manuscripts tradition see Galluzzi [1991]. By the same author, see also Galluzzi [1995b]. There have been critical editions for most 1400s' treatises, for a list see the bibliography in Galluzzi [1995a]; and Long [1997], which includes an interesting analysis, though limited to the importance and role of patronage for the authors of manuscripted treatises.

[9] See Luisa Dolza and Hélène Vérin, "Énigmes et raisons des théâtres de machines de la Renaissance," forthcoming, and Dolza [1999]; Vérin [1993] and [1998].

[10] The exact date of the first edition is controversial. François Beroald published the first dated edition with Latin commentary in 1578 and with an enlarged French commentary in 1579, both printed in Lyon by Barthèlémy Vincent. New Latin, Italian and French versions were published in 1582 by Giulio Pascali. Other French versions appeared in 1594 and 1596. A German edition was first printed in 1595, and a Spanish one in 1602. Besson is one of the few authors who received considerable attention in the literature; see in particular Hilliard [1979]; Keller [1973].

[11] Painting is wordless poetry, one of Simonides' most famous sayings (Simonides [1572] 524), was then used again by Leonardo da Vinci (Leonardo da Vinci [1890] paragraphs 14, 15 and 17). The same saying, quoted by Plutarcos, is the title of an emblem book (Mannasser [1630]).

[12] On emblems in the seventeenth century, see Praz [1946] and [1964]. Engravings in this work had already been used in a previous book of emblems, by the Jesuit David [1605]. See also Daly [1989]. On the issue of the role of illustrations and words, see Delégue [1990].

[13] More than twenty 1500s' texts that contain the term "theatre" in the title are books of emblems. As far as pagination is concerned, theatres followed books of emblems, so that captions inserted for each machine, more than explanations, resemble short maxims. Klein [1970].

[14] On hieroglyphs, see "Hiéroglyphes, languages chiffrés sens mystérieux au XVII siécle," XVII siécle, special issue, XL, 1988, no. 1. On the various editions of *Hieroglyphica*, see Drysdall [1983].

[15] According to Giovanni Pozzi and Lucia A. Ciapponi, the author of the tables was influenced by Mantegna and Bellini (see the critical edition of this work, Colonna [1980]). Others have argued that the influence is rather that of Botticelli, Carpaccio and Giovanni Bellini. Some scholars have argued that the author of the engravings was not a famous painter, but rather the same artist who realised the illustrations for Ovidius' Metamorphosis.

[16] The identity of the author remains to this day the object of much controversy; see Colonna [1979].

[17] On the reinterpretation of this work in terms of alchemy, see the critical edition. On its influence on Giulio Camillo, see Bolzoni [1984].

[18] A copy of Kerver's 1553 edition, translated by Martin, kept at the Bibliothéque Nationale de Paris as Rés. P.z. 641, is ex-libris of Viollet le Duc.

[19] For a discussion of the abundant literature on Alciati's emblems, see Scholz [1991].

[20] Letter dated 26 May 1528 to Amerbach, in Barni [1953] 73f.

[21] In 1534 he published Vegezio, Frontino and Bude with various tables. In 1543 he also published Jean Tagault, *de Chirurgica institutione libri quinque*.

[22] In the same years, Ramus published *Arithmetica libri tre*, see Ramus [1555]. On the printers of mathematical books in Renaissance Paris, see Margolin [1976].

[23] Concerning the various editions and their peculiarities, see Green [1871]; Duplessis [1884]; and Praz [1964] 248–252. On the publishing activity of Mathias (or Mathieu) Bonhomme Macé, see Baudrier [1964] 185–270. Guillaume Rondelet's *Histoire des Poissons* is considered Bonhomme's masterwork.

[24] According to various authors, Alciati was absolutely unconcerned with the illustrations in his own work. In the last edition that Alciati supervised, there were no plates, and he showed no interest in reprinting some tables that contained clear mistakes.

[25] On Alciati's emblems and the history of their publication, see Balavoine [1982].

[26] The contemporary Latin edition contains the same text, and the last sentence says: "ton eikonon picturam non inanem, quae oculos pascit."

[27] On Bernard Salomon, see Rondot [1897] 64–78.

[28] The theatres that most closely resemble Salomon's model are those by Agostino Ramelli and Vittorio Zonca. In both, figures are framed, and on the left, opposite page authors have inserted a short explanatory note. In other theatres, such as Besson's and Errard's, the page is designed like in Salomon's, but the explanation of the machines comes before the table.

[29] Giovio [1559]. See the new edition edited by Maria Luisa Doglio, Rome, 1978. On the life and works of Lodovico Domenichi, see Poggiali [1976] 221–225. Born in Piacenza, Domenichi, following a training in law, worked as proof reader in Venice in Giolittos' typography before spending a year in prison, accused by Doni of being the translator of Calvin's *Nicomediana*. Freed thanks to the intervention of Renata of France, Duchess of Ferrara, he worked for Cosimo I de Medici in Florence, where he died in 1564. On Paolo Giovio, see the proceedings of the congress "Paolo Giovio. Il Rinascimento e la Memoria." In *Raccolta, Storia*. Società storica comense, Como, vol. XVII, 1985.

[30] Della Porta's rule is completely different from Giovio's fourth rule, explained in the preface. Della Porta argued that in order to optimise the art of imagination and memory, it is necessary to use as *imagine agentes* only pictures of persons, characterised by either beauty or by physical elements that turned them into grotesque caricatures.

[31] See Brusatin [1980]. On Leonardo's "mechanical" enterprises, see Richter [1970] 387–392.

[32] Villard had drawn a similar machine in his *Livre de portraiture*, Table 62, which was supposed to produce a perpetual movement.

[33] Ruscelli [1555]. For some scattered information on his life see Bonino [1825] 223. On the importance of this book for the art of dyeing see Guareschi [1907] 437–439. Ruscelli's book had great success until the eighteenth century. In every major European library it is possible to find various editions of this book that used to be consulted in as late as the late 1700s. The French translation appeared in 1564 and was then published again in 1572 and in 1573. Ruscelli wrote his *Le imprese illustri*, published posthumously in Venice by Franco Patrizio in 1572, and added a *Ragionamento sopra i motti e disegni d'arme e d'amore* to some editions (Venice 1556 and 1560) of Giovio's work.

[34] Ruscelli [1556] 122. On symbolism in Ruscelli, member of the Venetian Academy, see Bolzoni [1995] 90–106.

[35] On Ruscelli see also Garin [1964].

[36] For reasons of both competence and opportunity, Bruno's thought on images cannot be analysed in this paper. However, it is possible to mention that Bruno's work on images constitutes an attempt to understand how images are created and function, something that in today's words could be called a semiotic research.

[37] On how both Gohory and Béroald interpreted the Polifilo according to alchemy, see Polizzi [1993].

[38] See Kahn [1995]. On Gohory see Hannaway [1973] 447f.

[39] Gohory [1572a] 1–5. Potum, or petum, is one of the indigenous names of tobacco, to whose knowledge Thevet contributed with his *Cosmographie universelle*, t. II, libro XX, Paris, 1575.

[40] Gohory took this motto from the *Avertissement aux lecteurs* of his 1571 translation of Macchiavelli's Discorso: "quant aux envieux, ils ne me font guéres de paour en cecy, d'autant que ma conscience me iuge, que ie n'ay vertu en moi qui les mérite, ne digne d'estre en vue qui les irrite, au contraire ie le désiderois volontiers à l'exemple de Themistocles, dont ie porte la devise: envie, d'envie, en vie." Gohory also translated Macchiavelli's *Il principe*. On the translations of Macchiavelli's work, see Balmas [1982]. The *Amadis* is a famous cycle of medieval Spanish novels, whose main author was Garcia Ordones de Montalvo, who wrote most of them in 1465. Published several times before 1577, Nicolas d'Herberay des Essarts's *Amadis* includes most of Montalvo's work. A manuscript of books X, XI and XIII was donated to Gohory, who translated and published them as a compendium to the work of Herberay. Secondary sources on the *Amadis* are numerous. On the role of Gohory and the meaning of his work, see Gorris [1996].

[41] "Au Seigneur Gohory sur la version du 13 d'Amadis." Remi Belleau, Jacques Brochier, Etienne du Tronchet, Dorat and Gilles Pinaut de Poitiers dedicate sonnets to Gohory's translation. See Hamy [1899], on the sonnets in particular see p. 11, note 3. To this day, this is the only study published on this academy.

[42] Gohory [1550] c. A 3v. Quoted and translated in Bolzoni [1995] 132.

[43] On this text and its interpretation, see Kahn [1995]. On the fortune of Paracelsus in France, see Debus [1991].

[44] See Ernest Theodor Hamy [1899] 2: "et une gravure sur bois, d'apres un dessin de M. Boucourt, montre quatre graves personnages, parmi lesquels on reconnait Ambroise Paré, causant dans un bosquet que domine une butte surmontée d'un moulin à vent."

[45] On Gohory's participation in the French translation of Colonna's *Polifilo*, testified by the marginal note in Latin on the 1554 and 1561 editions, see the section "Il Polifilio e dintorni" in Balmas [1992] 43–153.

Reference

Ackerman, J.
1985 "The Involvement of Artists in Renaissance Science."
In: *Science and the Arts in the Renaissance,* eds. J. W. Shirley and F. D. Hoeniger, 94–129. Washington.

Alciati, A.
1531 *Emblematur Liber.* Augsburg.
1549 *Les Emblemes.* Lyon.
1997 *Les Emblémes,* fac-simile de l'édition lyonnaise 1551, preface P. Laurens. Paris.

Aneau, B.
1549 "Preface." In: Alciati [1549].

Anonymous
1992 "Il Polifilo e dintorni." In: *Il Cinquencento visionario tra Italia e Francia,* 43–153. Florence.

Aristotle
Methaphysics.

Balavoine, C.
1982 "Les Emblemes de Alciat, sens et contresens." In *L'embleme a la Renaissance,* ed. Y. Giraud, 49–59. Paris.

Balmas, E.
1982 "Jacques Gohory, traduttore di Macchiavelli." In Saggi e studi sul Rinascimento francese, 23–73. Padua.
1992 *Cinquecento visionario tra Italia e Francia, Studi di letteratura francese vol. XIX,* ed. E. Balmas, 43–153. Florence.

Barni, G. L.
1953 *Le lettere di Andrea Alciato giureconsulto.* Florence.

Baroncini, G.
1996 "Note sull'Illustrazione Scientifica." *Nuncius* 11/2: 527–543.

Baudrier, H.
1964 *Bibliographie Lyonnaise, Recherches sur les imprimeurs, libraires, relieurs et fondeurs de lettres de Lyon ai XVI siécle,* vol. 10 (first edition, Lyon, 1895–1915). Paris.

Baxandall, M.
1972 *Painting and Experiences in Fifteenth-Century Italy. A Primer in the Social History of Pictorial Style.* Oxford.

Beretta, M.
1998 "Stampa, incisioni e terminologia nel *De re metallica* di Giorgio Agricola." In *Le forme della comunicazione scientifica,* eds. M. Galuzzi, G. Micheli, and M. T. Monti, 191–216. Milano.

Besson, J.
1567 *Le Cosmolabe.* Orleans.
1578 *Theatrum instrumentorum et machinarum.* Lyon.

Bolzoni, L.
1984 "Eloquenza e alchimia in Giulio Camillo, nuovi appunti." In *Il teatro della memoria. Studi su Giulio Camillo.* Padua.
1995 "I giochi della memoria." In *La stanza della memoria.* Turin.

Bonino, R.
1825 *Biografia medica Piemontese,* vol. 1. Turin.

Brusatin, M.
1980 "La macchina come soggetto d'arte." In *Storia d'Italia, Annali 3,* 31–81. Turin.

Chartier, R.
1983 "Histoire intellectuelle et histoire des mentalités, trajectories et questions." *Revue de synthése* 111–112: 277–303.

Colonna, F.
1979 *Bibliografie e opere,* eds. M. T. Casella and G. Pozzi, 2 vols. Padua.
1980 *Hypneromachia Poliphili.* Padua.

Daly, P. M.
1989 *Andrea Alciato and the Emblem Tradition.* New York.

David, J.
1605 *Occasio arrepta neglecta.* Antwerpen.

Debus, A. G.
1991 *The French Paracelsians: the chemical challenge to medical and scientific tradition in early modern France.* Cambridge.

Delégue, Y.
1990 *La perte des mots. Essai sur la naissance de la littérature au XVI et XVII siécles.* Strasbourg.

Dolza, L.
1999 *A gloria di Dio, a beneficio degli studiosi e servitio di vostra Altezza. I primi teatri di macchine nella cultura del tardo Cinquecento.* Ph.D. thesis in history of sciences, University of Florence.

Drysdall, D. L.
1983 "Filippo Fasanini and his explanation of sacred writings, text and translation." *Journal of Medieval and Renaissance Studies* 13/1: 127–159.

Duplessis, G.
1884 *Les emblems d'Alciat.* Paris.

Edgerton Jr., S. Y.
1985 "The Renaissance Development of the Scientific Illustration." In: *Science and the Arts in the Renaissance,* eds. J. W. Shirley and F. D. Hoeniger, 168–197. Washington.

Eisenstein, E.
1979 *The Printing Press as an Agent of Change: Communications and Cultural Transformations in Early-Modern Europe,* vol. 2. Cambridge.

Febvre, L.
1962 *Le probléme de l'incroyance au XVI siécle. La religion de Rabelais.* Paris.

Fumaroli, M.
1994 *L'école du silence. Les sentiment des images au XVII siécle.* Paris.

Galluzzi, P.
1991 "Le macchine senesi. Ricerca antiquaria, spirito di innovazione e cultura del territorio." In *Prima di Leonardo,* ed. P. Galluzzi, 15–44. Milano.
1995a *Prima di Leonardo,* ed. P. Galluzzi. Milano.
1995b "Dalle tecniche alla tecnologia." In *Storia delle scienze, Conoscenze scientifiche e trasferimento tecnologico,* ed. V. Marchis, vol. V 32–75. Turin.

Garin, E.
1964 *L'umanesimo italiano.* Bari.

Gingerich, O.
1986 "Copernicus's De revolutionibus: An Example of Renaissance Scientific Printing." In *Print and Culture in the Renaissance. Essays on the Advent of Printing in Europe*, eds. G. P. Tyson and Sylvia S. Wagonhein, 55–74. Newark and London.

Ginzburg, C.
1986 "Da A. Warburg a E. H. Gombrich. Note su un problema di metodo." In *Miti, Emblemi, Spie. Morfologia e storia*, 29–106. Turin.

Giovio, P.
1559 *Dialogo dell' imprese militari et amorose di monsignor Giovio vescovo di Nocera, con un ragionamento di Messer Lodovico Domenichi nel medesimo soggetto.* Lyon.
1574 *Dialogo dell' imprese militari et amoroso.* Lyon.

Gohory, J.
1550 *De usu et mysteriis notarum liber in quo vetusta literarum et numerorum ac divinorum ex Sibilla nominum ratio explicatur.* Paris.
1567 (Leo Suavius) *Theophrasti Paracelsi Philosophiae et Mediciane Utrisque Universae, Compendium ex potimis quisque ejus libris: cum scholiis in libros iiii ejusdem De Vita Longa, Plenos mysteriorum, parabolarum, aenigmatum, catalogus operum & librorum. Cum indice rerum in hoc opere singularium.* Paris.
1572a *Instruction sur l'herbe potum ditte en France l'herbe de la Royne ou Médicée, et sur la racine Mechiocan principalment avec quelques autres simples rares et exquis, exemplaire à manier philosophiquement tous autres Vegetaux.* Paris.
1572b *Libre de la fontaine perilleuse avec la Chartre d'Amour, autrement intitulé la Songe du Verger, ouvrage trés excellent de poesie antique, contenant la steganographie des mysteres secrets de la science minerale.* Paris.
1575 *Cosmographie universelle*, t. II, libro XX. Paris.

Gombrich, E. H.
1970 *Abi Warburg. An Intellectual Biography with a Memory of the History of the Library by F. Saxl.* London.
1972 *Studies in the Art of the Renaissance.* Oxford.
1982 *The Image and the Eye. Further Studies in the Psychology of Pictorial Representation.* Oxford.

Gordon, D. D.
1975 *The Renaissance Imagination.* Los Angeles.

Gorris, R.
1996 *Alla corte del principe. Traduzione, romanzo, alchimia, scienza, politica tra Italia e Francia nel Rinascimento.* Ferrara.

Green, H.
1871 *Alciati's Emblems, The Full Stream.* London.

Guareschi, I.
1907 *Storia della Chimica*, vol. II. Turin.

Hackmann, W.
1994 "Jan van der Straet (Stradano), and the Origins of the Mariner's Compass." In *Learning, Language and Invention*, eds. W. Hackmann and A. J. Turner, 148–179. Aldershot and Paris.

Hamy, E. T.
1899 "Un precurseur de Guy de la Brosse, Jacques Gohory et le Lycium philosophal de Saint Marceau les Paris, 1571–1576." *Nouvelles Archives du Museum*, IV série, I: 1–26.

Hannaway, O.
1973 "Gohory." In: *Dictionary of Scientific Biography*, vol. 5 447f. New York.

Havelange, C.
1998 *De l'oeil et du monde. Une histoire du regard au seuil de la modernité.* Paris.

Hilliard, D.
1979 "Jacques Besson et son *Theatre des instruments mathematique*." *Revue française d'histoire du livre* 48: 5–38.

Jay, M.
1993 *Downcast Eyes. The Denigration of Vision in Twentieth-Century French Thought*, Berkeley.

Kahn, D.
1995 "Le paracelsisme de Jacques Gohory." In: *Paracelse et les siens* (colloque des 15 et 16 decembre 1994 a la Sorbonne), *Aries* 19: 81–130.

Keller, A.
1973 "The missing years of Jacques Besson, inventor of machines, teacher of mathematics, distiller of oils and Huguenot pastor." *Technology and Culture* 14: 28–39.

Kemp, M.
1990 *The Science of Art, Optical Themes in Western Art from Brunelleschi to Seurat.* New Haven.

Klein, R.
1970 "La teoria dell'espressione figurata nei trattati italiani sulle imprese, 1555–1612." In *La forme et l'intellegible*, ed. R. Klein. Paris.

Leonardo da Vinci
1890 *Trattato della Pittura*, ed. G. Milanesi. Rome.
1883 *The Literary Works of Leonardo da Vinci*, ed. J.P. Richter. London.

Lindberg, D. C. and N. H. Steneck
1972 "The Science of Vision and the Origins of Modern Science." In: *Science, Medicine and Society in the Renaissance. Essays to Honor Walter Pagel*, ed. A. G. Debus, vol. 2, 29–45. London.

Long, P. O.
1997 "Power, Patronage, and the Authorship of Art: From Mechanical Know-how to Mechanical Knowledge in the Last Scribal Age." *Isis* 88/1: 1–41.

Mahoney, M. S.
1985 "Diagrams and Dynamics." In: *Science and the Arts in the Renaissance,* eds. J. W. Shirley and F. D. Hoeniger, 198–211. Washington.

Mannasser, D.
1630 *Poesis tacens, pictura loquens quibus occasio arrepta neglecta delineatur decantatur.* Dilinge.

Margolin, J.-C.
1976 "L'enseignement des mathematiques." In *French Renaissance Studies, 1540–70, Humanism and the Encyclopedia*, ed. P. Sharrat, 109–155. Edinburgh.

Mauri, D.
1992 "Il tableau di Beroalde de Verville: un viaggio attraverso i testi." In *Cinquecento visionario tra Italia e Francia, Studi di letteratura francese vol. XIX*, ed. E. Balmas, 367–381. Florence.

Mazzolini, R. (ed.)
1993 *Non-verbal Communication in Science prior to 1900*. Florence.

Montaigne
Essais.

Ossola, C.
1971 *L'autunno del Rinascimento*. Florence.

Poggiali, C.
1976 *Memorie per la storia letteraria di Piacenza*, vol. I. Piacenza.

Polizzi, G.
1993 "La fabrique de l'énigme: lecture 'alchimiques' du Poliphile chez Gohory et Béroalde de Verville." In *Alchimie et philosophie à la Renaissance*, eds. J.-C. Margolin and S. Matton, 265–282. Paris.

Pozzi, G.
1982 "Les hieroglyphes de l'hypnerotomachia poliphili." In *L'embleme a la Renaissance*, ed. Y. Giraud, 15–27. Paris.
1993 *Sull'orlo del visibile parlare*. Milano.

Praz, M.
1946 *Studi sul concettismo*. Florence.
1964 *Studies in Seventeenth-Century Imagery*, 2nd ed. Rome.

Ramus, P.
1555 *Eloquentia et philosophiae professoris regii, Arithmeticae libri tres, ad Carolum Lotharingum cardinalem*. Paris.

Richter, J. P. (ed.)
1970 *The Literary Works of Leonardo da Vinci*, vol. 1. London.

Rondot, N.
1897 *Bernard Salomon, peintre et tailleur d'histoires à Lyon au XVIe siècle*. Lyon.

Ruscelli, G.
1555 *Secreti del Rev. Donno Alessio Piemontese*. Venice.
1556 *Ragionamento di Mons. Paolo Giovio sopra i motti, e disegni d'arme, e d'amore*, commonly known as *Imprese*, by Girolamo Ruscelli. Venice.

Sarton, G.
1955 *The Appreciation of Ancient and Medieval Science during the Renaissance (1450–1600)*. Philadelphia.

Scholz, B. F.
1991 "The 1531 Augsburg Edition of Alciato's emblemata: a survey of research." *Emblematica* 5/2: 213–254.

Shirley, J. W. and F. D. Hoeniger (eds.)
1985 *Science and the Arts in the Renaissance*. Washington.

Simonides
1572 *Oeuvres morales*.

Vérin, H.
1993 *La gloire des ingénieurs. L'intelligence technique du XVIe au XVIIIe siècle*. Paris.
1998 "La réduction en art et la science pratique au XVI° siècle." *Raisons pratique* 9: 119–144.

Verville, B. de
1611 *Tableau des riches inventions couvertes du voile des feintes amoreuses, qui sont representees dans le Songe de Poliphile desvoilees des ombres du songe et subtilement exposees par Beroald*. Geneva.

II.
Theories of Matter Between Alchemy and Atomism and the Autonomy of Images

Alchemical Iconography at the Dawn of the Modern Age

The *Splendor solis* of Salomon Trismosin

Anne-Françoise Cannella

The example chosen to illustrate the problems bound with alchemical iconography is a sixteenth-century treatise called *Splendor solis oder Sonnenglantz*.[1] Written in German, this treatise describes the main steps of the philosophers' stone in a traditional way. It is divided in seven short treatises among which the third and the fourth are themselves separated into seven chapters or paragraphs, the fifth treatise being divided into four articles. The work is illustrated by a series of twenty-two remarkable illuminations which have contributed to its fame.

As soon as it appeared, the *Splendor solis* had a great reputation. The number of copies, manuscripts as well as printed books, from the middle of the sixteenth century up to the eighteenth century, proves it. The oldest exemplar known today is preserved at the Berlin Kupferstichkabinett, under the class mark Cod. 78 D 3; two of the illuminations are dated 1532 and 1535, respectively.[2] Several other copies were produced during the last third of the sixteenth century: Paris, National Library, ms. allemand 113;[3] Berlin, Staatsbibliothek, Cod. Germ. Fol. 42;[4] two versions dated from 1582: Nuremberg, Germanisches Nationalmuseum, ms. 146.766 and London, British Library, ms. Harley 3469; Kassel, Landesbibliothek, ms. Mus. 1954.76, 1584 and 1588;[5] finally, a last exemplar from the end of the sixteenth century is preserved in a not yet located private collection in Bern.[6]

Even though several editions of the *Splendor solis* had already been published since 1598, three more manuscripts were produced, one in 1613: London, British Library, ms. Sloane 3613; the other two in the first quarter of the eighteenth century, based on the French text of Charles Sevestre's edition: Yale, University Library, ms. Mellon 86,[7] Paris, National Library, ms. fr. 12.297.[8] Finally, three versions of the seventeenth and the eighteenth century are preserved at the Bibliotheca Philosophica Hermetica of Amsterdam, under the class marks ms. 5, 31 and 304,[9] not to mention the manuscripts and printed books that once could be discovered.

It should be pointed out that all of the above-mentioned manuscripts are anonymous. Indeed, Salomon Trismosin's name is not cited in any of the exemplars known up to now. It is only in the first edition of the treatise, in 1598, at Rorschach on Lake Constance, that the attribution to Trismosin appears in the title.

In the printed tradition, the *Splendor solis*' text is inserted in a score of others describing practical alchemical recipes, and attributed not only to Salomon Trismosin but also to Paracelsus and Korndorffer.[10] The whole is collected under the new title of *Aureum vellus oder Guldin Schatz und Kunstkammer* or *La Toison d'or.* It is well known that for the alchemists of the Renaissance, the Golden Fleece's conquest by Jason was one of the allegories of the philosophers' stone. The *Splendor solis* constitutes the third book of the *Aureum vellus*.

The printed book was republished several times by the end of the eighteenth century. A second edition, under the same title but enriched with a beautiful frontispiece, came out in Rorschach in 1599. The work was then republished in 1610, at the end of the *Promp-*

tuarium alchemiae of Joachim Tancke.[11] We also know the Basel edition of 1604. In 1612, the editor, Charles Sevestre, published *La Toyson d'or ou la Fleur des Trésors* in Paris. It is a French translation of only one of the German collection's treatises, the one that appears at the beginning of the third part of the first volume, under the title *Splendor solis*. More or less faithful to the original text but enriched with a number of paraphrases, this translation is attributed to a certain L. I. known only by his initials. Finally, the last edition of the *Aureum vellus* was printed in Hamburg in 1708.

As in the manuscript tradition, the printed books are enriched with a series of twenty-two (sometimes twenty-three) illustrations which here are woodcuts and sometimes touched up with colours.[12]

The first text of the *Aureum vellus*, published at Rorschach in 1598, is actually a kind of autobiography of Salomon Trismosin, or rather an account of his peregrinations. We come to know that Salomon Trismosin's passion for alchemy was born the day he attended a transmutation performed on lead by a miner called Flocker, who soon thereafter fell to the bottom of a mine well without having consented to reveal his secret. As early as 1473, Trismosin began to travel around looking for an artist who could teach him. He wandered through the Latin countries and finally encountered a shopkeeper associated with a Jew. Both transmuted tin into silver in large quantities. Trismosin was accepted as assistant until the day he left for Venice with the Jew and with part of the silver. He seized the opportunity of this trip to try the metal which resisted the tests. Disappointed, he found a job in an important laboratory managed by a German, Hans Tauler, and very well equipped and stocked with all kinds of products. Having won the foreman's trust by succeeding in his first transmutation, Trismosin became responsible for all kinds of operations and bound himself not to reveal the Art's secrets to anyone. But the owner (Trismosin calls him the "Centilon") drowned himself during a feast celebrating the wedding of the doge with the sea, and the laboratory was closed. Salomon left Venice and resumed his trip, during which he discovered cabbalistical and magical books written in Egyptian and had them translated into Greek and Latin. These books taught him the tincture methods of the greatest pagan kings such as Xophar, Sunsfor, Xogar, Xolophat, Lulaton,[13] which finally revealed to him transmutation's secrets. Emboldened by this secret, Trismosin asserted that with half a grain of the powder of youth, he had been able to make himself young again and that this "the red lion," could produce the same effect on ninety-year-old women so that they again would be capable of giving birth.

Since the first edition of the *Aureum vellus*, Salomon Trismosin had been considered to be Paracelsus' preceptor. Curiously, his autobiography does not mention this fact. The only note that refers to a meeting between both men, which happened in Constantinople in 1520, is found in the *Rosarium Novum Olympicuve m et Benedictum*, a treatise collecting the works attributed to King Solomon, to Salomon Trismosin and to Paracelsus, published by Benedictum Figulum in 1608. In his *Bibliographia Paracelsica*, Karl Sudhoff cites an extract from a treatise of Salomon Trismosin called "the red lion and its treasure":

> *I, Salomon Trismosin, have had nine disciples; none of them came to anything except an artful Swiss called Philippe Hohenheim. I declare that he will go very far in this art when he reaches the middle age; in this case, he will fill up the world with his marvels.*[14]

Salomon Trismosin is undoubtedly a mythical personage. Palma Cayet, in his *Chronologie Septénaire*, in 1605, qualifies his work as sublime and promises to produce a French translation of it.[15] But as early as 1697, Borrichius denounces the imposture in the *Conspectus Scriptorum Chemicorum* (Borrichius [1697] 27 § XXXIX). Indeed, the pseudonym of Trismosin has its place next to other pseudonyms from the same period such as Basile Valentin's. This name also has a symbolic connotation: from antiquity, Salomon was considered to be the great master in magical art; in the same way, Trismosin, which can be read as "three times Moses," constitutes a kind of pair with the Hermès Trismégiste which means "three times greater."

The sumptuousness of *Splendor solis'* illuminations makes it not only one of the most famous works of the hermetico-alchemical tradition, but also one of the monuments of illumination during the Renaissance.[16]

The German art historian, Gustav-Friedrich Hartlaub (Hartlaub [1937a], [1937b], [1954], [1955], and [1959]), was the first to be interested in this subject and to point out the different manuscripts existing. His studies focused mainly on the oldest manuscript of which he gave a succinct description. By means of a comparison with a missal of 1531 which had belonged to Albert of Brandenburg, the Bishop of Mayence, Hartlaub attributes the illuminations of this exemplar of the *Splendor solis* to Durer's circle in Nürnberg. More particularly, he sees in it Nikolaus Glockendom's style without ascribing the work to him with certainty.

René Alleau, in his *Etude iconographique du manuscrit de Berlin* (Alleau [1975]), discovered the etchings that were used as models for the astrological miniatures of the *Splendor solis*. These correspond to the series of inhabited vases and more precisely to the scenes surrounding the principal compositions, which themselves represent human categories settled by the god who dominates each of the miniatures.[17] It concerns seven engravings representing "the planets' children", attributed to Hans Sebald Beham. Alleau also considers him to be the author of *Splendor solis'* illuminations, with which he might have been helped by one of the Glockendom. Again, in keeping with the aforesaid missal, the manuscript would have been illuminated expressly for Albert of Brandenburg.

Jacques van Lennep (Lennep [1984] 110–129) adds to Hartlaub's recension the manuscript of Paris dated from 1577. He also completes the analyses of the German art historian and those of René Alleau. According to him, three artists had shared the task: Simon Bening, an excellent Flemish miniaturist,[18] had painted the naturalistic borders. Hans Sebald Beham, a "minor master" of Durer's circle,[19] was the author of the frames with character scenes. Indeed, many of the themes of Beham's etchings, such as the blazons, the battle scenes, the *putti*, and the ornaments, are similar to *Splendor solis'* illustrations; but those subjects appear in particular around the central compositions. Finally, contrary to what Hartlaub thought, in Lennep's analysis, the main figures were the work of Albrecht Glockendom[20] rather than Nikolaus. The names of these three artists are indeed associated with the missal illustrated for Albert of Brandenburg.

In spite of a fundamental concordance in the themes, considerable differences are perceptible, particularly in the choice of frames which surround the central composition. Those variances appear not only in a same manuscript but also between the several exemplars of the *Splendor solis*. Three types of frames can actually been distinguished: naturalistic frames, also called "seed-borders," are made of flowers, birds and insects. Remarkable for their extreme accuracy and for the variety of the inks used, they are entirely characteristic of the beginning of the sixteenth century's illumination, especially in books of prayers or "books of hours".[21] There is therefore no reason to look for any alchemical symbolism in those representations. Among naturalistic borders, some completely surround the main subject while others are limited to the lateral and the upper sides, the lower part of the picture being decorated with bas-reliefs in antique style, generally without any bearing on the evoked theme.

The second category is composed of monumental frames forming an altar-piece or a portico. They present all of the typical architectural elements of the Renaissance such as pilasters, pillars, medallions in relief, volutes, vases, basins, garlands, etc., frequent motives in sixteenth-century illumination[22] and painting.[23]

There exists an intermediate between these two types. In the seed-borders as well as in the monumental porticoes, the lower part of the frames shows deer and monkeys which sometimes extend beyond the composition and thus seem to have an independent existence. It is quite permissible to see in them a symbolic meaning, as the monkey, *simia naturae*, is a frequent image of the bad alchemist. From one manuscript to another, an identical theme therefore can be presented in two different kinds of framing, while keeping the same deer and monkeys in the bottom of the picture.[24]

Finally, the third type of frame is exclusively reserved for the series of inhabited vases. Usually set in a niche and sometimes surrounded by a seed-border, they inscribe themselves in vast compositions animated with scenes depending on the astrological divinity which dominates the whole.

The *Splendor solis* shows alchemical imagery's climax in the first half of the sixteenth century. It constitutes a sort of synthesis of

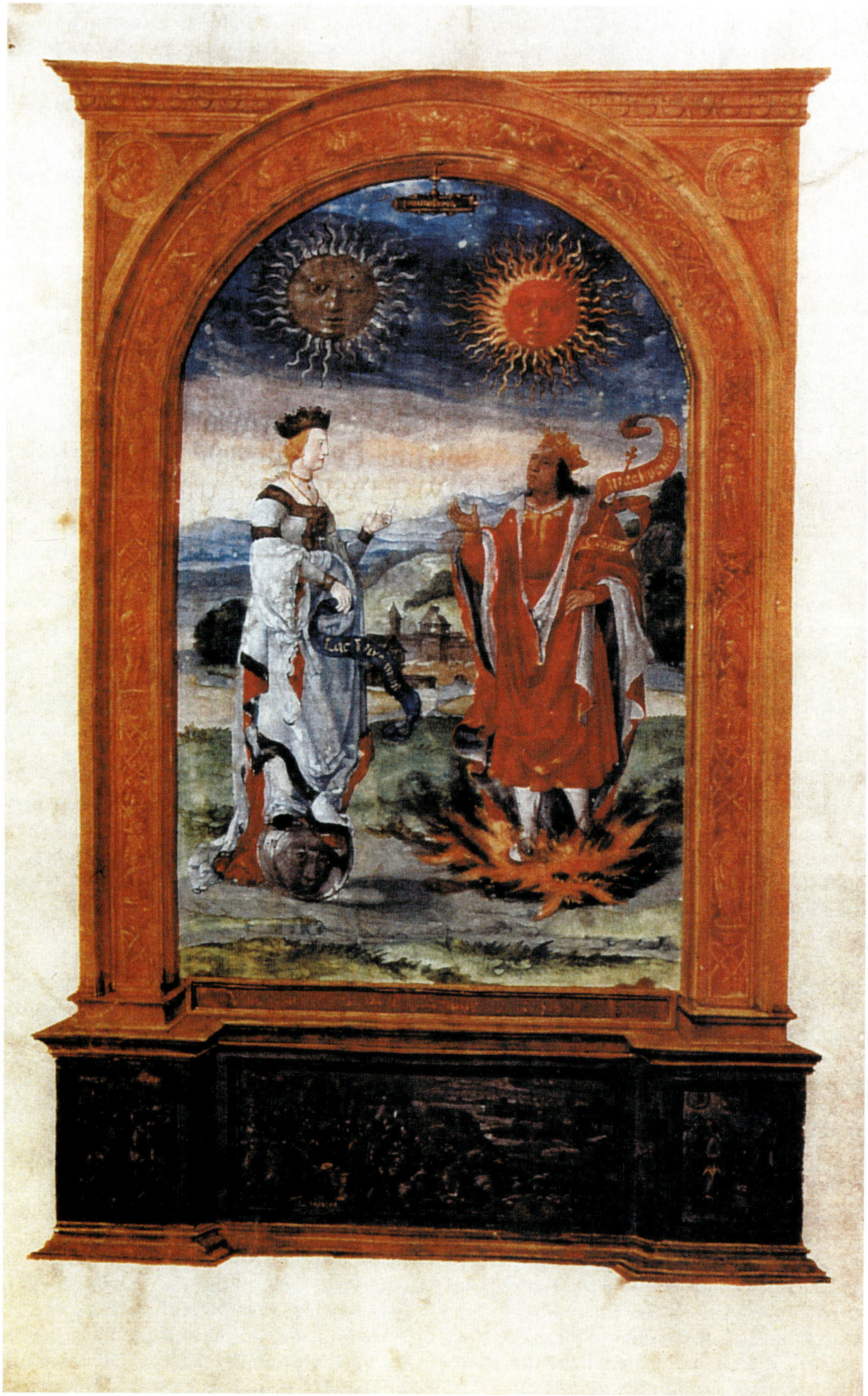

Figure 1
The alchemical wedding. This picture is one of the most specific in alchemical iconography. It symbolises the alchemical dualist vision of the universe. The philosopher's stone (often represented by the androgynous) is a 'double thing' issued from the wedding of two main opposed components. The queen stands for the feminine element; she is the moon, silver, mercury and water. The king is the sun, which also means gold, sulphur and fire. The wife holds a phylactery with the mentioned *lac virginum* (virgin's milk), whereas her husband bears the inscription *coagula masculinum*. Below of them, we can read *Via Universalis Particularibus Inclusis*.

the first illustrated alchemical works' iconography. Different categories of images cohabit in this iconography. Specific images are suited to alchemy and are recurring and conservative (fig. 1).

This is the case of the animated vases including a symbolical figure. Other images, called "of borrow," are inspired by religion, mythology or even from everyday life, such as the *Opus mulierum*, the *Ludus puerorum*, etc. Lastly, certain iconographical themes, apparently conventional, are peculiar to alchemy. The alchemical signification of the image is then revealed by certain details such as colours or unusual associations of elements.

A thorough study of one of *Splendor solis*' illuminations will permit a better understanding of the complexity of this kind of image (fig. 2).

Surrounded by a monumental portico in the typical "Renaissance" style, at the bottom of which are set two deer and two monkeys, the scene takes place in a traditional landscape of verdant hills, animated in the background with a few persons and groups of houses. A wood palisade separates the background landscape from the main subject. It may be viewed as a borderline between sacred and profane worlds, comparable with the *Hortus conclusus* in sacred iconography. In the foreground, a young winged lady, richly dressed and wearing many jewels, stands on the bank of a river of mud. She wears a flowery dress with a very elaborate bodice and yellow sleeves, and an azure blue cape. A golden shoe shows below the dress. The lady is wearing a golden crown topped by a silver star. Her wings are made of beautiful white and motley feathers similar to peacock's. Holding a crimson cloth, she is about to dry a strange character advancing from the muddy bog. This personage shows a curious morphology: it is a man, his body is black, his left arm white and his right arm and sphere-shaped head are red. His right hand is laid on his thigh and he holds his left hand out as if to grasp the cloth.

The concordance between text and image is especially evident. The text says:[25]

> *Menaldus[26] philosophus spricht also. Ich gebeut allen meinen Nachkömlingen daß sie die Cörper geistlich machen durch die Aufflösung und herwiderumb die geistlichen ding cörperlisch durch ein linde Kochung von deme redet Senior[27] also: Der Geist löset auff den Cörper und in diser Aufflösung zeucht er auß die Seel des Cörpers und verkeret den Cörper in die Seele und die Seel wird verwandelt in den Geist und der Geist soll wider zugefügt werden dem Cörper denn so ist er bestendig mit dem Cörper und herwiderumb der Cörper geistlich in Krafft deß Geistes. Das geben die Philosophi zuverstehen in einer solchen Signatur oder Figur: Sie sahen einen Menschen der war Schwarz wie ein Mor der stecket in einem Letten oder schwarzen unsaubern Schleim übel schmeckende. Dem kam zu hülff ein Junges Weib schön von Angesicht noch schöner am Leib und auffs hübschte geziret mit Kleidern die waren mancherley Farben. Sie war mit Weisen Flügeln auff ihrem Rucken geziret die Federen waren gleich den aller schönesten weisesten Pfauwen und hetten güldene Spigel und die Kisle waren geschmücket mit feinen Perlin. Sie hette eine Crone auff ihrem Haubt von reinem golde und auff der Cron einen Silbern Stern. Umb Ihren Halß hette Sie ein Halßband von feinem Golde darinnen versetzt den aller Edelesten Rubin den kein König vermag zubezalen. Sie hette an ihren Füssen güldene Schuhe. von Ir ging aus der aller Edeleste geschmack über alle Aromata. Sie bekleidet den Menschen mit einem Purpurgewande und bracht Ihn zu seiner höchsten Clarheit und füret ihn mit Ihr zu Himmel. Darvon redet Senior also: Es ist ein Lebendiges Ding das nicht mehr stirbet wan es ist mit einer ewigen Vermerung geübt.*

Although the text does not describe the tricolour personage, it abounds in details concerning the female angel who can be seen as the Saving Virgin also called "Wisdom" in the alchemical tradition. Her golden crown symbolises the accomplished transmutation.

The curious morphology of the man to be saved must be read from the lower to the upper part of the picture. At first, the man goes out of the mud which sucks him in. This "fetid earth" is of course the image of putrefaction. The liberation from the mud means that for transmuting, it is necessary to extract the spiritual principle from the corporeal.

The illustration can also be understood in light of other treatises of the alchemical tradition. One of the passages of the *Aurora consurgens*,[28] a treatise which is considered to be one of the principle sources of Salomon Trismosin's inspiration for the *Splendor solis*, contains a text very close to our illumination. Indeed, in the "Conversation of the well-beloved," we can read:

> *Tournez-vous vers moi de tout votre cœur et ne me méprisez pas si je suis noire et sombre, car c'est le soleil qui m'a ôté mon teint, et les abîmes ont recouvert mon visage, ma terre a été empoisonnée et corrompue dans mes oeuvres. Car les ténèbres se sont établies sur elle, parce que je suis plongée dans la vase du fond et ma substance n'est pas ouverte.* [...] *Je chercherai pour l'épouser une vierge chaste, belle de visage, plus belle de corps, très belle de vêtements, afin qu'elle roule la pierre de l'entrée de mon sépulcre, et elle me donnera des ailes comme celles de la colombe; je volerai avec elle dans le ciel et je dirai alors: Je vis à jamais et je reposerai en elle , parce que la reine se tient à ma droite en vêtements dorés, parée de couleurs variées.* [...] *et je serai revêtu par toi d'un vêtement pourpre, et je sortirai de moi comme l'époux sort de sa chambre nuptiale, car tu m'entoures de gemmes brillantes et étincelantes, et tu me revêtiras de vêtements de salut et de joie pour vaincre les nations et tous les ennemis. Tu me pareras aussi d'une couronne d'or marquée du signe de la sainteté, tu m'entoureras du vêtement de justice, tu te fianceras à moi à l'aide de ton anneau et tu me chausseras de chaussures d'or* (Franz [1982] 140–145).

The black legs and body are the representation of the *nigredo* or putrefaction, the white arm symbolises the *albedo* through which it is necessary to pass to reach the *rubedo* or transmutation. In his *Dictionnaire mytho-hermétique* of 1758, Pernety, who had probably the *Splendor solis'* illustration before his eyes writes, at the article "red head": *que ce qui a les pieds noirs, le corps blanc, et la tête rouge, est le magistère. C'est-à-dire que l'œuvre commence par la couleur noire, passe ensuite à la blanche, et finit par la rouge* (Pernety [1758] 489f.). But here, not only the feet are red but the whole body. About this image, Jung quoted a passage from of Albertus Magnus who explained how: "il faut purifier la matière jusqu'à ce que la tête noire offrant une ressemblance avec un Ethiopien ait été bien purifiée et ait commencé à blanchir" (Jung [1963]). The *Aurora consurgens* mentions the Ethiopians as an allegory of the nigredo: *"Alors les Ethiopiens se prosterneront devant moi et mes ennemis lècheront ma terre"* (Franz [1982] 74 f.).

As we have seen, colours have a very important meaning in alchemical iconography. The purple in which the angel wants to dress the putrid human being is thus the image of the Salvation or of the philosophers' stone, symbol of both material and spiritual perfection, which can be generated only after a putrefaction and which takes on several different colours before becoming red. This picture, unparalleled,[29] can thus be considered as a summary of the whole alchemical work.[30]

The representation of the alchemical process by the sequence of animated vases also has contributed to the fame of the *Splendor solis*. The theme of the vase, also called "mother" or "matrix," is one of the specific subjects of alchemy. Appearing in the treatises of Constantine and Gratheus in the last quarter of the fourteenth century, then developed in the *Donum Dei* of Georges Aurach in the last quarter of the fifteenth century, it is re-interpreted and systematised in the Aurora consurgens and the Splendor solis. Generally round, it figures the cosmos, the universe which symbolically constituted the receptacle for alchemical research (fig. 3).

Seven vases follow one after the other in the *Splendor solis*. Each subject, enclosed in a crowned and luted vase, appears as one of the major symbols of the philosophers' stone. In operative alchemy (in and after the seventeenth century), the number seven signifies "as many times as it is necessary to obtain the requested result"; there are seven operations before obtaining the stone.

Each animated vase is also put through an astrological theme. Indeed, seven corresponds to the seven astrological divinities which people the alchemical Olympus. The god symbolising the planet and the corresponding metal are portrayed at the top of the miniature. Character scenes which develop the theme actuate the borders.[31] Saturn, for instance, designates the basest metal, lead,

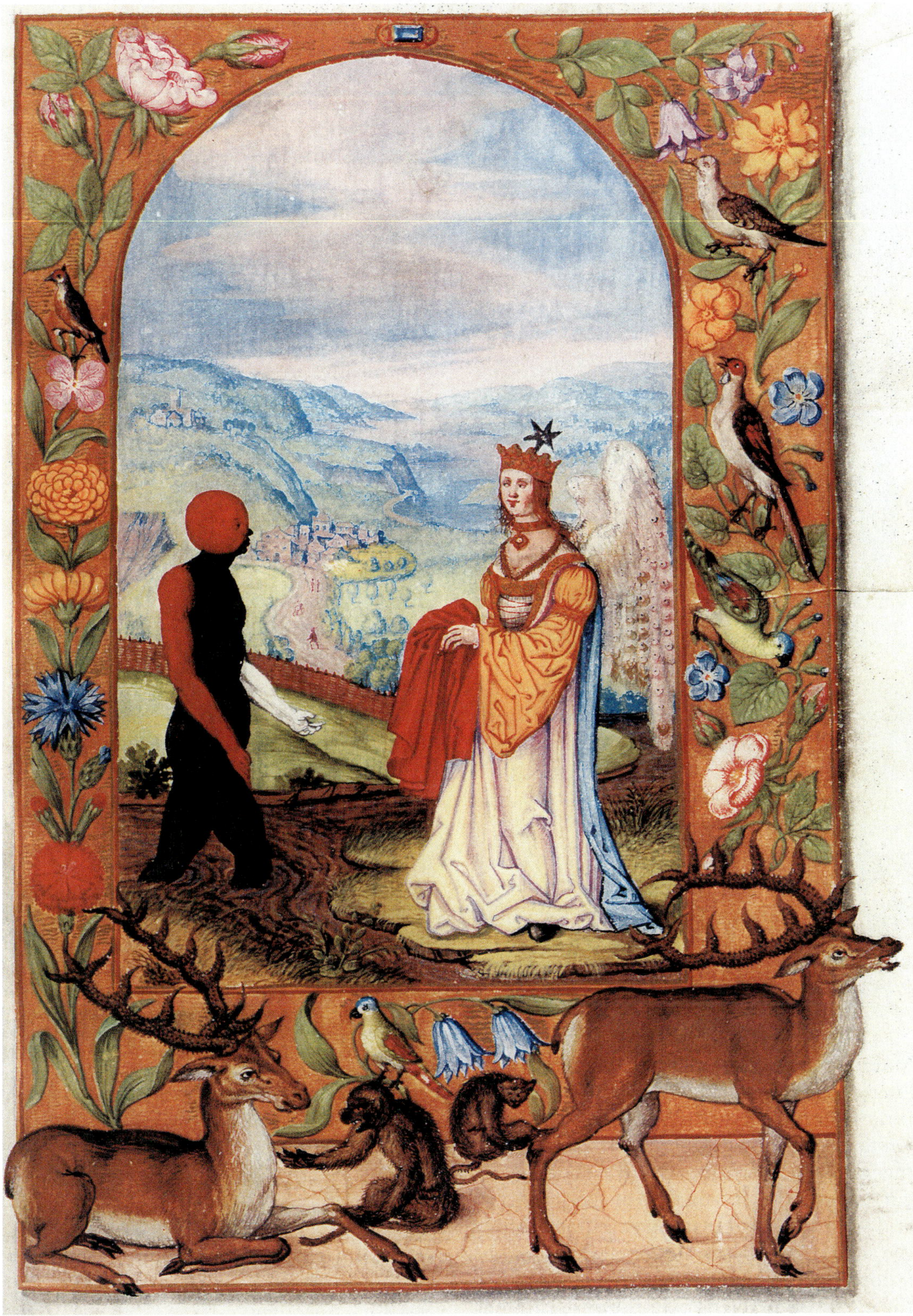

Figure 2
The woman angel comforting a strange muddy character (explanation in the text).

and thus the *nigredo*. It is associated with the vase which contains a *putto* pouring out the contents of a flask into a winged dragon's mouth, and which, with the other hand, sets a pair of bellows in action towards this monster. Jupiter corresponds to tin, to sublimation and to heat thanks to which metals take form in the bowels of the earth. It floats above the vase in which three birds with the stone's colours (black, white and red) are fighting.

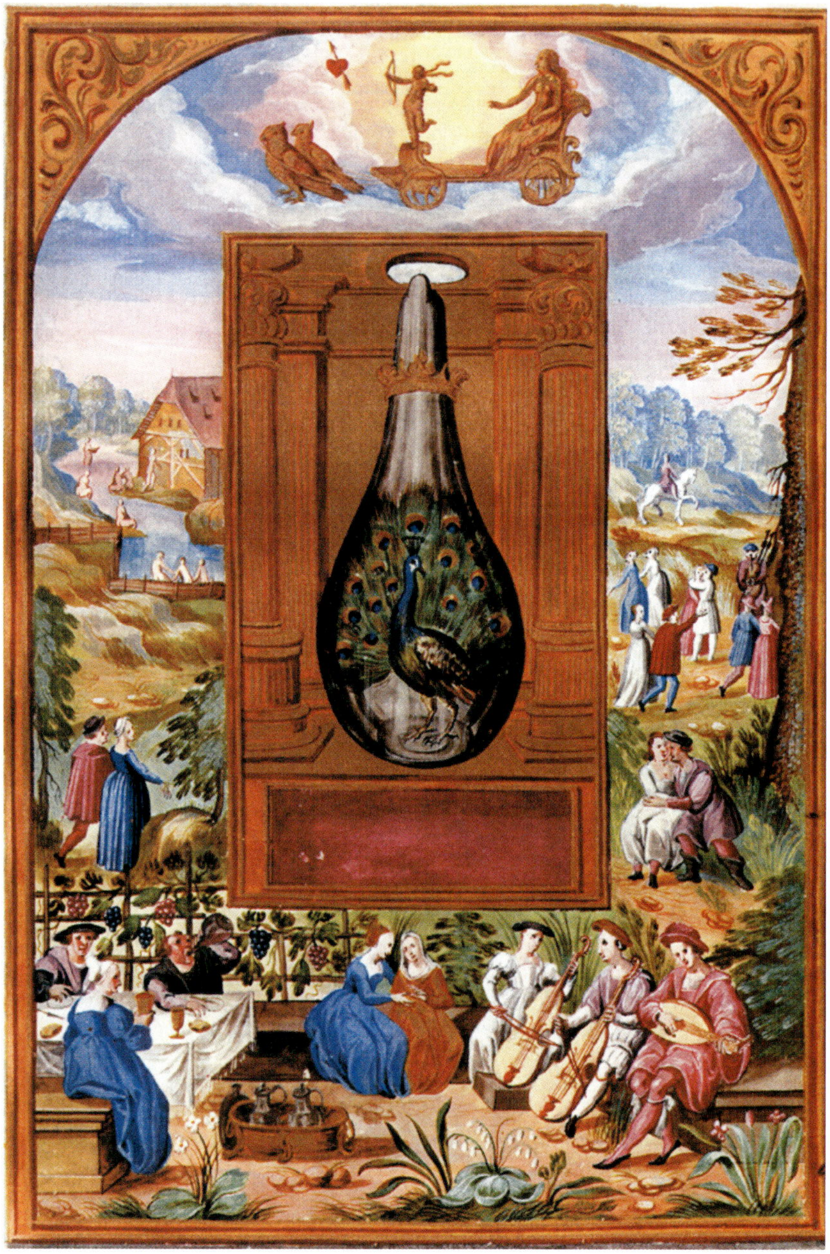

Figure 3
The peacock in the vase. The theme of the vase, also called 'mother' or 'matrix,' is one of the specific subjects of alchemy. The peacock is here associated with Venus, corresponding to copper. The peacock's tail symbolised one of the stages of the alchemical process, i.e., the moment when the substance takes on all kinds of colours. Venus inspires a few couples which give themselves up to the pleasures of the bath, walking, the palate and the senses.

Mars, the god of iron, designates the assimilated substance cooked at a certain temperature. It dominates a crowned, three-headed white angel. Venus, here associated with the peacock, corresponds to copper. Mercury, symbol of the duality of matter (fixity and volatility), dominates a vase imprisoning the mercurial queen. Diana, goddess of the moon, is associated with silver, designating the *albedo*. She reigns over the vase showing the fire-haloed king who stands on a crescent of moon. Finally, Apollo, the sun, gold, is represented above the vase keeping a green winged monster with three heads: black, white and red. Each illustration corresponds to a temperature of heat or cooking through which it is possible to obtain the philosophers' stone. Nevertheless, it seems hazardous to look for a direct junction of the astrological themes and the alchemical subjects. Moreover, in contrast to the illumination previously described, here there is not any evident concordance between the image and the text.[32]

A veritable masterpiece of the alchemical iconography of the Renaissance, the *Splendor solis* of Salomon Trismosin is, at the same time, both a synthesis and the starting point of several themes of fundamental meaning for the following centuries. After the *Aurora*, it reveals the independence of alchemical iconography and consequently its importance to the text it accompanies.

University of Liège

Notes

1 In the alchemical tradition, the sun is considered an allegory of gold.

2 The work is not complete; it contains only 19 illuminations instead of the traditional 22.

3 The miniature of folio 29 shows the date of 1577. According to a printed note added in the eighteenth century on folio 1 recto, this manuscript must have belonged to the Emperor Rodolph I; we can read: "*v. Schönfeldsches Museum der Technologie, zum Vortheil des Gewerbstandes errichtet im Jahre 1799. Kaiser Rudolphs des II Goldmacherbuch*." It should be observed that no less than forty years separate the first two manuscripts.

4 Dated from the end of the sixteenth century by Gustav-Friedrich Hartlaub. The manuscript has been reproduced in facsimile.

5 The dates of 1584 and 1588 appear on two folios. The text of the Splendor solis is included there in a series of other alchemical treatises.

6 According to G.-F. Hartlaub, this work contains 24 illustrations instead of the usual 22 illuminations.

7 It is a French translation of the compilation printed in 1598 under the title of "Aureum vellus" (see below). The manuscript is dated of 1720 for the texts; the illustrations that faithfully reproduce the sixteenth century manuscripts' illuminations could have been added around 1740.

8 Although faithful to German manuscripts' iconography for the main subjects, the remarkable illustrations are adapted to the reigning fashion.

9 We are still waiting for the microfilms.

10 Bartholomaeus Korndorffer lived around the 1520's. He is known only by the titles of his works; he is more particularly the author of a treatise on precious stones "*Beschreibung der Edelsteine*." See Ferguson [1906] 478.

11 Leipzig, in Verlag Henning Grossn. Besides medicine and chemistry treatises, Joachim Tancke (1557–1609) has also translated and published different books of Ewald van Hohenlande, Roger Bacon, Bernard le Trévisan, etc.; see Ferguson [1906] 426–429.

[12] This fact is significant enough when we know the importance of colours in alchemical iconography.

[13] The kings' names are associated with the tincture treatises that form part of the *Aureum vellus*.

[14] Sudhoff [1958] 483: *"Ich Trismosin hab neun Discipulos gehabt, ist keinem gerahten dan nur einem, Philips Hohenheimer, ein geborner Schweytzer, ein verschmitzter Mann: Ich sage, Er wird gross werden in dieser Kunst, wan er etwa in sein mitler Alter kompt, da wird die Welt gross Wunder von jhm hören, und viel Wunder an Tag bringen".*

[15] Cayet [1605] 459: *"Mais voyons tout de suite un nouveau thresor d'un paracelsiste dont la Boheme jouyt maintenant. Il y a long temps que d'Allemagne s'est travaillée et travaille après la pierre philosophale et en a recherché les livres des roys anciens d'Egypte, entre autres le Viatolon du Trismosin avec son Sarouadoap auri, et la teinture du Geroton, avec les livres excellents qu'il a faict du Suforeton, et aussi son Canganiveron et le Pareseton. Item le Moratosan de l'aigle noir, et le Nefolon de l'aigle rouge. Item les teintures de Xophares, roy de Silons, terre d'Aegypte, la Soronella de Crinot et les teintures du Petrumosin, qui sont toutes inventions sublimes de très grands roys et anciens philosophes; desquels noms et tiltres nous avons rendu les interprétations en la traduction par nous faicte en François de Guldin Schatz und Kunst Kammer en allemand; qui est à dire le thresor doré ou la chambre de la science".* The work that Palma Cayet has had the opportunity to see is the Rorschach's edition of 1598, but breaking his promise, he never translated it.

[16] The printed versions of the *Splendor solis* are also illustrated by a score of woodcuts, sometimes coloured, quite faithful to the manuscripts' illumination for the themes but of a much lower quality in execution.

[17] For the description of the animated vases, see below.

[18] Born around 1483–1484 in Gand – died around 1561. An excellent miniaturist, he presented his subjects following the typical Dutch model for frames, making a sort of altar-piece, or for naturalistic borders, using fruit, flowers and insects, or borders animated with character scenes.

[19] Hans Sebald Beham was born in Nuremberg in 1500; his work reveals Durer's influence. In 1525, he was banished from the city along with his brother and another Durer's student, Georg Pencz, because they had voiced subversive opinions.

[20] Albrecht Glockendom was considered to be a good illuminator, but also a poet who confined himself to colouring the etchings and to adapting verses to them.

[21] The comparison with one of the frames of ms. 3591 at the Library of the University of Liège, "book of hours" illustrated for Gerard van de Stappen, abbot of Saint-Laurent from 1520 to 1558, is a good example of this tradition.

[22] For this type of frame, it is interesting to compare our manuscript with the ms. 29 of the Wittert Collection at the University of Liège Library, a sixteenth-century "book of hours".

[23] Jacques van Lennep compares them with the *"constructions romanisantes des tableaux de van Orley, de Blondeel ou de Scorel qui, par leur tempérament germanique encore empreint de gothicisme, conférèrent un accent maniéré aux inspirations antiques de leurs oeuvres"* (Lennep [1966] 52).

[24] This fact supports the hypothesis of the symbolic role of the animals.

[25] Splendor solis, third treatise, fourth "similitude." B. Husson, (Husson [1975]) provides a French translation of this extract: Le philosophe Menaldus parle ainsi: "Je veux, dit-il, et entends que tous ceux qui désirent ensuivre la piste que nous y avons tenue fassent en sorte que les choses spirituelles se corporalisent, et que les corporelles se spiritualisent par une légère et seule décoction." Imitant encore en ceci Senior: "l'esprit, dit-il, délivre le corps et par cette délivrance l'âme se tire hors des corps, puis on réduit ces mêmes corps en l'âme: l'âme donc se change en un esprit et l'esprit de nouveau se fait corps." Car s'il demeure ferme au corps et qu'il rende de nouveau les corps de soi terrestres, massifs et grossiers, spirituels par la force de ces esprits, c'est le but de notre œuvre.
Un être humain aussi noir qu'un Maure se trouvait enlisé dans un marécage ou limon noir et sale et dégoûtant et une jeune femme venait à son secours. Cette dame était belle par excellence de corps et de face, enrichie de superbes habits de diverses couleurs, ayant de belles plumes blanches bigarrées comme celles d'un paon qui s'étendaient également sur son dos, à la merci d'un vent bénin et zéphire favorable, les ailerons en étaient d'or entrelacés de belles petites perles. Sur son chef bien agencé, elle avait une très belle couronne d'or, et sur icelle une étoile d'argent; à l'entour de son col elle portait un carcan d'or, dans lequel était richement enchâssé un précieux rubis d'excellent artifice, le plus juste prix et la valeur duquel n'eût pas su payer le plus grand revenu de quelque puissant roi: elle avait aussi des souliers dorés aux pieds et d'elle s'épandait une suave et très odoriférante odeur. Cette dame s'avance émue de compassion, et le retirant bénignement d'une telle infection, elle le nettoie pur et net, lui fait présent d'un bel habit de pourpre, et l'emmène jusqu'au ciel avec elle. Senior en parle tout de même, traitant de ce sujet, voire encore en termes bien plus clairs: "il y a, dit-il, une chose vivante qui n'est plus mortelle, ayant une fois été confirmée et assurée de sa vie par une éternelle et continuelle multiplication."

[26] Menaldus is one of the authors of the *Turba Philosophorum*.

[27] Senior: Mohammed-ibn-Umail; author of the *Tabula chemica*, an extremely allegorical treatise translated from the Arabic.

[28] The *Aurora consurgens* is a treatise dating from the beginning of the fifteenth century that tends to show alchemy as a quest for divine wisdom. It is also an alchemical interpretation of the Song of Songs.

[29] None of the former alchemical cycles, neither the *Aurora consurgens*, the *Liber Trinitatis* nor the *Donum Dei* proposes a comparable allegory, with the possible exception of an illustration taken from the *Aurora* which presents minor similarities. A woman, Alchemy, gives a bunch of herbs to an invalid who has a silver head and trunk and is sitting on a mattress. (This hypothesis should be reexamined.)

[30] As we have already said, the illustrations are different from one exemplar to another, not only in the sequence but also in certain details. The clothes and the ornaments worn by the angel, for instance, exhibit a few variances: in certain manuscripts, the dress is flowery, in others, it is entirely white; the bodice is sometimes fringed with fur and sometimes trimmed with a golden ribbon; the necklace

is not the same in all the versions. In the Berlin manuscript, (1582), the angel wears a black shoe rather than the golden one in the other copies. In the same manuscript, the lower part of the face of the tricolour personage (the nose, the mouth and the neck) is black and only the upper part and the spherical skull are red. The events which happen in the background are also quite different. A searching examination of the manuscripts also allows differences in the execution style of the paintings to be disclosed, ranging from a supple model (Nuremberg) to a more geometrical and calligraphic style (Berlin, London).

[31] To Saturn correspond the representations of tanners, hermits, farm-labourers, butchers and beggars. The scene also evokes death and tortures (the wheel and hanging) associated with this god. On Jupiter depend the princes, the treasurers and, here also, the alchemists. Mars governs war expeditions and marches past. Venus inspires couples who are enjoying a bath, a walk or a meal. Mercury appears here as the divinity of musicians, geographers and sculptors. Diana reigns over fishers, hunters and washerwomen. Lastly, with Apollo are associated games, wrestling and sword fencing.

[32] Indeed, if each subject contained in the vase always corresponds to the same astrological theme, the sequence of the illustrations is not consistent from one manuscript to the other.

References

Alleau, R.
1975 "Etude iconographique du manuscrit de Berlin." In: Trismosin [1975] 264–284.

Borrichius, O.
1697 *Conspectus Scriptorum Chemicorum.* Copenhagen.

Ferguson, J.
1906 *Bibliotheca Chemica. A catalogue of the alchemical, chemical and pharmaceutical books in the collection of the late James Young of Kelly and Durris,* vol. 1. Glasgow.

Franz, M.-L. von (ed.)
1982 *Aurora Consurgens. Le lever de l'aurore.* Paris.

Hartlaub, G.-F.
1937a "Arcana Artis. Spuren alchemistischer Symbolik in der Kunst des 16. Jahrhunderts." *Zeitschrift für Kunstgeschichte* VI: 289–324.

1937b "Signa Hermetis." *Zeitschrift des deutschen Vereins für Kunstwissenschaft* IV/3, 93–144.

1954 "Chymische Märchen. Naturphilosophische Sinnbilder aus einer alchemistischen Prunkhandschrift der deutschen Renaissance (*Splendor Solis*)." *BASF Werkzeitung 'Die BASF,'* Jg. 1954 Heft II: 47–51, Heft III: 103–107.

1955 „Chymische Märchen. Naturphilosophische Sinnbilder aus einer alchemistischen Prunkhandschrift der deutschen Renaissance (*Splendor Solis*)." *BASF Werkzeitung 'Die BASF',* Jg. 1955 Heft I: 4–9.

1959 *Der Stein der Weisen. Wesen und Bildwelt der Alchemie.* Munich.

Husson, B.
1975 'Commentaires des illustrations.' In: Trismosin [1975].

Jung, C. G.
1963 *Psychologie et alchimie,* transl. O. Raesvski. Paris.

Lennep, J. van
1966 *Art et Alchimie.* Brussels.

1984 *Alchimie. Contribution à l'histoire de l'art alchimique.* Brussels.

Palma Cayet, P. V.
1605 *Chronologie septénaire de l'histoire de la paix entre les rois de France et d'Espagne.* Paris.

Pernety, A.-J.
1758 *Dictionnaire mytho-hermétique. Dans lequel on trouve les allégories fabuleuses des poètes, les métaphores, les énigmes et les termes barbares des philosophes hermétiques expliqués.* Paris.

Sudhoff, K.
1958 *Bibliographia Paracelsica. Besprechung der unter Hohenheims Namen 1527–1895 erschienenen Druckschriften.* Graz.

Trismosin, S.
1972 *Splendor solis. Das gegenwertige Büchlein wird genandt Splendor Solis oder Sonnenglantz teilet sich in siben Tractat durch welche beshriben wird die künstliche Wirkunge des verborgenen Steins der Alten Weissen ...,* ed. G. Höhl (Faksimile-Ausgabe der Prunkhandschrift der Berliner Staatsbibliothek). Wiesbaden.

1975 *La Toison d'or ou la Fleur des Trésors. Texte de l'édition française de 1612. Introduction, traduction inédite du texte allemand de 1598.* Paris.

The Invention of Atomist Iconography

Christoph Lüthy

Introductory Puzzlement

For centuries now, particles of matter have invariably been depicted as globules. These globules, representing very different entities of distant orders of magnitudes, have in turn been used as pictorial building blocks for a host of more complex structures such as atomic nuclei, molecules, crystals, gases, material surfaces, electrical currents or the double helixes of DNA. Maybe it is because of the unsurpassable simplicity of the spherical form that the ubiquity of this type of representation appears to us so deceitfully self-explanatory. But in truth, the spherical shape of these various units of matter deserves nothing if not raised eyebrows.

How is it to be explained that the same type of illustrations should have survived unperturbed the most profound conceptual changes in matter theory? One needn't agree with the Kuhnian notion that revolutionary breaks dissect the conceptual evolution of science into incommensurable segments to feel that there is something puzzling about pictures that are capable of illustrating diverging "world views" over a four-hundred year period.[1] For the matter theories illustrated by the nearly identical images of fig. 1 have almost nothing in common. But what implications does this observation have for the relation between theory and illustration?[2]

It might be suggested that such globular drawings are theory-independent and ahistorical for some *a priori* characteristics which free them from all specific referentiality. But a quick look at twentieth-century debates over the nature of such imagery will quickly persuade us that this way out is barred and that there is nothing ahistorical, self-explanatory, or otherwise *a priori* about them. While Niels Bohr's atom model of 1913 was the expression of the belief in an empirically and mathematically verifiable correspondence between the solar system and the atomic nucleus with its surrounding spherical electrons,[3] the protagonists of the new quantum physics have urged, ever since 1924/5, that all *Anschaulichkeit* had been lost and that drawings of spatio-temporally bounded, material corpuscles were not only inadequate, but truly misleading (cf. Miller [1984] ch. 4). And yet, the old globular imagery has not only managed to survive its official refutations, but it has found its most paradoxical demonstration in the "pictures" that have, over the past few

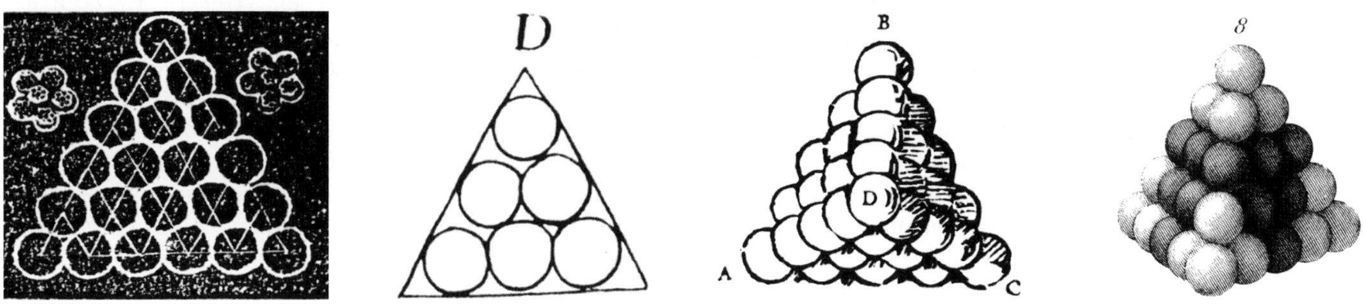

Figure 1
How many theories can be illustrated by a single image?

From left to right:
1a: From Giordano Bruno: *De triplici minimo et mensura* (Frankfurt 1591).

1b: From Robert Hooke: *Micrographia* (London 1665).

1c: From Christian Huygens: *Traité de la lumière* (Leiden 1690).

1d: From a paper of William Wollaston in *Philos. Trans.* 1813.

years, been produced by tunnel, field ion, and atomic force microscopes.[4] The relation between the depicted spherical particle on the one hand, and the various theories it happens to illustrate on the other, is thus seen to fluctuate so wildly that we must abandon all hopes of finding a single semiotic theory to define it. In fact, over the course of history, we find that the status of the illustrated globule oscillated between what, in the language of theory, might variously be described as "symbol," "sign," "model," "abstraction," "reproduction," or "map."

Faced with the globule's stubborn unwillingness to disappear from physics, one might also be tempted to view it as the product of some necessary mental operation, corresponding, for example, to the schematic operations of Kant's *transzendentale Einbildungskraft.*[5] But this hypothesis is also easily refuted. For as will be shown in the following, the globular atom is an invention of the late sixteenth century. Neither did it exist before, nor did its invention seem very useful at first. Instead, the globular particle of matter is a strange outgrowth of Renaissance speculation which required decades of reinterpretion before it began to seem useful here and there as a possible tool for the explanation of certain natural phenomena.

The many conceptual discontinuities that separate today's physics from the natural philosophy of the late sixteenth century render the survival of a stable atomist iconography a perplexing puzzle indeed. Although in this essay I limit myself to the task of providing a first and preliminary sketch of how the globular view of matter was first developed, and of how it managed to integrate itself into the emerging seventeenth-century view of nature, I hope that the results of this investigation will be able to contribute to a larger investigation of the oscillating relations between theories and images of matter.

Aristotle's Anti-atomism

While it is in most cases idiotic to ask why a certain phenomenon did not exist before the time when it first occurred, this question happens to be meaningful in our case. For there existed specific factors explicitly inhibiting the production of atomist images, factors that are mainly to be sought in Aristotle's natural philosophy. In his critique of the atomism of Leucippus and Democritus, Aristotle had argued that a world made up only of particles and void would be a mere "heap" and that such a "heap" could not explain the cohesion of natural bodies, let alone life and organic development. Apart from its other deficiencies, the atomists' model was also flawed, in Aristotle's eyes, by its sense-sceptical implications. For in an atomist cosmos, material substances merely *appeared* to be continuous bodies, while in truth they were made up of discontinuous particles of different types and of gaps between them. To a keen-eyed fellow such as the mythological Lynceus, an alloy such as bronze would, for example, not appear as a homogeneous, continuous substance, but as a mere juxtaposition of copper and tin particles. In other words, natural bodies

> *would only be mixed according to the standard of sense-perception, and the same thing would be a mixture to a man, who has not sharp sight, whereas to the eyes of Lynceus it would not be mixed.*[6]

Leucippus and Democritus had in fact assumed that an invisibly small micro-world underlay the macro-world of sense experience, but such a fragmentation of reality was anathema to Aristotle. Natural bodies, he insisted, not only appeared continuous and homogeneous, but, thanks to the agency of specific substantial forms, did indeed possess these qualities.[7] And although these Aristotelian "forms" (μορφή; morphé; or εἶδος, eidos) sound very "morphological," "eidetic" and hence picturable, they are not. In fact, they denote mainly (teleo-)logical principles, and they must strictly be distinguished from *figura* (σχῆμα, schema), a term of marginal importance in Aristotle's natural philosophy (cf. Emerton [1984] ch. 2; Des Chene [1996] ch. 4.3; Lüthy et al. [1997]). In contrast to most of his predecessors, Aristotle was no visual thinker, although one might have expected something different from such an outspoken sense realist. The rejection of material substructures was only a minor aspect of his conceptual, anti-visual natural philosophy, which did not think of the world *geometricè* even where its argument proceeded *more geometrico.* The strikingly pictureless nature of natural philosophy during Europe's scholastic centuries is the direct result of this style of reasoning.[8]

At the most basic level of Aristotelian matter theory, we encounter, under the name of "prime matter," a purely logical principle denoting the substratum underlying the specific forms of elements and other substances. This *materia prima* cannot subsist by itself and is neither imaginable nor indeed depictable. At the next higher level, we encounter the four elements which, however, possess none of the geometric qualities ascribed to them by the atomists or by Plato. Though they can be sensually experienced, they are not easily represented with ink on paper. Apart from theory-independent allegorical and semiotic means ("water" being represented by a lady pouring water out of a jar or by a wavy line indicating an ocean surface), the Aristotelian tradition has engendered only two types of theory-related graphic conventions. The first of these follows *De generatione et corruptione*, where the elements are defined in terms of the primary qualities of hot, cold, wet and dry in such a way that elements and qualities are found to overlap as in fig. 2.

While this circular arrangement of the elements illustrates merely logical relations, there is also a second convention which seems at first sight to possess more developed mimetic qualities. It is based on *De caelo*, where the four-element theory is derived from the central tenet that "all natural bodies and magnitudes are capable of moving themselves in space," (Aristotle [1945] 268b 15–17) and where the elements are defined by their respective tendencies to move to certain "natural places," rising "above everything else" (fire), or falling "below of everything else" (earth), or striving to the two intermediate positions (water and air) (ibid., 312a 27–29) (cf. fig. 3a).

This "onion-ring model" of elemental layering has a certain realistic appeal: experience gives us a world in which fires rise, winds blow over oceans, and lakes cover valleys. This realist interpretation was enhanced by Christian authors who combined the Aristotelian onion rings with the creation story of Genesis 1, where God formats the world, as it were, before introducing the creatures into this framework (fig. 3b). And yet, these drawings present a merely counterfactual situation. For although Aristotle's physics is

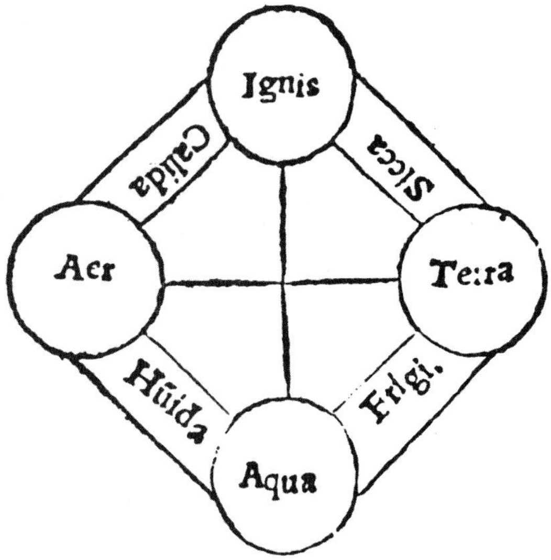

Figure 2
The logical relations between elements and qualities. From Charles de Bouelles: *Liber de generatione* (Paris 1510).

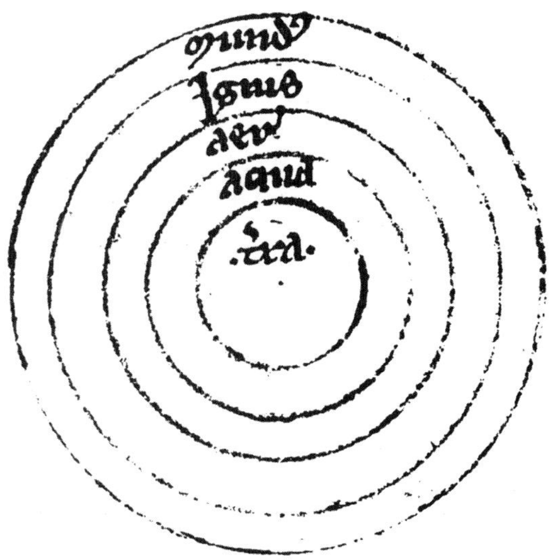

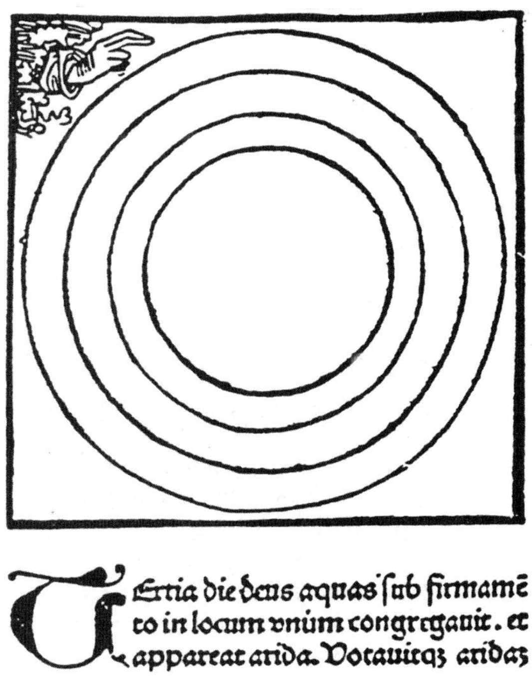

Figure 3
The onion-ring model of natural places.

3a (*left*): Fourteenth-century diagram from William de Conches: *Dragmaticon* (Ox. Bl. Digby, 1007).

3b (*below*): From Hartman Schedel: *Chronicarum Libri* (Nuremberg1497).

generally ruled by a converse law of entropy in which processes strive towards homogeneity, the accomplishment of homogeneity would in this case imply the death of the cosmos. For when the elements reach their proper sphere, they "grow together and become passive; but only where they meet, they act and react upon one another" (Aristotle [1980] 212b 32f.). In other words, only as long as the world does not correspond to the onion-ring model will generation and corruption, which depend on the continued mixing and unmixing of the elements, continue.[9]

These two Aristotelian pictorial conventions do thus not illustrate the elements themselves, but only the logical relations that hold between them. It is worth insisting once more that this limitation is not an accidental by-product of Peripatetic physics, but the explicit counterpart to Aristotle's laughter about his predecessors' suggestion that "the first simple bodies have any certain figure" (cf. Aristotle [1945] 303a 3–103b 8, 304a 10–15, and 306b 32–307a 34).

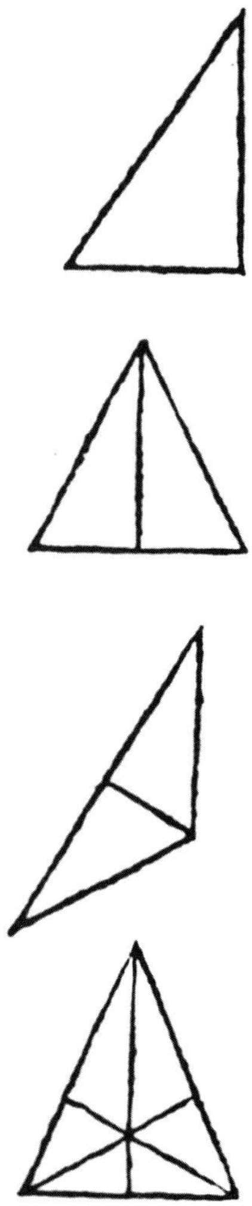

Figure 4
"Illustrating" Plato's *Timaeus*. From *Omnia divini Platonis opera* (Basel 1532).

Platonic Imagery

The reasons why scholastic Aristotelians had no use for microstructures and geometrical particles thus should be obvious. But then, the later Middle Ages and the Renaissance knew not just Aristotle, but also the thinkers against whom he had been arguing. From the early decades of the fifteenth century onward, Plato's *Timaeus*, Diogenes Laërtius' detailed summary of Epicurus' atomism (in *Lives of the Philosophers*, book 10) and Lucretius' equally Epicurean *De rerum natura* were all once more available to the erudite reader. Must we not assume that these texts represent the cradle of our modern atomistic imagery? Surprisingly enough, the answer is a fairly clear "no."

As for the *Timaeus*, let us recall that Plato there associates the elements with the shapes of the regular solids: earth particles are cubic, fire has the form of a pyramid, air comes in octahedra and water in icosahedra. These solids in turn are made up of the triangles that form their surfaces: while earth is made up of 45° isosceles, the other three elements are composed of half-equilateral triangles and can therefore be transformed into each other (Plato [1959] 53C–56C). About 60% of the *Timaeus* is dedicated to the playful application of these triangles and polygons to all kinds of natural phenomena.[10] Their geometrical shapes are notably used to explain our sensory experience: their impact on the human organs is said to be responsible for our specific perceptions of the world, an explanation that allows for the reduction of secondary (sensory) qualities to primary (geometrical) qualities. Democritus' conventionalism ("By convention sweet, by convention bitter, by convention hot, by convention cold, by convention colour: but in reality atoms and void")[11] was richly elaborated by Plato, who substituted "atoms and void" with "particles and interstices" of various shapes.

In opposition to Aristotle's matter theory, Plato's is thus essentially visual. Since the primary qualities of the elements are geometrical, they can be imagined and drawn with precision. And indeed, once the Renaissance had regained access to the corpuscularian parts of the *Timaeus* (which previously had been lost), these regular solids received widespread attention. But interestingly enough, it was not in the realm of matter theory that their visuality was exploited. While Renaissance painters became inordinately fond of displaying their perspectival skills in the construction of the regular solids and mathematicians explored their perfect geometry, no comparable enthusiasm can be observed among natural philosophers.[12] As for the editions of Plato's *Timaeus*, there are hardly any that even bother to depict the solids, and when they do, they limit themselves to very primitive drawings (fig. 4).

None of them seems to have attempted to illustrate how the regular solids combined into natural bodies or how they interacted with the sense organs.[13] Their reluctance to put these images to work was in this case not due to pictorial difficulties, and certainly not to the absence of pictorial conventions, but seems to be related to the generally dismissive view they took of the foundations of Plato's doctrine of the elements. Kepler's case is here particularly illuminating. Though this author invoked the Platonic solids very centrally in his explanation of the spacing of planetary orbits, he felt at the same time that their attribution to the elements relied on a weak and unconvincing *analogia* between geometrical shapes and elementary qualities.

Worse yet, it was based on a misunderstanding: Kepler was convinced that the Pythagoreans and Plato had in fact not spoken of the elements, but of the planets.[14]

If we speak of this general lack of enthusiasm for Plato's elementary shapes, we must also mention the one telling exception, the pyramidal fire particle, which was the only Platonic solid to have gained widespread popularity among early modern natural philosophers, physicians, and alchemists (e.g., Scaliger [1557]; Basson [1621]; Sennert [1636]). The reason for this phenomenon lies in the fact that it was the only Platonic element whose geometrical shape stood in an intuitively convincing relation to its physical and sensory qualities. Its wedge-shaped figure seemed to represent and thereby to explain those qualities of cutting and destroying that we associate with fire, in the sense that its "acute angles" appeared to explain the sensation of "acute pain" it caused. The other regular solids had no comparable advantage. Their ability to fill space without leaving any interstitial voids was of little importance to early modern alchemists and iatromechanicists who either allowed for such vacua or filled them with ether. The early modern disciples of a Platonising view of sense experience and physiology therefore swapped Plato's regular solids for irregular, but intuitively more useful, shapes.[15]

The Shapes of Democritus' Atoms

By contrast to Plato, who had at least proposed specific, if not necessarily convincing, shapes for the elements, the contribution of the ancient atomists to the development of atomist imagery turns out to have been even more marginal. Aristotle reports that Democritus and Leucippus had allowed for four types of differences among their particles (specific shape, magnitude, orientation in space, and relation to each other) (Aristotle [1928] 985b 4ff., [1980] 188a 23ff., 213b 1ff.), but that they had been reluctant to specify these shapes. Aristotle, who repeatedly criticises them for their vagueness, writes in *De caelo* that they had postulated "an infinite number of shapes" and hence "an infinite number of simple bodies," but without ever

> *defining or characterising the shape of each element, except to assign the sphere to fire. Air and water and the rest they distinguished by greatness and smallness only, as if it were their nature to be some ,seed-mixture' of all the elements* (Aristotle [1945] 303a 3ff.).

But, as Aristotle was quick to point out, the atomists' choice of the globular shape for fire contradicted Plato's: "For instance, because fire is the most mobile and has the faculty of heating and burning, the one school made it spherical, the other pyramidal"; but in fact, "it is absurd to assign to fire its shape for the purpose of division only" (Aristotle [1945] 306b 32–307a 34). Thus the only particle with which Plato managed to impress early modern thinkers had the misfortune of colliding with the description of the only elementary particle defined by the atomists. That the attributions of shapes on the basis of arbitrarily selected qualities easily led to contradictions was also pointed out by early modern critics of corpuscularianism. Hermann Conring, for example, used this argument in 1643 when attacking the attribution of the pyramidal shape to fire particles (Conring [1643] 68).

Although he was indirectly responding to Aristotle's criticism, Epicurus was no more precise on the shapes of particles, limiting himself to the identification of "soul atoms" with the "smoothest and roundest atoms" and claiming (*pace* Democritus) that these were nonetheless "quite distinct from fire atoms."[16] The same non-committal attitude is found in Lucretius' *De rerum natura*. Though this Roman Epicurean insists on the importance of the atomical *figurae* for physics in general and for our sense experience in particular (Lucretius [1992] II 333–729), his description is limited to the poor set of four predicates "round," "sharp-angled," "hooked," and "intertwined," predicates that furthermore suffer from the double defect of describing qualities, not substances, and of explaining at the same time a number of incompatible qualities.[17]

In sum, the reasons for why classical atomism did not engender its own iconographical tradition are these: first, in contrast to the Platonic model, the ancient atomists had left the relation between shapes and substances strongly underdetermined.[18] Second, in con-

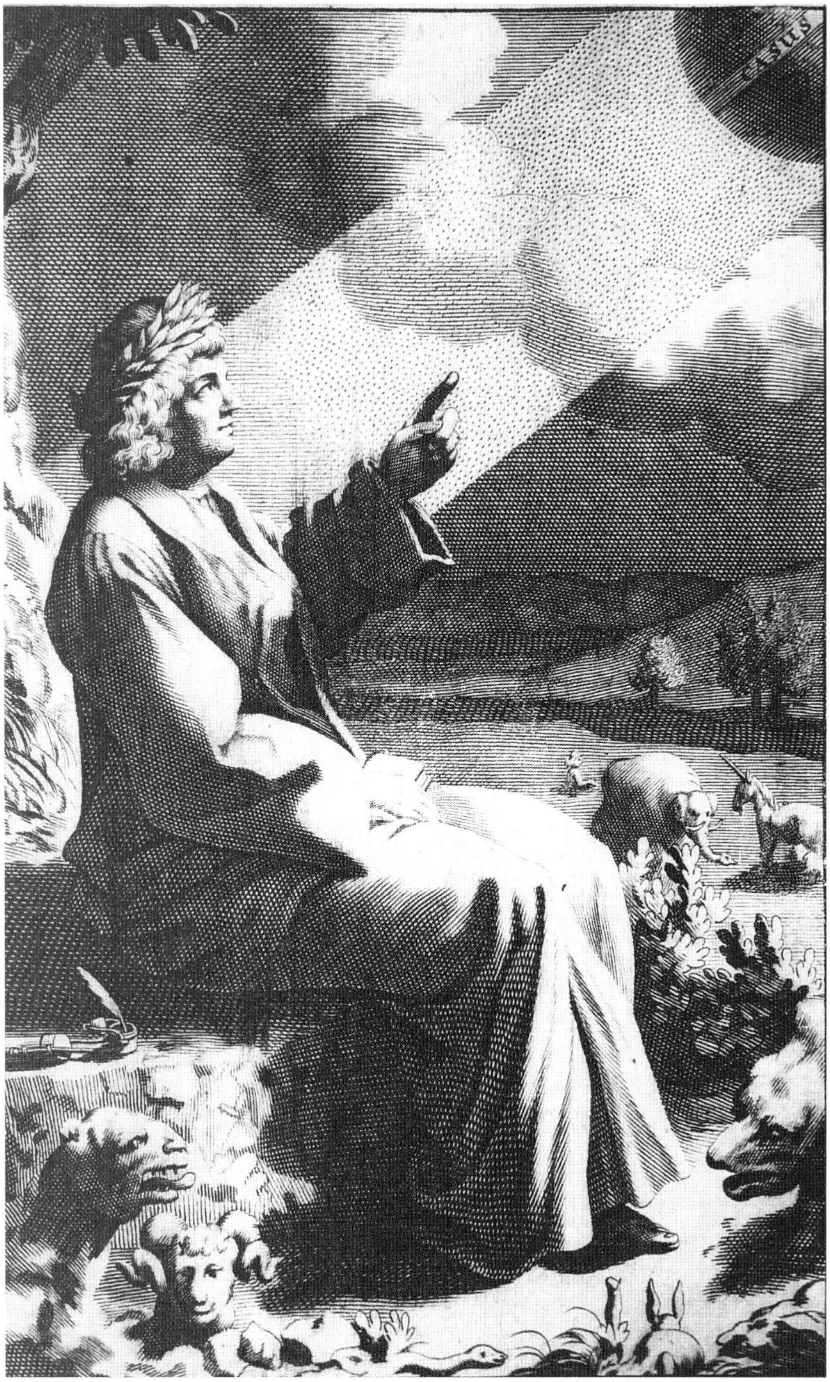

Figure 5
The only "atoms" depicted in a seventeenth-century edition of Lucretius (London 1683).

trast to the Aristotelian diagrams, which possessed at least some modest didactic and mnemonic usefulness, any attempt to depict Democritean micro-structures would have gone beyond the theory itself, without, however, serving any obvious ends.

Given what has so far been said, it cannot be considered a coincidence that Pierre Gassendi's thousands of pages in defense of Epicurean natural philosophy remained pictureless, despite the many attempts of this author to describe atomic shapes verbally (Gassendi [1658]). It also explains the disappointing result of an examination of the 79 editions of Lucretius' *De rerum natura* which were printed between the *editio princeps* of 1473 and the beautiful Leiden edition of 1725. Although quite a number of them contain illustrations and elaborate frontispieces, we find on them only dramatic scenes (depicting, e.g., the Athenian plague of book VI) or allegorical figures celebrating the triumph of philosophy. In these first 250 years of Lucretian editions, it appears that the only graphic representation of atoms is to be found on the frontispiece of the third edition of Thomas Creech's English translation of 1683 (fig. 5) (cf. Gordon [1962] 175; 233f.).

But what an ambivalent illustration this is! For if the dots seen descending from the sun refer to the motes dancing in shafts of light (mentioned in book II of *De rerum natura*), then they are explicitly not atoms, but merely *rei simulacrum et imago,* aids to help us imagine atoms (Lucretius [1992] II 112). Given Creech's frequent references to Descartes, it is also possible to understand these dots as an illustration of the Cartesian doctrine of the generation of particles in the sun. Again, however, these dots would be no atoms, since Descartes' corpuscles are always divisible. Finally, the word *CASUS* inscribed on the sun alludes to the atomists' denial of final and efficient causality in nature and to the generation of natural bodies out of the chance collisions of swerving atoms. This explanation, ridiculed already by Cicero, was perceived as Epicurus' weakest point even by seventeenth-century sympathisers.[19] Seen in this light, Creech's frontispiece begins to look almost like a criticism of the doctrines discussed in his book.

The Hesitation of Early Seventeenth-Century Atomists

This double handicap, that is to say, the lack of depictable corpuscularian models and the conflicting views presented in the inherited Platonist and atomist sources, had a stifling effect on the first neo-atomists of the early modern period. Authors such as Nicholas Hill, Daniel Sennert, Galileo Galilei, or Jean Chrisostome Magnen remained vague on the shape of the issue of atomic shapes, and they nowhere attempted to illustrate the corpuscles they promoted. David Gorlaeus explains his hesitation with these words:

> *Maybe the figure of the atom is quadratic. Thus no vacuum will be caused. But maybe there exist diverse figures. However this be, the shape of the atom is too small to be captured by our sense, and our intellect hardly can capture it. Hence we shall leave this question in suspense* (Gorlaeus [1620] 244).

While Gorlaeus' indecision is caused by conflicting criteria, Sébastien Basson's is also caused by a wish to reconcile the ancient corpuscularians for his fight against Aristotle:

> *We do not know whether* [the atoms] *are surfaces as in Plato, or corpuscles, as in Democritus. But let us say that these various surfaces or corpuscles, for their diverse shapes, are the matter of fire, air, water, and earth.* [...] *In this way, we have reconciled Democritus, Plato, and Empedocles.*[20]

But then, on rare occasions, Basson does employ graphic means (fig. 6).

But let us take note of the fact that these dots are not representations of what atoms look like, but only of how they are arranged in space. They thus address merely the last of the four Democritean predicates of atoms (i.e., shape, size, orientation in space, and reciprocal position). Basson – very much like the so-called medieval atomists[21] – was in fact convinced that our mind could form no mental image of indivisible entities: "Since nothing is in the intellect which was not first in the sense, how could the intellect manage to form such points, whose image the sense obtained nowhere?" (Basson [1621] 414). In fact, Basson's dot images are medieval in origin. They had always accompanied discussions of the mathematical arguments against atomism, which is also the context in which Basson uses them.[22] Libertus Fromondus' anti-atomist *Labyrinthus sive de compositione continui* of 1631 uses the same dots, but once again only to show why atomism leads to mathematically absurd consequences, such as the redefinition of diagonals in terms of broken lines (fig. 6b).

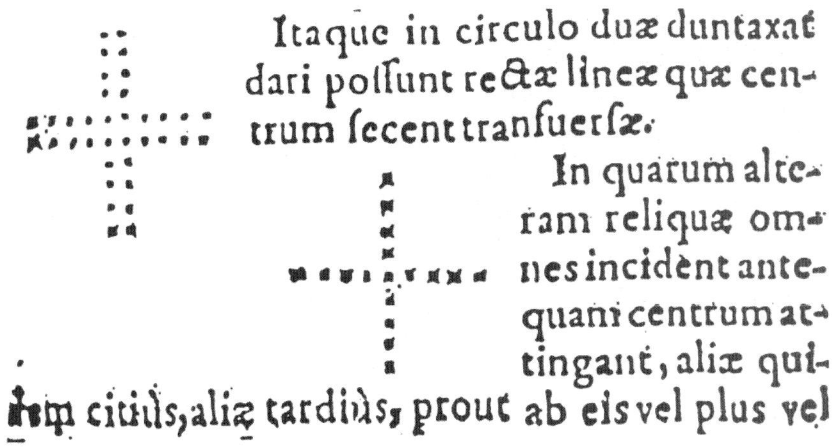

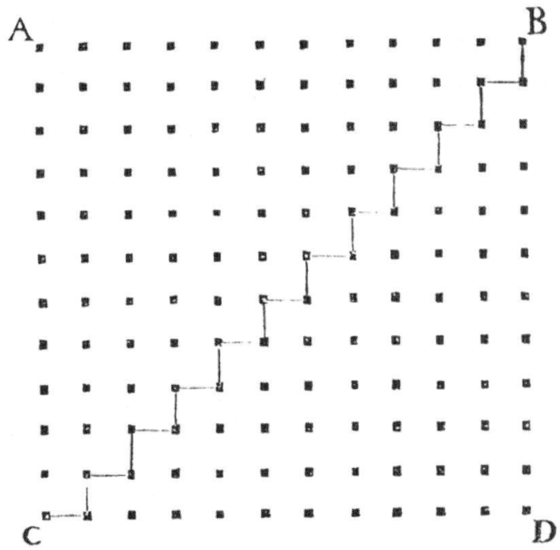

Figure 6
Discussing the mathematical arguments against atomism.

6a (*above*): From Basson: *Philosophiae naturalis* (Geneva1621).

6b (*left*): From Fromondus: *Labyrinthus sive de compositione continui* (Antwerpen 1631).

The Introduction of the Spherical Atom: Giordano Bruno and His Sources

When we look at the wood cuts personally prepared by Giordano Bruno for his *De triplici minimo et mensura* of 1591, we encounter a very different world. Here, we find ourselves finally in the presence of those agglomerated globules for which we have so far been looking in vain. And as if to show that these drawings really do refer to minimal particles of matter, we find the names of "Democritus" and "Leucippus" attached to them and the word "atom" occurring in the accompanying text.[23]

As far as I have been able to ascertain, our modern atomist iconography really does manifests itself for the first time in Giordano Bruno's work. This fact alone does not render this philosopher either the first or even a particularly important atomist. It has already been stressed repeatedly that we must divorce the *fortuna* of scientific images from the *fortuna* of the theories they happen to represent in a given (con-)text. In fact, the necessity of separating the history of a type of imagery from the theory that first engendered it could hardly be documented better than with reference to Bruno's own case. For as we shall now attempt to show, his images – though now for the first time applied to the realm of matter theory – are considerably older and come with a complex historical burden.

This story is most easily told if we focus on a particularly central image (fig. 7).[24] This *archetypus*, as Bruno calls his *Area Democriti*, has as its stated purpose to demonstrate how atomic units naturally assemble into further circular structures (Bruno [1591] 50). In a first step, when six spheres are added to a central sphere, one obtains once more a circular figure. The point is, as the chapter title announces, that "the minimum is again visible in the great and the largest."

It must strike us that Bruno is here mixing the dimensions in an illicit manner. While the "1 + 6" rule holds true for the two-dimensional case of seven circles of equal diameter, it is certainly false for the three-dimensional case of globules. In fact, twelve spheres touch a central sphere, the resulting figure being neither spherical nor of course circular.

Surprisingly enough, however, this is not the only place where Bruno uses the specific figure of the *Area Democriti* in combination with his dimensionally mixed geometry. We encounter in fact the same archetype twice in Bruno's astronomical defense of infinite worlds (fig. 8). The first occasion on which it is employed occurs in the *De l'infinito, universo et mondi* of 1584, where it is introduced with the following words:

> [...] *no more than six worlds can be contiguous to this one: because without penetration of bodies, no more than six spheres can be contiguous to one, just as no more than six equal circles can touch another circle unless their lines intersect* (Bruno [1956] 448).

Unshaken by such a blatant confusion of two-dimensional with three-dimensional reasoning, Bruno insists that his drawing demonstrates the beginning of a dynamically expanding series of further worlds: "We imag-

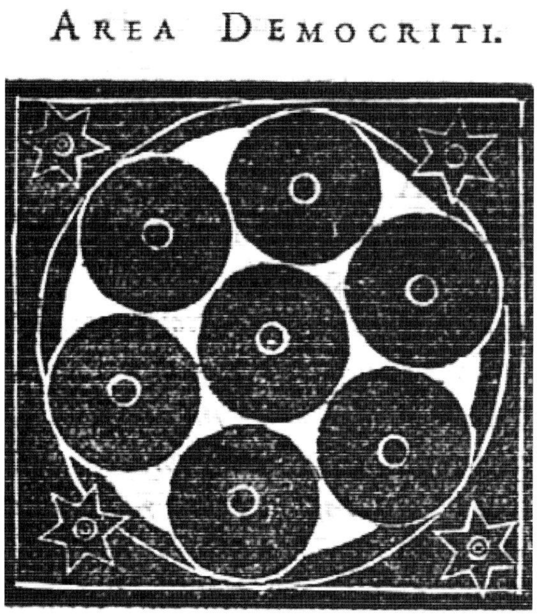

Figure 7
Giordano Bruno's "Area Democriti." From Bruno: *De triplici minimo et mensura* (Frankfurt 1591).

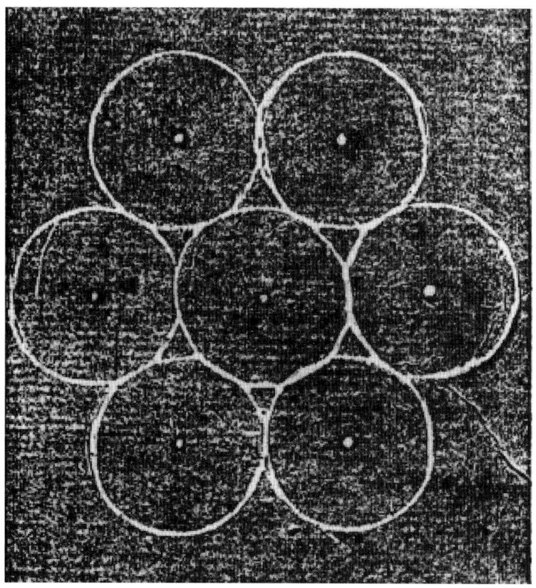

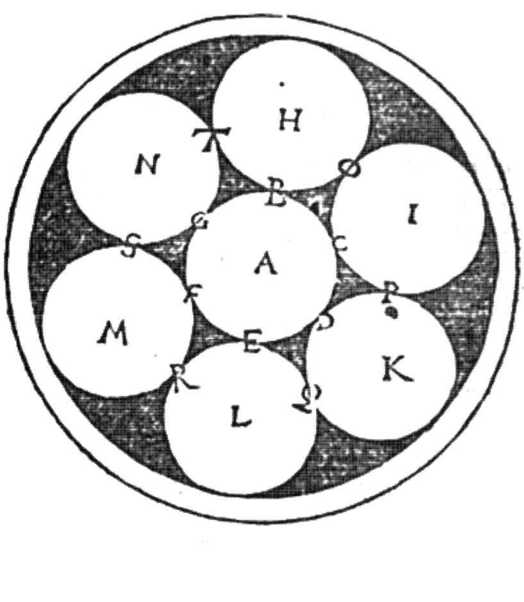

Figure 8
Giordano Bruno's celestial "gnomons."

8a (*right*): From *De l'infinito, universo et mondi* (London 1584).

8b (*below*): From *De innumerabilibus, immenso et infigurabili* (Frankfurt 1591).

ine many more of these, similarly spherical and likewise mobile." The seven circles of his archetype are thus merely an infinitely small visual sample of the innumerable worlds that we must imagine swimming in "an ethereal region which is, as we have said, immense, and in which moves, lives, and vegetates everything" (Bruno [1956] 461, 463).

The strange use to which Bruno puts his *Area Democriti* becomes understandable only once we understand its iconographical background, which is to be sought in the "gnomons" (or symbolical constellations) of a numerological tradition which, though very remote from our century's mind, was still very present in Bruno's thought.[25]

We must begin our explanation with Boethius' *De Institutione Arithmetica*, incidentally one of the Middle Age's preferred quadrivial manuals, which declares, in a professedly "Pythagorean" manner, that all of nature is constituted "by reason of number" (Boethius [1966] I 2). The central idea is that the principle of unity (or "1"), though itself not a number, "has the potential of a point, the beginning of interval and longitude" (ibid., I 4). The reason for this "potential" resides in the fact that "1," when multiplied by itself, remains "1," because 1 x 1 x 1 = 1. This mathematical truth is taken by Boethius and his followers to imply that "1" contains in itself all possible dimensions, as "unity in both power [i.e. 1^2] and force [i.e. 1^3] is a circle and a sphere."[26] In the Boethian tradition, then, "1," which remains identical in all dimensions, can serve as the foundation stone not only of arithmetic (by being the source of all subsequent numbers), but also of geometry (by being the source of all spatial unfolding).

This understanding of number stands behind Boethius' method of defining the character of the basic numbers in terms of specific geometrical patterns or gnomons.[27] For our understanding of Bruno's *Area Democriti*, we must know that Boethius claimed that from an ethical point of view, the number "6," being the sum of its dividends (6 = 1 + 2 + 3), is a "perfect number" not only in the modern and merely metaphorical sense, but also in a literal and ethical sense, representing virtue. The two-dimensional expression of its perfection was the triangle (Boethius [1966] I 20; II 7; I 15).

I
I I
I I I

In the Middle Ages and the Renaissance, the Pythagorean number symbolism was reinterpreted in a Christian light. In the process, the "perfect" number "6" received two new meanings. In his *De harmonica mundi* of 1525, for example, Francesco Giorgio Veneto linked "6" to the hexaemeronic account of creation. But he also asked his reader to contemplate the triangle made up of 1 + 2 + 3 units as a potent symbol for the mystery of the Trinity's oneness and the binary principles of matter and form through which creation had taken place (Giorgio Veneto [1525] ch. 11).

Importantly, however, there exists a different geometrical interpretation of the perfection of "6," which was proposed in Thomas Bradwardine's fourteenth-century *Geometria speculativa*. There, the proof that "six equal circles touch an equal circle on the outside" is followed by a demonstration that the gnomon of fig. 9 represents the number "6" in three different ways:

> *And in all these three conclusions, "six" attests the perfection of the circle, for in the first we have six of points, which are the extremities of the three lines, in the second six of lines, and in the third six of circles.*[28]

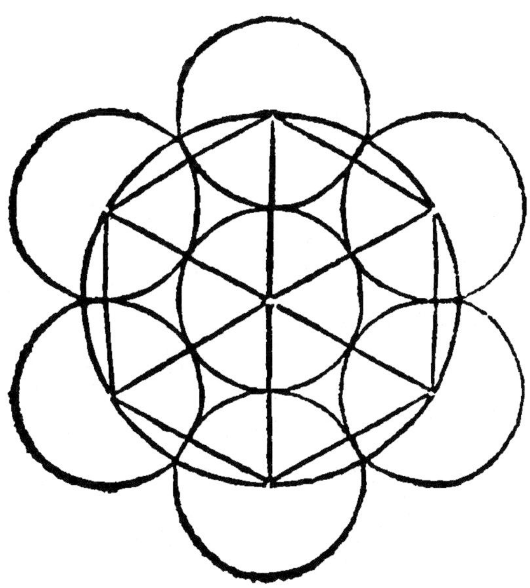

Figure 9
Bradwardine's symbol of perfection. From Bradwardine: *Geometria speculativa* (Paris 1495) [f. 8r].

This means that the medieval prototype of Bruno's *Area Democriti* served as a visual symbol of the perfection of "6." It soon became the preferred Christian symbol of the number "6," because it combined numerical perfection ("6") with geometrical perfection (the circle).

Two more authors must be mentioned before we can return to Bruno's *Area Democriti*. The first is Nicolaus of Cusa, whose *De ludo globi* (ca. 1462) praises the globular shape as the deepest symbol of the divinity. Very much like Boethius, whom he cites elsewhere in his text, Cusanus conceives of the unit-point as capable of "uncoiling" or "evolving itself" into atom, line and body, and as being present in all things.[29] Cusanus' well-known doctrine of the *coincidentia oppositorum* implies that the largest and the smallest, that is to say, God as the all-embracing measure of all things and the invisibly small unit-point, coincide formally. The Cardinal, always keen to exemplify his theology geometrically, proposed the sphere as the adequate symbol to symbolise both the divine Everything and the atomic almost-nothing: for when it shrinks, the sphere contracts into a point, and when it reaches its maximum extension, it becomes a straight surface. According to this view, the minimum and the maximum mirror each other as the extreme measures to all things and the image of the Creator (cf. Cusanus [1967] 579f. and 590ff.).

The second influence to be added is that of Ramon Lull, whose "art" for the demonstration of truths relied, in its most basic form, on the alphabetic notations of God's attributes or "dignities" on rotating wheels (*rotae*), with the letter "B" denoting, for example, the Goodness of God (*bonitas Dei*). The wheels could be turned into various constellations with other attributes, predicates, or signs (cf. fig. 10).[30]

The way in which all these influences could interact in the late sixteenth century is seen in a numerological treatise called *Mysticae numerorum significationis liber* by Bruno's contemporary Pietro Bongo.[31] In the chapter that deals with the number "6," we encounter three drawings whose relation to the *Area Democriti* is obvious (fig. 11).

Now, the fact that the number "6" is associated with Bradwardine's figure of the seven circles implies that the central circle is not being counted. In fact, Bongo views it as the divine generative principle and as the monad from which all others emanate. The meaning of the letters in fig. 11b, which are spelled out in fig. 11c, demonstrate this dynamic aspect: they refer, in a Lullian manner, to aspects of the created world in terms of divine projections. Bongo's archetype of "6" is thus an emblem of creation which does not intend to represent any actual spatial relations within the physical world.

Let us now return once more to Bruno, who, in his *De monade, numero, et figura* of 1591, offers his own numerology, also with strongly Boethian, Lullian, and Cusanean overtones. Thus he defines the circle as a "figure which belongs to the monad" and which replicates itself into numerologically significant patterns or gnomons (Bruno [1884] 335). The number "6," in turn, is praised in the usual Pythagorean manner. Bruno first describes it in terms of a set-up that will strongly remind us of Bongo and Lull; he writes that "matter, species, time, place, efficient cause and necessary cause are placed, in a mnemonically useful *imago*, around a common centre" (Bruno [1884] 419f.). As in Bongo, the centre is said to radi-

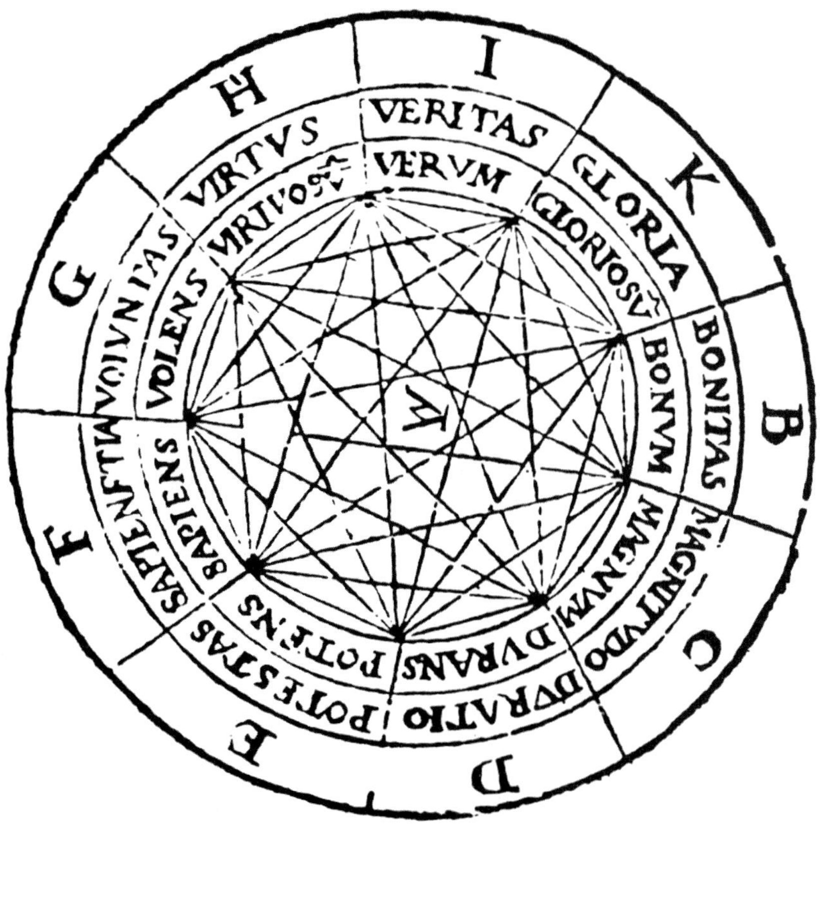

Figure 10
Divine Dignities on Lullian Wheels. From Lullus: *Opera* (Strasbourg 1617).

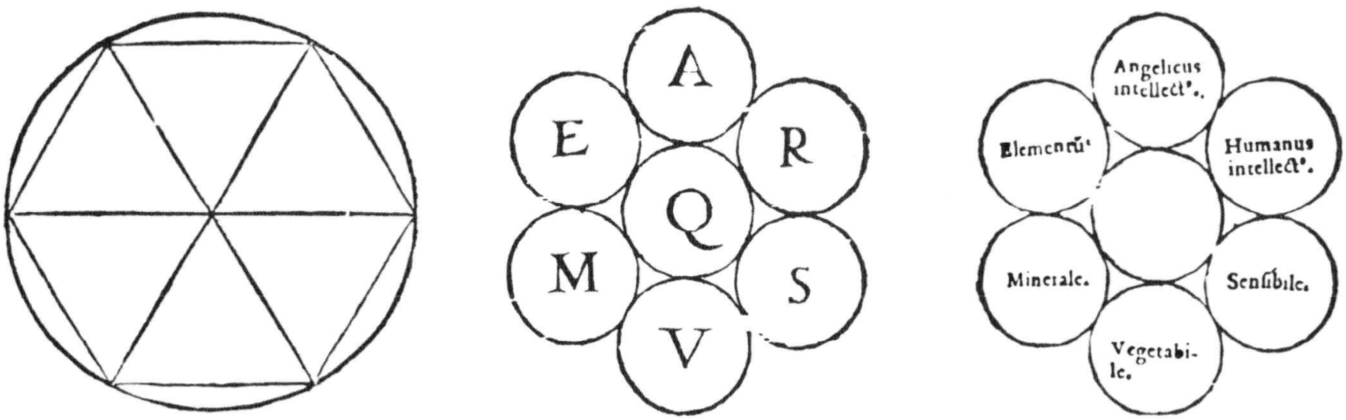

ate out to the periphery, forming a figure that "grows" (*crescit*) into various stellar and hexagonic shapes which, however, can time and again be "inscribed" into and "circumscribed" "by a circle" (*ibid.*) (fig. 12a).

A further image, called *Schala Comprensionis* (fig. 12b), which is related to both the gnomon of "6" and to the *archetypus* of the *Area Democriti*, brings us back to our point of departure. For the various little globular clusters in the four corners of this figure show how, for Bruno, the expansively self-replicating star is related to the basic composition of the "1 + 6" monads.[32] As Bruno writes in his explanation of the *Schala Comprensionis*, the circle and the triangle are the two fundamental shapes which in a process of dialectical mutual conditioning form the entire cosmos "like matter and form, potency and act, [...] minimum and maximum": three circles form a triangle, and six triangles form a hexagon, which can be inscribed again in a circle, and so forth (Bruno [1591] 189f.).

But whereas for the entire numerological tradition up to Bruno, such considerations had remained at the level of theological and cosmic symbolism, they assume in Bruno, for the first time, a physical meaning as well. As we have already seen, Bruno, who otherwise retains all of the traditional meanings, applies the inherited gnomonic imagery to spatial conditions holding in both the maximal or celestial, and the minimal or atomic, domains. At the same time, these circles are now also endowed with a mimetic significance.

Figure 11
Pietro Bongo's numerological exegesis of the number "6".
From Bongo: *Mysticae numerorum significationis liber* (Bergamo 1585).

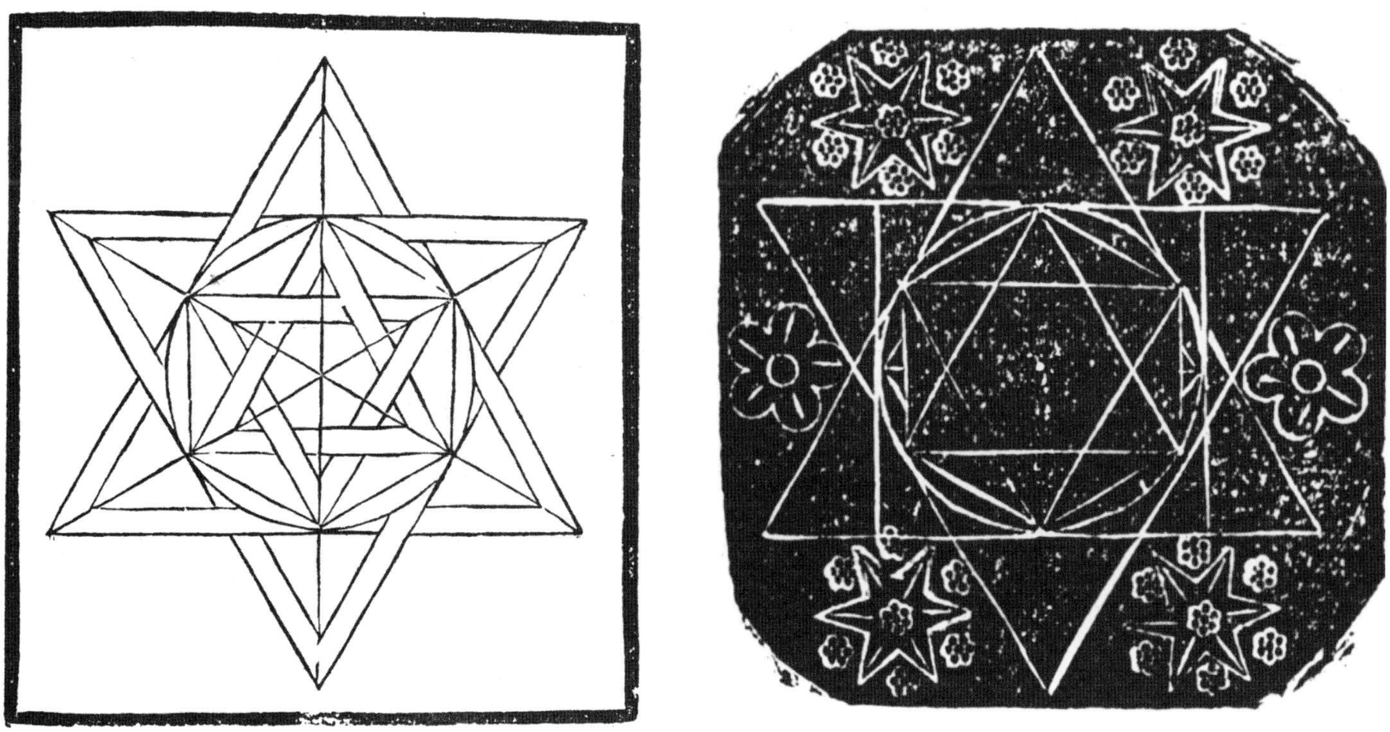

Figure 12
Giordano Bruno's *figuratio* of "6," the *Schala Comprensionis*, and their relation to the *Area Democriti*.

12a (*left*): From *De monade, numero, et figura* (Frankfurt 1591).

12b (*right*): From *De triplici minimo et mensura* (Frankfurt 1591).

The continuing gnomonic and numerological charge of these images explains why Bruno so obstinately ignored their inapplicability to the three-dimensional case. While Bradwardine, from whose writings the archetype of the *Area Democriti* ultimately derives, had even prepared a wax model so as to demonstrate to himself that "one sphere is contingent to twelve equal spheres placed around it" (Boethius in Molland [1989] prop. 4.42), Bruno's desire to have physics, mathematics, theology and epistemology spring from the same set of images led our author to suppress the distinctions between the various dimensions.

But we must not complain: the birth of our atomist imagery is due just to this – to us inacceptable – conviction that the Pythagorean gnomons demonstrate the patterns of divine world-making at all levels of magnitude. That the same numerological gnomon could equally apply to infinite worlds and to atomic aggregates was due to the belief that the same divine patterns were at work at all levels, in a manner that transformed Cusanus' *coincidentia oppositorum* into a veritably pictorial *coincidentia figurarum*.[33] "We attribute the same figure to the maximum and to the minimum," Bruno states in *De triplici minimo et mensura* (Bruno [1889b] 181), and he introduces the *Area Democriti* with the title: "It is evident that the minimum appears again in the great and the largest" (Bruno [1889b] 182). Let us here not forget that Democritus had not just been an atomist, but also the champion of infinite worlds. The title of our *Area Democriti* must therefore be understood as an intended *double entendre*.[34]

Now, although Bruno rejected Democritus' vacuum and replaced it with ether, he did accept indivisible and incorruptible atoms.[35] This becomes clear, for example, in the central question of the nature of mixtures. We recall that Aristotle had chastised the atomists for their inability to account for "true mixtures," calling the structures they proposed mere "heaps" (*acervi*) (Aristotle [1982] 328a 13ff.). "Heaping" (*coacervare*), however, is precisely the word Bruno uses to describe the formation of bodies out of atoms (Bruno [1889b] 222). At least in this sense, the *Area Democriti* and related archetypes are, *inter alia*, veritable atomist images. This fact becomes even more significant when we take into consideration that the words *acervus* and *coacervare* had long before been used to describe gnomons. In his ornate edition of Boethius' *De institutione* of 1521, Girardus Ruffus had, for example, emphatically distinguished between metaphysical, mathematical, and natural-philosophical entities (*entia*), reserving the term "number of aggregates and of heaps" (*numerus aggregatorum coacervatorumque*) strictly to gnomons, that is, to the discontinuous figures produced by mathematical entities (Ruffus [1521] f. 85r). In doing so, he accepted Proclus' distinction between physical *continua* and numerical *acervi* (Boethius in Molland [1989] prop. 1,1). Bruno, however, by applying numerical *acervi* to describe physical conditions, explicitly rejected this ontic distinction. In fact, it is a key aspect of his proposed reform of physics and mathematics that for him, the ensouled and dynamically expansive point-units were always also considered physical and corporeal enough to develop beyond their abstract arithmetical status into the concrete realm of physical existence. In other words, images such as the *Area Democriti* represent, from an iconographical point of view, the identification of the gnomonic *acervi* of the mathematicians with the classical atomists' physical *acervi*, which Aristotle had rejected.

The Physicalisation of the "Gnomons"

Bruno uses his globular pictures in support of views so unique and strange as to provoke goose pimples on the skin of modern readers. They represent the bastard offspring of theology, mathematics, cosmology, metaphysics and psychology, and they are almost nowhere directly employed to solve physical questions.[36] And where they are, they appear inadequately applied to the three-dimensional case, notably in the discussion of globular stacking of the *Articuli adversus mathematicos* (cf. Bruno [1588]). This is why it is obvious that our story of the development of atomist iconography cannot end with Bruno. We will necessarily have to show, first, that (and why) Bruno's images managed to influence later authors; second, we must explain how they were being redefined so as to become useful scientific tools.

Since an essay of this length cannot possibly cover the increasingly rich iconographic domains of seventeenth-century physics, chemistry and iatromechanical medicine, my final pages will be limited to providing a sketchy proof of Bruno's influence; and to equally sketchy remarks concerning the factors contributing to the eventual victory of the globular unit of matter.

Using Bruno's Globules: The Cases of Kepler and Jungius

Direct and indirect traces of Bruno's atomist iconography in subsequent authors do not exist in abundance, but in sufficient number to allow us to document a continuous chain of influence.

It would seem that the earliest printed appearance of Bruno's globules in the context of a physical explanation is to be found in the *Strena seu De nive sexangula*, Kepler's 1611 study on the snowflake (Kepler [1611]) (fig. 13).

Kepler's attempt to understand the hexagonal structure of snowflakes in terms of the packing of globules has been hailed as a revolutionary anticipation of molecular models, and its effect on Peiresc, Gassendi, Mersenne, Descartes, the Bartholinus brothers, Rossetti, Grew, Hooke, and on the general history of crystallography is well documented.[37] With the exception of Heusler-Keßler, no one has, however, recognised Bruno as Kepler's (unacknowledged) source of inspiration. But not only is fig. 13a quite close to two drawings in Bruno's *Articuli adversus mathematicos*,[38] in which two diverse ways of packing globules are addressed, but we also know that Kepler had been reconsidering Bruno's astronomical views precisely in the months in which the *Strena* was composed.[39]

In the context of our present interest in the physicalisation of Bruno's archetypes, it is worth mentioning that there is a further, somewhat amusing aspect, to this presumed connection between Bruno's theological mathematics and the snowflakes that fell on Kepler's sleeve. For in one of his outbursts against Euclid, Bruno had claimed that a circle could not be divided into more than three diameters, because no central point could touch more than six globules and hence six radii. He added as an illustration fig. 14, in which he showed the six admissible radii plus a seventh, inadmissible radius (Bruno [1889b] 245).

This image, which has been chastised for its childish mathematical realism and which is of course a mere variation on the theme of our *Area Democriti*, happens to resemble a snowflake.[40] Kepler's overall familiarity with Bruno's geometrical beliefs would in fact have easily allowed him to perceive the snowflake's structure as a paradoxical piece of evidence in favour of the Italian's impossible mathematics. Though such an association cannot be demonstrated, a number of Kepler's musings strongly remind us of Bruno, notably when he wonders whether the hexagon represents an "archetype of beauty" because of its resemblance to the circle, and whether this shape is the result of a "formative power" which radiated out from the centre (Kepler [1611] 6, 17, 22f.). Kepler's own views implicitly discredit both Bruno's geometry and the causal explanations that underlie it.[41] Though Kepler left the question open – "The matter is not yet clear to me" (ibid., 23) –, he did end up convincing himself that the snowflake was made up of condensed vapour globules, a view that was to have a considerable effect on the development of mechanistic models of matter in general, and of crystallic forms in particular. This, then, is one of the ways in which corpuscular *acervi* entered the world of physical explanation.

While Bruno's imagery made but a fleeting, though highly influential, appearance in Kepler, it was to be a central source of

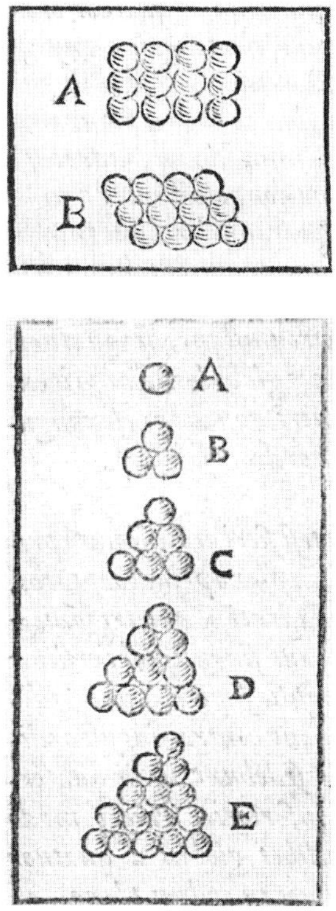

Figure 13
Kepler's reflections on the geometry of the snowflake. From Kepler: *Strena seu De nive sexangula* (Frankfurt 1611).

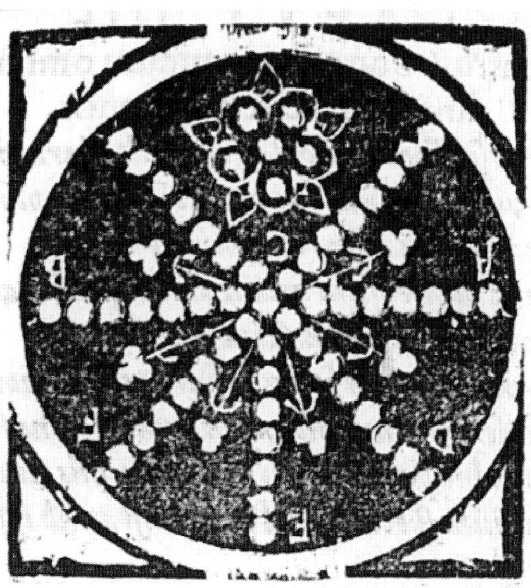

Figure 14
Giordano Bruno's argument that a circle cannot be divided more than three times. From Bruno: *De triplici minimo et mensura* (Frankfurt 1591).

inspiration to Joachim Jungius, who, over a twenty-year period from 1630 to 1650, made extensive annotations and calculations on Bruno's geometry.[42] The large bundle called *Circuli locum replentes* is correctly characterised by Meinel as "notes on the repletion of space by spheres with respect to a corpuscularian structure of matter, mostly in response to Giordano Bruno."[43] Most of its 259 *schedae* contain in fact reflections on the globular models of Bruno's *De minimo*, which fascinated Jungius because of the attempt they seemed to represent of developing a *mathesis universalis* and to apply it to the physical world.[44] In his notes, Jungius tries to capture Bruno's doctrines in a net of rules and theorems, to check their validity, and to apply them to his own "syndiacritical method," which relied on the assumption that the *syncrisis* and *diacrisis* (gathering and separation) of indestructible and unchanging first bodies was the real cause of material change.[45]

Among Jungius' earliest Brunian annotations of April, 1630, we encounter a "figure which Bruno calls *Area Democriti*" (fig. 15) (Jungius, *Circuli*, f. 9 and 11).

Our archetype appears under the very Brunian title of "The circle capturing circles"; it is followed by a long set of calculations, of which the last dates from June, 1647 (ibid., *Circulus circulos capiens*), and which concern the ratio between circumference and surface of the original circle and those of the resulting circle. Jungius here attempts to calculate the surface of the empty segments and to discover the general rules defining the growth of Bruno's circle by means of the reiterated addition of six units (1 + 6 + 6 + 6 + etc.). He discovers, early on, the grave errors resulting from Bruno's analogous treatment of point, circle and sphere, correctly identifies its Boethian sources, and offers the necessary corrections.[46] He notes, in particular, that more than six spheres touch a central sphere; that a circle made up of globules does not grow in a hexagonal manner; and that the empty spaces between spheres stacked three-dimensionally are by no means pyramidal.[47]

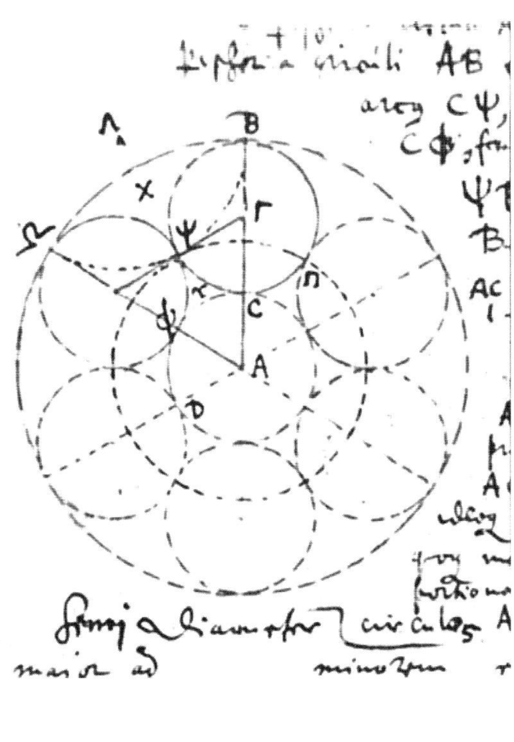

Figure 15
Joachim Jungius' use of Bruno's "Area Democriti." From Jungius: *Circuli locum replentes* (Codex Hanseaticus IV p6, NL J. Jungius Pe 67).

The necessity of such corrections did not, however, affect Jungius' overall interest in Bruno's idea. What interested him was the notion that an identical number of identical material corpuscles could be rearranged in such a manner as to take up a variety of volumes which, in turn, might be said to possess a variety of physical properties.[48] Jungius, who knew Kepler's *Strena*, employed such spatial *metaschematismata*, for example, in his description of the transformation of vapours into snowflakes.[49] It also appeared to him that a range of physical qualities such as "transparency, opacity, density, rarity, hardness, softness, etc." were due to the configuration of the interstitial pores.[50] A serious application of figures such as the *Area Democriti* to an atomist matter theory required therefore a mathematical exploration of the interrelation between *schemata*, their geometrical qualities, and their respective physical properties (cf. Jungius, *Circuli*, f. 164ff.).

Jungius' use of Bruno's imagery remained, however, devoid of numerological, mnemonic, or religious overtones. His own globules never denoted any cosmic unity, nor did they vaccillate in their dimensionality. They now referred to the solid, three-dimensional building blocks of natural substances only (cf. Jungius [1642a] §§ 58 and 76).

A Doubly "Ideal Shape": Two Reasons for the Victory of the Globular Atom

Kepler's influential modelling of the snowflake and Jungius' vast bundle of unpublished notes do not only document the debt of later authors to Bruno's globular patterns, but also indicate in which way these had to be reinterpreted before they could become the representatives of physical bodies.

But first, note that neither Kepler nor Jungius subscribed to the view that *all* atoms were of globular shape (cf. Jungius [1642a] §77). The ancient atomists had, in a very successful simile, compared the atomic shapes to an alphabet of letters (Meinel [1988] esp. 2–5). Bruno's atomism implied, by contrast, that the alphabet consisted of one single letter only. Such a proposal could certainly not satisfy mechanistic philosophers, who needed an array of shapes to explain the variety of natural forms and of sense experience. To let everything depend on one shape meant to hide once again all specific properties of an atom *inside* its surface, rather than have them expressed by its primary, geometric quality. It is therefore not surprising to find seventeenth-century mechanistic corpuscularians as different from each other as Pierre Gassendi, René Descartes or Giovanni Alfonso Borelli develop corpuscularian alphabets consisting of a variety of shapes.

If the spherical corpuscle nonetheless emerged victorious from that century, this is due to the double sense in which it happened to possess an ideal shape. The first sense in which it was "ideal" is due to the fact that the triumphs of the new mathematical sciences all involved material globes.

The most ancient science dealing with spheres is of course astronomy (fig. 16). But mathematical optics, too, relied on globular models. Theoderic of Freiberg's famous explanation of the rainbow (ca. 1310) postulates a multiple refraction of light rays in globular rain drops (fig. 17), a view that was to become once more popular among seventeenth-century authors. It is, in this context, no coincidence that atomism found some of its first disciples among early modern students of optics. Thus Thomas Harriot, who otherwise remained rather ambivalent with respect to corpuscularian matter theories, shows himself at his most atomistic in his correspondence with Kepler on optical questions.[51] The same holds true of the corpuscularian drawings found in Isaac Beeckman's *Journal*, which repeatedly relate to optical questions (e.g., Beeckman [1939-1953] I 211 (Sept. 1618)). René Descartes' optics, finally, is essentially a science of globules, as his explanation of the luminosity of a comet's tail demonstrates (fig. 18).

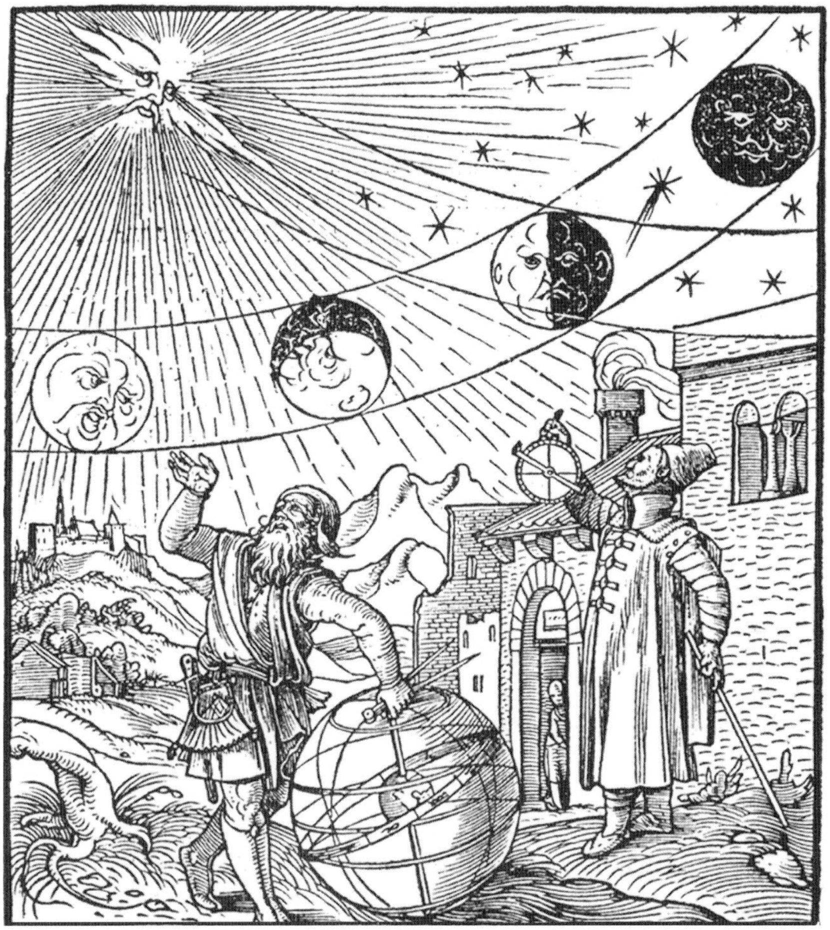

Figure 16
Astronomy, the oldest "globular science." From Münster: *Canones super novum instrumentum luminarium* (Basel 1534).

But not just the bodies of astronomy and optics, but also those of mechanics were globular. The objects of ballistics, Galileo's accelerating balls and his pendula, as well as the bodies demonstrating the law of shock – in short, all those bodies whose behaviour was found to follow mathematically calculable laws – happened to be spherical (fig. 19).

Given the general aspiration of the mechanistic philosophy to explain the behaviour of macroscopic bodies in terms of the behaviour of microscopic corpuscles, it will thus not surprise us if the globular shape obtained priority among all other forms.

But there is also a second, and historically posterior, sense in which this shape was found to be "ideal." As is well known, the attempt to reduce all natural phenomena to the interaction of colliding corpuscles had encouraged natural philosophers to propose a rich variety of often highly implausible and *ad hoc* shapes, none of which managed to generate a consensus. The success of Newton's universal force of attraction, however, rendered the need for this type of geometrical reductionism (as we might call it) far less

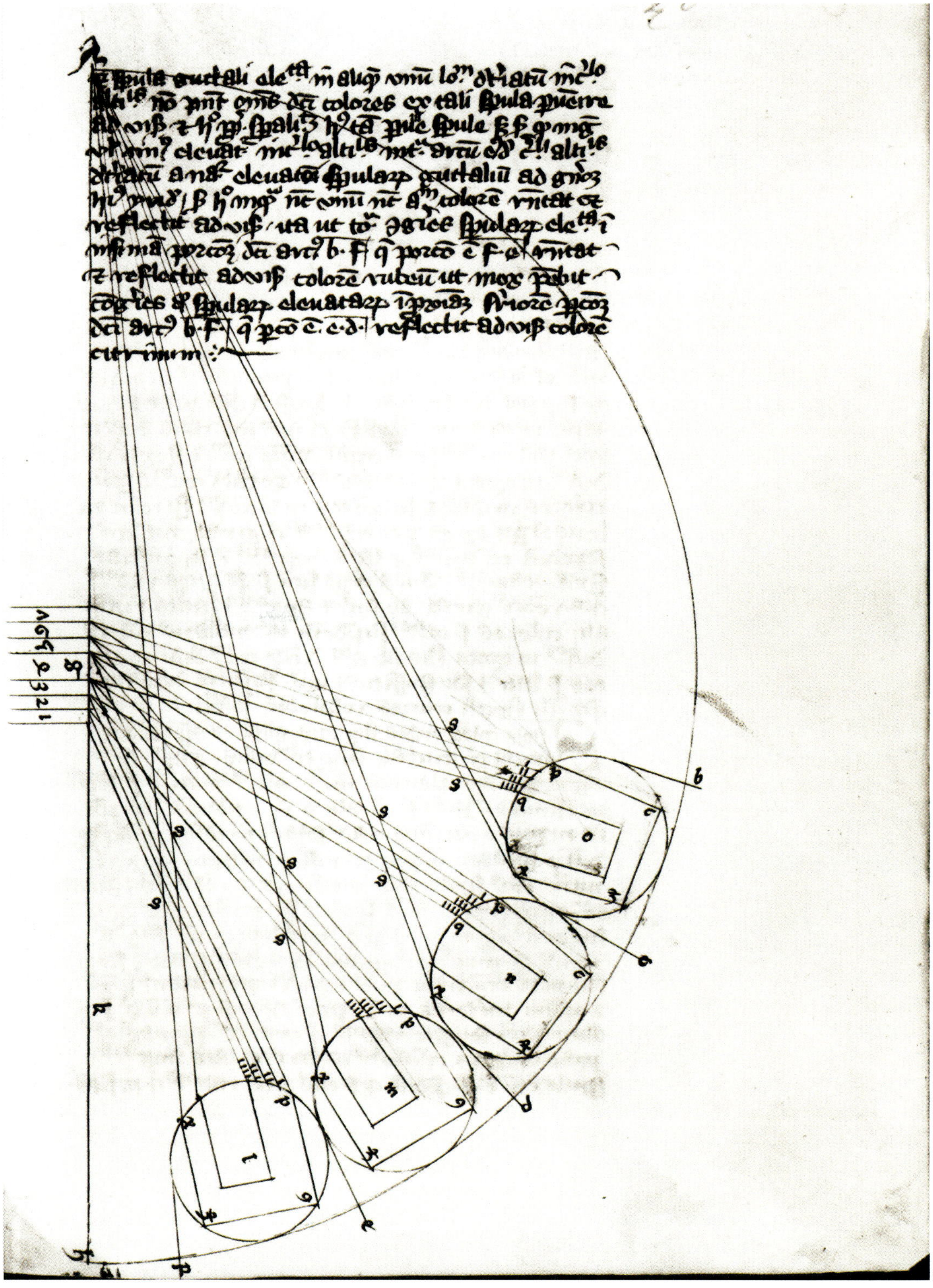

Figure 17
Drop model to explain the rainbow. From Theodericus de Friburgo: *De iride et radialibus impressionibus* (ca. 1310) (Universitätsbibliothek Basel, F IV 30, 2r).

pressing. This explains the growing nonchalance of eighteenth-century corpuscularians. A chemist such as Stahl, though continuing to believe "that all particles have certainly and in truth some figure, but not any shape such that we might know well and assuredly," began to search for alternative ways of classifying substances (Stahl [1744] 52). The Oxford physicist John Keill, in turn, opened the *Preface* of his lectures of 1700 with an attack on what he felt to be a false need for visuality, saying of the corpuscularian models of the mechanical philosophers that "there is scarce anything mechanical to be found besides the name. Instead whereof, the Philosophers substitute the Figures, Ways, Pores and Interstices of Corpuscles, which they never saw" (Keill [1726] iii).

The natural consequence of this growing unwillingness to hunt after the shapes of corpuscles was that the globule, the privileged figure of physical demonstration, now became the token representative of *any* unit of matter. In other words, while for Bruno, the globule had represented the one and only possible figure, and for the mechanistic philosophers the best-understood shape, it became for the eighteenth-century physicist the symbol for any invisibly small particle of matter, of whatever kind. As has been mentioned in the introduction, however, this agnosticism was but temporary and was later to give way again to both realistic interpretations and to outright rejection.

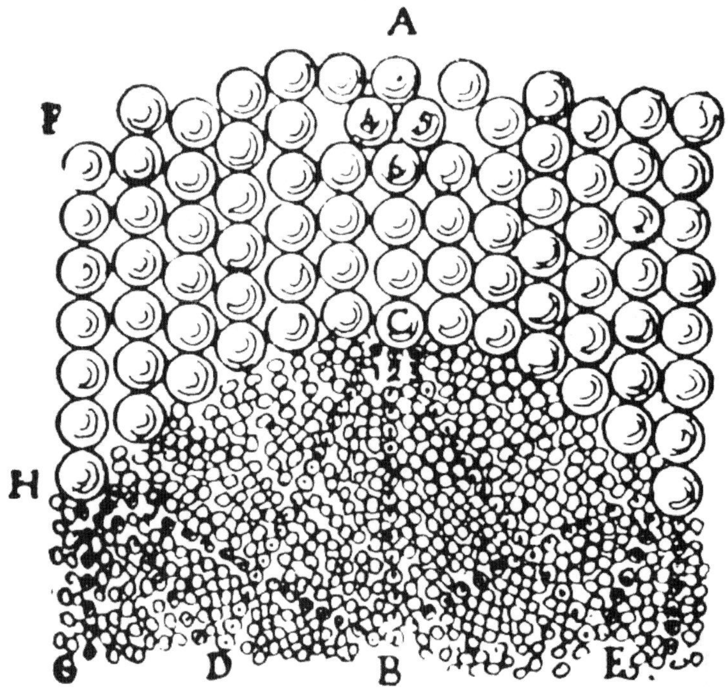

Figure 18
Descartes' explanation of the luminous ray of a comet's tail. From Descartes: *Principia philosophiae* (Amsterdam 1644).

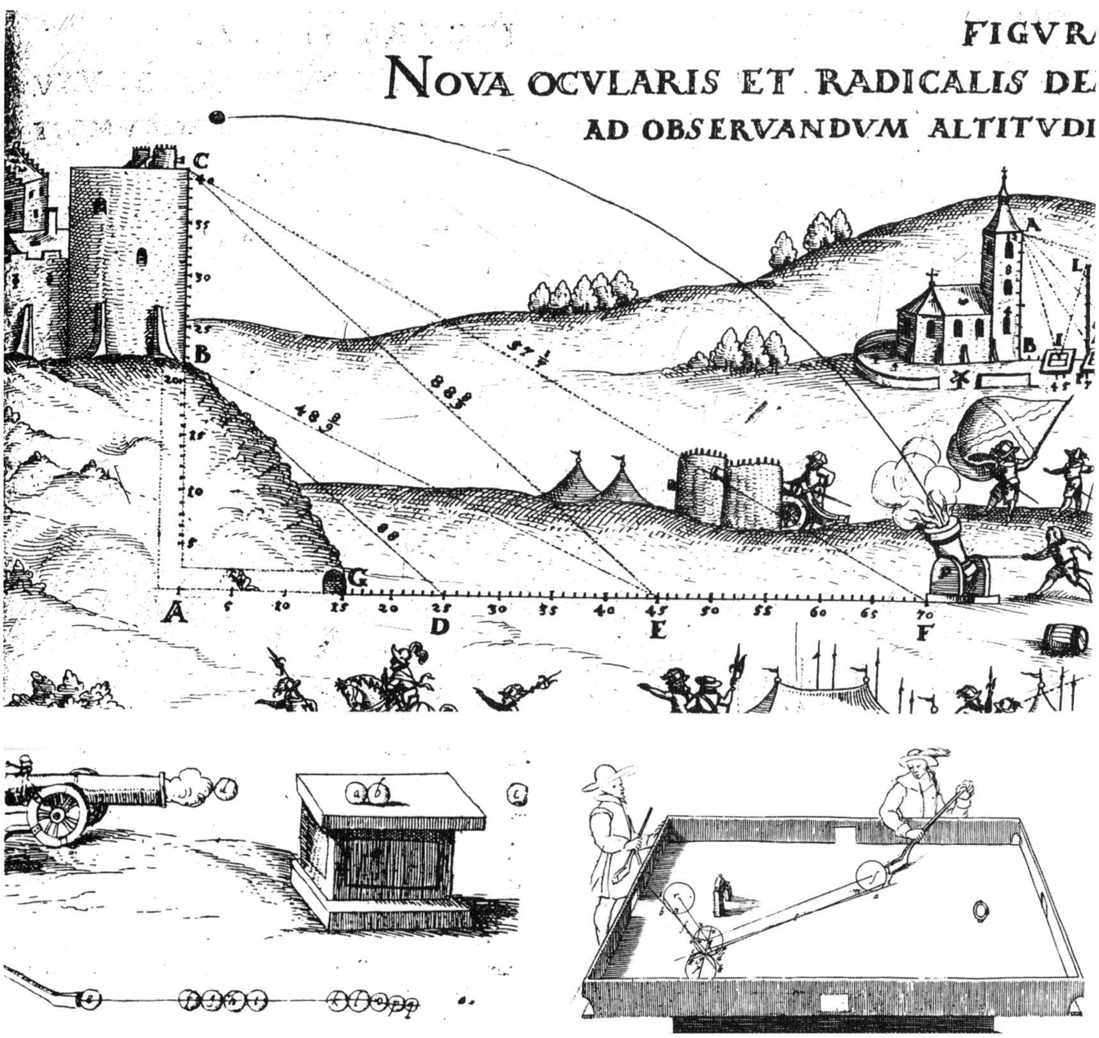

Figure 19
The globular objects of early modern mechanics.

19a (*above*): From Hulsius: *Tractatus primus instrumentorum mechanicorum* (Frankfurt 1609). [detail]

19b (*left*): From Marcus Marci: *De proportione motus* (Prague 1639). [detail]

19c (*right*): From Marcus Marci: *De proportione motus* (Prague 1639).

Conclusion: Old Images for Ever New Theories

The history of the globular particle in the history of physics demonstrates just how independent the *fortuna* of images is from that of scientific theories. We have observed, in particular, with what peculiar historical baggage Bruno's *Area Democriti* entered the world of matter theory at the end of the sixteenth century, and how it became gradually physicalised in the works of later natural philosophers. Unaffected by such reinterpretations, the same "archetype" migrated from author to author, appearing at regular intervals in seventeenth and eighteenth-century works. Thus we encounter it, for example, in the *Physica* of the Jesuit Honoré Fabri (1670), where the spherical shape represents the element of earth and at the same time the heaviest of all particles.[52] Fabri's *Area Democriti* (fig. 20) serves in this case to demonstrate how we have to image the densest possible mixture of elements (i.e., the closest packing of earth particles with minimal interstitial spaces for other elements) (Fabri [1670] vol. 3, tract. V, lib. II, prop. 173, p. 141).

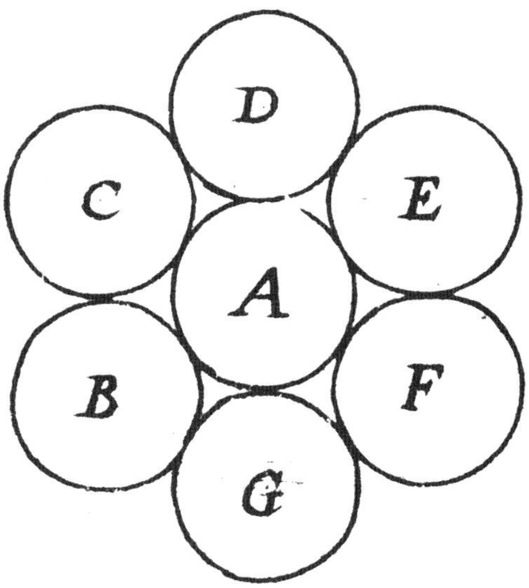

Figure 20
Earth atoms gathering to form the densest possible body. From Fabri: *Physica* (Lyon 1670).

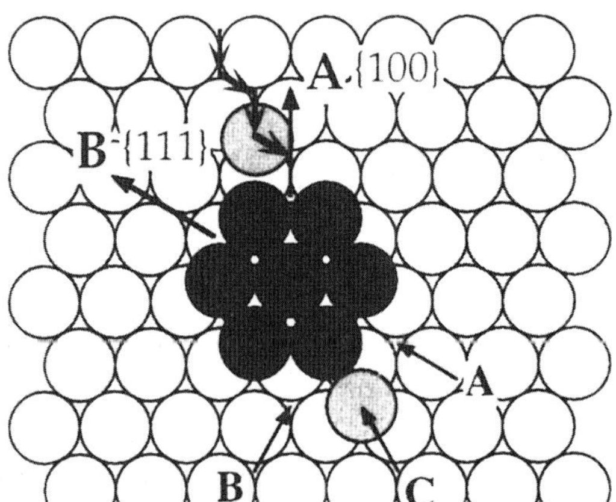

Figure 21
Atomic surface processes, 1998. From J. V. Barth and H. Brune: "Atomare Prozesse an Oberflächen." *Physik in unserer Zeit* 29 (1998): 256.

Given the simplicity of this figure, one might be led to assume that it is a mere coincidence that Fabri hits upon the same figure to illustrate his physics. But Fabri himself implicitly admits that he is borrowing a figure he has seen elsewhere, and he also shows awareness of its inadequacy for the present purpose: "Just how many spheres can be placed around an equal sphere is something known to geometers, but is of no concern here" (Fabri [1670] vol. 3, tract. V, lib. II, prop. 172, p. 141). For the idea Fabri wanted to convey, the old two-dimensional graph happened to be sufficient.

The appearance of a descendant of our *Area Democriti* in a contemporary journal of physics is of course all the more astonishing once we learn that the seven black disks represent "a heptamer of 7 silver atoms" which are "growing in the direction preferred by snowflakes" on top of a "hexagonally close-packed surface" (Barth et al. [1998] 255f.). But such unexpected transhistorical reminiscences aside, there is of course almost nothing that today's experimental physicist of low-temperature surfaces shares with the natural philosophy of either a Bruno or a Kepler.

What thus remains mysterious, at the end of this article, is the force with which the globular conception of atoms, developed four hundred years ago and in common use since the late seventeenth century, has managed to survive the profound conceptual transformations that have affected physics since then.

In a recent book, entitled *Image et Science*, we find various ways of visually representing the structure of morphine. The globular representation is there described as "an image more legible" than other types (Bajard et al. [1985] 105). But if it is true that the visual is to be measured in terms of its legibility, one wonders even more strongly what renders this ridiculously underdeveloped atomic alphabet with its one single letter so irreplaceable.

Centre for Medieval and Renaissance Natural Philosophy, University of Nijmegen[53]

Notes

[1] Cf. in particular Kuhn [1962] ch. 10: "Revolutions as Changes of World View."

[2] Emerton [1984], whose work has left strong traces in the present article, is one of the few scholars to have drawn attention the visual aspect of the corpuscularian revolution of natural philosophy.

[3] On the development of Bohr's views on the representability of physical theory, cf. Chevalley [1973]; on the relation between theory and imagery in twentieth-century physics, cf. Falkenburg [1995] and Galison [1997].

[4] On the "manufacturing" of microscopical images, cf. Hacking [1983] ch. 11, and Hacking [1985]. According to Hacking (personal communication), the paradoxical side of these new types of microscopic "pictures" of atoms has not yet received sufficient scholarly attention.

[5] For a Kantian (but historically accurate) understanding of atomism, cf. Lasswitz [1890].

[6] Aristotle [1982] 328a 13ff. On Lynceus as a protomicroscopist, cf. Lüthy [1996].

[7] On the different visual implications of an atomist and an Aristotelian explanation of rarefaction, cf. Des Chene [1996] 107.

[8] Murdoch [1984] x, who, when speaking of Aristotle's non-visual style of philosophizing, adds that "the illustrations are notoriously few in the manuscript copies of Aristotle and Galen and of their medieval translations and seemingly endless commentaries."

[9] Maier [1943] 12 emphasises "daß tatsächlich keine dieser [elementaren] Substanzen je in reiner unvermischter Form angetroffen wird, daß also alle Aussagen über sie, und alle Regeln, die für sie aufgestellt werden, nur für den lediglich in abstracto zu denkenden Grenzfall der elementa pura gelten."

[10] Plato speaks of his hypotheses as "an innocent pleasure" undertaken for the "sake of recreation," and he adds that their truth cannot be put "to a practical test," because only gods know how "to resolve the many into one," Plato [1959] 59C–D; 68D.

[11] Democritus of Abdera, fragment 9, quoted from Kirk et al. [1983] 410.

[12] Kemp [1990] part I, contains various examples of such painterly obsession with the solids (notably pp. 56; 62f.; 68; 77). Cf. in addition Leonardo da Vinci's drawings for Luca Pacioli's *De divina proportione* (Milan, Ambrosiana) and Jacopo de' Barbari's "Portrait of the mathematician Fra Luca Pacioli with an unknown youth" (Naples, Museo e Gallerie Nazionali di Capodimonte).

[13] Plato [1534] I: 473-500 (Plato) and II: 1-348 (Proclus) are both entirely without illustrations.

[14] Kepler [1610] 82ff; 120ff; Kepler's famous "*Tabula orbium planetarum dimensiones, et distantias per quinqueregularia corpora geometrica exhibens*," is found in [1596].

[15] The only seventeenth-century drawing I have so far seen in which the regular solids are used for a physiological explanation occurs in Giuseppe Zambeccari's *Idea glandulae* (Bibl. naz. Florence, Ms. II.IV.363), reproduced in Belloni [1963]. All other iatromechanicists tried to find more useful shapes, cf. notably the illustrations in Craanen [1689].

[16] Diogenes Laertius [1925] "*Letter to Herodotus*" 54 and 66. That Democritus had identified soul atoms with round fire atoms is testified by Aristotle [1957] 404a 1–9.

[17] Thus Lucretius claims, for example (Lucretius [1992] II 402–407), that sweet things are round, and a little bit later (II 450ff.) that what is fluid must be made up of round particles. How, then, would one have to explain the cohesion of sugar atoms?

[18] Martianus Capella [1969] II 46G offers graphic descriptions of a number of ancient philosophers, including a Democritus surrounded by atoms. It would be interesting to know whether there existed (or still exist) any illustrations of this theme.

[19] Cicero [1998] I vi, 18–21. Cf. Galilei [1890–1909] I 28 and Charleton [1652] 40f. for repetitions of this criticism.

[20] Basson [1621] 426. On Basson's eclectic atomism, cf. Lüthy [1997].

[21] Zoubov [1959] has shown that even the medieval finitists, who had argued against Aristotle that magnitudes were only divisible into a finite number of parts, had on the whole denied that these parts had any geometrical properties, for fear of thereby undermining the foundations of geometry. Cf. John Wyclif, *Tractatus de logica*, III 60; Nicolaus d'Autrecourt, *Exegit ordo executionis*.

[22] On the mathematical arguments against atomism, cf. Fromondus [1631] chs. 8–15; here 44. Basson [1621] 417 employs his dot figures in fact to demonstrate just this: that not even God could introduce an unbroken diagonal into the orthogonally formatted world. On medieval atomism, cf. Zoubov [1959]; Murdoch [1974].

[23] My pictures are taken from the original editions, here Bruno [1591]. The modern editions, here Bruno [1889b], all schematise and thereby falsify the meaning of Bruno's woodcuts.

[24] For a more detailed examination of this drawing, cf. Lüthy [1998].

[25] It was Mulsow [1991] esp. 235-240, who directed my attention to this numerological background.

[26] Boethius [1966] II 30. Our expressions "square" and "cubic numbers" are an echo of the notion that numbers can possess geometrical dimensionality.

[27] This "gnomonic" tradition was reinforced by Euclid's "monadic" understanding of numbers. Cf. Bertius [1599] § VII: "Numerus est multitudo ex unitatibus collecta, τὸ ἐκ μονάδων συγκείμενον πλῆθος [Euclid, Elements VII Def. 2.], ut ternarius, quaternarius, &c. quæ propterèa discreta dicimus, quia binarius cum binario nullo communi terino conjungitur ad constituendum quaternarium."

[28] The translations follow in outline Molland [1989] 78f.

[29] While Cusanus [1967] 578 speaks of the "evolution of the point, where 'to evolve' means that the point explicates itself," Boethius [1966] II 4 had spoken of the unit ("1") "uncoiling itself towards all numbers."

[30] For a brief introduction, cf. *Prologus ad Lectorem* in Lullus, 1744; for a survey, cf. Yates [1966] ch. 8.

[31] Bongo [1585]. The relation of Bruno's drawings to Bongo's has been noticed by Mulsow [1991] 237.

[32] Bruno's fondness for self-replicating figures is evident in various writings of his, cf. Bruno [1889b] 247.

[33] Bönker-Vallon [1995] 108f. justly defines the *Area Democriti* as the expression of a *Gestaltungsgesetz*.

[34] For a key passage linking the microscopic and the macroscopic *globi* physically, cf. Bruno [1889b] 176f.

[35] Bruno [1889b] 140: "The vacuum and atoms alone are not sufficient to us; for we require a certain matter which glues [the atoms] together," i.e., the ether.

[36] A rare mention of an actual physical shape is found in a reference to the regular shape of the diamond; cf. Bruno [1889b] 196f.

[37] Kepler [1611]. For an appraisal of Kepler's proposal from a scientific point of view, cf. Mason [1966]. For a historical appraisal of the *Wirkungsgeschichte*, cf. esp. Halleux [1975]; Emerton [1984] 40ff.

[38] Heuser-Keßler [1991]. Cf. the reproductions in Bruno [1889a] 81f.

[39] Galileo's celestial discoveries of 1610 had led Kepler to reconsider Bruno's theory of the multiple worlds, cf. Kepler [1610] 2f., 22, 24f., 30, 32. The addressee of the *Strena*, J. M. Wackher von Wackenfels, was furthermore a staunch defender of Bruno's cosmological views.

[40] Védrine [1976] 246: "Toute construction géométrique se réduit à une analyse de la composition des minima, exactement comme dans un jeu d'enfant. On imagine les résultats: par exemple, le centre d'un cercle ne peut entrer en contact qu'avec six minima, donc il n'est pas possible de mener réellement plus de six rayons."

[41] Cf. Kepler [1611] 16, where Kepler demonstrates his awareness of the distinctness of the two-dimensional and three-dimensional cases.

[42] Staats- und Universitätsbibliothek, Hamburg, Codex Hanseaticus IV p6, NL J. Jungius Pe 67.

[43] Meinel [1984] 75f. also gives an index of the subheadings of this *fascis*.

[44] Cf. Jungius [1929], for his views on the role of mathematics in pedagogy and the sciences.

[45] These are some of Jungius' subtitles: "Situs atomorum laxior [vel] pressior, et hinc figurarum aequalitas vel inaequalitas. Ad Jordanum Brunum" (Jungius, *Circuli*, f. 102); "Arithmetica ad Jordanum Brunum pro physica" (f. 182); "Geometria ad Jordanum Brunum pro physica" (f. 202). For a more extensive report on Jungius' notes on Bruno, cf. Lüthy [1998] 83-90. I have recently received a copy of Wolfgang Neuser's hitherto unpublished "Bruno, Jungius. Vorstellungen von Raum und Atom im 16./17. Jahrhundert," which I should also like to acknowledge.

[46] Jungius, *Circuli*, f. 119f. and 128ff. Jungius criticises Boethius' doctrine of "circular and spherical numbers" (f. 182f.; cf. Boethius [1966] II 30), redraws Bruno's *Area Democriti* and concludes: "aliud est numerus cyclicus Boethio quam nobis circularis."

[47] Jungius, *Circuli*, f. 164ff. demonstrates that Bruno is wrong in assuming that the space between four touching spheres is pyramidal, adding: "Dubito an talis pyramis quantum sibi imaginatus est Jord. Brunus, geometricè ... verè formari possit."

[48] On the analogy between chemical and mathematical resolution, cf. Jungius [1642b] §§ 62–64.

[49] Jungius [1662] pars II, sec. i, cap. 12. ass. 18. That Jungius knew Kepler's *Strena* is documented by Meinel [1992] 144f.

[50] Jungius [1642a] § 77. In Jungius, *Circuli* f. 203, he tries to explain the robustness of substances in terms of "durabiles compositiones."

[51] Cf. on this issue Shirley [1983] 385–388 and Gatti [1993] 18f. In the context of Bruno's iconographic legacy, it is interesting to note that one of Harriot's only references to Bruno (Brit. Library, Add. Ms. 6788, f. 67v) is accompanied by an image of a pile of globules. Cf. Gatti [1989] 52.

[52] Fabri [1670] vol. 3, tract. V, lib. II, prop. 162, p. 138. For an overview of his elemental geometry, cf. *Scholion*, 154.

[53] Research for this contribution was made possible by a fellowship at the Max Planck Institute for the History of Science (Berlin) and by the Netherlands Organisation for Scientific Research (NWO), grant 200-22-295. This article is based on a 1997 lecture.

References

Aristotle
1928 *Metaphysics*. Oxford.
1945 *De caelo*. Cambridge MA.
1957 *On the Soul. Parva Naturalia. On the Breath*. Cambridge MA.
1980 *The Physics*. Cambridge MA.
1982 *De generatione et corruptione*. Oxford.

Bajard, C., C. Saint Martin et al.
1985 *Image et Science*. Paris.

Barth, J. V. and H. Brune
1998 "Atomare Prozesse an Oberflächen." *Physik in unserer Zeit* 29: 251–260.

Basson, S.
1621 *Philosophiae naturalis adversus Aristotelem libri XII*. Geneva.

Beeckman, I.
1939–1953 *Journal tenu par Isaac Beeckman de 1604 à 1634*. The Hague.

Belloni, L.
1963 "Schemi e modelli della macchina vivente nel Seicento." *Physis* 5: 259–298.

Bertius, P.
1599 *Theses philosophicae de quantitate*. Leiden.

Boethius
1966 *De institutione arithmetica libri duo*. Frankfurt.

Bongo, P.
1585 *Mysticae numerorum significationis liber*. Bergamo.

Bouelles, C. de
1510 *Liber de generatione*. Paris.

Bönker-Vallon, A.
1995 *Metaphysik und Mathematik bei Giordano Bruno*. Berlin.

Bradwardine, T.
1495 *Geometria speculativa recoligens omnes conclusiones geometricas studentibus artium et philosophiae valde necessarias*. Paris.

Bruno, G.
1584 *De l'infinito, universo et mondi*. Venice [= London].

1588 *Articuli centum et sexaginta adversus huius tempestatis mathematicos atque philosophos.* Prague.

1591 *De triplici minimo et mensura ad trium speculativarum scientiarum & multarum activarum artium principia libri V.* Frankfurt.

1884 "De monade, numero et figura." In: *Jordani Bruni Nolani Opera Latine Conscripta*, ed. F. Fiorentino, vols. I, II 323–484. Naples.

1889a "Articuli adversus Mathematicos." In: *Jordani Bruni Nolani Opera Latine Conscripta*, eds. F. Tocco and H. Vitelli, vols. I, III. Florence.

1889b "De triplici minimo et mensura." In: *Jordani Bruni Nolani Opera Latine Conscripta*, eds. F. Tocco and H. Vitelli, vols. I, III. Florence.

1956 "De l'infinito, universo et mondi." In: *Opere di Giordano Bruno e di Tommaso Campanella*, eds. A. Guzzo and R. Amerio. Milan.

Charleton, W.
1652 *The Darknes of Atheism Dispelled by the Light of Nature. A Physico-Theologicall Treatise.* London.

Chevalley, C.
1991 "Introduction. Le dessin et la couleur." In: N. Bohr, *Physique atomique et connaissance humaine*, ed. C. Chevalley, 7–140. Paris.

Cicero
1998 *De finibus bonorum et malorum libri V.* Oxford.

Conring, H.
1643 *De fermentatione exercitationes.* Frankfurt.

Craanen, T.
1689 *Tractatus physico-medicus de homine.* Leiden.

Creech, T.
1683 *Titus Lucretius Carus His Six Books of Epicurean Philosophy, Done into English Verse.* London.

Cusanus, N.
1998 *Dialogus de ludo globi.* Berlin.

Della Porta, G. B.
1610 *Elementorum curvilineorum libri III.* Rome.

Des Chene, D.
1996 *Physiologia. Natural Philosophy in Late Aristotelian and Cartesian Thought.* Ithaca.

Diogenes Laertius
1925 *Lives of Eminent Philosophers.* London.

Emerton, N. E.
1984 *The Scientific Reinterpretation of Form.* Ithaca.

Fabri, H.
1670 *Physica, id est, scientia rerum corporearum.* Lyon.

Falkenburg, B.
1995 *Teilchenmetaphysik. Zur Realitätsauffassung in Wissenschaftsphilosophie und Mikrophysik*, 2nd ed. Heidelberg.

Fromondus, L. (Froidmont)
1631 *Labyrinthus sive de compositione continui.* Antwerp.

Galilei, G.
1890–1908 *Le opere di Galileo Galilei.* Florence.

Galison, P.
1997 *Image and Logic. A Material Culture of Microphysics.* Chicago.

Gassendi, P.
1658 *Opera omnia.* Lyons.

Gatti, H.
1989 *The Renaissance Drama of Knowledge. Giordano Bruno in England.* London and New York.

1993 *The Natural Philosophy of Thomas Harriot.* Oxford.

Giorgio Veneto, F.
1525 *De harmonia mundi totius cantica tria.* Venice.

Gordon, C. A.
1962 *A Bibliography of Lucretius.* London.

Gorlaeus, D.
1620 *Exercitationes philosophicae post mortem auctoris editae.* Leiden.

Hacking, I.
1983 *Representing and Intervening. Introductory Topics in the Philosophy of Natural Science.* Cambridge.

1985 "Do We See through a Microscope?" In: *Images of Science*, ed. P. M. Churchland, 132–152. Chicago.

Halleux, R.
1975 *Johannes Kepler: L'Étrenne ou la neige sexangulaire, traduit du latin, avec une introduction et des notes.* Paris and Vrin.

Heuser-Keßler, M.-L.
1991 "Maximum und Minimum. Zu Brunos Grundlegung der Geometrie in den 'Articuli adversus mathematicos' und ihre weiterführende Anwendung in Keplers 'Neujahrsgabe oder Vom sechseckigen Schnee.'" In: *Die Frankfurter Schriften Giordano Brunos und ihre Voraussetzungen*, ed. K. Heipke, 181–197. Weinheim.

Jungius, J.
Circuli locum replentes. Codex Hanseaticus IV p6, NL J. Jungius Pe 67.

1642a *Disputationum de principiis corporum naturalium prima ... respondente Johanne Hogio.* Hamburg.

1642b *Disputationum de principiis corporum naturalium altera ... respondente Jodoco Slaphio.* Hamburg.

1662 *Doxoscopiae physicae minores, ed. Fogelius.* Hamburg.

1929 "Über den propädeutischen Nutzen der Mathematik für das Studium der Philosophie (1629)," trans. J. Lemcke and A. Meyer. In: *Beiträge zur Jungius-Forschung. Prolegomena zu der von der Hamburgischen Universität beschlossenen Ausgabe der Werke von Joachim Jungius (1587–1657)*, ed. A. Meyer. Hamburg.

Keill, J.
1726 *An Introduction to Natural Philosophy: or, Philosophical Lectures Read in the University of Oxford, Anno Dom. 1700.* London

Kemp, M.
1990 *The Science of Art. Optical Themes in Western Art from Brunelleschi to Seurat*. New Haven.

Kepler, J.
1596 *Mysterium cosmographicum*. Tübingen.

1610 *Dissertatio cum Nuncio Sidereo*. Prague.

1611 *Strena seu De nive sexangula*. Frankfurt.

Kirk, G. S., J. E. Raven, and M. Schofield
1983 *The Presocratic Philosophers*. Cambridge.

Kuhn, T.
1962 *The Structure of Scientific Revolutions*. Chicago.

Lasswitz, K.
1890 *Geschichte der Atomistik vom Mittelalter bis Newton*. Hamburg and Leipzig.

Lucretius
1992 *De rerum natura*. Cambridge MA.

Lullus, R.
1744 *Logica Nova, iam Valentiae impressa anno 1512*. Palma.

Lüthy, C.
1996 "Atomism, Lyncaeus, and the Fate of Seventeenth-Century Microscopy." In: *Early Science and Medicine* 1: 1–27.

1997 "Thoughts and Circumstances of Sébastien Basson. Analysis, Micro-History, Questions." In: *Early Science and Medicine* 2: 1–73.

1998 "Bruno's 'Area Democriti' and the Origins of Atomist Imagery." In: *Bruniana & Campanelliana* 4: 59–92.

Lüthy, C. and W. R. Newman
1997 "'Matter' and 'Form': By Way of a Preface." In: *The Fate of Hylomorphism. 'Matter' and 'Form' in Early Modern Science*, eds. C. Lüthy and W. R. Newman. Leiden.

Maier, A.
1943 "Die Struktur der materiellen Substanz." In: A. Maier, *An der Grenze von Scholastik und Naturwissenschaft. Studien zur Naturphilosophie des 14. Jahrhunderts*, 7–140. Essen.

Martianus Capella
1969 *De Nuptiis Philologiae et Mercurii libri VIII*. Stuttgart.

Masi, M.
1983 *Boethian Number Theory. A Translation of the "De Institutione Arithmetica" with Introduction and Notes*. Amsterdam.

Mason, B. J.
1966 "On the Shapes of Snow Crystals." In: J. Kepler, *The Six-Cornered Snow-Flake*, ed. C. Hardie, 47–56. Oxford.

Meinel, C.
1984 *Der handschriftliche Nachlaß von Joachim Jungius in der Staats- und Universitätsbibliothek Hamburg, Katalog*. Stuttgart.

1988 "'Das letzte Blatt im Buch der Natur.' Die Wirklichkeit der Atome und die Antinomie der Anschauung in den Korpuskulartheorien der frühen Neuzeit." In: *Studia Leibnitiana* 20: 1–18.

1992 *Die Bibliothek des Joachim Jungius. Ein Beitrag zur Historia litteraria der frühen Neuzeit*. Göttingen.

Miller, A. I.
1984 *Imagery in Scientific Thought, Creating Twentieth-Century Physics*. Boston and Basel.

Molland, G.
1989 *Thomas Bradwardine, Geometria speculativa*. Stuttgart.

Morellus, F.
1609 *De Numerorum Historia*. Paris.

Mulsow, M.
1991 "Kommentar." In: G. Bruno, *Über die Monas, die Zahl und die Figur als Elemente einer sehr geheimen Physik, Mathematik und Metaphysik*, ed. E. von Samsonow, 181–269. Hamburg.

Murdoch, J.
1974 "Naissance et développement de l'atomisme au bas Moyen Age latin." In: *Cahiers d'Études Médiévales II: "La science de la nature: théories et pratiques,"* 11–32.

1984 *Album of Science. Antiquity and Middle Ages*. New York.

Münster, S.
1534 *Canones super novum instrumentum luminarium*. Basel.

Plato
1532 *Omnia divini Platonis Opera*. Basel.

1534 *Ἅπαντα Πλάτωνος μεθ' ὑπομνημάτων Πρόκλου εἰς τὸν Τίμαιον*. Basel.

1959 *Timaeus*. New York.

Ruffus, G.
1521 *Boetii Arithmetica*. Paris.

Scaliger, J.C.
1557 *Exercitationum exotericarum liber XV de subtilitate, ad Cardanum*. Paris.

Sennert, D.
1636 *Hypomnemata physica*. Frankfurt.

Shirley, J.W.
1983 *Thomas Harriot: A Biography*. Oxford.

Stahl, G. E.
1744 *Einleitung zur Grund-Mixtion derer Unterirdischen mineralischen und metallischen Cörper*. Leipzig.

Védrine, H.
1976 "L'obstacle réaliste en mathématique chez deux philosophes du XVIe siècle: Bruno et Patrizzi." In: *Platon et Aristote à la Renaissance* (= XVI Colloque International de Tours), 239–248. Paris.

Yates, F.
1966 *The Art of Memory*. Chicago.

Zoubov, V. P.
1959 "Walter Catton, Gerard d'Odon et Nicolas Bonet." In: *Physis* 1: 261–278.

III.
The Classification of Life and the Interaction Between Images and Texts

Image and Text in Natural History, 1500–1700

Brian W. Ogilvie

Introduction

The sixteenth century witnessed an explosion of natural history images. From the publication of the *Herbarum vivae eicones* in 1532 to that of Carolus Clusius' *Rariorum plantarum historia* in 1601, European presses churned out thousands of copies of hundreds of illustrated botanical books. While fewer zoological titles flowed from the presses, they too were lavishly illustrated. The Renaissance naturalists who wrote these books had at their disposal new graphic techniques, woodcut and engraving, that permitted the easy reproduction of identical pictorial images (Ivins [1953]). Woodcut in particular allowed writers and publishers to easily juxtapose text and image, presenting both written descriptions and artists' renditions of strange and familiar flora and fauna (fig. 1). By the end of the century, both image and text were necessary components of a marketable natural history book.

Art historians, bibliographers, and historians of science have long known about this pictorial explosion, which they have followed through the succeeding centuries down to the present day.[1] But few scholars have examined the relation between image and text in early modern natural history books.[2] As a result, they have failed to notice a curious phenomenon: the disappearance of images from botanical books in the seventeenth century. That is a deliberate exaggeration. Nevertheless, important botanical books published in the seventeenth century often omitted images, whereas earlier they had been practically obligatory. The images themselves did

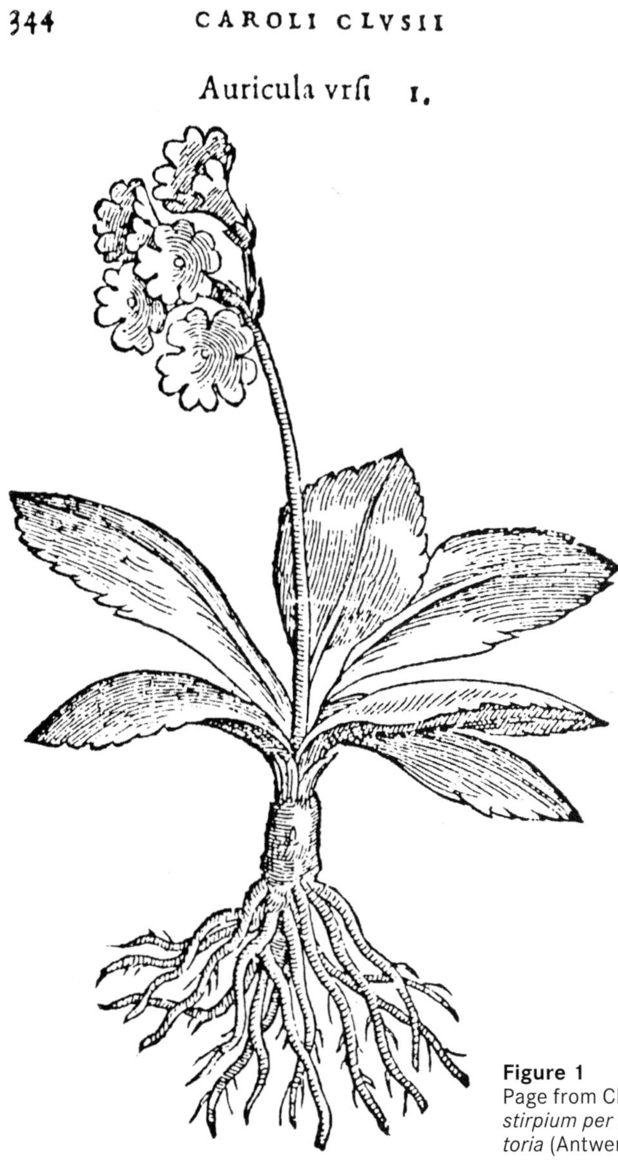

Figure 1
Page from Clusius: *Rariorum stirpium per Pannoniam historia* (Antwerp 1583).

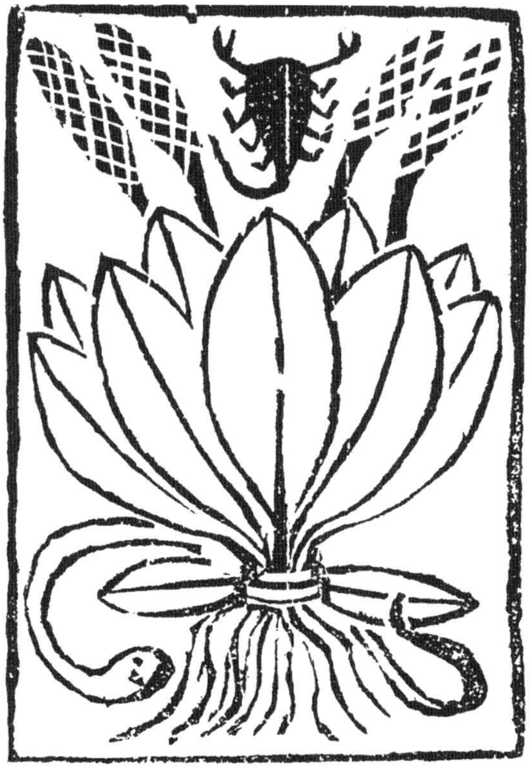

Figure 2
Plantago. From *Herbarium Apulei Platonici* (ca. 1481).

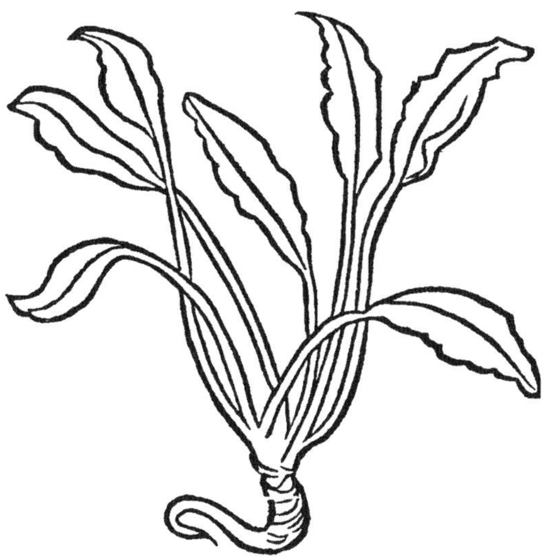

Figure 3
Serpentaria. From *Hortus sanitatis* (1484).

not disappear, of course: generations of scholars cannot have been so misled as to imagine what did not exist. And zoological books continued to be elaborately illustrated. But by and large botanical illustrations migrated into a new genre of book, the *florilegium*, in which texts were secondary appendages and artists employed all their skills of verisimilitude. Texts remained behind, in books written for scholars who had little need of illustration, for they had developed a precise technical language that, to the initiated, conveyed more than woodcuts or engravings could. The continuous existence of natural history illustrations from the sixteenth century to the present has masked the history of the relation between image and text in natural history books.

In this study I sketch this history, focusing on the stylistic features and communicative functions of images and texts in sixteenth- and seventeenth-century natural history. I will restrict my remarks primarily to botanical images, the area in which I have the greatest expertise, and the bulk of my remarks concern the period from 1530 to 1630. This will thus be a partial account from one perspective. I have also had to omit any consideration of the manuscript natural history illustrations that also proliferated in this period. Nonetheless, I think my results are valuable. They illuminate how image and text served similar roles in sixteenth-century natural history and how the increasing importance of classificatory systems in the seventeenth century rendered texts more useful to scholarly naturalists than images. They thus hint not only at the specific, changing relation between word and picture but also the broader scientific context in which that relation takes on meaning. Finally, this story reminds us that the development of modern science in the seventeenth century sometimes involved not the adoption of images but their renunciation.

Educating Athena

Renaissance botanical illustration emerged fully formed from the Strasbourg press of Hans Schott, like Athena from the head of Zeus. In 1532, Schott issued the *Herbarum vivae eicones*, or "living pictures of plants," with illustrations by Hans Weiditz accompanying a text edited by the reformist minister and physician Otto Brunfels. Recently, scholars have stressed the role of Schott and Weiditz in the production of this work, relegating Brunfels to the role of a hired pen who added descriptions (compiled, I might add, from classical writers) to the illustrations already prepared by Weiditz (see Arber [1986] 55). Brunfels also added, in appendix, excerpts and epitomes from important recent works on botany, including the late fifteenth-century exchange between Niccolò Leoniceno and Pandolfo Collenuccio about the alleged botanical errors of Pliny the Elder. Brunfels was hence not ignorant of the learned tradition in botany and its contemporary classics, but it does seem that his role in the production of the work was secondary.

Hans Weiditz was a pupil of Albrecht Dürer, and he took full advantage of his master's innovations in woodcut when preparing the illustrations for the *Herbarum vivae eicones.*[3] The book's title – not "pictures of living plants," but rather "living pictures of plants" – emphasises Weiditz' skill and the naturalism of the book's woodcuts. Contemporaries were stunned by the contrast between Weiditz' illustrations and those used in earlier printed herbals, as Konrad Gesner testified (fig. 2, 3, 4, 8).[4] Earlier illustrations had been for the most part schematic products of second-rate artists; Weiditz and Schott produced a masterpiece.

But however striking Weiditz' illustrations were, they did not satisfy some naturalists. Certain elements in the Dürer school's naturalism sat uneasily with the demands placed on illustrations by Renaissance naturalists. Weiditz' woodcuts represented individual plants, often with withered or insect-eaten leaves, stems, or flowers, and they were often distorted, as if drawn from a pressed specimen (fig. 4). Many parts are foreshortened or obscured by others – a realistic representation of the specimen, and a testament to the artist's skill, but also a hindrance to determining the form of those parts.

Brunfels' contemporary Leonhard Fuchs, whose *De stirpium historia* appeared a decade later, in 1542, alluded to these failings in the preface to his own herbal:

> *As far as concerns the pictures themselves, each is positively delineated according to the features and likeness of the living plants, we have taken peculiar care that they should be most perfect, and, moreover, we have devoted the greatest diligence to secure that every plant should be depicted with its own roots, stalk's leaves, flowers, seeds and fruits. Furthermore we have purposely and deliberately avoided the obliteration of the natural form of the plants by shadows, and other less necessary things, by which the delineators sometimes try to win artistic glory: and we have not allowed the craftsmen so to indulge their whims as to cause the drawing not to correspond accurately with the truth.*[5]

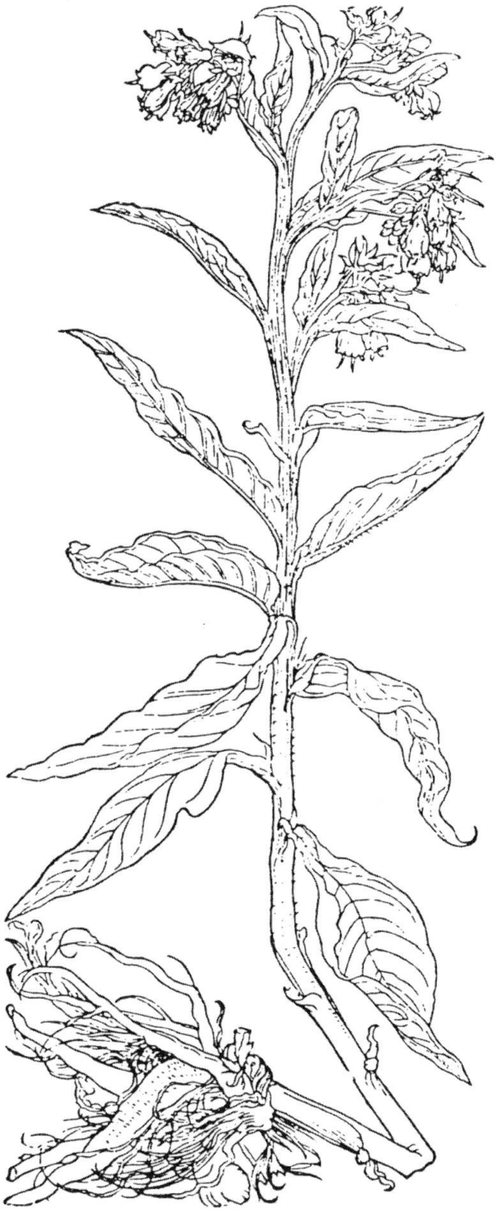

Figure4
Walwurtz Männlein. From Brunfels: *Herbarum vivae eicones* (Strasbourg1532).

Fuchs recognised that artistic verisimilitude, and the search for "glory," did not always correspond with the naturalist's truth.[6] The perspectival eye of the artist, who sought to render an object from a particular point and model it in space with light and shadow, was not that of the naturalist. Fuchs and his successors developed an alternate set of conventions to suit their particular needs. Rarely artists themselves, they trained their artistic collaborators to see with the eye of a naturalist.

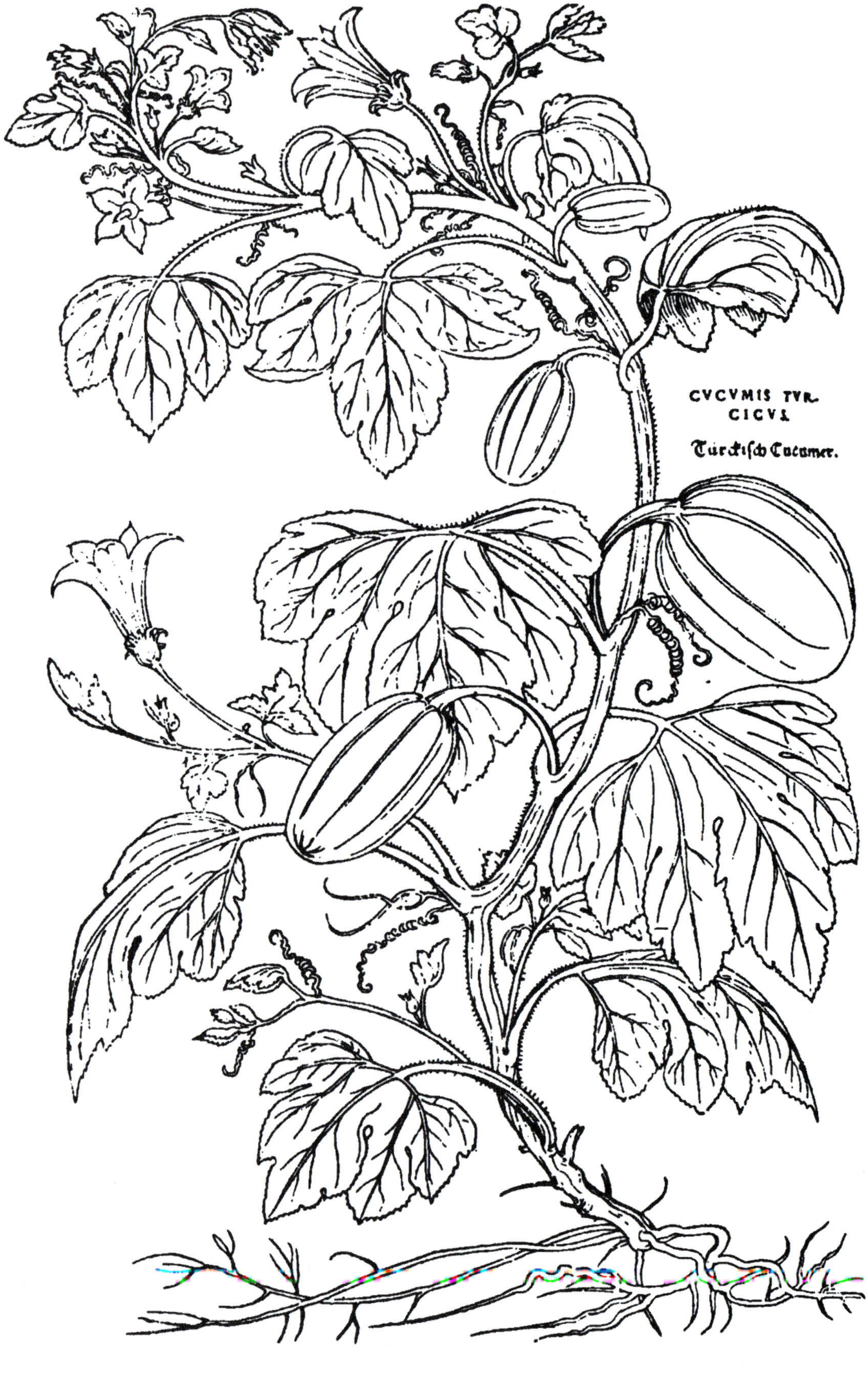

Figure 5
Cucumber. From Fuchs: *De stirpium historia* (Basel 1542).

What did characterise the naturalist's eye? In the first instance, Fuchs often had plants depicted in ways they would never be seen in real life: for example, plants bearing both flowers and fruit at the same time (fig. 5). The artist strove to capture a particular plant at a given moment of its existence; the naturalist wanted to illustrate the plant in general, as it might be observed at different times. If the plant changed greatly over the course of its life cycle, this could require two illustrations, as with the figures of the *Scilla hispanica* published by Carolus Clusius (fig. 6, 7).

Aside from eliminating foreshortening and modelling, natural history illustration was characterised by a number of other conventions. Plants were usually portrayed flattened, with their stems, leaves, flowers, seeds, and roots spread out in ways they would not be found live (fig. 8, 9). By the 1550s, some illustrations contained enlargements of flowers and seeds (fig. 10). Another convention developed around the same time was cutting away twigs and stems to reveal other parts more clearly (fig. 11). Not all contemporary illustrations followed this convention – for instance, the woodcuts to Pierandrea Mattioli's commentaries on Dioscorides are often lush and visually confusing (fig. 12) – but it was the prevailing mode. With these conventions, Fuchs and his successors established a visual style for botanical woodcuts which reigned for almost a century – and still has noticeable affinities with present-day botanical illustrations (see Zweifel [1988]).

All of these conventions, which the nineteenth-century botanist L. C. Treviranus considered the *sine qua non* of proper botanical woodcuts, served the same purpose (Treviranus [1855]).They aimed to make the woodcut express, as clearly as possible, individual parts of the plant depicted: hence naturalists rejected foreshortening and shadow, which might distort or obscure certain parts; they eliminated superfluous vegetative parts, which could obscure others; and they presented enlargements of certain characteristic elements. At the same time, however, these conventions distorted the overall appearance of the plant, what later naturalists would call its habit. The plantain from Fuchs' *Historia stirpium*, for example, has its leaves pointing up, rather than strewn flat on the ground,

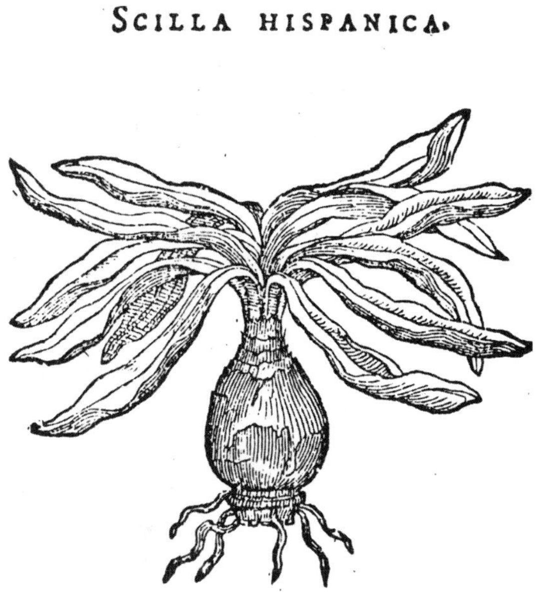

Figure 6
Scilla hispanica. From Clusius: *Rariorum stirpium per Hispanias historia* (Antwerp 1576).

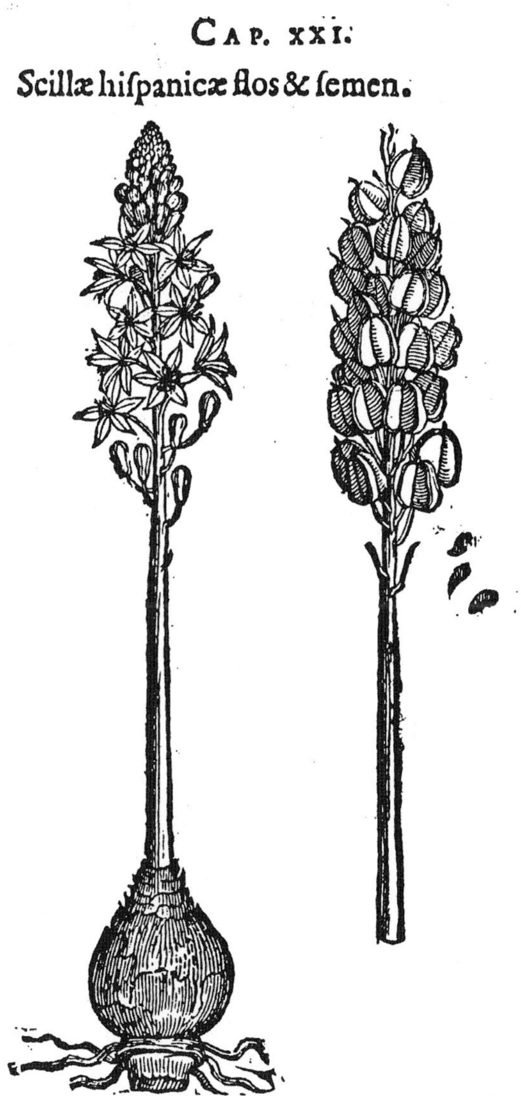

Figure 7
Scilla hispanica. From Clusius: *Rariorum stirpium per Hispanias historia* (Antwerp 1576).

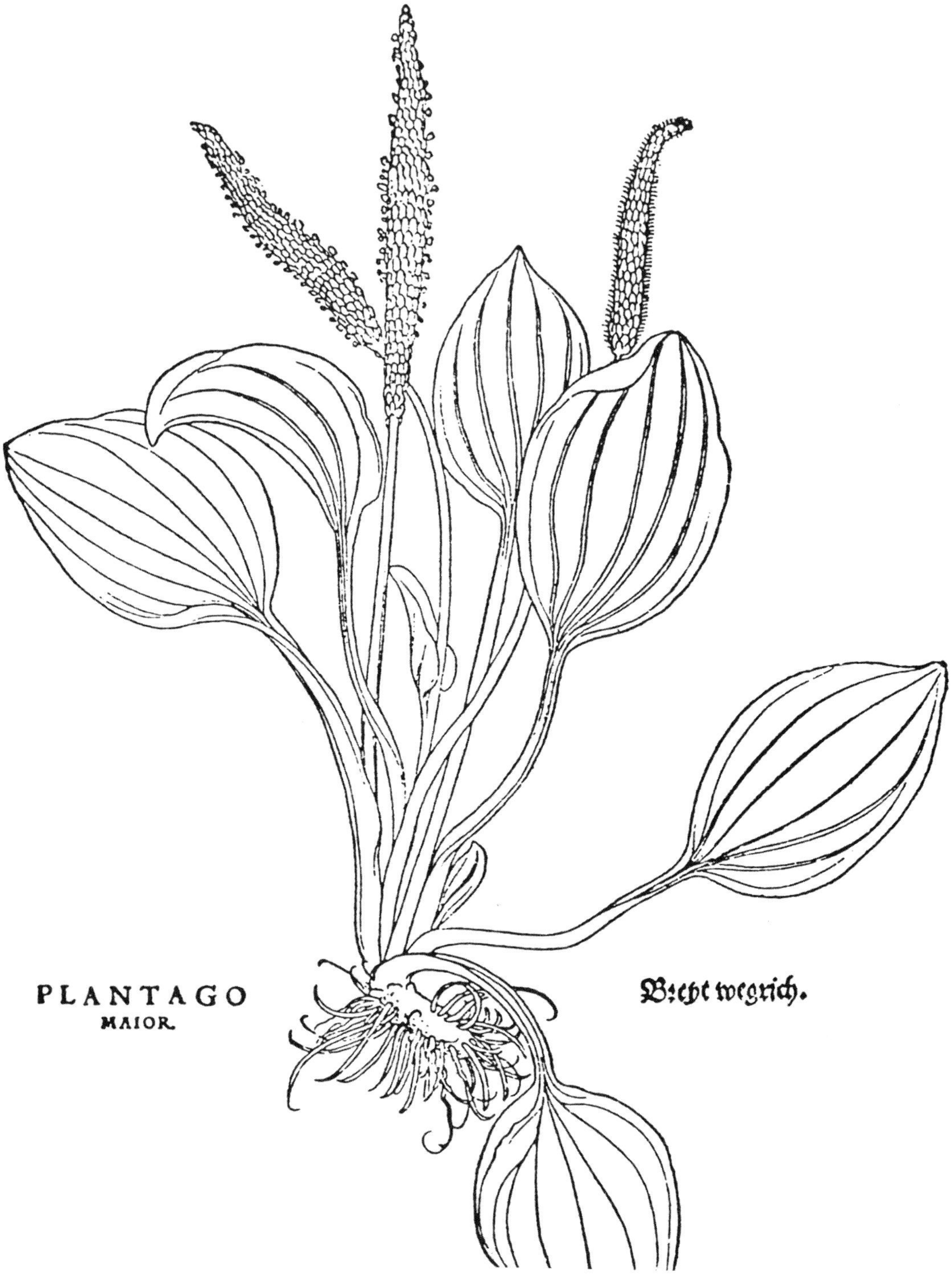

Figure 8
Plantago. From Fuchs: *De stirpium historia* (Basel 1542).

which is how the plant is found in the wild (fig. 8, 9). Instances in which a plant seems to have been drawn from a pressed specimen are uncommon. Nonetheless, sixteenth-century illustrations generally convey an overall impression of spreading and flattening.

Thus, after a brief fling with Renaissance artistic naturalism, Renaissance naturalists reconsidered the charms of Dürer's school and found them wanting. In order to serve their ends, naturalists had to educate artists in a new set of conventions, conventions that rejected perspective and modelling of an individual object in favour of rendering clearly and distinctly the individual parts of an idealised specimen. This style of botanical illustration can be characterised as "analytic": it broke down the plant into distinct elements,

Plantago Maior.

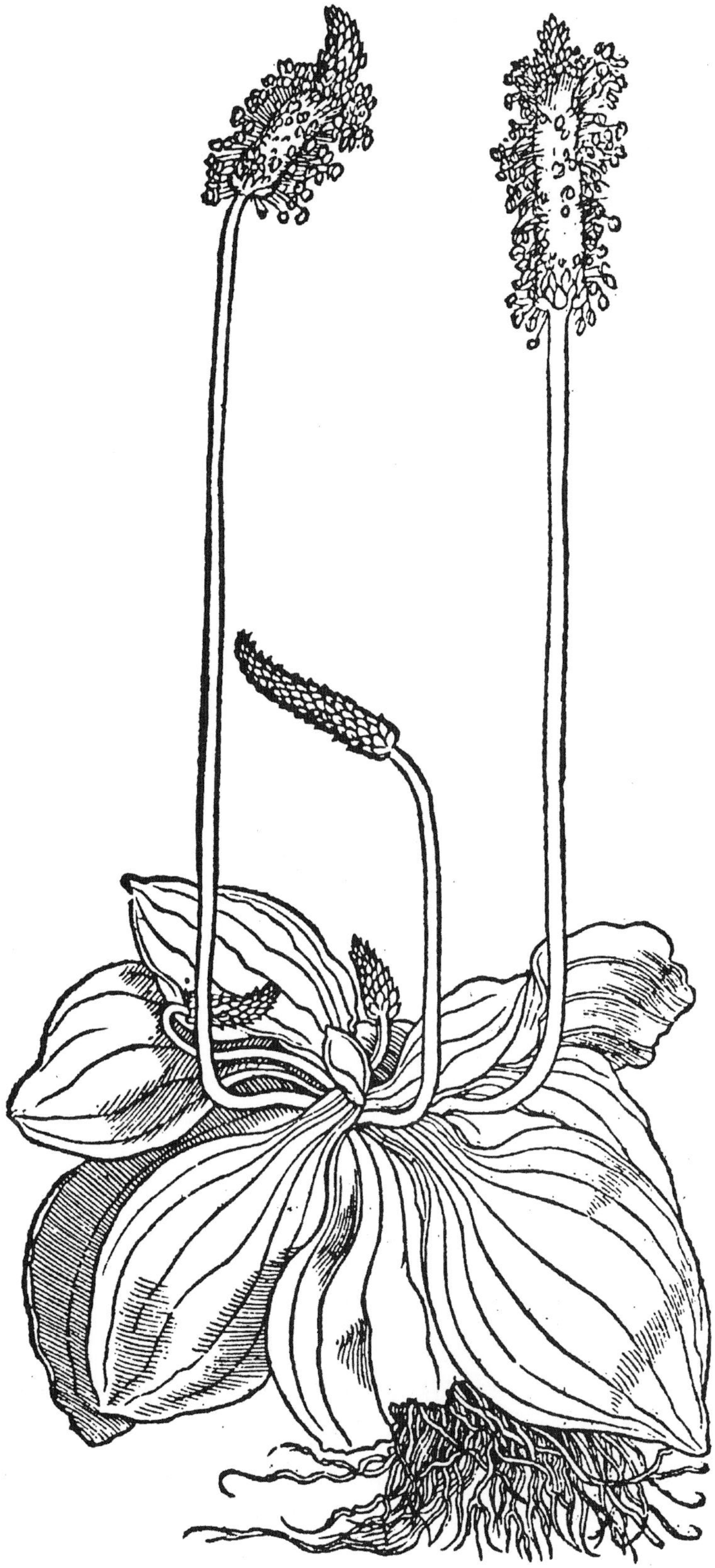

Breyter Wegrich.

Figure 9
Plantago. From Brunfels: *Herbarum vivae eicones* (Strasbourg 1532).

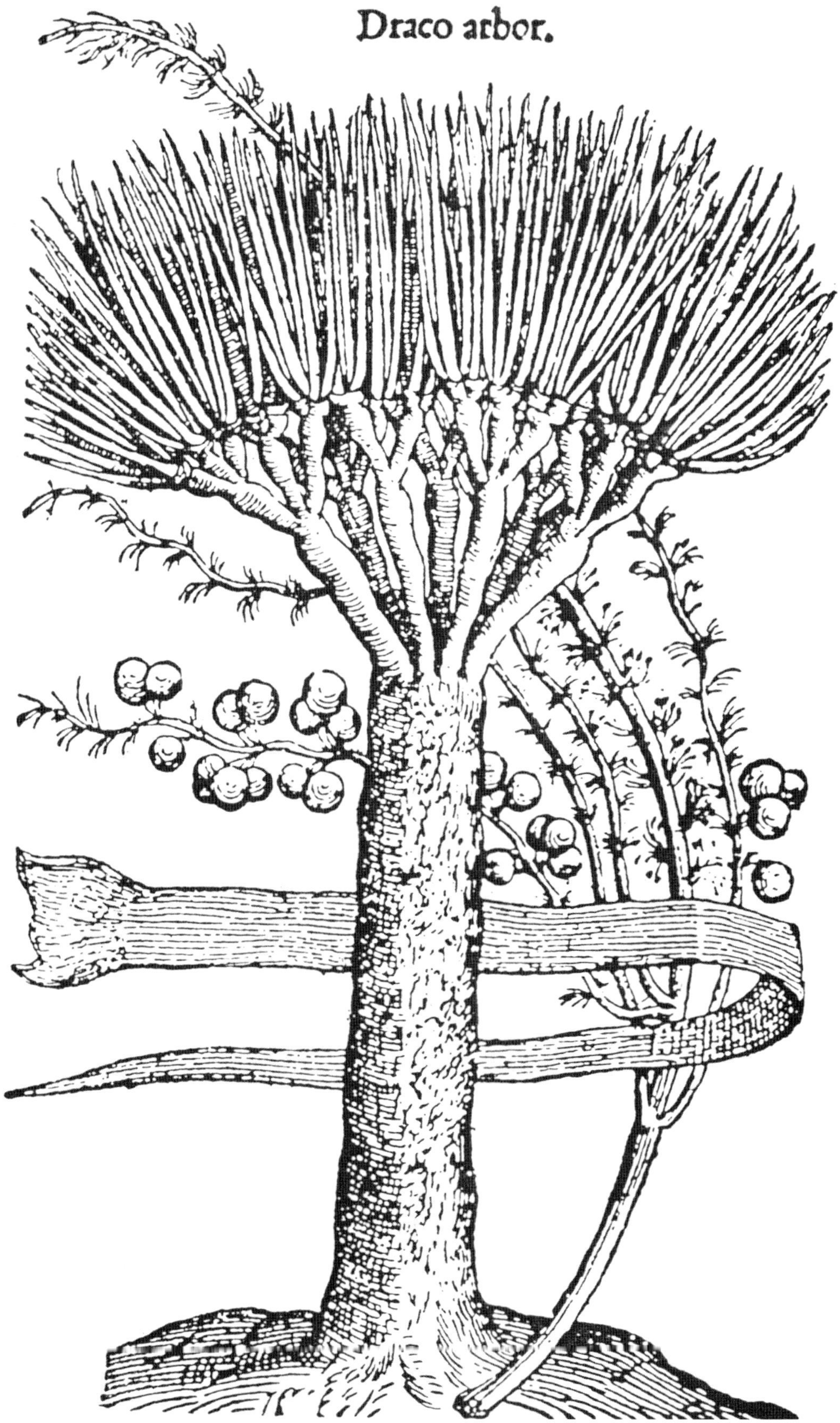

Figure 10
Draco arbor. From Clusius. *Rariorum stirpium per Pannoniam historia* (Antwerp 1583).

and its goal was to represent those elements precisely rather than to convey a realistic overall impression of how the plant looked. In the seventeenth century, Renaissance naturalism would return to botanical illustration with a vengeance. At this point, however, I would like to turn from image to text and trace a parallel development in the verbal description of plants.

Ut pictura prosa

Sixteenth-century phytography was modelled on the descriptions of the ancients: in particular, those of Pedanius Dioscorides of Anazarba, the author of five books *De materia medica*. Dioscorides' descriptions were often maddeningly brief, and in the case of common plants he usually omitted the description entirely. Nonetheless, he provided a convenient model for phytographic descriptions; as a result, his method was adopted, with increasing detail, by his sixteenth-century successors.[7]

Otto Brunfels, as I have noted, contented himself with reproducing or epitomising descriptions drawn from Dioscorides and other ancient sources. Leonhard Fuchs had adopted a similar method in his 1542 Latin herbal (though he often disguised his sources, as later writers complained), but in the 1543 German version (Fuchs [1543]), Fuchs attempted to be more precise in his descriptions. For example, in his description of the Akelei (*Aquilegia*, or columbine), Fuchs provided his own description of the plant's stem, leaves, flowers, fruit, and root. A similar pattern was adopted by Fuchs' contemporaries Jerome Bock, whose *New Kreutterbuch*, in German, was issued in 1539 (Bock [1539]) partly through the offices of Brunfels, and Valerius Cordus, whose *Historia Plantarum*, in Latin, was largely complete by the young man's death in Rome in 1544, though it was first published by Konrad Gesner some fifteen years later; Cordus [1561a]).

When Cordus described the Black Hellebore, for example, he began with the stems which grow out of the ground, followed by the leaves, flowers, and seeds. He ended the chapter by describing the plant's root – as if he had uprooted the plant after first watching its progress from sprout to mature plant and noting the places and times in which it grew. Cordus occasionally varied from this pattern to emphasise a point; for instance, he often began descriptions of Ranunculae by noting where they grew, since similar plants in this family have very different habitats (Cordus [1561b] fol. 109v–121v). But for the most part, Cordus' descriptions represented the natural development of a growing plant (Greene [1983] I 375–377).

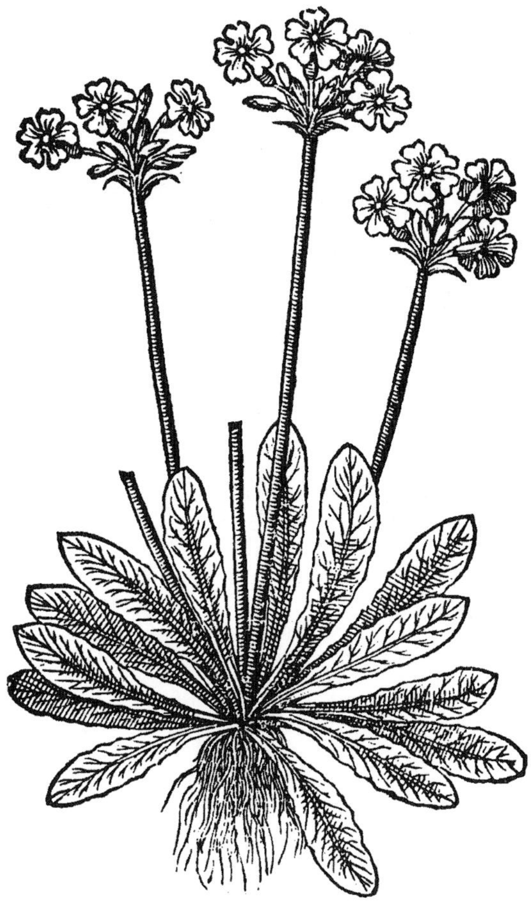

Figure 11
Primula. From Clusius: *Rariorum stirpium per Pannoniam historia* (Antwerp 1583).

For Cordus and his sixteenth-century successors – and, significantly, "popular" nature writers through the present[8] – this vicarious element was crucial to description. The reader followed a description just as if he or she were observing the plant itself in a garden or in the wild. Such vicarious description characterises natural histories published between the middle and the end of the sixteenth century, even when writers drew on books rather than their own experience to describe a plant. Konrad Gesner combined his own observations with published accounts and correspondents' reports for his description of the "Lunaria arthritica" (or "Auricula ursi," a kind of primula), but his description reduced

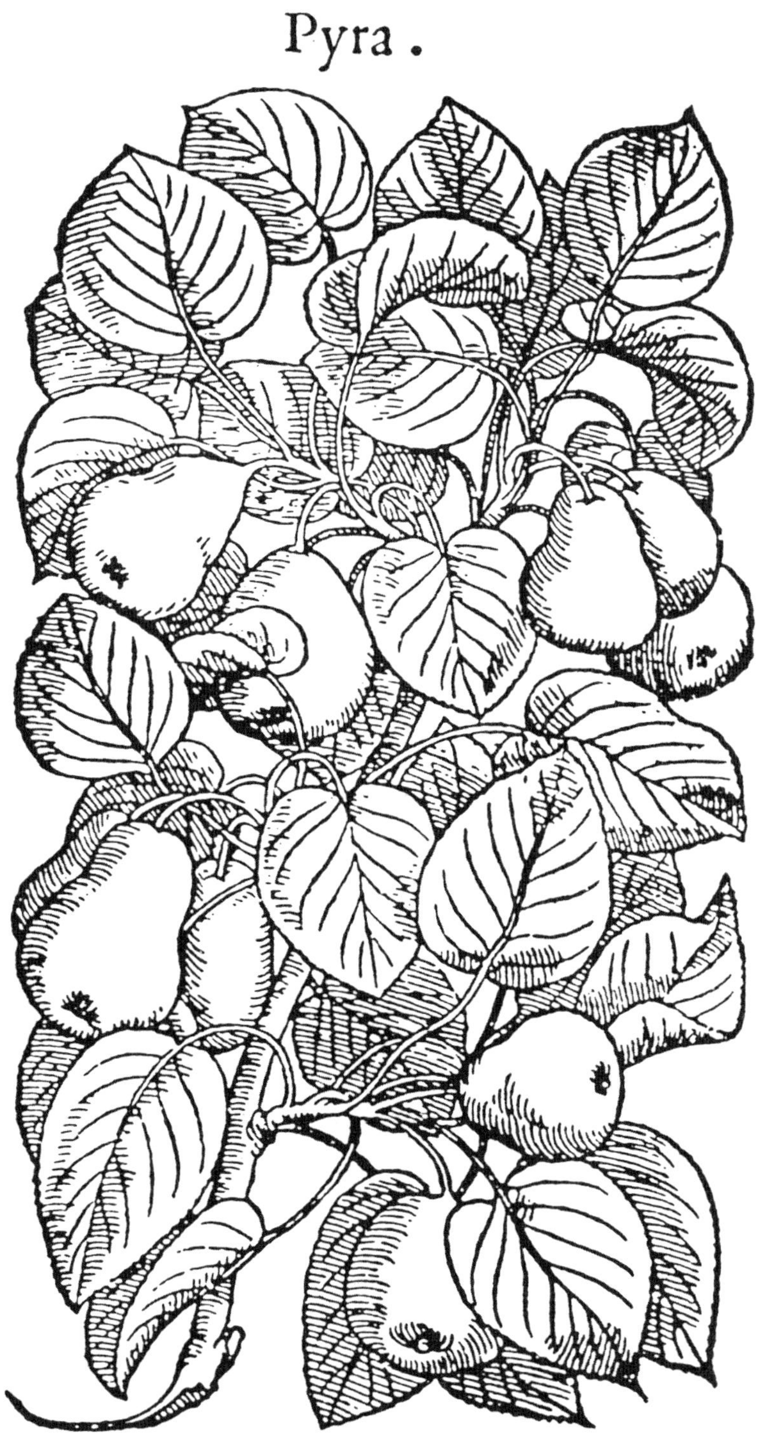

Figure 12
Pyra. From Mattioli: *Commentarii* (Venice 1560).

these disparate sources into one dynamic narrative. He began with the leaves, noting their varying forms, and continued with the small, yellowish or pale flowers, "which, in our gardens, flower in April and early May, with a very pleasant odor."[9] After noting that the leaves do not fall off, Gesner described the root before turning to the alleged medicinal uses of the plant and the ease of cultivating it in a garden.

The vicarious description of *naturalia* reached its high point in the works of Carolus Clusius, which for the same reason are some of the most delightful sixteenth-century natural history texts to read. Clusius' two masterworks, his accounts of Spanish and Austro-Hungarian plants, are particularly important in the development and dissemination of vicarious descriptions (Clusius [1576], [1583)]). Each of these works was the fruit of long travel and observation. Although Clusius spent less than a year travelling through Spain, his descriptions in the Spanish flora rely on observations made in gardens and the wild over a much longer period (e.g. Clusius [1576] 16f., 289, 304, 321–325). Similarly, Clusius repeatedly prolonged his stay in Vienna in order to continue gathering material for his Austro-Hungarian work.[10] The descriptions contained in them thus reflect the process of distilling years of careful observation into brief descriptions.

In his Spanish flora, Clusius referred frequently to his activity as "digging up" plants. He named one type of plant *Iris bulbosa*, or bulbous iris, because "though they had the flower of an Iris or Xyris, when I dug them up I discovered that they had a bulbaceous root."[11] He dug up another plant, the *Scilla Hispanica*, at several stages in its growth: as its stem emerged, when it flowered, and when its leaves grew out (Clusius [1576] 289f.).When describing these and other plants, Clusius followed the pattern mentioned above: his description imitated the plant's life history. In some cases, the history was incomplete, as with the *Pancratium*, which he had seen only when it was putting out leaves (Clusius [1576] 292). In others, he complemented observations made during his eight-month journey through Spain and Portugal with information and material from correspondents. When he saw the dragon tree near a monastery outside of Lisbon, it was not flowering – and indeed the monks claimed that it never did. But the next year a friend sent him a new shoot with ripe berries, which he duly had illustrated and placed in his book (Clusius [1576] 12f.).

Ideally, then, naturalists in the late sixteenth century aimed to describe a plant's entire life cycle. Clusius' eight months in Spain did not allow him to achieve this ideal in every case. His observations of plants in Austria, Pannonia, and neighbouring provinces, on the other hand, were carried out over almost a decade before the book containing them went to press.[12] In this case he had ample opportunity to repeat his observations in the wild, and to grow plants from seed or bulb in his garden.

Clusius' description of the *Anemone silvestris II*, or *Anemone montana Pulsatillae folio*, provides a sterling example. He began with the bristly top of the root, from which first leaves, then a stem bearing a flower emerged. At the end, as usual, he described the root itself. The plant grew, he noted, in high mountain ridges, among rocks with a bit of soil, and since it sent its roots into cracks in the stone, it was very difficult to uproot. "It flowers in July, occasionally in August, depending on when the summer begins, that is, when the snows melt, for it is one of the first to flower."[13] Clusius tried to cultivate it in his garden, but it was one of those Alpine plants which could not bear the "milder air" of Vienna. The *Cotyledona altera taurica*, on the other hand, which was found in the Tauern (one of the Austrian alpine ranges), flowered there in July, but "in gardens, in May, like the other plants of that kind."[14]

To a twentieth-century reader, such vicarious descriptions seem almost like a verbal version of stop-action photography. However, they were the result of multiple observations, usually made with several plants in different places and times. They were a condensation of experience. Much as an illustration was intended to represent the characters of a species, not of an individual, the description had to suppress the peculiarities of individual plants. After all, as Aristotle had said, natural history deals with what happens "always or for the most part" (Aristotle [1965–91] 6.17, 571 a 25).

The preceding analysis demonstrates a clear affinity between illustrations and verbal descriptions in sixteenth-century botanical works. Both were analytic, oriented toward describing individual parts of a plant and their relative positions. In a word, they were done in the same style - understanding, by "style," not elements of the illustrator's art such as line, modelling, and so forth, but rather the cognitive approach behind both image and text. Renaissance scholars were fond of quoting Horace's dictum "ut pictura poesis," from the *Ars poetica*: in botanical illustration and description, the tag "ut pictura prosa" would be more appropriate. One is reminded of Johan Huizinga's analysis of stylistic similarities between fifteenth-century Flemish painting and poetry; but in this case, both illustrations and descriptions appeared in the same works, with the same object (Huizinga [1996] 329ff.).

Words Outstrip Pictures

As botanical descriptions focused ever more narrowly on morphology and the place and time of growth, those aspects which interested naturalists, they became increasingly more detailed. Bock, Fuchs, and their contemporaries wasted little ink on morphology. In some cases they did not even bother with a description: hence Hieronymus Bock's chapter on common garlic (Knoblauch) began with instructions for cultivation. Though later editions of Bock's text included a picture, his only mention of the plant's form was in reference to its seed, "set on a round, smooth, rush-like stem, like the head on onion-pipes."[15] Bock gave more detail in his description of the "Waldt Knoblauch," which he considered a distinct plant. This plant

> *emerges in the spring, about the same time that Aron first appears. It is not very common, and its usual place is in very dark and moist woods. It is a plant with one or two leaves, similar to those of the Meyenblumen. The leaves acquire stems like that of garlic; at the end of April they bear little white flowers, which are also like those of Meyenblumen. But they are not hollow, like the Meyenblumen, but star-shaped, and the leaves [of the flower, i.e. petals] are separated from one another. The roots of this flower are white and longish, like a small garden garlic plant which is not more than a month old.*[16]

Bock presumed that his reader was familiar both with common garden garlic and the Meyenblumen. He described the Waldt Knoblauch by comparing and contrasting its features with these two plants. This method, used already by Theophrastus, provides a powerful and economic means of describing a new object in terms of features already experienced by the observer, who draws on that previous experience to understand the novelty.

A similar method was employed by Rembert Dodoens some fifteen years later in his herbal. He distinguished three kinds of garlic: cultivated garlic, wild garlic, and "bear garlic" (the same as Bock's Waldt Knoblauch). He too based his descriptions of the first and third kinds on comparisons with known forms:

> *Garden garlic has the leaves of grass or onion, between which round, hollow stems grow in the second year; on these, the flowers and seeds grow, similar to the flowers and seeds of onion. The root is round, bulbous like the bulb of an onion, composed of several nodes joined together, beneath which hang small, spindly fibers.*[17]
>
> [...]
>
> *The third garlic, called bear garlic, usually has two large, broad leaves, very similar to the leaves of the "grand Muguet," between which come one or two stems, on which many small white flowers grow. The root is shaped like that of a new garlic, with a very strong smell and taste.*[18]

Unlike Bock, Dodoens described the common garlic, but the level of detail in his description of bear garlic is roughly equivalent to that in Bock's description of the same plant. The same level of detail characterized later sixteenth- and seventeenth-century herbals such as revisions of Dodoens' *Cruydeboek* and John Gerard's plagiarised translation of Dodoens, the *Herball, or general Historie of plants.* For physicians and apothecaries, it was just enough detail to serve the practical ends of identifying common plants.

By the late sixteenth century, however, this level of detail no longer sufficed for scholarly botanists, the *studiosi rei herbariae.* Carolus Clusius distinguished "several kinds of Garlic or Moly, which are distinguished either in their leaves, flowers, or *capita*. Most of them, I think, have not yet been observed by others."[19] Since these plants were found in the mountains of Austria and Pannonia, Clusius named them "Moly montanum," with brief morphological epithets to distinguish the kinds. He described the first, *Moly montanum latifolium*, as follows:

> *The first has a stem which exceeds two cubits. It is surrounded by leaves to the midpoint; the leaves are much broader than those of other genera of Allium, and oblong, very close to those of the leek. At the top of the stem, which is light and reed-like, sits a head composed of many bulbs, of a darkish purple and pretty thick, among which grow pale purple flowers on long pedicules. These flowers are succeeded by triangular vessels which contain the seed. The root is thick and bulbous, consisting of many nuclei, with many white fibers; moreover, little round bulbs adhere to the root, similar to those which grow in the head. The whole plant has a strong smell of garlic. It flowers and bears bulbs in June and July.*[20]

Clusius continued to describe four other kinds of *Moly montanum*, one of which he divided into two "species." In describing the second, he remarked that its leaves were narrower than the first, while his description of the third noted that its flowers were "similar in shape and size to those of the second, but whitish in colour, whose petals are marked by faint purple veins running lengthwise in the centre and on both edges."[21] He followed this chapter, in turn, with a description of two similar plants which he designated *Moly narcissinis foliis.*

Clusius' descriptions are clearly much more detailed than those provided by Bock and Dodoens. Some elements, for instance the observation that the leaves of the *Moly montanum latifolium* extend about halfway up the stem, did in fact convey an overall impression of the plant's form, a verbal picture. But Clusius' descriptions were still largely contrastive, both within and across generic boundaries. He compared the leaves of the different kinds of *Moly montanum* with those of leeks, grasses, and reeds, as an economical yet accurate means of conveying their appearance. Within this chapter, he explicitly compared elements of different plants, emphasising only those morphological elements which distinguished the plants from one another. In his Spanish flora, Clusius had used the same technique. After describing the *Aristolochia longa* which grew in southern France, he identified another:

> *A certain difference can be observed in the Aristolochia longa: for the flower of the Spanish kind (although similar to the preceding in almost every regard) is seen, by someone examining it closely, to differ a bit: the topmost part of the interior is purple, whereas it is green in the other one. Moreover, the end of the root is not slender, as in new plants of the other kind, but much more obtuse.*[22]

Clusius did not bother to describe the other features of this plant – since, after all, they were almost identical. In this case, the two items added to the previous description were truly *differentiae*: they were the features which distinguished these two species. Such contrastive descriptions within a genus, unlike those which compared one genus with another, did not necessarily require that the reader be familiar with the original plant, since its description was itself part of the text.

The growing detail of botanical description in the later sixteenth century was thus not due to a love of detail for its own sake. Instead, naturalists provided just enough detail for their descriptions to capture the distinctions between different kinds of plants. In Clusius' work, one can see this process within each chapter, as subsequent types of a plant are described with references to earlier descriptions. The same is true of Caspar Bauhin's lapidary morphological descriptions. In his *Prodromos Theatri Botanici* (1620), Bauhin described three forms of *Allium*. His description of the third, *Allium montanum radice oblonga*, was extremely short:

> *III. Allium montanum with oblong root: similar to the first, but with a taller stem and longer leaves, a little head, pale purple, with flowers consisting of six leaves and the same number of stamens.*[23]

As the vegetable world became better known, it was also more populous: descriptions had to keep pace precisely with the amount of knowledge naturalists required to distinguish its inhabitants.

Indeed, naturalists sometimes criticised one another for inadequate descriptions. Clusius included the *Cotyledon altera Matthioli* in his Austro-Hungarian flora, "although its history has been presented by Mattioli and others, since in my judgement they describe it with more brevity than they should."[24] Mattioli, in fact, had only repeated Dioscorides' description:

> *There is also another kind of Cotyledon which some call Cymbalum, having broader, & fat leaves, like little tongues, thick about ye root, describing as it were an eye in ye middle, like as ye greater Sempervivum, binding by ye tast; a little thin stalk, & upon it flowers, & seeds like to Hypericum, but ye root greater. It is good for ye same uses that Sempervivum is.*[25]

Mattioli added only that he had seen this plant in the garden of an Austrian physician and in many mountainous regions; all of Dioscorides' characteristics agreed with what he had seen (Mattioli [1598] 788). One of the others who had described this plant was Rembert Dodoens, in the herbal that Clusius had translated: Dodoens said that it "has broad, thick, fairly round leaves, spread out around the stem, as in the large Sempervivum, from the middle of which comes the tender stem, bearing small flowers."[26] Dodoens did not even mention the eye-shaped depression in the middle of the leaves. Compare Clusius' description:

> *This plant has numerous, densely-packed leaves, similar to oblong little tongues, rather thick, pointed, somewhat toothed on the edges, tending toward white, with an acid and somewhat astringent taste. From the centre, or middle of the leaves, rises up a stem eighteen inches long, or sometimes only a foot, hairy on the upper part, with shorter leaves here and there. At the top of the stem, it puts out little pedicules between the leaves, on which rest two or three white flowers composed of five petals, which on the inside appear to be spotted with drops of blood. They have a pallid, shiny navel [umbilico] which is surrounded by ten little stamens [staminula]; then in little capsules very small black seed. Underneath is a thin, black, fibrous root; many other plants are attached by*

the root to the mother plant, forming a circle; sometimes they form a sod, as with common sedum.[27]

This description is much more extensive than Dioscorides' or Dodoens', and it also serves as the basis for a contrastive description of a second variety of this plant, "with narrower and longer leaves which are not toothed but covered with a white wool, less white; its stem is shorter, thicker, and covered with many leaves, which are usually somewhat hairy on top and bottom. It has fewer flowers, scattered on the stems by the leaves."[28] By giving these two descriptions, Clusius judged that he had provided necessary information which his predecessors had omitted.

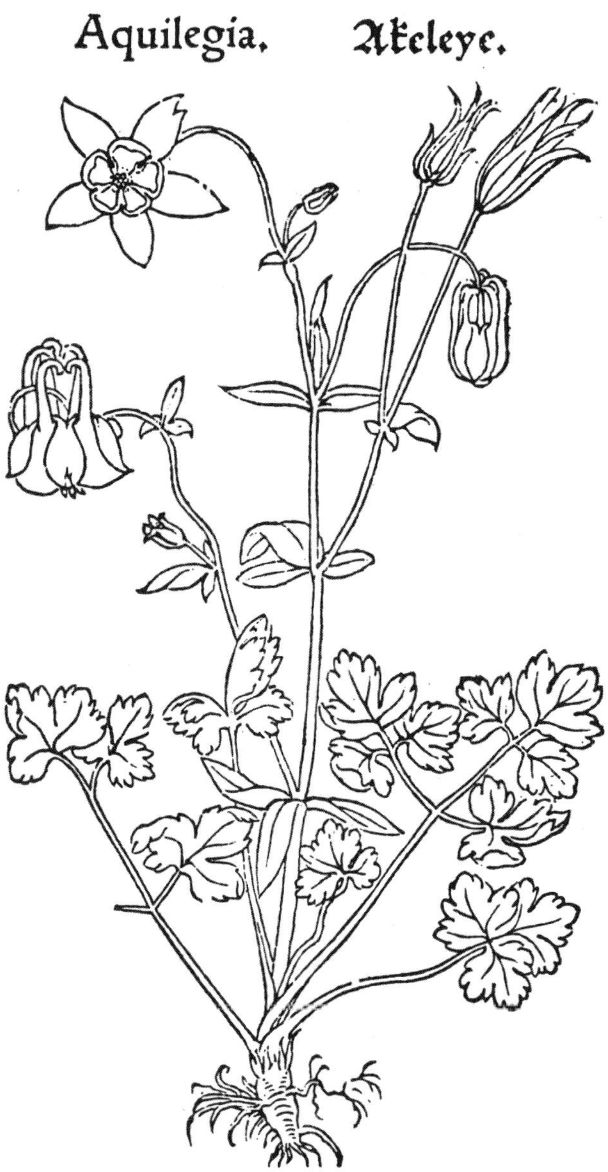

Figure 13
Akelei. From Fuchs: *New Kreueterbuch* (Basel 1543).

As the preceding discussion has emphasised, descriptive techniques in early modern natural history made sense only within a system of differences. It might have been possible to provide a verbal picture of a plant which sufficed to recreate it in the mind's eye, but naturalists did not usually attempt this. The goal of description was, rather, to provide enough detail to distinguish a species or variety from its congeners by emphasising *differentiae* in the shape, arrangement, and colour of parts. Like illustrations, descriptions focused on elements of a plant, not the whole.

Image and text embodied the same analytic style, but we can observe a marked difference in precision between the two forms, a difference that shifted with the development of descriptive language. The relative precision of image and text was rarely the same, but the relation inverted in the generation between Fuchs' work and Clusius'. At first, illustrations were much more precise. For example, in 1543 Leonhard Fuchs described the columbine as follows:

> *On this [plant] grow many pretty flowers, which are filled in two parts, and develop five curved points on the bells, almost like a knight's spurs. These flowers are commonly purple-brown, but one finds also entirely white ones, and some are red-brown.*[29]

If one compares this rather vague description with the illustration produced by Fuchs' artists (fig. 13), it is clear that the illustration conveys a more precise impression of the part. The illustrator had adapted a set of visual elements developed over the course of the past two centuries, aimed at naturalistic representation, and while Fuchs and his contemporaries modified the relationship of those elements to fit their notions of "realism," the artist still had the jump in the flexibility and range of his descriptions.

The development of botanical description had by the end of the sixteenth century reversed this relationship. As more subtle *differentiae* were introduced in order to separate similar but distinct species (as contemporaries saw them), the power of the image to convey these *differentiae* was eclipsed by the flexibility of words. We have seen that Clusius distinguished the Hispanic variety of *Aris-*

tolochia longa from the kind which grew in France by a difference in colour and a small difference in the shape of the root-tip. The latter difference could have been conveyed in an illustration, if the naturalist had the funds to pay for another woodcut and the patience to oversee its production, but the former would have been indistinguishable in a black-and-white line drawing.

Justifying the Illustrations

It is, however, incontestable that naturalists in the later sixteenth century continued to use woodcut illustrations in their books, and to go to great troubles to ensure that those illustrations were both accurate and, so far as possible, beautiful. One reason was economic: publishers found that illustrated books sold better. But illustrations continued to serve an intellectual function even as verbal descriptions became increasingly precise. For the rest of the sixteenth century, analytic images and descriptions would coexist on the pages of natural history books.

In the first instance, the book market after 1530 seemed to demand illustrations. Jerome Bock published the first edition of his *Kreutterbuch* in 1539, without illustrations. In the preface to the second German edition, published in 1551 with woodcuts by the Strasbourg artist David Kandel – mostly copied from earlier books – Bock was still lukewarm about their usefulness. Though it is sometimes alleged that he omitted illustrations because of their expense, in the preface to his work he minimised their importance:

> *But as far as illustrated herbals are concerned, it is clear that they are useful to a certain degree, when we do not have any living plants, or cannot get hold of recently-collected ones. But whoever has his own gardens and gardeners can plant many and various plants (as we read that, in former times, Antonius Castor did at Rome) and contemplate their living images. For such people there is no need of pictures, except for those plants which are truly foreign and which we cannot see recently collected in every place, and which completely refuse to be acclimated to our soil.*[30]

Bock's remarks remind us forcefully that words and pictures were not the only ways in which Renaissance naturalists learned to recognise plants – in fact, they took a back seat to visual inspection of the plant itself, whether in the wild or in a garden. Nonetheless, Bock acquiesced when his publisher insisted on illustrating the second German edition of his work (1551) and the Latin translation, issued a year later (1552). He may have found illustrations useless, but most book buyers disagreed.

Many of those buyers were wealthy bourgeois and noble book collectors, not practicing naturalists. Despite the descriptive, technical focus of sixteenth-century botanical illustrations, they were – and were considered to be – aesthetically pleasing. This is seen most clearly in the production of coloured exemplars. Leonhard Fuchs referred in the text of his herbal to the colours in the illustrations, and some extant copies are coloured to match Fuchs' text. Hans Weiditz's original watercolours for the *Herbarum vivae eicones* seem to have served as models for coloured copies (Arber [1940] 803). Konrad Gesner too thought that colour was an important part of the illustrations to his *Historia animalium*. Though the expense of hand-colouring prevented the publisher from selling every copy with coloured images, he told his readers that they could order copies coloured according to the original drawings, if they had the funds (Gesner [1551] sig. g1v). In Plantin's shop, it took three months to properly colour an herbal with some two thousand illustrations, and the work of colouring formed over ninety percent of the cost.[31] It is thus not surprising that most exemplars which have survived are not coloured or only partially coloured. Largely or completely coloured books were great gifts, comparable to the richly-illustrated medieval miniatures in effect if not, due to the technology involved, in cost. Plantin donated a coloured copy of the Flemish edition of Lobelius' herbal to the Duke of Prussia; the tome cost about a quarter of a university medical professor's annual salary.[32] Such coffee-table editions were probably rarely owned by working naturalists.

But market forces played only a partial role in the demand for illustrated books. Bock may have been largely indifferent to the illustrations in his work, but Konrad Gesner,

Carolus Clusius, and many of their contemporaries went to great lengths to ensure that their books, and those they edited, were published with accurate illustrations of plants. The reason, so far as it appears in contemporary texts, lies in the different ways in which picture and text were apprehended by the reader or viewer.

When Konrad Gesner edited Valerius Cordus' posthumous works, he took care to choose illustrations for the work, which Cordus had not provided (Gesner in Cordus [1561a] sig. a ii^{v}; fol. 85r–v). Since the book's publisher, Wendel Rihel of Strasbourg, also published Bock and owned the blocks for the illustrations in his herbal, Gesner adopted many of those. He also added several from his own collection and copied others from earlier books (Gesner in Cordus [1561a] fol. 85v; see Treviranus [1855] 18f.). Though he referred to the images as an "ornamentum," they had a more serious purpose: "Almost all will help the reader come more easily to a knowledge of unfamiliar things." On the other hand, Gesner omitted images in some cases, "for instance, where there is another picture of a plant of the same type which is almost identical to it."[33]

Gesner was aware thus of both the advantages and limitations of woodcut illustrations for natural history. On the one hand, he inverted Bock's reason for placing little value on illustrations. Their aim was not to add to descriptions of plants already known to the reader but to aid in understanding the descriptions of those which were previously unknown. Gesner implied that pictures could convey much more forcefully and successfully certain elements which were difficult to understand from a verbal description. On the other hand, they were useless for conveying subtle differences (especially when considerably reduced in scale, as was the case with most sixteenth-century natural history illustrations).[34] These *differentiae* could be conveyed adequately only in words – for example, the subtle distinctions between two varieties of *Aristolochia longa* described by Clusius (Clusius [1576] 321; see also 306 f.). Thus, for Gesner, the two forms of description were complementary, and each was required to complete the history of a plant.

Gesner articulated a distinction which most of his contemporaries accepted in silence. This attitude toward illustration best explains such monstrous results of late Renaissance botany as Carolus Clusius' illustration of the "*Ficus indica*." Clusius provided his contemporaries with a description and illustration of this plant at the beginning of his *Exoticorum libri*. He had never seen this tree, but he had heard about it from the "distinguished man, the well-travelled Fabrizio Mordente of Salerno (who lived for some years in Goa, and explored the areas around that city)."[35] Mordente visited Vienna during the last years of Maximilian's reign, where he met Clusius and described the tree.

From the woodcut illustration (fig. 14) and detailed description, the modern reader can easily recognise this "Indian fig" as the banyan tree. According to Fabrizio, the tree grows out of the ground and sends out branches, from which fibres dangle toward the ground. When they reach it, they root and thicken, eventually sending out their own branches with fibres which repeat the process. They also grow upwards, so that "out of one tree a dense forest grows." The foliage is so thick that the sun cannot penetrate it, and within the tree voices sometimes echo. Fabrizio claimed that he had camped with a force of 800 men under one of these trees, and that it could have sheltered three thousand. He added descriptions of the leaves and the fruit, of which elephants were especially fond.[36]

The woodcut itself is a strange amalgam of Fabricius' observations, Clusius' botanical knowledge, and an unnamed illustrator's skill. "I had this tree sketched, as elegantly as it could be, from Fabricius' narration, and he approved it."[37] This is one of the few instances in Clusius' works in which he included an illustration that was not drawn from life. Elsewhere in the *Exoticorum*, and occasionally in his other works, he included illustrations taken over from other books or from his correspondents, but these were themselves the result of firsthand observation.[38] Compared with European trees, the banyan was odd enough that even an inaccurate illustration would stress its peculiarities.

In fact, this woodcut deviates substantially from the analytic style used in Clusius' other works and by other contemporary naturalists. It is presented in perspective and modelled with strong highlights and shadowing, precisely those elements in Weiditz's pictorial style to which Fuchs had objected six decades earlier. Many of the other illustrations in Clusius' book – pictures of seeds, of strange woods and barks, and of unusual birds – use the same style, quite different from that of Clusius' *Rariorum plantarum historia* of four years earlier. It is tempting to see this work as part of a transition from the analytic style of later sixteenth-century botanical illustration to the lush florilegia of the early seventeenth-century. It is to this transition that I now turn.

Disappearing Pictures

Given the prevalence of illustrations in botanical works in the later sixteenth century, their decline in the seventeenth century gives one pause. Of course, it is not entirely proper to say that they declined, for this period saw the introduction of florilegia. These books included detailed copperplate engravings or etchings of flowers, often from particular gardens, such as Basilius Besler's *Hortus Eystattensis* (1613), which depicted the episcopal gardens in Eichstätt, or Crispijn van de Passe's *Hortus floridus* (1614) (fig. 15). The latter was the first illustrated bulb catalogue: the flower fancier could order those plants he or she liked for his or her private garden (MacDougall [1991], and Kenseth [1991] 371f.). These were only the first of dozens of picture-books catering to the *amateurs de fleurs* in Baroque Europe.

Simultaneously, more specialised, detailed studies of the plant world and its geography began to appear. The local floras of the early seventeenth century make a sober contrast to the richly illustrated florilegia. Mere lists of plants, sometimes with brief descriptions, they were aimed at medical students. The earliest included Ludwig Jungerman's catalogue of plants growing around Altdorf, and Caspar Bauhin's catalogue of plants found within a mile of Basel. Florilegia were large-format books, quartos or folios: these duodecimo catalogues, on the other hand, could be slipped into a student's pocket, and they were interleaved with blank paper for note-taking. These two genres instantiate the increasing differentiation between natural history as a means of investigating the world and collecting as a type of self-fashioning in the early seventeenth century.

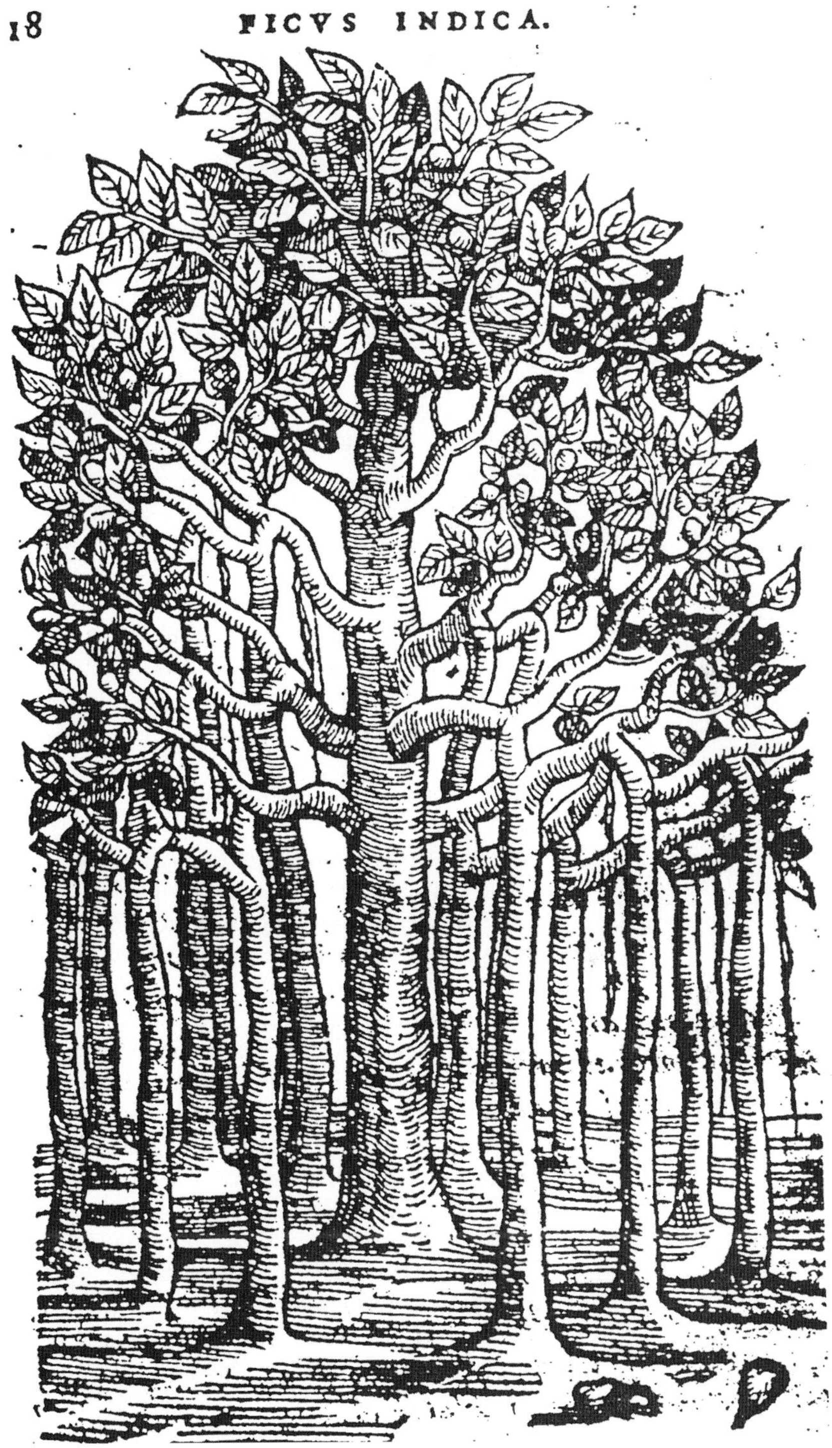

Figure 14
Ficus indica. From Clusius, *Exoticorum libri decem* (Leiden 1605).

Florilegia, in a sense, siphoned off the coffee table book market from the drier descriptive works in natural history. In the latter, even detailed engravings could no longer reflect the distinctions made by naturalists like Caspar Bauhin, always eager to find a new species or variety (the two terms were not yet clearly distinguished). Furthermore, as Agnes Arber observed long ago, the processes of

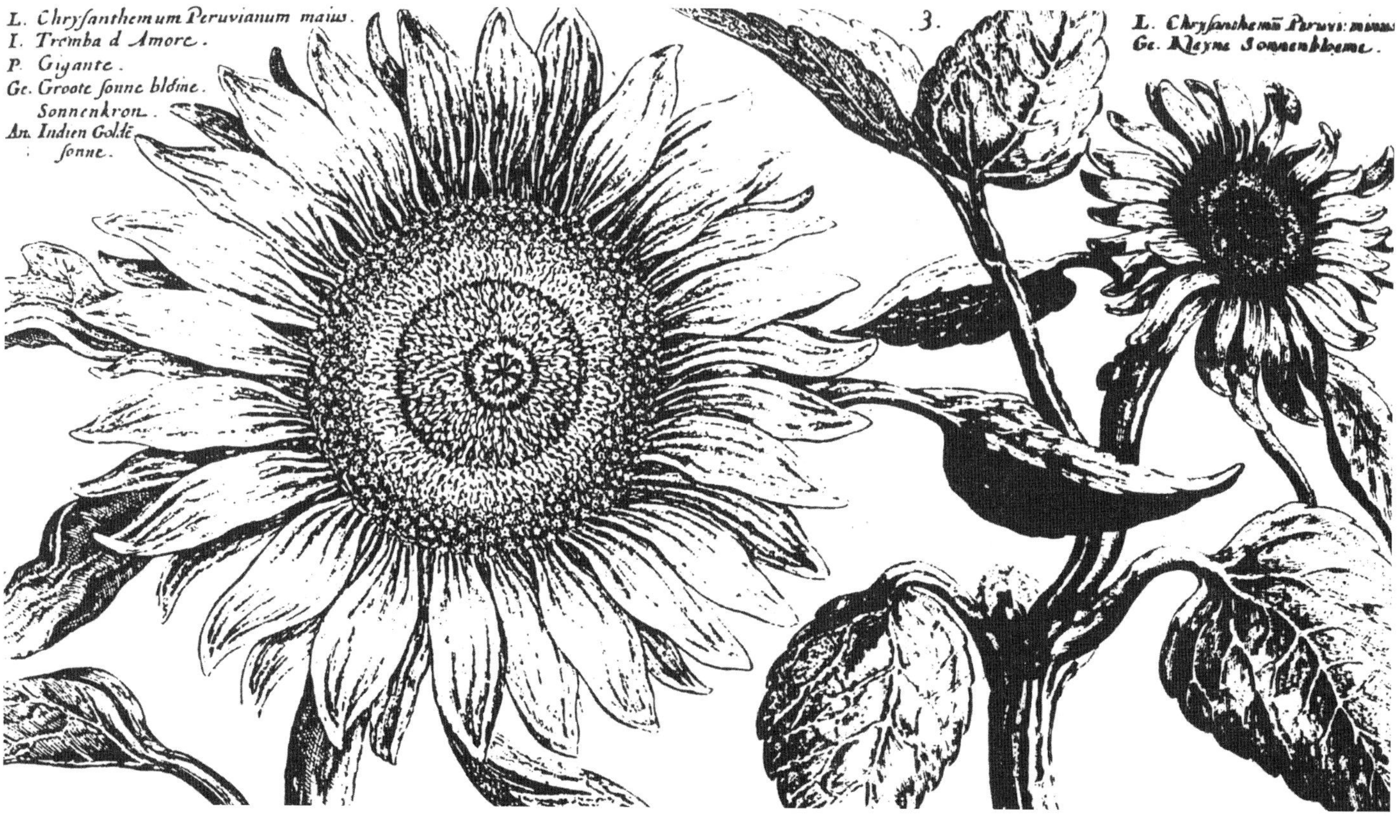

Figure 15
Sunflower. From Crispijn de Passe: *Hortus floridus (Arnheim* 1614).

engraving and etching encouraged a separation between image and text (Arber [1986] 245). Woodcuts could be printed on the same page as text with only one run through the press, but combining letterpress and copperplate required two runs and careful alignment. It was easier to print plates and text separately; and, in turn, to concentrate on plates.

The artists who illustrated florilegia did not work under such a watchful eye, and their pictorial idiom differed from the analytic style of sixteenth-century botanical woodcuts. Whether they focused on the plant's flower or provided an image of the entire plant, they did not analyse the plant into its component organs. The illustrations in the *Hortus floridus* and other florilegia were heavily modelled and often foreshortened (fig. 16).[39] They were intended to allow their viewers to experience nature vicariously, but that experience took place in a different mode. They conveyed a naturalistic representation of an individual plant taken as a whole, returning as they did so to those elements in Weiditz's naturalistic style that naturalists had suppressed.

The authors of florilegia realized that their works were not, in fact, aimed primarily at an audience of naturalists. The lavish illustrations of Jacob Breyne's 1678 florilegium far outdid the descriptive text, whose author admitted that he had tried to write "in a botanist's manner and style," but often with more prolixity than necessary and with some metaphoric descriptive turns (Breyne [1678] sig.)(**2v). At least Breyne wrote descriptions. His contemporary Dominique Chabrey turned sixteenth-century woodcuts into a florilegium of sorts by reprinting them with notes on the name, location, time, and virtues of the plant they depicted – but with no verbal description to complement and complete the woodcut. Though he claimed to have compressed an entire history of plants into one volume, it was a volume which would have been useless to a serious student of botany, unless he had access to the books from which Chabrey's illustrations had been drawn (Chabrey [1666]).

The development of florilegia and the introduction of more specialised, detailed works for serious botanists contributed to the decline of strictly botanical illustrations, but shifts in botanists' intellectual focus were equally important. Although late sixteenth-century botanical descriptions and illustrations shared a common style, the difference in the way

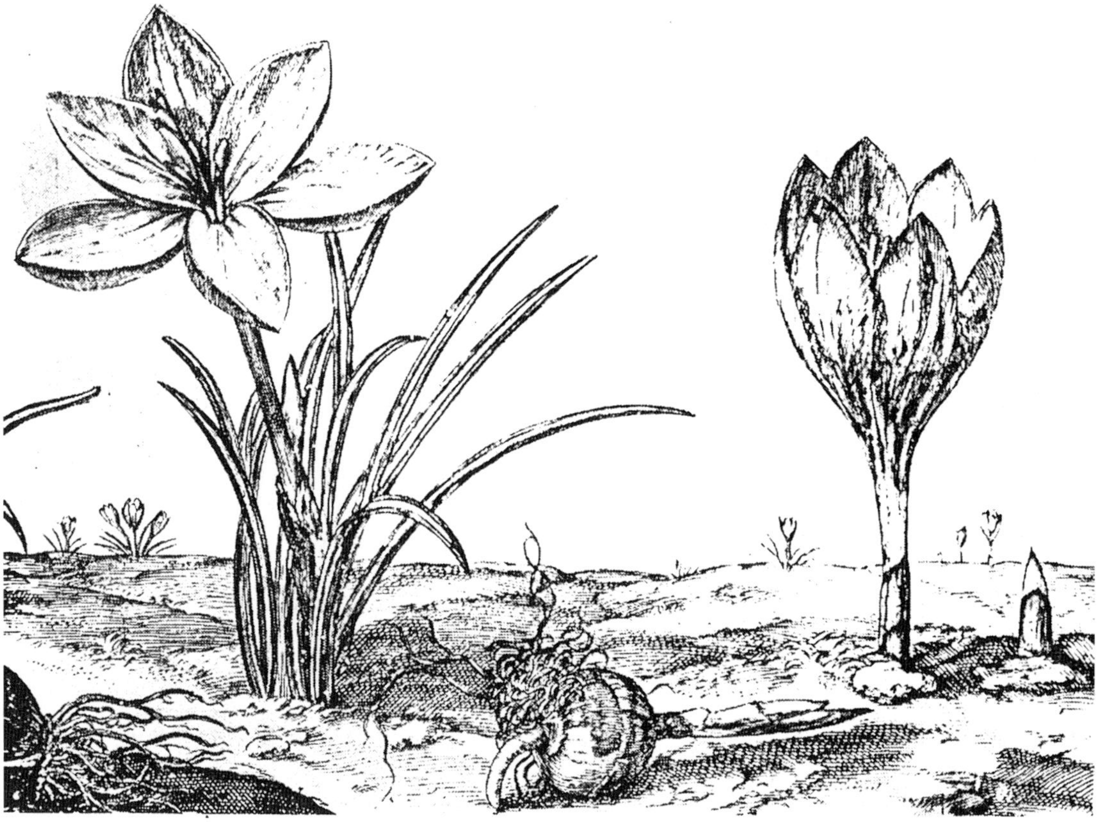

Figure 16
Crocuses. From Crispijn de Passe, *Hortus floridus* (Arnheim 1614).

they were apprehended would ultimately lead to the eclipse of botanical images. Technical and financial factors also played a role. Naturalists' verbal control over a description far outpaced their ability to control the rendering of a woodcut or an engraving, despite the best efforts of Fuchs and Clusius to oversee their artists. Furthermore, illustrations represented a tremendous financial outlay for authors or publishers. But these causes would not have sufficed to reduce botanical illustration to secondary importance had naturalists not become increasingly concerned with classification rather than description. With this shift, naturalists needed to control not only the analysis of a plant's description but its apprehension by the reader.

In this area as in others, Andrea Cesalpino anticipated the concerns of later generations. In his *De plantis* (1583), Cesalpino aimed not to describe individual plants but to reduce their innumerable species to a philosophical method: a feat not attempted since the days of Theophrastus, he claimed (Cesalpino [1583] sig. a4r). Cesalpino's contemporaries do not seem to have much appreciated his approach; even two decades later the young Adriaan van de Spiegel praised Cesalpino's descriptions while chastising him for using Tuscan rather than Latin names for plants (Spiegel [1606] 129). Nonetheless, his concern to find order in the chaos of species was shared by all the great botanists of the seventeenth century, from Caspar Bauhin to Joseph Pitton de Tournefort and John Ray.

Brevity of description was, Cesalpino claimed, one of the advantages of his system. By reducing plants to a natural order, on morphological grounds, "we are not compelled to repeat in each description those elements which are common to the genus." "Such certain knowledge comes from this brief description," he continued, "that a picture cannot make it more certain. A picture, unlike words, does not express all the *differentiae*" (Cesalpino [1583] sig. a4). Furthermore, verbal descriptions could be easily purged of the "fictions" that are often found in printed illustrations (Cesalpino [1583] sig. b1).

None of Cesalpino's contemporaries would have disagreed with these remarks; indeed, Clusius' contrastive style of description was implicitly based on economy of expression and grouping of similar plants. But Cesalpino's concern with systematisation forced him, as it did not force his contemporaries, to rely solely on the written word. His decision not to repeat the characters common to a genus required that he determine which characters those were. In other words, it required descriptive analysis of plants before they were classified. Sixteenth-century botanical illustrations were analytic, but they ultimately left it to the viewer to perform a final analysis – to identify all the significant characters, be they generic or specific. Cesalpino could not afford to leave this process to the reader: otherwise his audience would have to repeat, in effect, Cesalpino's own labour in developing a system. And, it may have occurred to him, they would also have material to challenge his systematic. Because illustrations did not adumbrate a hierarchy of characters, they could potentially subvert any classificatory system which laid some claim to be natural or philosophic, as Cesalpino's did.

Cesalpino's method was not adopted by any of his contemporaries or immediate followers. The closest was Caspar Bauhin, whose incomplete *Phytopinax* (1596) grouped plants by classes and subclasses. Bauhin's *Pinax Theatri Botanici* (1623) completed the work begun in the *Phytopinax*, but neither book attempted to provide descriptions of the plants they contained. Rather, they comprised indexes to the already copious botanical literature.[40] The great age of botanical systems would not begin until the last decades of the seventeenth century. I will conclude by briefly considering the way two of the greatest systematisers, Tournefort and Ray, conceived the relation between image and text.

In both Tournefort's and Ray's works, Cesalpino's basic principle of the subordination of specific to generic characters reappears. Tournefort's method, as set out in his *Elemens de botanique* (1694) would have struck Clusius or Gesner as irredeemably myopic. For Tournefort, the aim of botany was "to place each species within its true genus." In this task, "nothing matters other than the essential character which distinguishes one genus of plant from any other genus, and nothing is more important in botany than to separate completely this character from anything which might disguise or obscure it" (Tournefort [1694] sig. eii^{r}). Given the number of species known to botanists at the end of the seventeenth century, Tournefort even denied that it was his task to describe individual species. For him, certain knowledge of plants consisted of recognising the characters which defined a genus, and knowing whether some authority has named and described the genus' species (Tournefort [1694] sig. eii^{v}).

Unlike Cesalpino, Tournefort did not shrink from illustration. Indeed, two of the Elemens' three volumes consist of engraved plates. However, Tournefort's illustrations were not figures of *plants*. Rather, they depicted only "the parts which precisely define the character of each genus" (Tournefort [1694] sig. eii^{r}). He expressly excluded figures of individual species or entire plants, because they would occupy too much space – and because many features, such as leaves, vary too much within a genus to be depicted. The illustrations themselves follow this prescription. With a few exceptions, primarily plants with bulbs, they depict only the flowers and fruit of genera, along with leaves in the case of trees.[41] Like Gesner before him, Tournefort realised that illustrations could teach a beginner to recognise plants much more quickly than could words; but unlike sixteenth-century naturalists, Tournefort did not aim to convey the appearance of an entire plant. Rather, his illustrations served to teach students how to recognise parts; and the last series of plates in his *Elemens* comprised a visual dictionary of botanical terminology similar to those found in modern floras and field manuals (Tournefort [1694] III pl. 423–451).

For Tournefort, the botanists' task of classification in encaptic groups rendered the illustration of a particular plant superfluous or even misleading, as it had for Cesalpino a century earlier. But even naturalists who continued to focus on description, rather than classification, came to see illustrations as a luxury, not a necessity. Such was the case with John Ray's monumental *Historia plantarum* (1686–1704), arguably the most important seventeenth-century botanical work.

In a letter of 1684, Ray had expressed the hope of providing figures for the work – like Gesner, a century before, on pedagogical grounds. One of the reasons for writing the book, Ray said, was "to facilitate the learning of plants without a guide or demonstrator, by so methodising them that and giving such certain and obvious characteristic notes of the genera that it shall not be difficult for any man to find out infallibly any plant especially being assisted by the figure of it."[42] But Ray did not have the financial resources to have plates prepared. His friends thought woodcuts would be inadequate, and a plea to potential subscribers, who might underwrite the preparation of a series of plates, went unanswered (Raven [1950] 213f.).

Ray had wanted illustrations specifically for the learner, especially one without a teacher; another of his reasons for issuing the *Historia plantarum* was also for learners, those who could not afford many books, since Ray's work would contain descriptions of "all the species already described and published."[43] But his inability to prepare figures did not deter Ray from publishing his work; nor did it detract from the text's usefulness to other serious naturalists. Leonard Plukenet, by all accounts a difficult and jealous man, called Ray's history "the best medium to reach heaven, better than the Divinity of the Schools" (Raven [1950] 232).

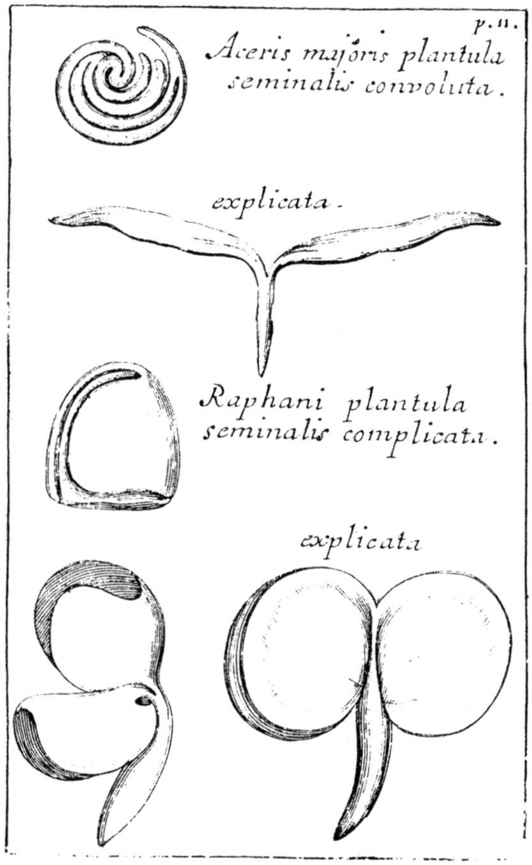

Figure 17
Seed-leaf figure. From Ray: *Methodus plantarum nova* (London 1682).

Ray's earlier *Methodus plantarum nova* (1683) had been illustrated - with exactly one plate, illustrating the different kinds of seed-leaves (fig. 17). This illustration was not intended to describe a particular plant; rather, it was a key to Ray's terminology. Once the reader understood this terminology, verbal descriptions would suffice. But though it did not include other plates, the *Methodus* contained a large number of synoptic tables – in fact, its main body consisted of little else (fig. 18). The table, rather than the illustration, was the characteristic image of late seventeenth-century botany. Illustrations continued to be used in books on exotic botany, such as Van Reede tot Drakenstein's *Hortus malabaricus* – books devoted to plants which had as yet been neither seen nor grown in European collections and gardens (see Heniger [1986]). But their ultimate fate was to be replaced by verbal descriptions.

The decline of illustrations in seventeenth-century botany has few contemporary parallels in other areas of natural history. Zoological works, such as they were, continued to be illustrated; the six volumes of John Johnston's natural history, copiously illustrated with engravings by Matthias Merian, are fitting successors in this regard to Gesner's and Aldrovandi's compilations. One of the most stunning productions of seventeenth-century natural history, Robert Hooke's *Micrographia* (1665), owes its interest to contemporaries and historians largely to the numerous engravings, to which Hooke's text continually refers. Hooke's description of a water-gnat, for instance, is keyed to his engraving by means of letters (fig. 19). Reading his description without following it on the plate is an exercise in frustration. Hooke's language was not up to providing an adequate description of the insect.

I think there are two distinct reasons why botanical illustrations lost importance in the seventeenth century. One of them has to do with the nature of plants, the other with the history of botany. In the first instance, plants are relatively simple in their gross morphology. Theophrastus divided their organs into root, stem, leaf, flower, and seed; all of his sixteenth-century successors found this enumeration to be sufficient. Each of these parts could vary widely between specimens, but it was possible for naturalists to construct a fairly simple grammar and syntax of plant description – simple due to the nature of the plants themselves. Animals, with a greater number of organs whose interrelations were more complicated, proved more difficult to describe adequately in words.

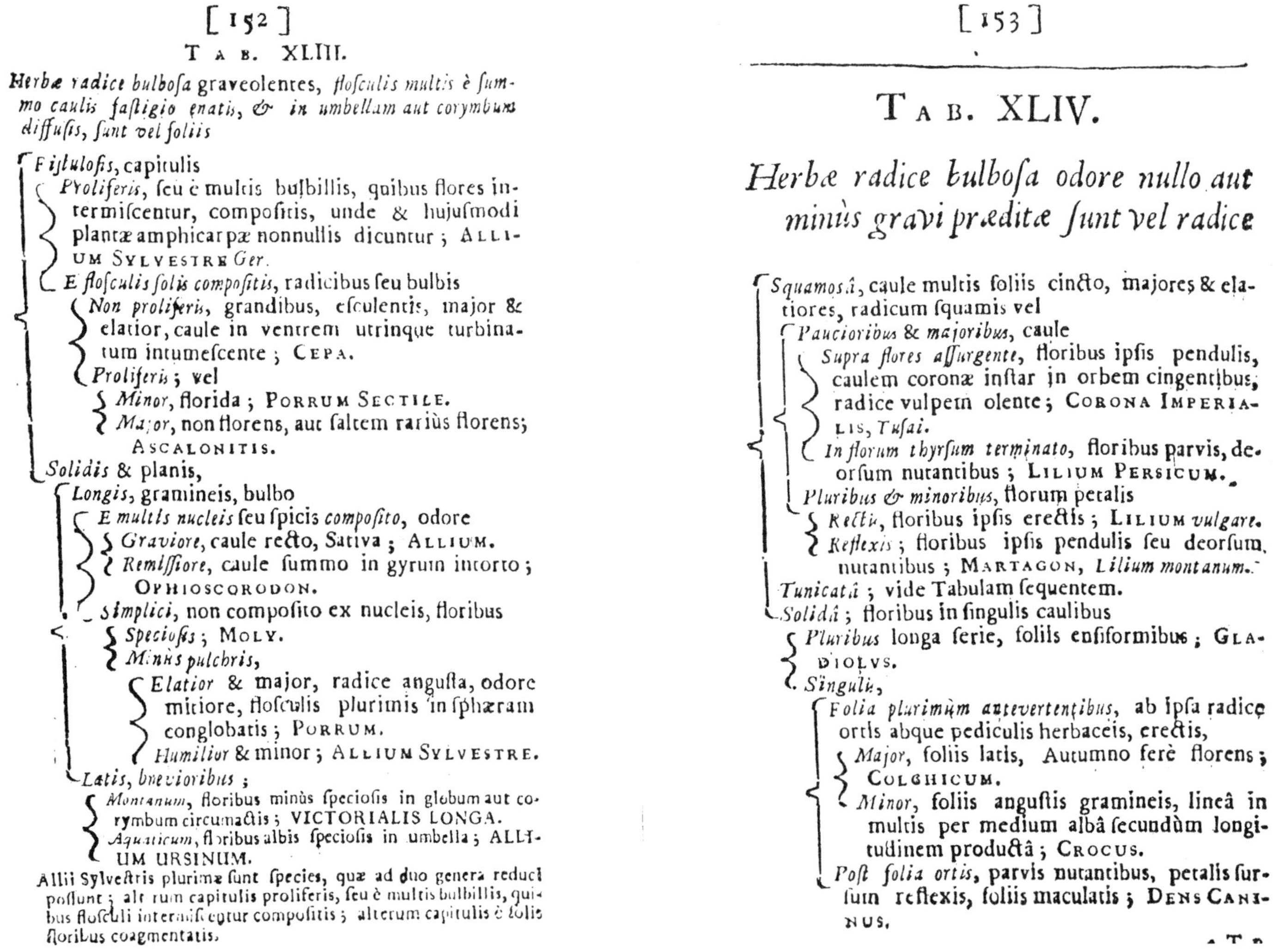
[152]

TAB. XLIII.

Herbæ radice bulbosa graveolentes, *flosculis multis è summo caulis fastigio enatis, & in umbellam aut corymbum diffusis, sunt vel foliis*

Fistulosis, capitulis
- *Proliferis*, seu è multis bulbillis, quibus flores intermiscentur, compositis, unde & hujusmodi plantæ amphicarpæ nonnullis dicuntur; ALLIUM SYLVESTRE *Ger.*
- *E flosculis solis compositis*, radicibus seu bulbis
  - *Non proliferis*, grandibus, esculentis, major & elatior, caule in ventrem utrinque turbinatum intumescente; CEPA.
  - *Proliferis*; vel
    - *Minor*, florida; PORRUM SECTILE.
    - *Major*, non florens, aut saltem rariùs florens; ASCALONITIS.

Solidis & planis,
- *Longis*, gramineis, bulbo
  - *E multis nucleis* seu spicis *composito*, odore
    - *Graviore*, caule recto, Sativa; ALLIUM.
    - *Remissiore*, caule summo in gyrum intorto; OPHIOSCORODON.
  - *Simplici*, non composito ex nucleis, floribus
    - *Speciosis*; MOLY.
    - *Minùs pulchris*,
      - *Elatior* & major, radice angusta, odore mitiore, flosculis plurimis in sphæram conglobatis; PORRUM.
      - *Humilior* & minor; ALLIUM SYLVESTRE.
- *Latis, brevioribus*;
  - *Montanum*, floribus minùs speciosis in globum aut corymbum circumactis; VICTORIALIS LONGA.
  - *Aquaticum*, floribus albis speciosis in umbella; ALLIUM URSINUM.

Allii Sylvestris plurimæ sunt species, quæ ad duo genera reduci possunt; alterum capitulis proliferis, seu è multis bulbillis, quibus flosculi intermiscentur compositis; alterum capitulis è solis floribus coagmentatis.

[153]

TAB. XLIV.

Herbæ radice bulbosa odore nullo aut minùs gravi præditæ sunt vel radice

Squamosâ, caule multis foliis cincto, majores & elatiores, radicum squamis vel
- *Paucioribus* & *majoribus*, caule
  - *Supra flores assurgente*, floribus ipsis pendulis, caulem coronæ instar in orbem cingentibus, radice vulpem olente; CORONA IMPERIALIS, *Tusai.*
  - *In florum thyrsum terminato*, floribus parvis, deorsum nutantibus; LILIUM PERSICUM.
- *Pluribus & minoribus*, florum petalis
  - *Rectis*, floribus ipsis erectis; LILIUM *vulgare.*
  - *Reflexis*; floribus ipsis pendulis seu deorsum nutantibus; MARTAGON, *Lilium montanum.*

Tunicatâ; vide Tabulam sequentem.

Solidâ; floribus in singulis caulibus
- *Pluribus* longa serie, foliis ensiformibus; GLADIOLUS.
- *Singulis*,
  - *Folia plurimùm antevertentibus*, ab ipsa radice ortis abque pediculis herbaceis, erectis,
    - *Major*, foliis latis, Autumno ferè florens; COLCHICUM.
    - *Minor*, foliis angustis gramineis, lineâ in multis per medium albâ secundùm longitudinem productâ; CROCUS.
  - *Post folia ortis*, parvis nutantibus, petalis sursum reflexis, foliis maculatis; DENS CANINUS.

Figure 18
Synoptic table. From Ray: *Methodus plantarum nova* (London 1682).

The other reason, however, lies in the intense effort which sixteenth-century botanists expended in observing plants and developing a vocabulary to express their *differentiae*. Hooke's pioneering work, like Brunfels' and Fuch's in botany, could not draw upon and refine the labours of predecessors. Eventually, descriptive languages for entomology and crystallography – to name only two modern sciences whose subjects Hooke addressed – were developed, languages adequate to express verbally the variety of nature. As those languages were refined, illustrations came to fill the same function they had filled for Konrad Gesner: they introduced beginners to an unfamiliar subject, but they could be dispensed with by skilled practitioners who could see in their mind's eye objects that were described in technical languages.

In the case of Renaissance botany, images played a vital role in the development of such a language. By allowing naturalists to portray those elements which they considered important but could not name, analytic botanical images in the sixteenth century served as an impetus to developing more refined words. When words could adequately describe the morphology of a plant, those images disappeared – leaving room for new images, like tables, serving new functions.

The story I have told, incomplete as it may be, suggests that the history of botanical illustration – and, by implication, all scientific illustration – must take its place within a broader historical narrative that treats both image and text seriously. The crucial ties between images and text are lost when either is treated in isolation, and the resulting gap impoverishes both the history of science and the history of art. Historians of science who see images as decoration fail to consider their real communicative function, while art historians construct a factitious continuity in the history of illustration that masks sig-

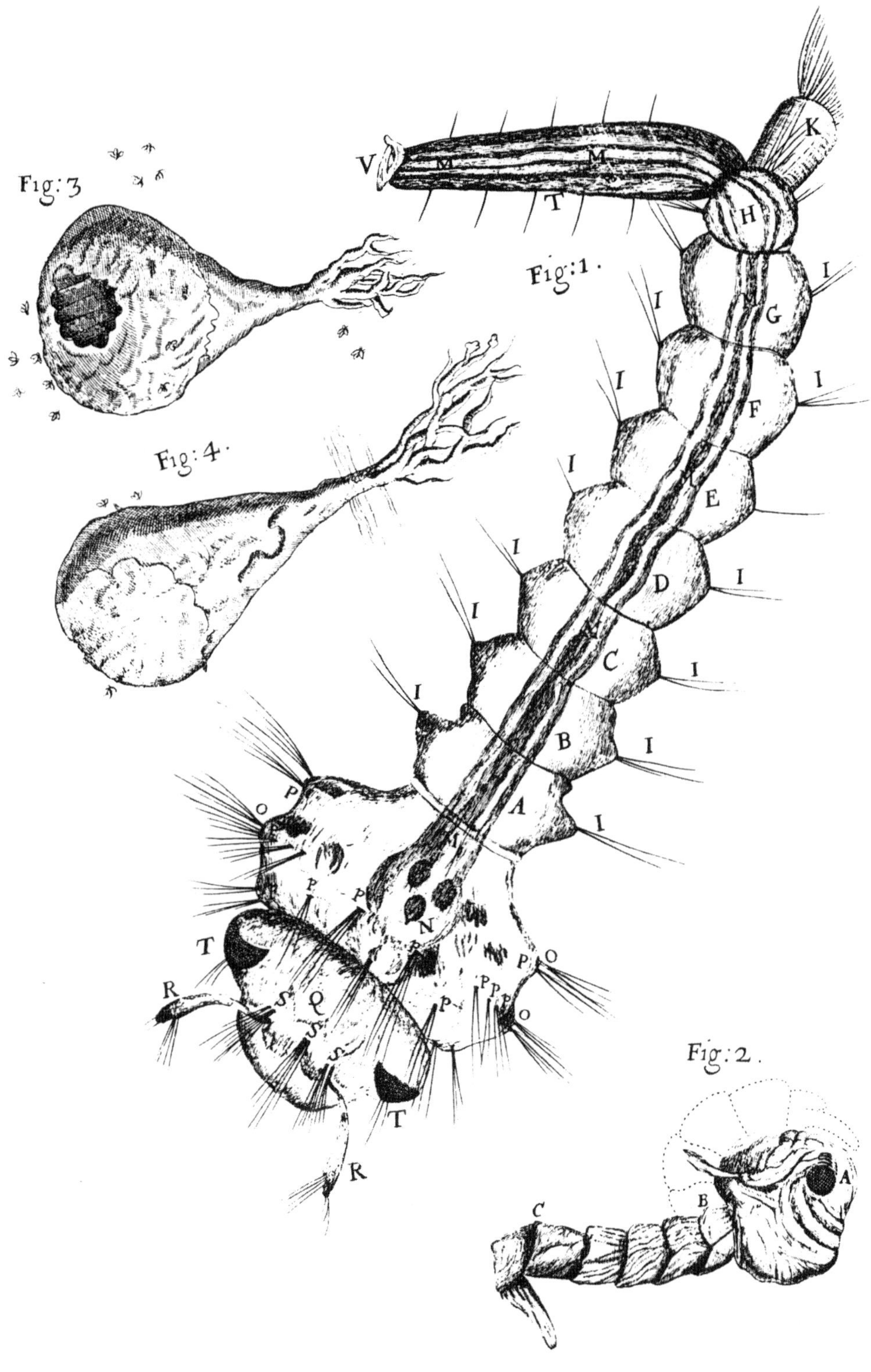

Figure 19
Water-gnat. From Hooke: *Micrographia* (London 1665).

nificant changes in its cognitive role. When images and text appear side-by-side in our sources, we must work side-by-side to fully understand them.

Department of History,
University of Massachusetts, Amherst

Notes

[1] A sampling of the literature includes: Treviranus [1855]; Arber [1986]; Nissen [1951]; Ackerman [1985]; Bridson and White [1990]; Blunt and Stearn [1994].

[2] Even Agnes Arber, for whom illustration was an integral aspect of Renaissance herbals, considered illustrations and descriptions in two different chapters of her seminal work, *Herbals.* Arber [1986].

[3] On the illustrations, rediscovered early in the twentieth century, see Rytz [1936].

[4] Konrad Gesner, preface to Bock [1552] sig. c iiii^r.

[5] Quoted in translation by Arber [1986] 219.

[6] On Fuchs' approach to illustration, see Kusukawa [1997].

[7] On Dioscorides, see Riddle [1985]; on his importance to early Renaissance botanists, see Reeds [1976].

[8] This is a broad generalisation. Writers like Gilbert White and John Muir arranged their observations in a temporal framework, noting down in chronological sequence what they saw, with "incidental" detail from a purely morphological standpoint.

[9] Gesner [1556] 24: "qui in hortis nostris mense Aprili & initio Maii florent, longe suavissimo odoris."

[10] Clusius to Joachim Camerarius II, 2 January 1578, 27 January 1579, and 24 July 1580, in Hunger [1927–43] II 354, 366, 377.

[11] Clusius [1576] 273: "quae cum flore Iridis aut Xyridis essent, erutas tamen bulbacea radice esse deprehendi."

[12] Clusius arrived in Vienna in the autumn of 1573 and made his first botanical expedition there in the spring of 1574 (see chapter 4). He sent the MS. of his *Rariorum stirpium per Pannoniam historia* to Plantin in Antwerp in the spring of 1582: Clusius to Camerarius, 22 May 1582 (Hunger [1927–43] II 388).

[13] Clusius [1583] 382–384 (quotation on p. 384): "Floret autem istic Iulio, interdum etiam Augusto, pro aestatis maturius aut serius incipientis natura, solutis videlicet nivibus: nam ex primis floribus est."

[14] Clusius [1583] 483: "in hortis etiam Maio, cum reliquis eius generis stirpibus."

[15] Bock [1577] fol. 265v: "auff einem runden glatten Bintzechten stengel gesetzt/ wie die köpff auff den Zwybel Pfeyffen."

[16] Bock [1577] fol. 266v: "laßt sich gegen dem Früling/ so bald der Aron herfür kreucht züm ersten sehen/ ist nit allenthalben gemeyn/ doch ist sein gewonliche Statt in den gantzen finstern feuchten unnd nassen Wälden. Ist ein kraut etwann nür mit einem blatt/ etwann mit zweyen blettern/ die vergleichen sich aller ding dem Meyenblümen kraut/ welche Knoblauch stöcklein aber zwey bletter gewinnen/ die tragen im ende des Aprillen weysse Blümlein/ auch den weissen Meyenblümlein gleich/ doch seind die blümlein an disem Knoblauch nit hol/ als der Meyenblümlein/ sondern gestirnt/ unn die blättlein underschiedlich von einander gesetzt. Diser Blümen wurtzel seind weiß/ unnd etwas lang/ ein jede anzüsehen als ein kleiner junger garten Knoblauch/ der nicht uber ein Monat im feld ist gestanden."

[17] Dodoens [1557] 443: "L'Ail de iardin a les fueilles de Gramen ou d'Oignon, entre lesquelles viennent à la seconde année de tiges rondes & creuses, sur lesquelles croissent les fleurs & semence, semblables aux fleurs & semence d'Oignon. La racine est ronde, bulbeuse comme le bulbe d'un Oignon, amassée de plusiers noyaux ioincts ensemble, soubs lesquelz pendent fibres petites & gresles."

[18] Dodoens [1557] 444: "Le tiers Ail appelé Ail d'ours, a communement deux fueilles grandes & larges, pres semblables aux fueilles du grand Muguet, entre lesquelles vient une tige ou deux, sur lesquelles croissent mout de petites fleurs blanches. La racine a la figure d'un Ail nouveau d'odeur & goust tresforte."

[19] Clusius [1583] 211–213: "Allii sive Moly genera, vel foliis, vel floribus, wel capitibus inter se discrepantia: quorum maximam partem ab aliis hactenus minime observatam fuisse existimo."

[20] Clusius [1583] 213: "Primum caulem habet duorum cubitorum altitudine excrescentem, quem ad medium usque ambiunt folia, multo quam Allii generibus latiora, oblonga, porraceis proxima: summo cauli, qui laevis & iunceis est, insidet caput e multis bulbis ex atro purpureis simul congestis satisque crassis consiatum, inter quos in oblongis pediculis nascuntur flores dilute purpurascetes [*sic*] quibus succedunt trigona vascula in quibus semen: radix crassa bulbosaque ex multis nucleis constans, numerosis albis fibris praedita est: sed et bulbuli rotundi radici adhaerent iis fere similes qui in capite nascuntur. Toti plantae Alii gravis odor inest. Iunio et Iulio floret & bulbos fert."

[21] Clusius [1583] 213–216: "flores superiori forma & magnitudine similes, exalbido tamen colore, quorum media foliola secundum longitudinem et utrasque oras vena obsolete purpurascens notat."

[22] Clusius [1576] 321: "Sed & differentia quaedam in Longa observari potest: nam Hispanicae (tametsi huic fere per omnia similis) flos diligentius inspicientibus nonnihil differre conspicitur, quippe cuius extima pars interiore parte purpurei coloris sit, qui in alia herbaceus est: radix item huius extrema parte non ut altera gracilis est in novellis plantis, sed plerumque obtusior."

[23] Bauhin [1620] 27: "III. Allium montanum radice oblonga: priori simile, sed caule altiore & foliis longioribus, capitulo rotundo pallide purpurascente, floribus ex foliolis sex, totidemque staminibus composito." Bauhin added that he had observed this plant in the Hortus Dei near Montpellier (see chapter 4).

[24] Clusius [1583] 484: "licet a Matthiolo et aliis eius historia sit prodita, quoniam meo iudicio brevior apud eos est quam oportebat."

[25] Dioscorides [1934] 489 (I have checked the translation against Mattioli's version); Mattioli [1598] gives the Latin translation of Jean Ruelle: "Est & alterum Umbilicum Veneris genus, cymbalion aliqui vocant, pinguibus foliis, latioribus, ligularum modo, densis circa radicem, medium oculi ambitum describentibus, ut in maiore sempervivo, adstringentis gustus: cauliculo tenui, in quo flores, & semina hyperici: radice maiuscula. Facit eadem, quae Sempervivum."

[26] Dodoens [1557] 29: "La seconde espece a les fueilles larges, espesses, asses rondes, espandues en large tout au tour de la tige, comme à la grandes Ioubarbe, du milieu desquelles provient la tige tendre, portant petites fleurs."

[27] Clusius [1583] 483: "Numerosa & conferta huic sunt folia, oblongis ligulis similia, crassiuscula, mucronata, nonnihil per ambitum crenata, candicantia, gustu acido & subadstringente: e quorum centro, seu medio foliorum exortu caulis plerumque cubitalis assurgit, vel pedalis duntaxat, superiore parte hirsutus, brevioribus foliis hinc inde septus, qui in summo ex singulis foliorum alis pedicellos emittit, quibus insident bini aut terni flores colore candidi, e quinque foliolis compositi, interne veluti sanguineis guttis aspersi, umbilico pallido & splendente, quem decem staminula ambiunt; semen deinde in capitulis perexiguum,

nigrum: radix subest nigra, tenuis, fibrosa, cui secundum matrem multae aliae in orbem adnascuntur plantae, ut integrum cespitem interdum tegant vulgaris Sedi modo."

[28] Clusius [1583] 484: "angustioribus & longioribus foliis est, minime dentatis, sed lanugine alba duntaxat hirsutis, minus incanis, cauliculo breviore, crassioreque & multis foliis septo, quae plerumque tam superne, quam inferne subhirsuta sunt: paucioribus floribus hinc inde secundum caulem ad singulas alas sparsis."

[29] Fuchs [1543] cap. XXXV, sig. H4r: "darauff wachsen vil schöner blümen/ die seind zweyfeltig gefült/ und gewinnen an den schellen fünff gebogne spitz/ fast wie die Rittersporn. Dise blümen seind gemeynlich purpurbraun/ doch findt man auch gantz weiß/ etlich seind rotbraun."

[30] Bock [1552] sig. d vii^r: "At quod ad herbarios depictos attinet, certum est nonnihil illos utilitatis tum adferre, cum vivas herbas, aut omnino non habemus, aut recentes nancisci non possumus. Quisquis vero hortos proprios & hortulanos habet is multas ac varias stirpes plantare, quod olim Antonium Castorem Romae fecisse legimus, earumq; vivas imagines contemplari poterit, ita ut picturis non sit opus, nisi in iis plantis quae maxime peregrinae sunt, quas nec recentes ubiq; locorum intueri licet, & quae nostro solo assuefieri omnino recusant."

[31] Christophe Plantin to Severinus Gobelius, 11 October 1581 (letter no. 954) in Plantin [1916] 315f.; see n. 3 for the costs. To keep costs down, Plantin employed women – at least two of them widows – to colour these illustrations: Arber [1940] 317f. in reprint.

[32] Plantin to Gobelius, 11 October 1581, in Plantin [1916] 315f. The book cost 113 florins coloured; Carolus Clusius received some 400 florins annually a decade later as professor at the University of Leiden.

[33] Gesner in Cordus [1561a] fol. 85v: "Lectorem pleraeq; omnes, ut ad rerum ignotarum cognitionem facilius perveniat, promovebunt;" "ut ubi proxime alia eiusdem generis & per omnia fere similis proponitur pictura."

[34] This point was made explicitly by eighteenth-century naturalists such as René-Antoine Ferchauld de Réaumur, Réaumur [1737] I.1 63f.: "Les Desseins disent bien plus vîte ce qu'ils ont à dire; ils ne peuvent pourtant pas toujours représenter tout ce qu'on voudroit qu'ils représentassent, mais ils soutiennent toujours l'imagination, & avec leur secours on lit & on entend des descriptions qu'autrement on n'entendroit ni ne liroit."

[35] Clusius [1605] 1: "ornatissimus vir, multisque peregrinationibus clarus, Fabricius Mordente, Salernitanus (qui aliquot annis Goae vixerat, & vicina illi urbi loca perlustraret)."

[36] Clusius [1605] 1f.: "unicam arborem densissimam silvam efficere."

[37] Clusius [1605] 2: "Huius arboris iconem, quam concinne fieri potuit, ex Fabricii narratione, accedente etiam ipsius iudicio."

[38] See the "Anser Magellicanus," or Penguin, Clusius [1605] 101, following the text of the narrative of Mahu and de Cordes' voyage: Wieder [1923] I 234–236. In his *Rariorum stirpium per Pannoniam* (Clusius [1583] 419f.), Clusius reproduced a description and illustration of a type of geranium, both of which had been sent by his English friend Thomas Penny.

[39] Contemporary manuscript illustrations also used perspective and modelling to depict plants, as in the seventeenth-century *vélins* of the Muséum d'Histoire Naturelle, Paris. Laissus [1967].

[40] I have developed this idea further in my essay on "Encyclopædism in Renaissance botany: From *historia* to *pinax*," Ogilvie [1997].

[41] Examples where entire plant, or much, is illustrated are pl. 181, Colchicum; pl. 183, Crocus; pl. 185, Narcissus (with stem and flower); pl. 189, iris with bulbous root. pl. 291, Bistorta, also includes the roots.

[42] Ray [1848] 160f., quoted by Raven [1950] 213.

[43] Ray, same letter, quoted in Raven [1950] 213.

References

Ackerman, J.
1985 "The involvement of artists in Renaissance science." In: *Science and the Arts in the Renaissance*, eds. J. W. Shirley and F. D. Hoeniger. London and Toronto.

Arber, A.
1940 "The colouring of sixteenth-century herbals." In: *Nature* 145. Reprinted in Arber [1986] 315–318.

1986 *Herbals: Their Origin and Evolution; a Chapter in the History of Botany, 1470-1670.* Cambridge.

Aristotle
1965–91 *History of Animals*, trans. A. L. Peck and D. M. Balme. Cambridge MA.

Bauhin, C.
1620 *Prodromos Theatri Botanici.* Frankfurt.

Blunt, W. and W. T. Stearn
1994 *The Art of Botanical Illustration.* London.

Bock, H.
1539 *New Kreütter Buch von underscheydt, würckung und namen* ... Strasbourg.

1552 *De stirpium, maxime earum, quae in Germania nostra nascuntur, ... Commentariorum libri tres*, trans. D. Kyber. Strasbourg.

1577 *Kreütterbuch. Darin underscheidt, Namen und Würckung der Kreütter, Stauden, Hecken und Beumen, sampt ihren Früchten, so inn Teutschen Landen wachsen* ..., ed. Melchior Sebizius. Strasbourg.

Breyne, J.
1878 *Exoticarum aliarumque minus cognitarum plantarum centuria prima.* Danzig.

Bridson, G. D. R. and J. J. White
1990 *Plant, Animal and Anatomical Illustration in Art and Science: A Bibliographical Guide from the Sixteenth Century to the Present Day.* Detroit.

Cesalpino, A.
1583 *De plantis libri XVI.* Florence.

Chabrey, D.
1666 *Stirpium icones et sciagraphia.* Geneva.

Clusius, C.
1576 *Rariorum aliquot stirpium per Hispanias observatarum Historia, libris duobus expressa.* Antwerp.

1583 *Rariorum aliquot stirpium, per Pannoniam, Austriam, & vicinas quasdam provincias observatarum historia, quatuor libris expressa.* Antwerp.

1605 *Exoticorum libri decem: quibus animalium, plantarum, aromatum, aliorumque peregrinorum Fructuum historiae describuntur.* [Leiden].

Cordus, V.
1561a *Annotationes in Pedacii Dioscoridis Anazarbei de Medica materia libros V* [with other of Cordus' works appended], ed. C. Gesner. Lyon.

1561b "Historia plantarum." In: Cordus [1561a].

Dioscorides
1934 *The Greek herbal of Dioscorides*, trans. J. Goodyer, ed. R. T. Gunther. Oxford.

Dodoens, R.
1557 *Histoire des plantes*, traduite par C. de L'Ecluse. Anvers.

Fuchs, L.
1543 *New Kreüterbuch* ... Basel.

Gesner, C.
1551 *Historiae animalium lib. I. de quadrupedibus viviparis: Opus philosophis, medicis, grammaticis, philologis, poetis, & omnibus rerum linguarumq; variarum studiosis, utilissimum simul iucundissimumq; futurum.* Zurich.

1556 *De raris et admirandis herbis, quae sive quod noctu luceant, sive alias ob causas, Lunariae nominantur, Commentariolus* ... Zurich.

Greene, E. L.
1983 *Landmarks of Botanical History*, ed. F. N. Egerton. Stanford.

Heniger, J.
1986 *Hendrik Adriaan van Reede tot Drakenstein (1636–1691) and Hortus Malabaricus: A Contribution to the History of Dutch Colonial Botany.* Rotterdam.

Huizinga, J.
1996 *The Autumn of the Middle Ages*, trans. R. J. Payton and U. Mammitzsch. Chicago.

Hunger, F. W. T.
1927–43 *Charles de l'Escluse (Carolus Clusius) Nederlandsch kruidkundige, 1526–1609*, 2 vols. 's-Gravenhage.

Ivins, W. M. Jr.
1953 *Prints and visual communication.* Cambridge MA.

Kenseth, J. (ed.)
1991 *The age of the marvelous.* Hanover NH.

Kusukawa, S.
1997 "Leonhart Fuchs on the importance of pictures." In: *Journal of the History of Ideas* 58: 403–427.

Laissus, Y.
1967 *Les vélins du Muséum [Palais de la Découverte].* Paris.

MacDougall, E. B.
1991 "A paradise of plants: Exotica, rarities, and botanical fancies." In: *The Age of the Marvelous*, ed. J. Kenseth. Hanover NH.

Mattioli, P. A.
1598 *Opera quae extant omnia, ed. Caspar Bauhin.* Frankfurt.

Nissen, C.
1951 *Die botanische Buchillustration: Ihre Geschichte und Bibliographie.* Stuttgart.

Ogilvie, B. W.
1997 "Encyclopædism in Renaissance botany: From historia to pinax." In: *Pre-modern Encyclopædic Texts*, ed. P. Binkley, 89–99. Leiden.

Plantin, C.
1916 *Correspondance*, ed. J. Denucé, vol. 6. Antwerp and 's-Gravenhage.

Raven, C. E.
1950 *John Ray, Naturalist: His Life and Works.* Cambridge.

Ray, J.
1848 *Correspondence.* London.

Réaumur, R.-A. F. de
1737 *Mémoires pour servir à l'histoire des insects.* Amsterdam.

Reeds, K. M.
1976 "Renaissance humanism and botany." In: *Annals of Science* 33: 519–542.

Riddle, J. M.
1985 *Dioscorides on Pharmacy and Medicine.* Austin.

Rytz, W.
1936 *Pflanzenaquarelle des Hans Weiditz aus dem Jahre 1529.* Bern.

Spiegel, A. van de
1606 *Isagoges in rem herbarium libri duo.* Passau.

Tournefort, J. P. de
1694 *Élémens de botanique, ou Méthode pour connoître les plantes.* Paris.

Treviranus, L. C.
1855 *Die Anwendung des Holzschnittes zur bildlichen Darstellung von Pflanzen, nach Entstehung, Blüthe, Verfall und Restauration.* Leipzig.

Wieder, F. C. (ed.)
1923 *De reis van Mahu en De Cordes door de straat van Magalhães naar Zuid-Amerika en Japan.* 's-Gravenhage.

Zweifel, F. W.
1988 *A Handbook of Biological Illustration.* Chicago.

Notes on the Function of Early Zoological Imagery

Allan Ellenius

When writing a review for *ISIS*, the Journal for the History of Science, published some years ago, I started by taking the reader to the Florentine Baptistery, to pause in front of the famous bronze doors by Lorenzo Ghiberti. Framing the biblical scenes on the door of Paradise are a number of plants and animals portrayed with painstaking naturalism and thus in accordance with the artist's own pledges in his *Commentarii*. The animals are easily identified, except for a few that still defy attempts at finding their equivalents in nature. Although non-existent, they seem to have been inspired by living specimens. Illusion plays a double trick on the spectator. The pictorial idea may originally have been inherited from manuscript illustrations of the fifteenth century where recordings of the biblical stories were combined with naturalistic digressions in the margins, often displaying an extraordinary degree of empirical observation. To Ghiberti, a general character of naturalistic imitation was enough; the function of the decorative framework did not include more rigorous demands of illusion of the kind that was soon to be developed by artists such as Antonio Pisanello and Leonardo.

In a contribution to a symposium at Uppsala university in 1983, called The Natural Sciences and the Arts, professor James Ackerman gave a lecture on *Early Renaissance Naturalism and Scientific Illustration* (Ackerman [1985] 2). He made the important point that naturalism in the meaning of truth to nature could only be accurate within a particular context agreed upon by the creator as well as the receiver. The ways in which artists contributed to the aims of science in the early Renaissance were described by Renaissance writers already as *natura naturans*, meaning a kind of naturalism that did not adjust visual evidence to an ideal concept of order which was, in turn, called *natura naturata*.

Accordingly, there is no gradual development towards some kind of perfect illusion; rather on the contrary, sometimes backlashes occur, to the extent that we adopt latter day concepts of naturalism as standards when describing the images and their relation to visual reality. Images are always embedded in words that provide the spectator with ideas of interpretation that may originate from different kinds of sources.

What makes this problem an interesting challenge is that the symbiosis of art and science was taken for granted, a situation that was to last up to the second half of the seventeenth century. Moreover, this cross-fertilisation of art and scientific ambitions was wedded to the idea of progress. The search for truth introduced a new dynamics which at times made older formulae obsolete; by making *dimostrazioni*, artists proved their skill in solving new and often intricate problems of composition, movement and action. It has been maintained, and with good reason, that the prime mover of the development was the emergence of scientific discoveries, as can be seen from important achievements in the field of optical sciences. Visual conquests were thus due to a competitive impulse.

Iconographical systems with mythological and allegorical motifs occasionally included contributions that corresponded to far-reaching naturalistic ambitions. We shall take one example as a starting point for a discussion of the budding science of ornithology.

Among the commissions carried out by Raphael and his workshop was the garden loggia of Villa Farnesina in Trastevere (1518–1519), produced at the initiative of Agostino Chigi and, in the centre of the ceiling, depicting the wedding of Amor and Psyche.[1] The loggia is an illusionistic arbour with decorative trellises that imitate a garden bower, thus forming a smooth transition from the ground-loggia to the garden. Against the painted sky that shines through the painted bower, flying birds are to be seen in the vaultings. Some of them clearly announce their origin in studies of nature and most probably were carried out by Giovanni da Udine, who also painted the garlands with flowers, fruits and vegetables that are such a conspicuous part of the ceiling. His skill in rendering *cose naturali* was praised by Vasari:

> *Giovanni in questo genere di pitture ha passato tutti coloro che in simili cose hanno meglio imitata la natura, percìo che, oltre all'altre cose, insino i fiori del sambuco, del finocchio e dell'altre cose minori vi sono veramente stupendissimi* (Caneva [1992] 32).

An extensive kind of naturalism was accepted within iconological systems that were otherwise dominated by classicising and idealising ambitions. The art of describing conformed very well to the hierarchy that emanated from Alberti and was further elaborated by the academies.

The program of the loggia di Psiche was inspired by Apuleius and his *Golden Ass*. Interpretations have been offered ever since the beginning of this century, mostly emphasising the function of an allegory of love in the neo-Platonic sense of the word. One argument has been the inclusion of three women with Cupid that were not to be found in the classical source; they have been interpreted as the three Graces, who played an important role in neo-Platonic thought as expounded by Marsilio Ficino. According to Erwin Panofsky, Cupid is entreating the three Graces to protect Psyche (Panofsky [1960] 191, n. 3). A more recent interpretation departs from the function of the suburban villa and accentuates the character of sensuality; the display of classical mythology would then be explained by the enthusiasm for love and beauty in man and nature, without far-reaching statements of iconological subtleties (Scalabroni [1983]). It should be remembered that Italian *villeggiatura* created a particular kind of leisure environment that may have been regarded as central to the commission. In 1491, Niccolò da Correggio published the first edition of his celebrated poem on Psyche in which he describes a garden that included an aviary, a fruit garden and an animal park. Early visitors to the Villa Farnesina recalled the famous gardens of classical antiquity when describing the natural environment, and Bellori spoke of "la vaghezza naturale del sito." I would point to the fact that ancient mythology around 1500 was wedded to an increasing sense of naturalism that strengthened the realistic message.

Returning to the birds, the plumage and the general character reveal a couple of kingfishers (*Alcedo atthis ispida*), ingeniously related to their natural surroundings (figs. 1, 2). They are combined with a putto holding a trident, the well-known attribute of Neptune; associations with the sea are confirmed by the landscape below which gives an atmospheric impression of reflecting waters, probably indicating marshes as well. The painter may have been acquainted with stuffed birds; but he was not satisfied with repeating formulae of profiles like those to be found in medieval manuscripts and model books. Flight silhouettes vary and suggest the impression of movement. Compared to the heraldic eagle and the thunderbolt as attributes of Jupiter, seen at the left, the kingfishers reveal a less conventional rendering. The mythology of the kingfishers, or halcyons, goes back to classical antiquity and is in fact one of the oldest tales in Western poetry. They were assumed to breed in mid-winter when Zeus miraculously calmed the waves; their mutual love was proverbial, to the point of the female partner carrying the old and disabled male on her back.

The experienced ornithologist may have noticed that the breasts are not painted in the deep red colour that is typical for the species. Young birds, however, have a pale reddish colour that is gradually transformed into the appearance of the adults. In the same vaulting, there is another bird, depicted from above and with the beak turned downwards. There can be no doubt that this is another

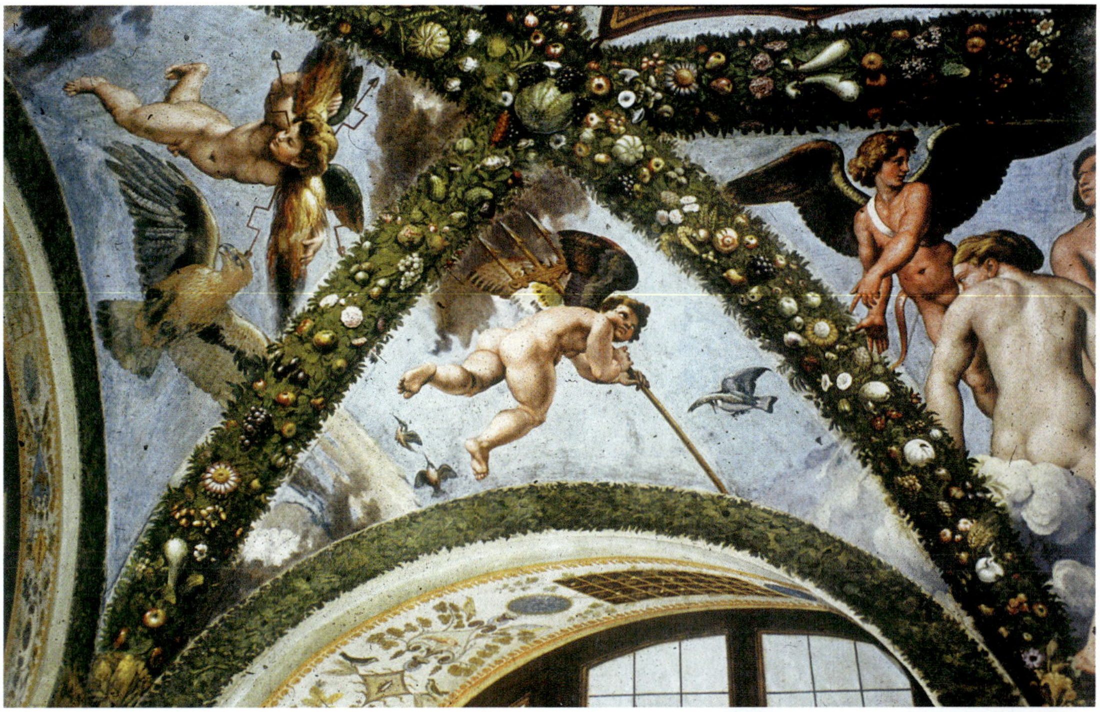

Figure 1
Fresco in the Loggia of the Villa Farnesina, Rome, by Raphael and his workshop(1518–1519), detail. (Photo: Ellenius.)

Figure 2
Fresco in the Loggia of the Villa Farnesina, Rome, by Raphael and his workshop (1518–1519), detail. (Photo: Ellenius.)

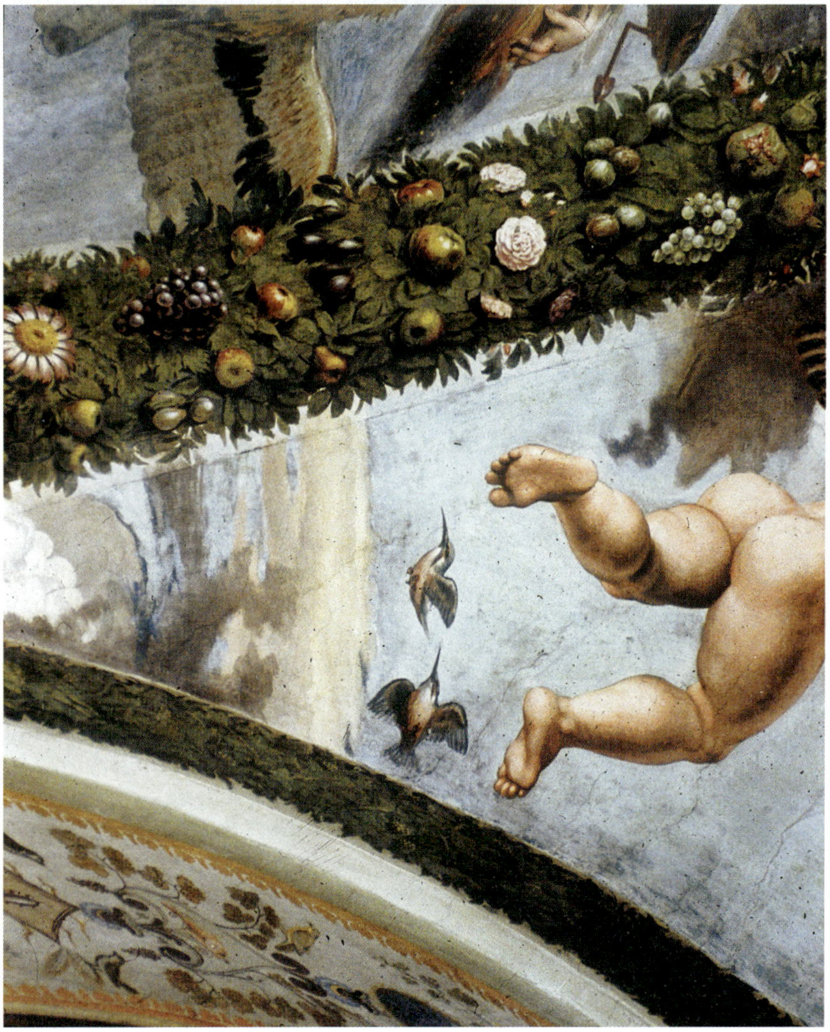

indication of the maritime environment; the plumage reveals a grey kingfisher (*Ceryle rudis*).

According to Vasari, Giovanni da Udine had accompanied his father on hunting and bird-catching, thereby acquiring first-hand knowledge of the animals (Caneva [1992] 31). He was among those who belonged to the team that decorated the Vatican loggias, almost contemporary with the loggia di Psiche. On one of the pilasters there he painted a young farmer, or rather, a rambling naturalist, an ornithologist if you like, positioned close to the trunk of an oak and watching birds that are sitting or flying around the upper part of the tree. Moreover, it is a striking fact that images from the natural world usually crept in, as it were, from the margins; the process is well known from the emergence of genres such as landscape, still-life, flower and animal painting in the Netherlands during the sixteenth century.

In the loggia di Psiche, the function of the imagery seems to explain the oscillation between the classicising rendering of the ancient mythology on the one hand, and the naturalistic images of the birds that can be combined with the painstaking naturalism of the garlands on the other. The spectator came to dwell in a space between different kinds of illusion; deciphering them demanded knowledge from such different sources as the literary heritage of classical antiquity, spiced with the lingering tradition of medieval chivalry and courtly love, and a fresh experience of visual reality. The allegorical theme, with or without neo-Platonic love permeating earth, may well have paved the way for a new rational and emotional approach to natural phenomena. If we accept the interpretation of Cupid calling the attention of the Graces to the divine beauty of Psyche on Earth, then the function of the scene nearby seems quite logical; his gesture may be said to open up a sphere of secular art that responded to the demands of *invenzione* as well as to a meticulous kind of observation that reveals the mind of a naturalist. At this point, it is close at hand to recall the achievements of Leonardo and his empirical studies of plants and animals; although acquainted with neo-Platonic thought, he was of course primarily the defender of sensory knowledge. But there is a strong echo of neo-Platonic idealism when he speaks of the lover who "is moved by the thing beloved as the senses are by the sensible object, and is united with it and become one and the same" (Kemp [1981] 106).

Reliable accounts of animals were, in other words, not necessarily the result of a current liberation from a mythological and literary past. Rather, in order to give a convincing expression of the mythological message, it was of advantage to gain a deeper knowledge of animals, their appearance and habits. Artistic skill became an *ancilla scientiae*, without consciously aiming at the function of acuity within the framework of institutionalised science. This, again, would be a reminder that the search for truth in natural history could be carried out far from the centres of natural history science in Zurich and Bologna, with their pioneering achievements of Konrad Gesner and Ulisse Aldrovandi. High standards of truth were aimed at by painters like Hieronymus Bosch, and by others who certainly did not share the rationalist ideas of those naturalists who had the ambition of performing a *catalogazione della natura*, to borrow the subtitle of a recent work by Professor Giuseppe Olmi (Olmi [1992]). As we shall see, the encyclopaedic ambitions also included the ways in which birds nest.

It is a paradox that the woodcut illustrations generally conveyed a fairly coarse and summary treatment of the shape of animals that had originally been recorded in naturalistic watercolour paintings, made under the supervision of the naturalists themselves. The team of artists working for Ulisse Aldrovandi produced a great number of animal portraits, as can be seen from the rich Aldrovandi collection in the university library of Bologna. The function of the images in *Historia animalium* was restricted to give a general idea of the animal; much depended on how well the reader was acquainted with the linear syntax adopted by those who carried out the woodcuts. Apart from the pioneering observations made by William Ivins, no systematic account seems to have been given of the ways and methods in which plumage and other details of ornithological portraits were recorded (Ivins [1953]); the relation between creator and viewer was certainly of another kind than that in cases where the viewer saw and experienced the water-coloured original. This situation was brought to a head in the early eighteenth century when dreams of an exact pictorial language arose among naturalists who were also prominent artists.

The step from myth to reality was complicated due to the psychological implications. Prior knowledge could produce extraordinary situations, even to masters of empirical observation. When Andreas Vesalius performed his dissections in the chancel of S. Francesco at Bologna in 1540, he opened the abdomen and described the vermiform appendix and the appendix; but he did so not with the visual acuity that could be expected, but in accordance with Galen's observations that had once been made at dissections of animals (Eriksson [1959]). In the *Fabrica* of 1543, however, this anomaly disappeared.

Biological imagery obviously has to be discussed in relation to the art of describing; the inventories of natural history phenomena that were made up by early naturalists aimed at cataloguing outward shapes and, occasionally, anatomical and physiological details. But difficulties tended to emerge even for the most sharp sighted and meticulous naturalists, especially when descriptions of the shape had to be completed by accounts of the colours.

Different ways of confronting the issue can be traced back as far as Aristotle.[2] Straightforward attempts at description were made by resorting to the range of local colours, mingled with metaphors and aesthetic judgements. According to Aristotle, the kingfisher is blue, green and slightly purple-coloured; the colours are not distinctly visible, the result being a shimmering effect. The next step was taken by Albertus Magnus, to whom the green and blue look like sapphire when exposed to the radiating light of the sun, while the breast is compared to glowing coal. New metaphors are mobilised by Gesner, to whom the colours imitate the rainbow and even remind him of the colouristic effect of silk; he has noticed the same effect in lapwings and ravens. Subjectivism is given free rein when he speaks of the blue colour as creating a glaring impression to the point of jarring the eyes. In the German edition of his ornithology, he tries a characterization that is subtle indeed when describing the wings as "grün, aber der innertheil an denselbigen zeucht sich purpurfarb, mit violfarb vermischt, wie auch die blawe Federn in den mitten auff dem Rücken" (Gesner [1550] 57). Aldrovandi, finally, speaks of the blue colour as approaching and being mixed up with green colour which creates an effect that is said to be most delightful to the eye (*visu iucundissima*) (Aldrovandi [1599-1603] 519). In other words, description can not eliminate the problems produced by the psychology of perception. It is fair to say that painters recording birds in their environments at the time under discussion were hardly aware of the effects of changing sources of light. Discoveries of the interplay of colour and natural surroundings had to wait until the nineteenth century.

The empirical objective had even more complications. Leonardo's empirical approach was coloured by the existence of visual schemes that emanated from visual hypotheses; an experiment yielded a result, which, in turn, seldom was taken as a starting point for further enquiries and successive changes for the purpose of generating new visual recordings. This kind of empiricism was, of course, at variance with modern methods in the empirical sciences. Turning to the images illustrating Gesner, Aldrovandi and other sixteenth century naturalists, the situation was more or less the same. The empirical approach produced descriptive images; these did not necessarily, however, lead to further questions as to the shape and physiology of he animals. To take one conspicuous example.

A sensation was caused in the middle of the sixteenth century when stuffed birds of paradise appeared in Europe and, as has been shown by Wolfgang Harms, immediately were described by Girolamo Cardano (Harms [1985]). The feet had been removed, however, which brought about the idea that the birds were *apodes*, without feet; they were assumed to live all their life on the wings which created problems when it came to explain their breeding. Myth and emblematic meaning took over the explanation and the birds became symbols of virtue and soaring ambitions. Aldrovandi referred to emblematic sentences like *Sine pondere sursum* and *Sic animus petat alta* (Aldrovandi [1599–1603] 809). In his para-

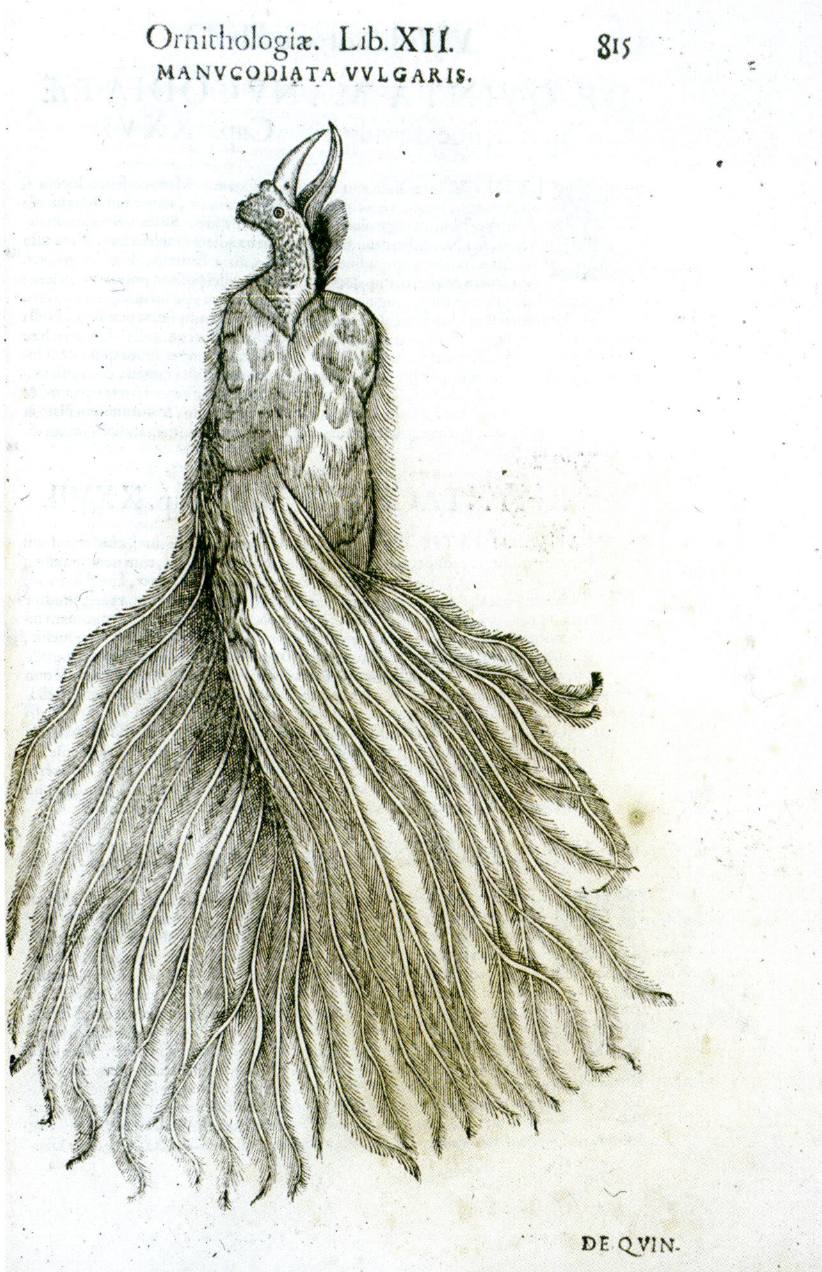

Figure 3
Bird of paradise. From Aldrovandi: *Ornithologiae libri XX* (Bologna 1599–1603). (Photo: Uppsala University Library.)

graph on the birds of paradise he published no less than five portraits, all of them showing flying birds, occasionally in cloudscapes (fig. 3). The artist had no doubt been inspired by stuffed birds in front of him and took the opportunity to indulge in the ornamental pattern created by the tail feathers. It did not occur to the naturalist that making comparative observations of kindred species would have been a means of questioning conditions that were in fact unique. There is a difference between these images and those of other birds illustrated in Aldrovandi's treatise; no ready-made formulae existed that could be used when depicting the birds of paradise, the result being a remarkable freshness of observation.

The step from these sixteenth-century images to a watercolour by Rembrandt showing a bird of paradise is not as big as could be expected (fig. 4). Rembrandt had in his private *kunstkammer* a stuffed specimen, preserved as an exotic miracle, along with sponges, shells, minerals and weapons (Freedberg [1991] 416 ff.). He had apparently been captured by the beauty of the bird, without asking any questions about anatomical anomalies. But they are painted rather as objects on display than as living creatures, as can be seen from the parallel strokes of the pen near the bird to the left. The descriptive character prevails, in spite of the masterly handling of the medium.

Figure 4
Birds of paradise. Watercoloured drawing by Rembrandt. Louvre, Cabinet des Dessins, Inv. RF 4687.

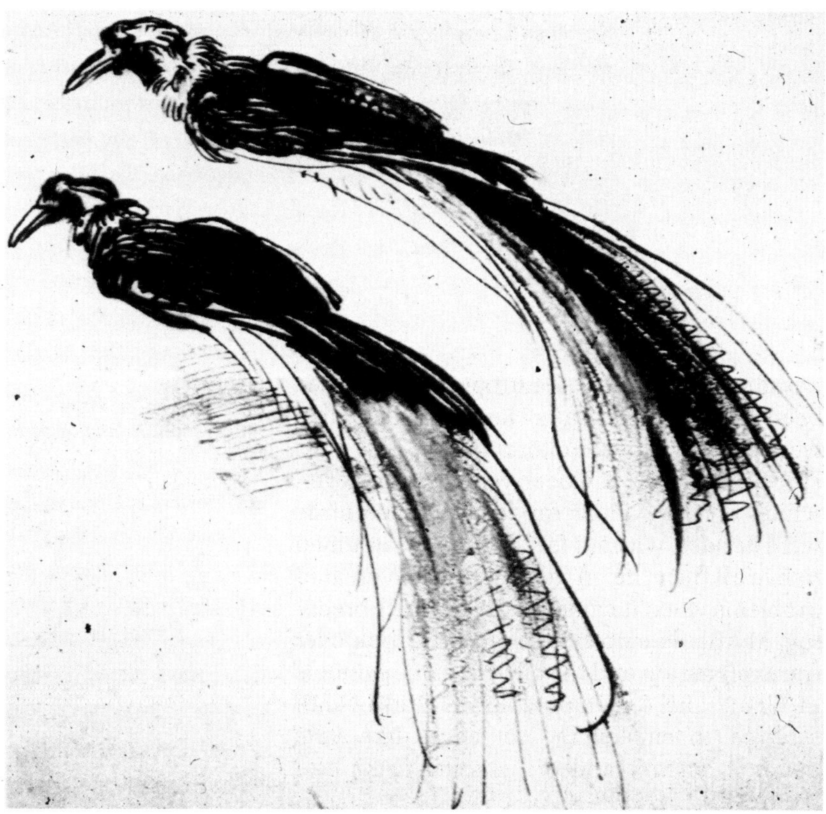

Let us now return to the kingfishers, to see whether the budding science of zoology could contribute to the extension of knowledge by introducing further aspects of the bird, its appearance and habits. It will be necessary again to recollect the mythical past in order to decipher the illustrations that were combined with the literary digressions.

Great interest was devoted to the nesting and breeding of the halcyons. Gesner gave a first and very detailed description of the nest taken from a series of authors dating back to Aristotle. He confesses that nobody had seen the nest; yet, he embarks upon a reconstruction of the maritime nest (fig. 5). Andrea Alciati had introduced the breeding bird in his *Emblemata* (1531), together with the lemma *Ex pace ubertas* (Alciati [1600] 609). The halcyon days in midwinter when Zeus calmed the winds to make possible the survival of the bird belonged to the mythological heritage. Although sceptical to the idea of the nest floating on the waves, Gesner paradoxically remained faithful to the literary tradition and even submitted an illustration. In the German translation of his *Historia animalium*, he reported on the halcyon and its remarkable nesting activity:

> *Darum was Weissheit und Kunst sie an ihrem Nest zumachen/ brauchen/ ist nur ein Wunder zusagen. Dann der Alcyon macht mit keinem andern Werkzeug dann allein mit seinem Schnabel sein Nest/ ja er zimmert diss als ein Schiff/ dieweils es ein Werck ist/ das von den Wällen nicht vmgekehret/ noch ertrenckt/ mag werden/ dann er flechtet leine Fischgrät alls ein Wupp ineinander/ also/ dass er/ etliche/ gleich als den Zettel/ gerad leget, etc.* (Gesner, Thierbuch, 30).

In any case, Gesner refers to Alciati and accepts the idea of the nest as being fastened to a rock, illustrated by Alciati in a way reminiscent of paintings showing the pelican feeding her young. He also accepted the idea of the nest as built on the shores. One of the literary descriptions of the nest links up with the lines already quoted and speaks of an oblong shape with narrow openings that prevented water from pouring into the nest; moreover, the openings were perfectly adjusted to the shape of the birds. According to Basilius in his *Hexaemeron*, the nest was built in the middle of the sea, and with admirable skill. Pliny gave the most detailed description by speaking of an oblong shape reminiscent of a sponge. Gesner refrains, however, from having an illustration made, probably well aware of the uncertainties as to the real shape.

This cautious attitude disappeared when Aldrovandi turned to the traditions collected by Gesner (Aldrovandi [1642] XIII 720). The idea of the nest floating on the waves apparently appealed to him when he had it illustrated in his *Historia animalium*. His illustration comes very close to Pliny's description and also seems to have been inspired by his identification of the halcyon with a bird he calls *ispida*. When coming to the relation between halcyon and ispida, a topic much discussed among sixteenth-century naturalists, Aldrovandi referred to Gesner who argued that they were in fact one and the same species. As can be seen from the woodcuts, however, the linear syntax is by no means the same; the fairly detailed description in the case of Gesner has been replaced by Aldrovandi's synthetic rendering of the shape (figs. 6, 7). Textual information was indispensable as a necessary link to the visual image. The shape of the nest provides another key to the fantasy of the illustrator which came to have its impact on ideas about halcyons up to the end of the seventeenth century. To take just one example from the outskirts of seventeenth century Europe.

In 1682, the young student Erik Humbla defended his dissertation on the halcyon at the university of Åbo in Finland, at that time part of the kingdom of Sweden (Humbla [1682]). The topic is announced not only in the title but also by means of a woodcut showing the kingfisher, combined with fragments of a poem that immediately indicates the tradition to which this dissertation belongs. The text implies that the author himself even carried out the illustrations. It may well be questioned whether he had ever seen a kingfisher; he says that the bird is unknown in his own country. As producing a specimen of learning, he throughout links up to the tradition we have already discussed. The image on the title page has some likeness with older illustrations; yet it seems to have a character of its own. The author repeatedly refers to his illustrations when he follows Aldrovandi by identifying the halcyon bird with the species called ispida. In discussing the nest, the author again refers to Aldrovandi who reports that a nest had been brought to him and then had been illustrated in his work. The author takes over this image but nevertheless maintains a sceptical attitude; the illustration does not seem to correspond with the varying testimonies from the historical writers concerning the true shape of the nest (fig. 8). The conclusion is: *dubium sane est.* As to the place of the nest in the maritime

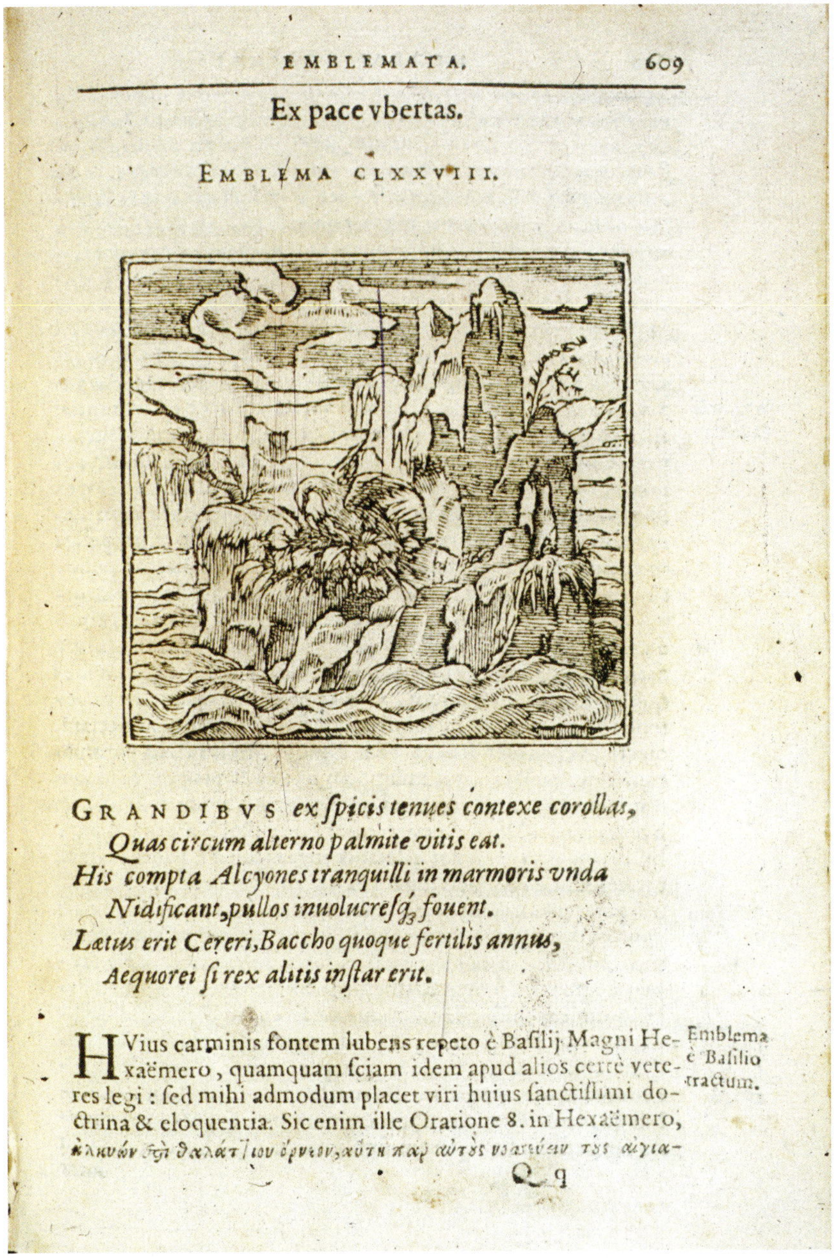

EMBLEMATA. 609

Ex pace vbertas.

EMBLEMA CLXXVIII.

GRANDIBVS ex ſpicis tenues contexe corollas,
Quas circum alterno palmite vitis eat.
His compta Alcyones tranquilli in marmoris vnda
Nidificant, pullos inuolucreſq; fouent.
Lætus erit Cereri, Baccho quoque fertilis annus,
Aequorei ſi rex alitis inſtar erit.

HVius carminis fontem lubens repeto è Baſilij Magni Hexaëmero, quamquam ſciam idem apud alios certè veteres legi: ſed mihi admodum placet viri huius ſanctiſſimi doctrina & eloquentia. Sic enim ille Oratione 8. in Hexaëmero, ἀλκυὼν ἐστὶ θαλάττιον ὄρνεον, αὕτη παρ' αὐτοὺς νοσσεύειν τοὺς αἰγια- (Emblema è Baſilio tractum.)

Q q

Figure 5
"Ex pace ubertas." Engraving from Alciati: *Emblemata* (Lyon1600). (Photo: Dept. of Art History, Uppsala University.)

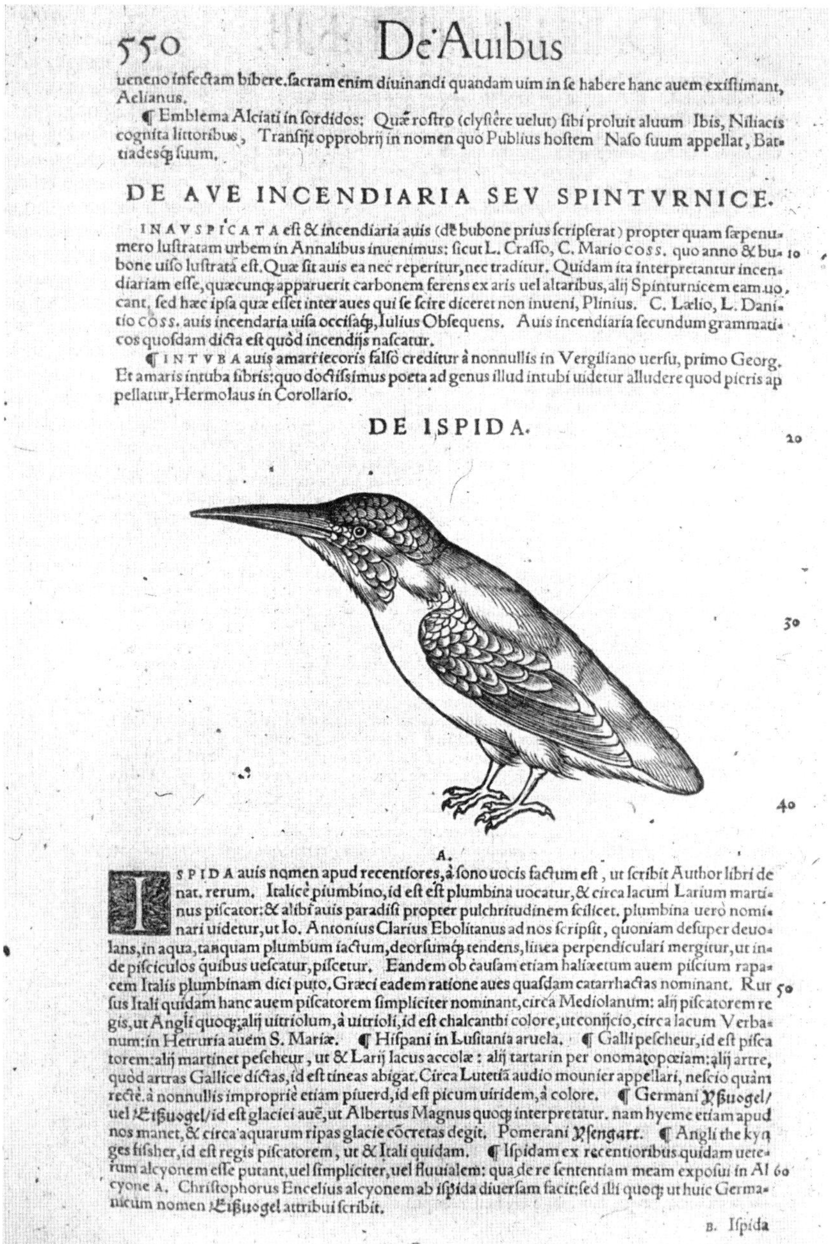
550 De Auibus

ueneno infectam bibere. sacram enim diuinandi quandam uim in se habere hanc auem existimant, Aelianus.

¶ Emblema Alciati in sordidos: Quæ rostro (clystere uelut) sibi proluit aluum Ibis, Niliacis cognita littoribus, Transijt opprobrij in nomen quo Publius hostem Naso suum appellat, Battiadesq; suum.

DE AVE INCENDIARIA SEV SPINTVRNICE.

INAVSPICATA est & incendiaria auis (de bubone prius scripserat) propter quam sæpenumero lustratam urbem in Annalibus inuenimus: sicut L. Crasso, C. Mario COSS. quo anno & bubone uiso lustrata est. Quæ sit auis ea nec reperitur, nec traditur. Quidam ita interpretantur incendiariam esse, quæcunq; apparuerit carbonem ferens ex aris uel altaribus, alij Spinturnicem eam uocant, sed hæc ipsa quæ esset inter aues qui se scire diceret non inueni, Plinius. C. Lælio, L. Danitio COSS. auis incendaria uisa occisaq;, Iulius Obsequens. Auis incendiaria secundum grammaticos quosdam dicta est quòd incendijs nascatur.

¶ INTVBA auis amari iecoris falsò creditur à nonnullis in Vergiliano uersu, primo Georg. Et amaris intuba fibris: quo doctissimus poeta ad genus illud intubi uidetur alludere quod picris appellatur, Hermolaus in Corollario.

DE ISPIDA.

A.

ISPIDA auis nomen apud recentiores, à sono uocis factum est, ut scribit Author libri de nat. rerum. Italicè piumbino, id est plumbina uocatur, & circa lacum Larium martinus piscator: & alibi auis paradisi propter pulchritudinem scilicet. plumbina uerò nominari uidetur, ut Io. Antonius Clarius Ebolitanus ad nos scripsit, quoniam desuper deuolans, in aqua, tanquam plumbum iactum, deorsumq; tendens, linea perpendiculari mergitur, ut inde pisciculos quibus uescatur, piscetur. Eandem ob causam etiam haliæetum auem piscium rapacem Italis plumbinam dici puto, Græci eadem ratione aues quasdam catarrhactas nominant. Rursus Itali quidam hanc auem piscatorem simpliciter nominant, circa Mediolanum: alij piscatorem regis, ut Angli quoq;; alij uitriolum, à uitrioli, id est chalcanthi colore, ut conijcio, circa lacum Verbanum: in Hetruria auem S. Mariæ. ¶ Hispani in Lusitania aruela. ¶ Galli pescheur, id est piscatorem: alij martinet pescheur, ut & Larij lacus accolæ: alij tartarin per onomatopœiam: alij artre, quòd artras Gallice dictas, id est tineas abigat. Circa Lutetiã audio mounier appellari, nescio quàm rectè. à nonnullis impropriè etiam piuerd, id est picum uiridem, à colore. ¶ Germani Ysuogel/ uel Eißuogel/ id est glaciei auẽ, ut Albertus Magnus quoq; interpretatur. nam hyeme etiam apud nos manet, & circa aquarum ripas glacie cõcretas degit. Pomerani Ysengart. ¶ Angli the kynges fissher, id est regis piscatorem, ut & Itali quidam. ¶ Ispidam ex recentioribus quidam ueterum alcyonem esse putant, uel simpliciter, uel fluuialem: qua de re sententiam meam exposui in Alcyone A. Christophorus Encelius alcyonem ab ispida diuersam facit: sed illi quoq; ut huic Germanicum nomen Eißuogel attribui scribit.

B. Ispida

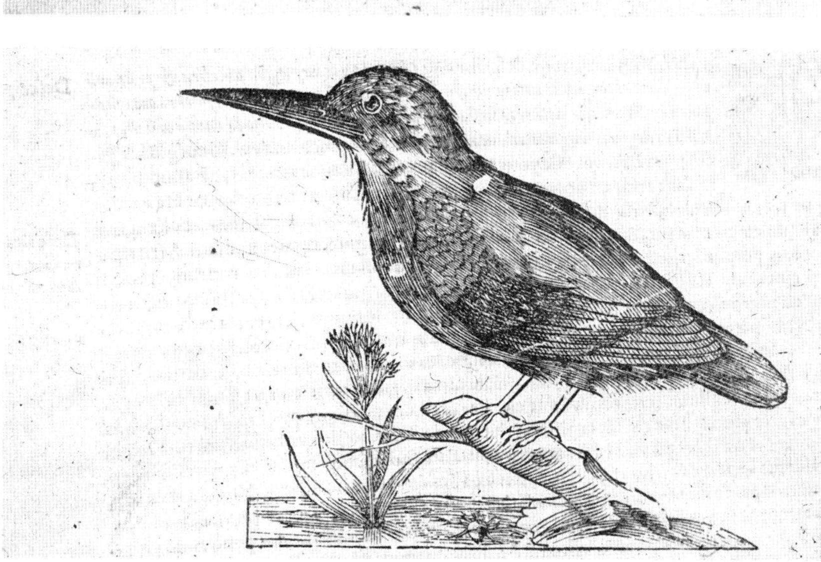

environment, the author seems to accept the idea already illustrated by Alciati.

The illustrations in Gesner, Aldrovandi and others generally tend to show the specific character of the birds by repeating profiles. When it comes to describing the plumage, the character of the visual language changes, due in part to the intentions of the naturalist writing. The particular character of a specimen occasionally could produce images from life and habits, as seen in the example of the halcyon. There are, however, depictions of breeding birds that may have been started off by interest in a mythological past but then turned into careful studies of a reality close at hand. Drawings of the nests built by house sparrows, martins, swallows and the nightingale are reproduced by Aldrovandi and seem to be the first depictions to be found in zoological literature. This would mean that the systematic account of the birds and their relation to the tradition also included a budding interest in visually rendering of the life and habits of birds and other animals. A source of knowledge of the conditions of the living birds was an influential work written by a Swedish refugee in Rome, the former archbishop Olaus Magnus who published his *Historia de gentibus septentrionalibus* in 1555.

Basic to Magnus's conception when writing about the life of the northern peoples are climate and nature, which he believes affected the life of men and animals in a fundamental way. Like his brother and fellow refugee, Johannes Magnus in his book on the genealogy of the Gothic kings, he was deeply inspired by the kind of nationalist ideas that belonged to the humanistic legacy. One should be aware that both of these works were published in Italy as national propaganda, intended for an international audience. By sorting out the specific climat and nature as hallmarks of the country in the north, Olaus Magnus also described the indigenous fauna which may have looked exotic to foreign readers whose knowledge of the northern countries was practically non-existent. Visual documentation was apparently considered necessary as a means of convincing people, as indeed had already been shown by Olaus Magnus in his *Carta marina*, edited in Venice in 1539; style and iconography of the map recur in Magnus's work on the history of the northern peoples.

Olaus Magnus was among the sources cited when Gesner published his *Historia animal-*

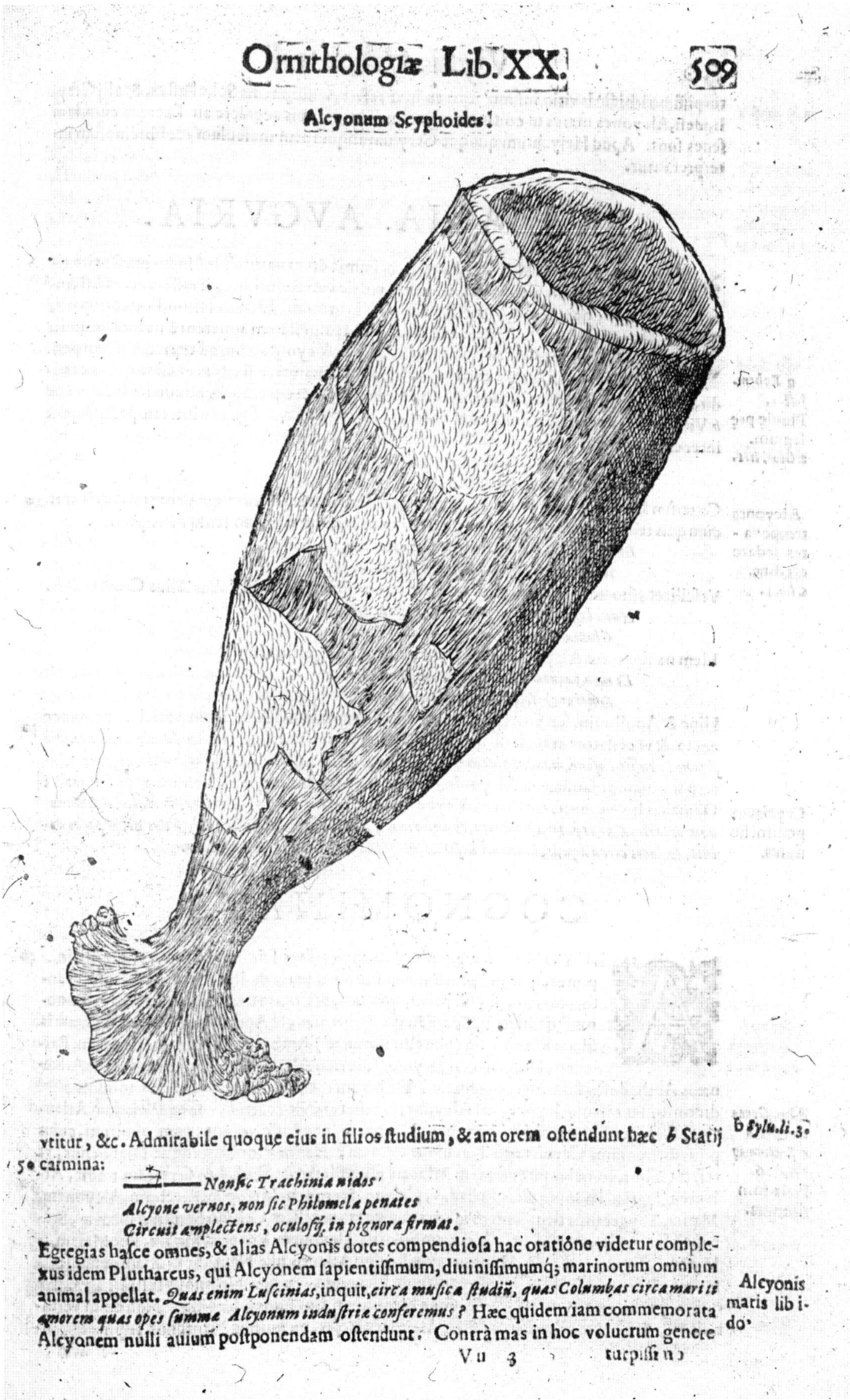
Ornithologiæ Lib. XX. 509

Alcyonum Scyphoides.

vtitur, &c. Admirabile quoque eius in filios studium, & amorem ostendunt hæc b Statij carmina: b Sylu.li.3. 5.

——— Non sic Trachinia nidos
Alcyone vernos, non sic Philomela penates
Circuit amplectens, oculosq́; in pignora firmat.

Egregias hasce omnes, & alias Alcyonis dotes compendiosa hac oratiône videtur complexus idem Plutharcus, qui Alcyonem sapientissimum, diuinissimumq; marinorum omnium animal appellat. Quas enim Luscinias, inquit, circa musica studiũ, quas Columbas circa mariti amorem quas opes summæ Alcyonum industriæ conferemus? Hæc quidem iam commemorata Alcyonem nulli auium postponendam ostendunt. Contrà mas in hoc volucrum genere turpissi- Alcyonis maris libido.

Vu 3 turpissi m

Figure 6 (*opposite, above*) Kingfisher. From Gesner: *Vogelbuch* (Frankfurt 1550). (Photo: Uppsala University Library.)

Figure 7 (*opposite, below*) Kingfisher. From Aldrovandi: *Ornithologiae* (Bologna 1599–1603). (Photo: Uppsala University Library.)

Figure 8 (*left*) Kingfisher's nest. From Aldrovandi: *Ornithologiae* (Bologna 1599–1603). (Photo: Uppsala University Library.)

Figure 9
Hunter aiming at capercailye. From Magnus: *Historia de gentibus septentrionalibus* (Rome 1555). (Photo: Dept. of Art History, Uppsala University.)

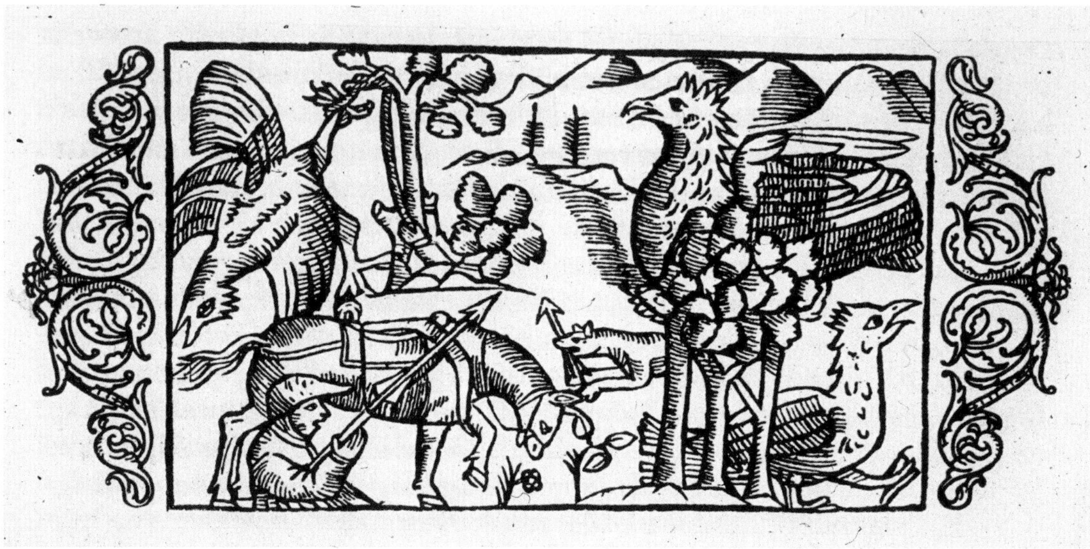

Figure 10
Capercailye. From Gesner: *Vogelbuch* (Frankfurt 1550). (Photo: Uppsala University Library.)

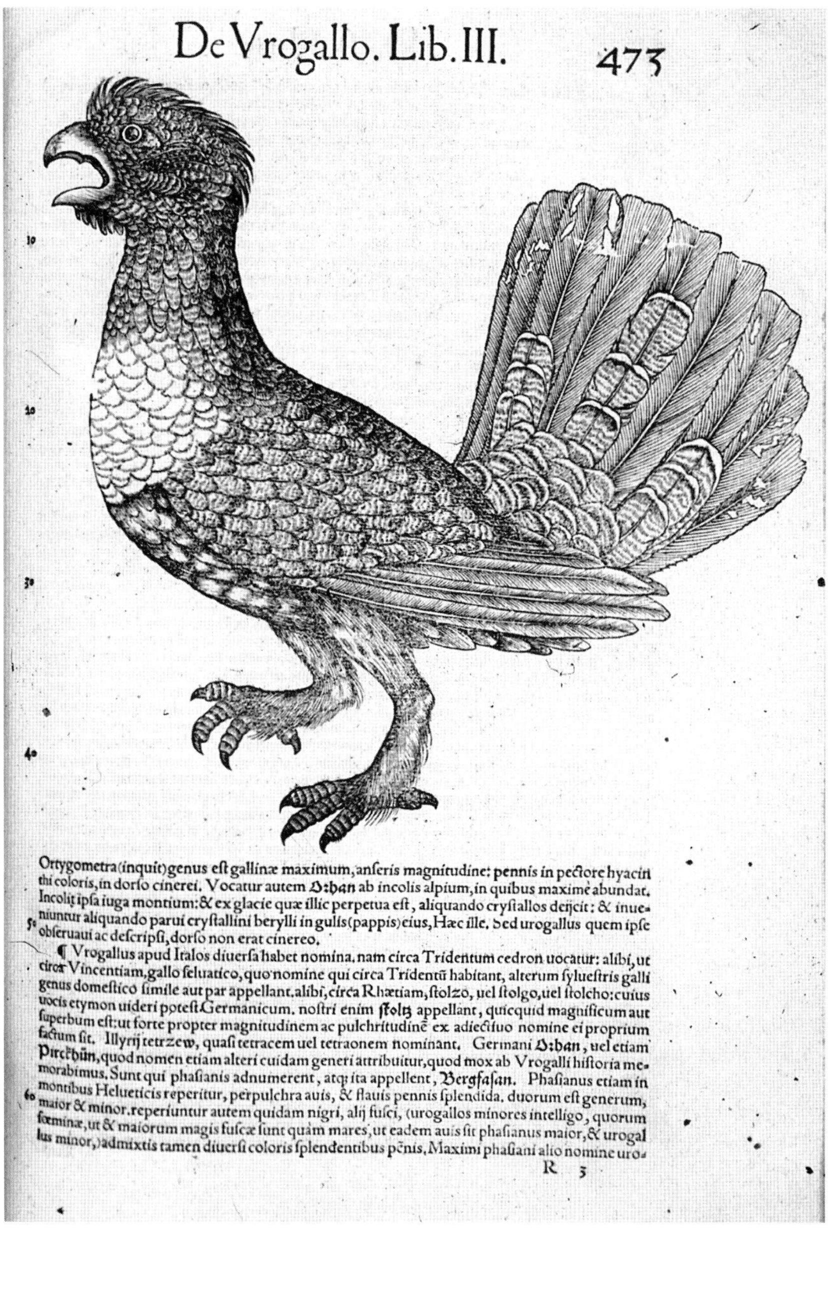
De Vrogallo. Lib. III. 473

Ortygometra (inquit) genus eſt gallinæ maximum, anſeris magnitudine: pennis in pectore hyacinthi coloris, in dorſo cinerei. Vocatur autem Orhan ab incolis alpium, in quibus maximè abundat. Incolit ipſa iuga montium: & ex glacie quæ illic perpetua eſt, aliquando cryſtallos deijcit: & inueniuntur aliquando parui cryſtallini berylli in gulis (pappis) eius, Hæc ille. Sed urogallus quem ipſe obſeruaui ac deſcripſi, dorſo non erat cinereo.

¶ Vrogallus apud Italos diuerſa habet nomina. nam circa Tridentum cedron uocatur: alibi, ut circa Vincentiam, gallo ſeluatico, quo nomine qui circa Tridentũ habitant, alterum ſyluestris galli genus domeſtico ſimile aut par appellant. alibi, circa Rhætiam, ſtolzo, uel ſtolgo, uel ſtolcho: cuius uocis etymon uideri poteſt Germanicum. noſtri enim ſtoltz appellant, quicquid magnificum aut ſuperbum eſt: ut forte propter magnitudinem ac pulchritudinẽ ex adiectiuo nomine ei proprium factum ſit. Illyrij tetrzew, quaſi tetracem uel tetraonem nominant. Germani Orhan, uel etiam Pirckhũn, quod nomen etiam alteri cuidam generi attribuitur, quod mox ab Vrogalli hiſtoria memorabimus. Sunt qui phaſianis adnumerent, atq; ita appellent, Bergfaſan. Phaſianus etiam in montibus Helueticis reperitur, perpulchra auis, & flauis pennis ſplendida. duorum eſt generum, maior & minor. reperiuntur autem quidam nigri, alij fuſci, (urogallos minores intelligo, quorum fœminæ, ut & maiorum magis fuſcæ ſunt quàm mares, ut eadem auis ſit phaſianus maior, & urogallus minor,) admixtis tamen diuerſi coloris ſplendentibus pẽnis. Maximi phaſiani alio nomine uro-

R 3

ium. We do not know whether visual material had been brought by Magnus from Sweden to Rome, nor do we know about his contacts with the artists who submitted illustrations for the work. In this situation, it is no wonder that many images take on a fanciful character that had to be complemented by the text.

In any case, the visual imagery of Magnus's *Historia* was to affect generations of naturalists, all the way to the eighteenth century. Myths told the tale of swallows hibernating on the sea floor or reported about the greed of the wolverine which made him look for trees growing close to each other to be able squeeze out the faeces from his body. The scenes were illustrated in the *Carta marina* as well as in the *Historia*.

One of the chapters in Magnus's work is devoted to bird life of the north (Magnus [1555]). Much legendary material is collected and illustrated; the images caught the interest of Gesner and his artists. This can be seen from the example of the capercailye, portrayed by Magnus as sitting in the top of a tree and aimed at by a hunter (fig. 9). Technically reduced to a few marks it is certainly a most approximative portrait of the bird. Gesner accepted Magnus's portrait of the capercailye, as taken from *Carta marina*. But later he changed his mind after having seen the bird himself (*a me conspectus*). As can be seen from the woodcut in *Historia animalium*, there can be no doubt that Gesner and his artists had watched the bird when mating in spring. So he abandoned the scheme that had been taken from Magnus and produced the first authentic portrait of the mating bird (fig. 10). Gesner's encounter

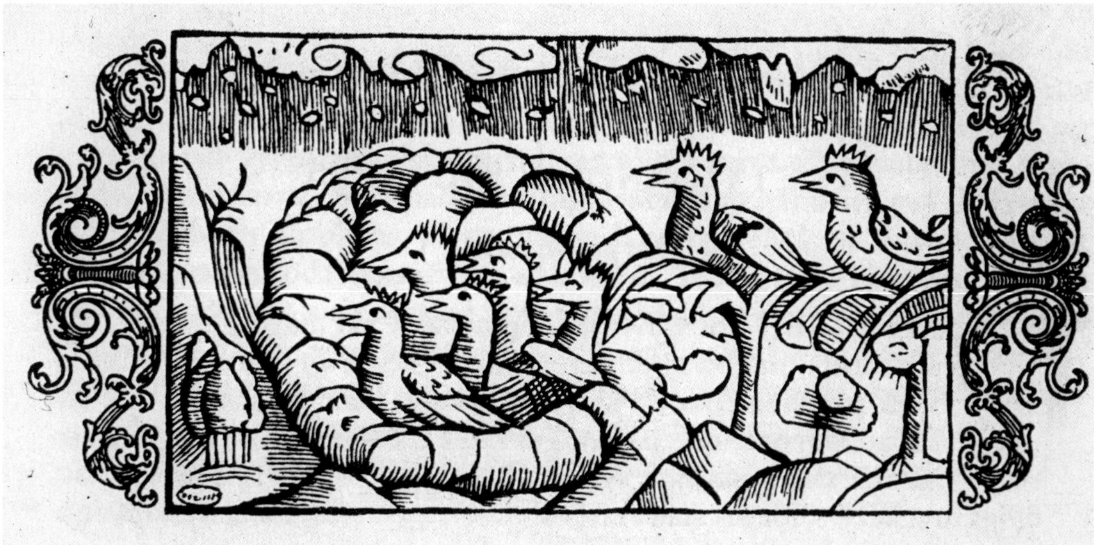

Figure 11
Capercailye's nest. From Magnus: *Historia de gentibus septentrionalibus* (Rome 1555). (Photo: Dept. of Art History, Uppsala University.)

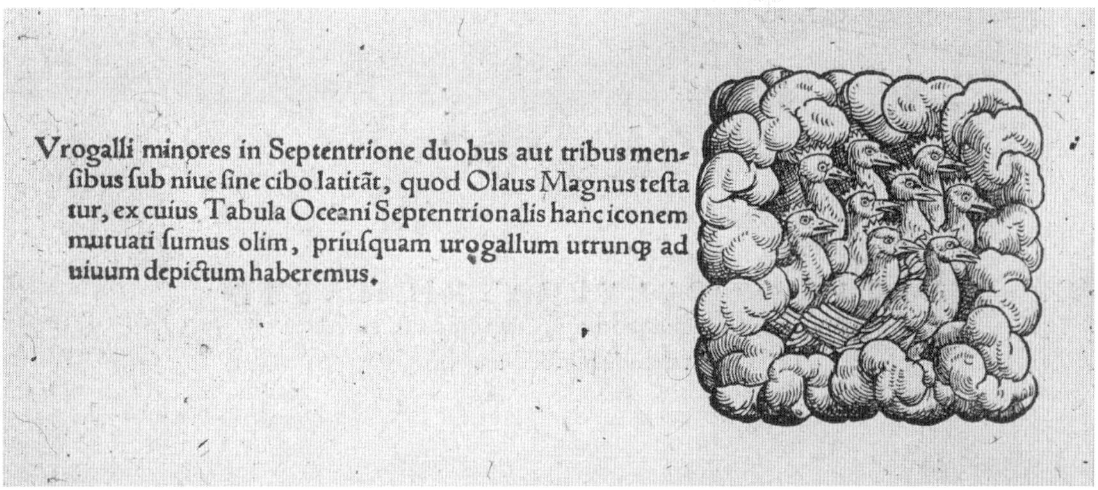

Vrogalli minores in Septentrione duobus aut tribus menſibus ſub niue ſine cibo latitãt, quod Olaus Magnus teſtatur, ex cuius Tabula Oceani Septentrionalis hanc iconem mutuati ſumus olim, priuſquam urogallum utrunq; ad uiuum depictum haberemus.

Figure 12
Capercailye's nest. From Gesner: *Vogelbuch* (Frankfurt 1550). (Photo: Dept. of Art History, Uppsala University.)

with the capercailye, which also yielded fairly precise notes on the colours, introduced a momentary impression of the bird; the ordinary profile as recorded from a bird that had been shot and studied in the artist's workshop had been replaced by a report from the hunting ground.

Gesner was the foremost authority to Aldrovandi who called him simply, the Ornithologist. But when it came to the capercailye, however, he produced a most simplified version which cannot possibly refer to an experience of the bird in its natural surroundings. Readers had to rely on the interplay of text and image.

To extend the field of ornithological knowledge as recorded in images, there was an increased interest in the life cycle of the birds, including the nest-building. In emblematic literature, nests of swallows were occasionally depicted as part of the moralizing message. Sixteenth-century prints in the Netherlands and a late painting by Pieter Bruegel, called *The peasant and the nest robber* belong to the same category of images (Boström [1949]). When Gesner ventured into the domestic habits of birds, he found that Olaus Magnus had an illustration showing the nest of the capercailye (Gesner [1555]). According to Magnus, these birds leave their shelter in the snow and turn to the mountains for hatching and breeding. A woodcut shows the nest with five young in a wintry landscape, with the parents sitting in trees nearby while snow falls from the sky (fig. 11). Gesner repeats the idea, now with nine young and maintains that the nest had been portrayed *ad vivum* (fig. 12). At any event he received the initial impulse from Olaus Magnus. And this may also have stimulated Aldrovandi to include in his work a series of drawings showing nests of house sparrows, martins, swallows and the nightingale. In his presentation of the house sparrow, Aldrovandi explicitly refers

Figure 13
Nightingale's nest. From Aldrovandi: *Ornithologiae* (Bologna 1599–1603). (Photo: Uppsala University Library.)

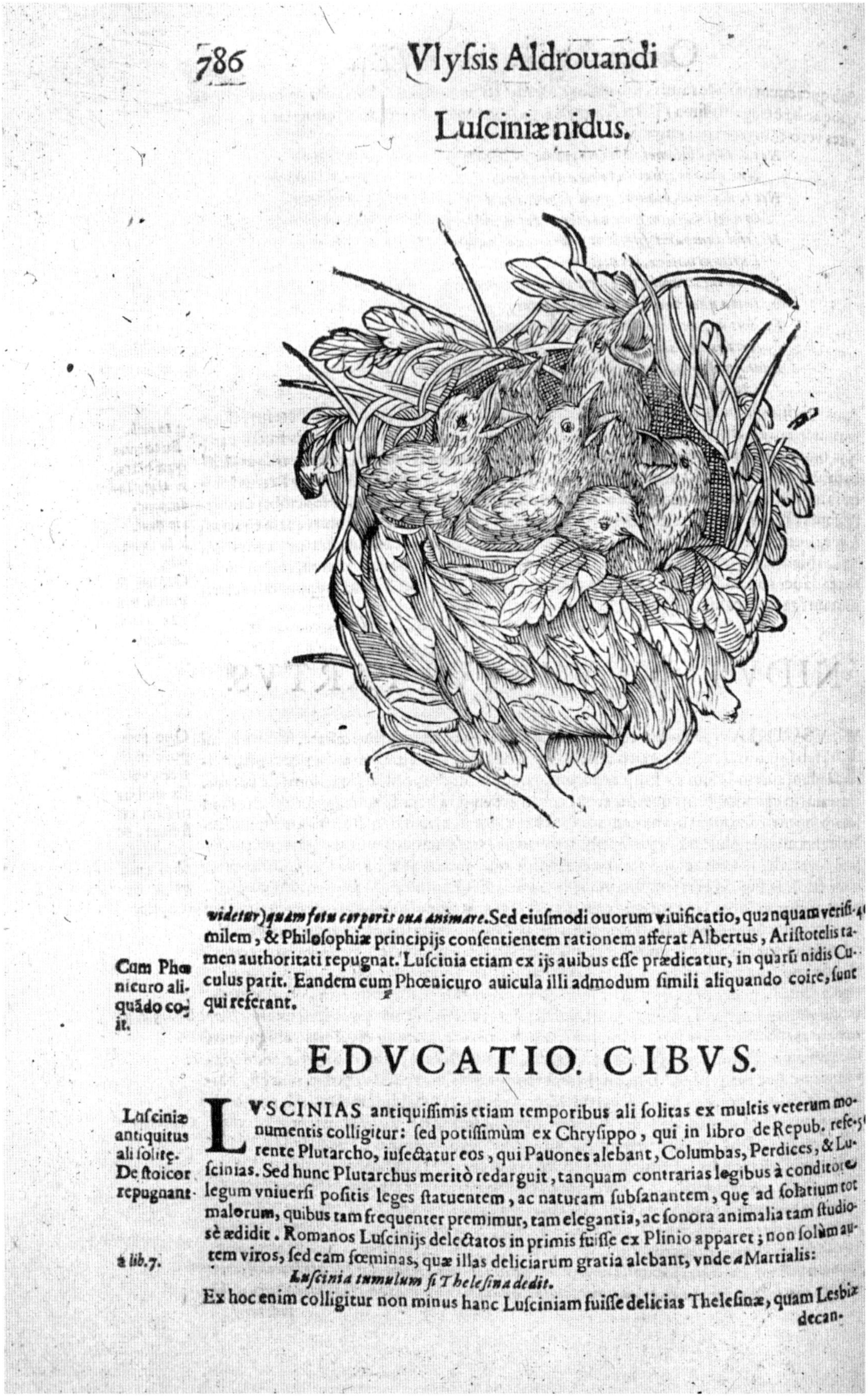

786 Vlyſsis Aldrouandi

Luſciniæ nidus.

videtur) quam fetu corporis oua animare. Sed eiuſmodi ouorum viuificatio, quanquam veriſimilem, & Philoſophiæ principijs conſentientem rationem afferat Albertus, Ariſtotelis tamen authoritati repugnat. Luſcinia etiam ex ijs auibus eſſe prædicatur, in quarũ nidis Cuculus parit. Eandem cum Phœnicuro auicula illi admodum ſimili aliquando coire, ſunt qui referant.

Cum Phœnicuro aliquãdo coit.

EDVCATIO. CIBVS.

LVSCINIAS antiquiſſimis etiam temporibus ali ſolitas ex multis veterum monumentis colligitur: ſed potiſſimùm ex Chryſippo, qui in libro de Repub. referente Plutarcho, inſectatur eos, qui Pauones alebant, Columbas, Perdices, & Luſcinias. Sed hunc Plutarchus meritò redarguit, tanquam contrarias legibus à conditore legum vniuerſi poſitis leges ſtatuentem, ac naturam ſubſanantem, quę ad ſolatium tot malorum, quibus tam frequenter premimur, tam elegantia, ac ſonora animalia tam ſtudioſè ædidit. Romanos Luſcinijs delectatos in primis fuiſſe ex Plinio apparet; non ſolùm autem viros, ſed eam fœminas, quæ illas deliciarum gratia alebant, vnde a Martialis:

Luſcinia tumulum ſi Theleſina dedit.

Ex hoc enim colligitur non minus hanc Luſciniam fuiſſe delicias Theleſinæ, quam Lesbiæ decan-

Luſciniæ antiquitus ali ſolitę. De ſtoicor repugnant.

a lib. 7.

to frequent observations in his country house and gives a fairly accurate rendering of the nest, with the young faithfully depicted, their mouths open and the material of the nest carefully sketched out (fig. 13). An interest in the nests of the swallows possibly also had its roots in the myths related to these birds; classical authors indulge in praising the skills with which the nests were constructed. The nest built of clay is recorded by Aldrovandi in drawings showing the semicircular form, *cuius iconem hic habes expressam*.

The drawings of the nests may seem fairly uncomplicated, as simple reports of visual reality. But as so often, the artist had to make use of formulae that could oscillate between the principle of *natura naturans* and the adjustment of the image to some sort of established norm. Aldrovandi observes that two nests of sparrows had been combined, the result being a form that reminds him of the human thorax (fig. 14). In his *Fabrica* (1543), Vesalius had combined the bowels of the human body with a scheme taken from *Torso de Belvedere* (Ackermann [1985] 13).

After the pioneering achievements of the sixteenth century, conventional formulae dominated up to the beginning of the eighteenth century. Attempts were now made to question the elder tradition of ornithological imagery by having the bird portrayed on a 1:1 scale and emphasising the mimetic character by using the technique of coloured copper engraving. A critical approach to the mythological tradition gained ground in the vein of ideas emanating from Francis Bacon. But the scientific ambition was due not only to immanent causes. The tradition from Olaus Magnus and his accounts of life in the north inspired Olof Rudbeck the Younger in producing his *Book of Birds* (1695–1707), a patriotic project that was sponsored by Charles XI who explicitly spoke of the work as contributing to the fame and glory of the country.[3] Taken together, the result was a new aim and direction of ornithological illustration, a new concept of visual truth.

Uppsala University

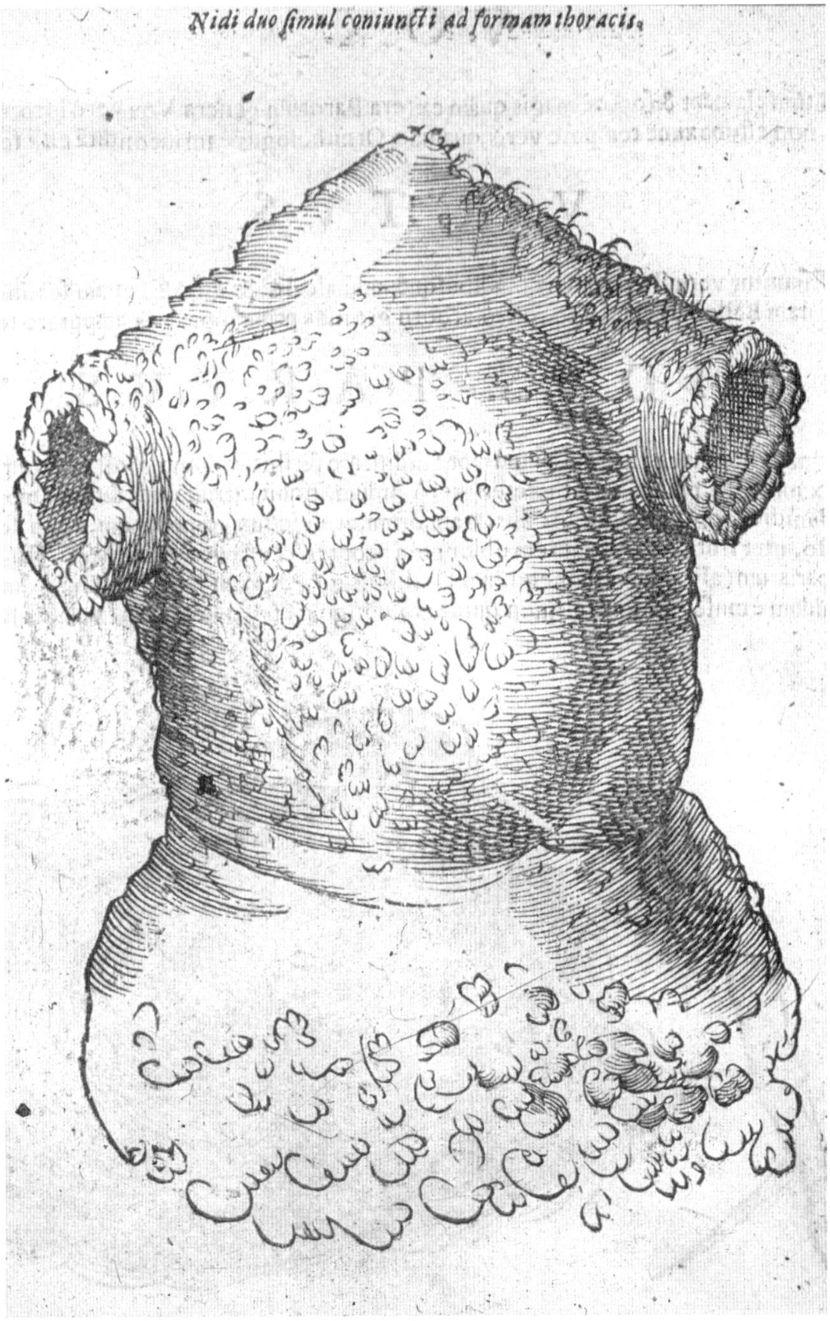

Figure 14
Sparrow's nest. From Aldrovandi: *Ornithologiae* (Bologna 1599–1603). (Photo: Uppsala University Library.)

Notes

[1] For a recent interpretation of the paintings, see Caneva [1992].

[2] For a survey, see Aldrovandi [1599–1603] 502 ff.

[3] Ellenius [1986]. See also Ellenius [1997]. To Rudbeck, Aldrovandi was a well-known precursor, a general source of inspiration and, most probably, also of competition when he embarked upon his patriotic project; Aldrovandi's images were, however, of minimal importance when it comes to explaining the innovative character of those created by the Swedish ornithologist.

References

Ackerman, J. S.
1985 "Early Renaissance Naturalism and Scientific Illustration." In: *The Natural Sciences and the Arts. Aspects of Interaction from the Renaissance to the 20th Century. An International Symposium*, ed. A. Ellenius. Motala.

Alciati, A.
1600 *Emblemata*. Lyon

Aldrovandi, U.
1599–1603 *Ornithologiae libri* XX. Bologna.

1642 *Historia animalium*. Bologna.

Boström, K.
1949 "Das Sprichwort vom Vogelnest." In: *Konsthistorisk Tidskrift XVIII.*

Caneva, G.
1992 *Il Mondo di Cerere nella Loggia di Psiche. Villa La Farnesina Sede dell' Academia Nazionale dei Lincei*. Rome.

Ellenius, A.
1986 "Olof Rudbeck the Younger and the Pictorial Tradition in Natural History." In: *Olof Rudbeck's Book of Birds,* 49–68. Stockholm.

1997 "Ulisse Aldrovandi as a Source of Olof Rudbeck's Book of Birds." In: *Sidereus Nuncius & Stella Polaris. The Scientific Relations between Italy and Sweden in Early Modern History*, eds. M. Beretta and T. Frängsmyr. Canton MA.

Eriksson, R. (ed.)
1959 *Andreas Vesalius' first public anatomy at Bologna 1540. An eyewitness report* ... Uppsala.

Freedberg, D.
1991 "Science, Commerce, and Art: Neglected Topics at the Junction of History and Art History." In: *Art in History, History in Art. Studies in Seventeenth-century Dutch culture*, eds. D. Freedberg and J. de Vries. Santa Monica CA.

Gesner, K.
Thierbuch. Von den Vögeln.

1550 *Vogelbuch*. Frankfurt am Main.

1555 *Historia animalium Liber III qui est de Auium natura*. Zurich.

Harms, W.
1985 "On Natural History and Emblematics in the 16th Century." In: *The Natural Sciences and the Arts. Aspects of Interaction from the Renaissance to the 20th Century. An International Symposium*, ed. A. Ellenius, 67–83. Motala.

Humbla, E.
1682 *De halcyone*, praes. A. Wanochius. Abo.

Ivins Jr., W.
1953 *Prints and Visual Communication*. Cambridge MA.

Kemp, M.
1981 *Leonardo da Vinci. The Marvellous Works of Nature and Man*. Singapore.

Magnus, O.
1555 *Historia de gentibus septentrionalibus*. Rome.

Olmi, G.
1992 *L'Inventario del mondo. Catalogazione della natura e luoghi del sapere nella prima età moderna*. Bologna.

Panofsky, E.
1960 *Renaissance and Renascences in Western Art*. (= The Gottesman Lectures, Uppsala University, vol. 11). Uppsala.

Scalabroni, L.
1983 "Raffaello Sanzio e scuola. La Loggia di Amore e Psiche." In: *Luoghi di Raffaello a Roma*, 55–61. Rome.

Elephant, Mammoth, Unicorn, or What?

Notes on the Interrelations of Pictures and Texts in Leibniz

Alexandre Métraux

Introduction

Picturing new discoveries made by natural philosophers was already a well established practice when Gottfried Wilhelm Leibniz began to study the history of the earth. In the reconstruction of this savant's geological endeavours, however, historians of science (and likewise their colleagues in the departments of philosophy) generally have preferred to lend their inner ear to the messages conveyed by Leibnizian texts and to visually ignore the iconic information, as if the sole function ascribed to the latter was that of illustrating propositions believed to be sufficiently telling in themselves. This iconophobic bias is rooted in the belief that iconic information constitutes the redundant apanage of theoretically sound texts. Yet, taking pictures to be eye-catching illustrations turns out to be detrimental to understanding the strategic role which Leibniz attributed to iconic information in the construction of the geological theory he outlined in the (posthumously published) treatise *Protogaea*.[1] The aim of the present essay is to restore the balance between text and image in Leibniz' geological research.

On the History of the History of Fossils

A few weeks after Duke Johann Friedrich died on December 28, 1679, Leibniz displayed a pronounced eagerness to study the history of the house of Hanover-Lüneburg. But several years elapsed before the polymath's historiographical propensity met the favourable response of Duke Ernst August, the deceased's brother and successor. It is known that on May 28, 1685, Otto Grote informed Ernst August that Leibniz was, indeed, ready to sit down in order to study the relevant sources and to act as the official historiographer of the duke's house if, in return, he were granted a pension *ad vitam*. On August 10 of that same year, the philosopher finally received his patron's commission to prepare a frescoe-like history of the family of Guelphs.

Just another five and a half years later, Leibniz outlined the work in progress in a memorandum submitted to the duke in mid-January 1691. Having explained why collecting, deciphering, and critically evaluating a great variety of documents scattered over many localities was necessarily so strenuous and time-consuming, he indicated, "I first will have to address the most remote antiquities of these lands, showing thereby that these very lands were, as it seems, covered by seaways extending closely to the slopes of the Harz mountains; that so-called 'tongues of vipers' have been found at Lüneburg, like on the isle of Malta, which are but teeth of some marine marvels; that bones of unknown creatures occur in the Baumann cave, and in the excavation of Scharzfeld; I, too, have collected such bones in the Baumann cave; that entire lakes have been filled and have turned into rock, where effigies of fish remain (like those of flies in amber); and that metallic matter has filled these holes; and I shall examine more such things which are likely to decorate the work."[2]

Pleasing erudition was thus the aim which Leibniz sought to achieve. The treatise, both erudite and pleasing, on the most pristine traces of life on the lands of the house of Hanover-Lüneburg, was not brought to completion until September 1715. The busy and unsteady philosopher unquestionably did not disdain fossilised fish, petrified teeth, and bones of both known and mysterious species, but other matters of erudition obviously mattered more to him in the twenty-four years which separate the memorandum of 1691 and the letter to Andreas Gottlieb von Bernstorff (1649–1726) announcing the *Protogaea*'s completion.[3]

It may be suitable to mention some episodes in Leibniz' biographical record relating to the gathering of evidence concerning the fossil record in order to fill some of the gaps between the two aforementioned dates.[4]

The first mention (or put with some caution: what seems to be the first mention) of the geological monograph in slow progress appeared on September 6, 1692, in Leibniz' letter to Henri Justel (1620–1693).[5] It so happened that six days later, a gigantic petrified tooth or a large piece of a prehistoric skeleton - elephant, mammoth, unicorn, or what? - was found near Wolfenbüttel; Leibniz saw to it that the specimen be properly preserved. In January 1693, the *Acta eruditorum* published a sketch of the *Protogaea* (cf. Leibniz [1693]). On September 1, 1696, Leibniz met Wilhelm Ernst Tentzel (1659–1707) at Wolfenbüttel, where both men examined the relics of a fossil elephant (possibly the piece found in 1692). In 1706, a memoir on stones containing animals and plants was read before the Academy of Science in Paris (cf. Leibniz [1989a]),[6] followed in May 1710 by a short publication on the same topic in the *Miscelana Berolinensia* (cf. Leibniz [1989b]).

The available evidence suggests, however, that the period of geological research reached its apex in the years 1685 to 1690. During that time, Leibniz visited the collection of Friedrich Lachmund (1636–1676) at Göttingen as well as the cabinet of natural history at Kassel, examined the Baumann cave in the Harz (mentioned in the outline quoted above), undertook mineralogical and chemical studies of geological interest in Bohemia and Saxony, etc.

Though completed before its author's death, the *Protogaea* remained unpublished until 1749, when the original Latin version, accompanied by the engravings diligently designed by Nikolaus Seelander under Leibniz' supervision, was edited by Christian Ludwig Scheid and published at Göttingen, followed by the publication of the German translation (cf. Leibniz [1749]) in the same year, and finally by the French translation in 1859.

Ontological Reflections on Fossil Traces of Animals and Plants

Why did Leibniz intend to have the history of the family of Guelphs preceded by a treatise on events that occurred in the voiceless silence of antedeluvian times? For a scholar of Leibnizian width and depth, an enterprise of this sort, from which considerations concerning what he would call *geographia historica*[7] were absent, was inconceivable indeed. Not because the *savant* might have felt the urge to flatter his patron by enumerating the natural resources that made up the wealth of the territory to be described, but rather because the encyclopaedic scope of Leibnizian research reflected his commitment to a worldview according to which all things present and past were mutually related by way of a pre-established, logically transparent harmony. The universe thus being conceived as the incarnation of logic – as Aron Gurwitsch aptly put it (cf. Gurwitsch [1973] 240–255) –, propositions on present historical phenomena implied that at least some references be made to the sum of all compossible states of the world. There would, of course, have been no need to mention this metaphysical (and physical) belief if arguments drawn from the logic of compossibility and the rational order of things hadn't played a role in the philosopher's reflections on the distinctive marks of stony and bony traces of animal life (as will be shown below).

The examination of the preparatory studies done for the historiographer's work on the family of Guelphs shows that, as far as the *geographia historica* was concerned, Leibniz had a certain preference for two thematically related fields:

(a) the history of the earth and the ancient forms of life (how did the earthly globe transform itself in the course of time until it attained its actual physiognomy, what did these transformations entail mainly for animal species?); and

(b) the identification of specimens of the fossil record (what do such petrified items stand for and which species do they represent?).

Among the productions of nature which captured the philosopher's attention more vividly than other antiquities were petrified traces of animals, some of which would look familiar to any knowledgeable amateur of natural history, whereas others could embarrass even widely travelled and learned polymaths. But in both circumstances, the crucial methodical step consisted in ascertaining *that* these traces were natural, and *what* they represented. Leibniz tried to come to grips with both issues. In doing so, he mobilised various resources – ontological reflections on the properties of natural imprints, visual perception, direct and/or indirect quotations from published and unpublished zoological and paleontological works, iconic representations, copies of such representations, etc.

Some brief comments on these resources are likely to illustrate how iconic and linguistic information came to bear upon Leibniz' analysis of the fossil record.

In *De natura fossilium*, Agricola lists several minerals to which nature has given "various shapes and forms" (Agricola [1955] 14). Some such minerals

> *imitate* [...] *well known objects. Ammonis cornu imitates a horn; tephrites, a new moon; alum, asbestos, and native silver, hair.* [...] *Certain others represent effigies of things, for example, enorchis, the testes; diphysis, the genitals of both sexes; enthrochos, a wheel.* [...] *Going back to minerals, many with lines of various colors running through them resemble various objects as leucophthalmos resembles the human eye; aegophthalmos, a goat's eye;* [...] *lapis eislebanus, pike, perch, various marine fish, and even a cock and salamander* (Agricola [1955] 14f.).

The author of this passage does not (yet) distinguish specimens which imitate known objects by material inclusion (such as an ammonite embedded in solid matter) from specimens which represent known objects by iconic imitation (such as the eye-agate). But Leibniz addresses this distinction from the outset. The issue is, indeed, methodologically relevant on several counts. On the one hand, it matters whether a configuration recognisably refers by *itself* and in any way to a *real* object or whether it is, so to speak, *superimposed as such* onto rocks and stones by the eyes (and the imagination) of the beholder. On the other hand, it also matters whether the configuration looks like an already known object of nature, or whether it represents (rather than imitates by mere chance) an *unknown*, but presumably *real* object of nature.

At Eisleben in the region of Mansfeld as well as at Osterode in the Harz, Leibniz writes, black and foliated schist is being extracted, in which "one often sees the properly and artfully delineated forms of fish, as if some artist had inserted into the black rock a sharply cut piece of metallic matter; it is also referred to as 'ichthyomorphum.' [...] I own a piece of this rock which is decorated on both sides with effigies of fish, but they are different from one another."[8] The rock to which the *Protogaea* alludes is obviously of the *same* (or of the same *sort*) as Agricola's *lapis eislebanus*. In both cases, however, the effigies of fish (see fig. 1) are self-referential; they appear to be traces of what they represent – individual organisms –, not artefacts depicting marine animals by imitation, nor the outcomes of a sport of nature (Leibniz uses the conventional term *lusus naturae* (cf. Leibniz [1949] 64)). To put it differently, these configurations stand for themselves as well as for the species to which they belong.[9]

The question is then: On what grounds does Leibniz reject explanations which do not refer to a natural scenario by which specimens contained in the fossil record are linked to the existence of individual organisms having lived in the most remote times? Why are plastic faculty,[10] equivocal generation (cf. Leibniz [1949] 92–94), and/or the propensities of some supernatural genius no suitable candidates to represent the *explanans* responsible for the visible, palpable, recognisable specimens pertaining to the domain of the *explanandum*?

Leibniz' theory of fossil traces rests on two major propositions:

(1) By comparison with natural organisms, even the most artfully designed imita-

Figure 1
Various petrified fish. From Leibniz: *Protogaea* (Leipzig 1749) plate 2.

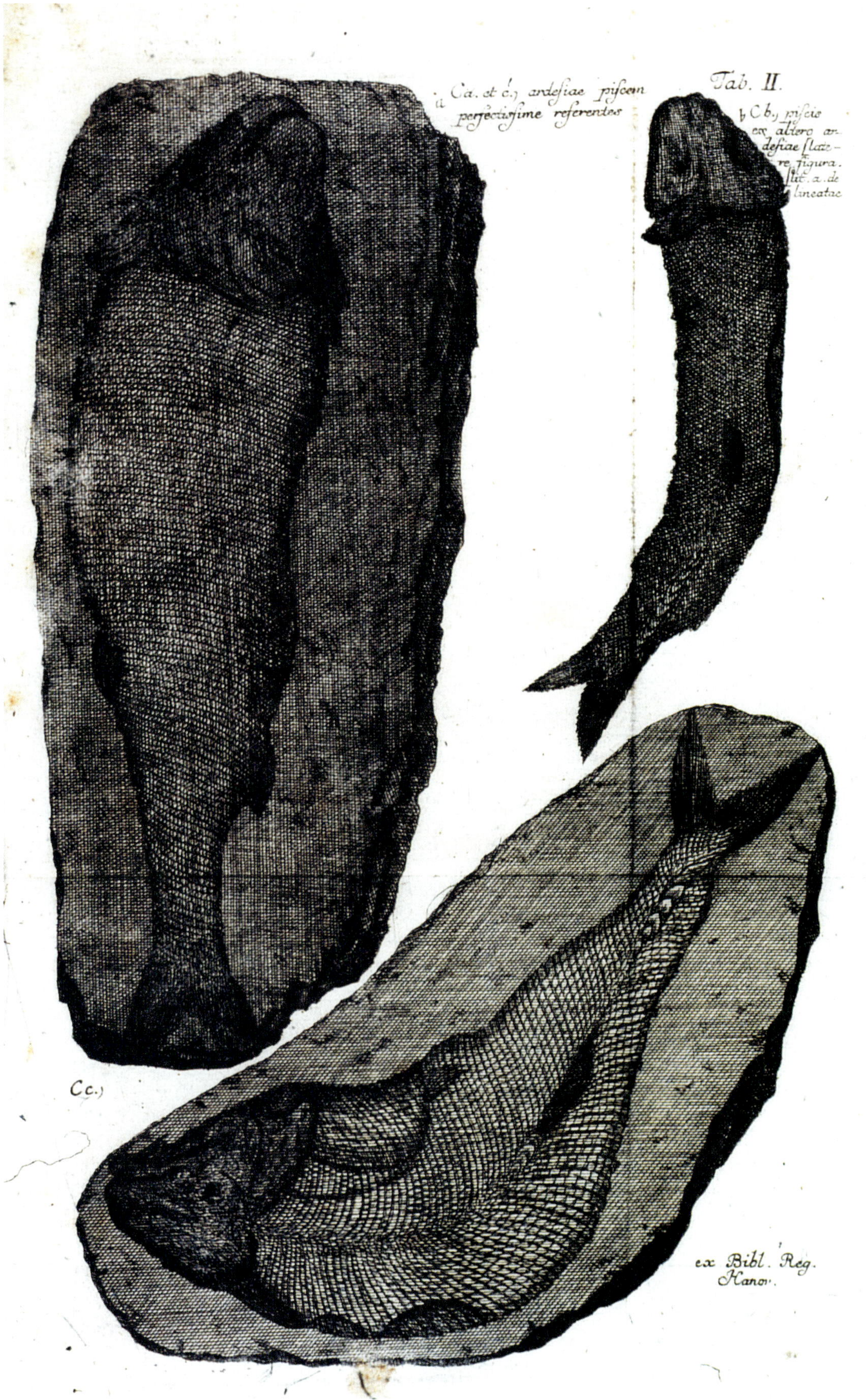

tions of animals and plants cannot reach the same degree of perfection (cf. Leibniz [1949] 94–96).

(2) Similarly, by comparison with natural bodies, imitations of animals, plants, crystals, or other configurations resulting from chance are far from being as perfect as their models.[11]

The dissimilarity between artefacts and natural bodies concerns morphological features which can be perceived, described, depicted, measured, or otherwise objectified. If a shell found in some outcrop looks like a body identified as specimen of a determined species, the degree of match between the trace and a living exemplar is indicative of the 'naturalness' of the petrified specimen. Hence, morphological disparity between any two (or more) bodies is likewise indicative that one or the other is either the outcome of human art or of mere chance. The latter notion conveys the idea that configurations, though resembling specimens of known species, still result from deterministic (causal) processes, but not from equally deterministic biological processes (Leibniz' rejection of equivocal generation in fact bars bodies begotten by chance, plastic power and the like, from the biological order).

The methodologically plausible rule of disparity between natural and artificial configurations of organisms contrasts with another rule spelled out in the *Protogaea*. Leibniz asserts that experimental manipulation is a helpful device for gaining insight into processes pertaining to the realm of historical geography. "I believe that he, who will compare with greater care the works of nature excavated from the subterranean depth with the products of the laboratory [...], will deserve merit for his pains, for often a marvellous similarity between things natural and artificial shows up."[12] Leibniz mentions several examples of laboratory studies dealing with similarities between natural and artificial productions, among which the production of cinnabar from quicksilver and sulphur[13] and the production of a kind of ruby by submitting orpiment to the action of fire.[14]

Is the contrast – or even the mutual exclusion – of these two rules as radical as it seems at first reading? I do not believe so. Upon closer consideration, these rules turn out to complement each other and to constitute a powerful strategy that allows the assessment of both the physical and the biomorphological aspects of petrified antiquities. As long as physical states and changes are at stake (processes of fossilisation, of dislocation of specimens from one spot of the terrestrial surface to another spot in the course of time, of tectonic metamorphosis, and the like), analogies between small-scale artificial events and large-scale natural events are not only helpful, but theoretically illumining. However, whenever configurations of a special kind, viz. of animals and plants, are at stake, such analogies are not only misleading, but theoretically untenable. The hypothesis that any existing species could have been duplicated in virtue of some plastic faculty, of some benevolent or malevolent genius, or of equivocal generation would contradict the logic of compossibility (cf. supra). Two kinds of reasoning thus merge in the *Protogaea*, and each of these entails its own procedures of testability. One procedure is that of top-down physics *plus* artificial duplication under controlled experimental conditions; the other procedure is that of bottom-up observation *plus* inductive generalisation on the basis of mutually corroborating perceptual evidence. In the case of configurations of organisms in rock, sand, amber, and other such material containers of traces of ancient life, the overarching physics of natural geography paradoxically confers more – and not less – significance to morphological considerations, hence to iconic information. This confirms the thesis formulated above that pictures are not just redundant by-products pleasing the mind's eye, but rather the medium by which the eye's mind is informed.

Iconological Strategies

The eye's mind needs methods by which it identifies petrified traces of organic life, such that these traces be properly classified, evaluated, and possibly used in the theoretical construction of the earth's transformations.

The first methodical step to which Leibniz alludes aims at establishing the degree of similarity between specimens (tokens) and species (types) by vision. Reflecting on his own experience as a natural historian and casual collector, the author of the *Protogaea* asserts that "most often you recognize the

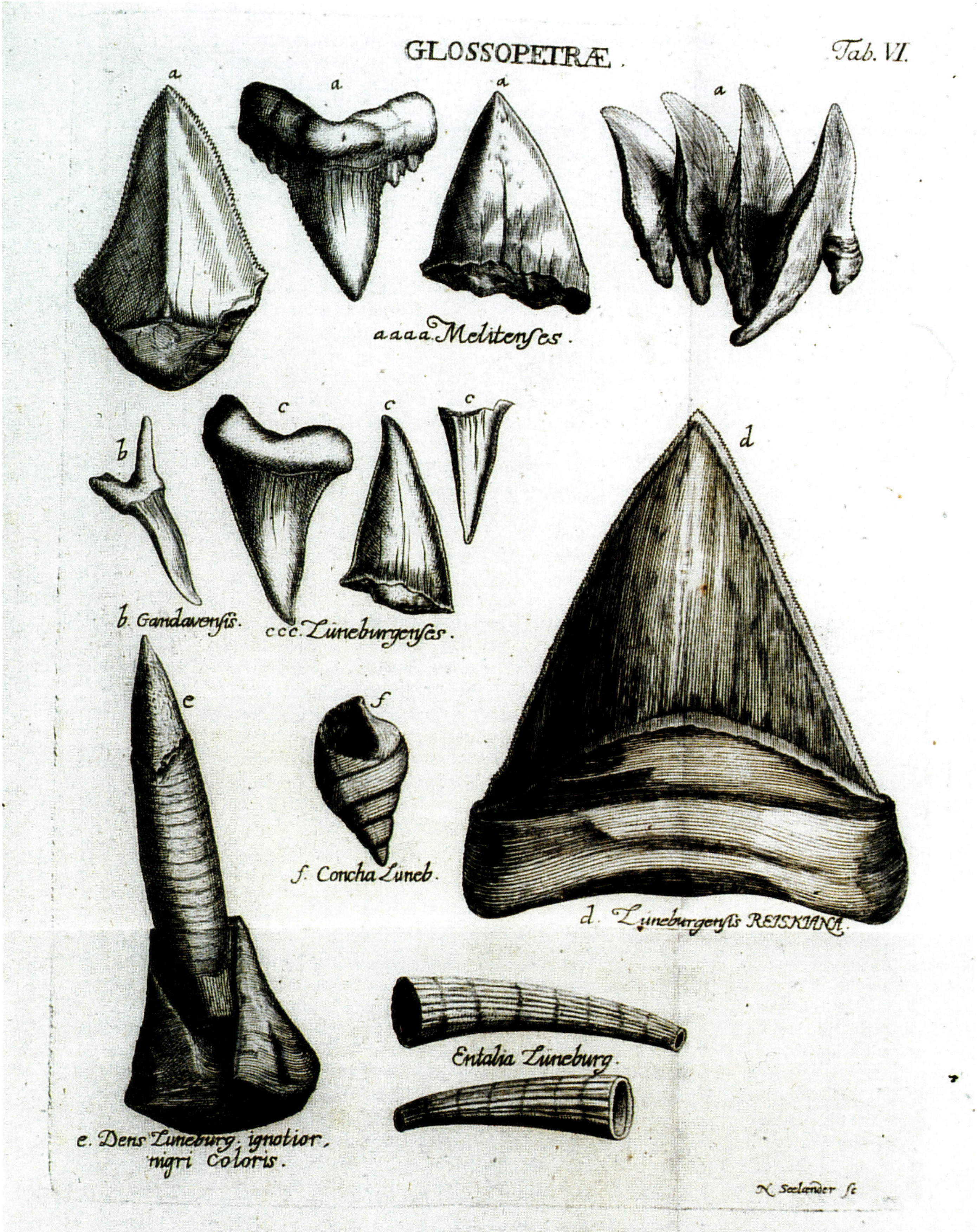
GLOSSOPETRÆ.
Tab. VI.
a
a
a
a
aaaa. Melitenses.
b
c
c
c
d
b. Gandavensis.
ccc. Luneburgenses.
e
f
f. Concha Luneb.
d. Luneburgensis REISKIANA.
Entalia Luneburg.
e. Dens Luneburg. ignotior, nigri Coloris.
N. Seelander sc

species of the fish at first sight [...]" and that "the animal never diverges from its own symmetry, nor does it not possess its own size."[15] Leibniz also reports to have seen and examined specimens of ancient life himself: "I have held in my hands a barbel, a perche, and a whiting, which were imprinted on rock."[16]

In a second methodical step, Leibniz compares his own visual and tactile evidence with testimony by other authors which either is explicitly quoted (as in the case Lachmund, cf. infra) or alluded to in few words (as in the case of Guericke, cf. infra). The instances of 'I have seen' are correlated thus with, and corroborated by, a variety of instances of 'They have seen.'[17]

What others report to have seen appears to be relevant for settling questions of classification, of nomenclature, of the places where fossils have been found, i.e. questions natural historians and historical geographers have to deal with. But such information is of limited value as far as biomorphological features of fossilised organisms are concerned. Iconic information thus adds to textual information what the eye's mind and the mind's eye need for rounding up what texts and pictures cannot convey by themselves. In other words, pictures are supposed (1) to visually stabilise prototypes of fossil traces[18] and (2) to inspire confidence that textual messages really deal with the objects they claim to deal with.

Two examples are likely to illustrate the function of iconic information in Leibniz' effort to stabilise prototypical visual patterns of the traces of ancient life. The first example concerns the topic of *glossopetrae* and pictorial quotations; the second example concerns the visualisation of an unknown species in the absence of available iconic information.

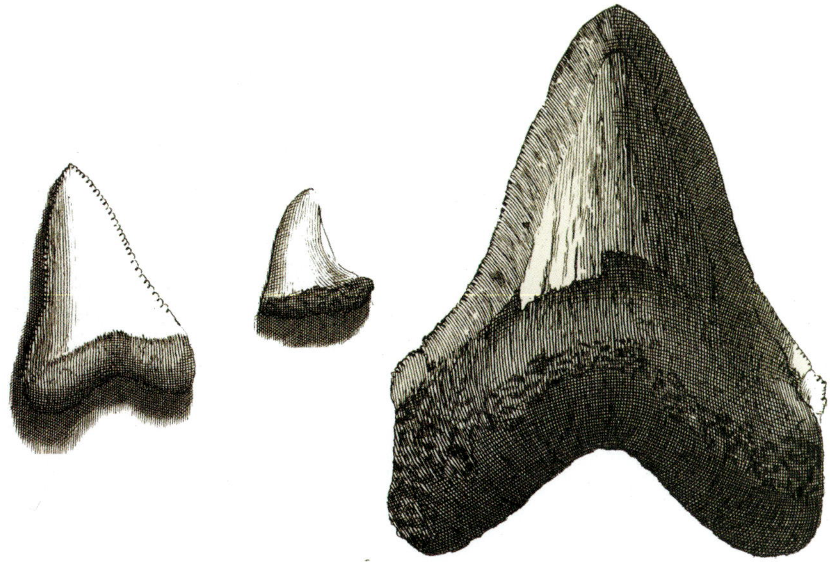

Figure 2 (*opposite*)
Various *glossopetrae* from Lüneburg and from Malta
From Leibniz: *Protogaea* (Leipzig 1749) plate 6.

Figure 3
Pictures of *glossopetrae*. From Hooke: *Posthumous Works* (London 1705) – rotated.

Leibniz asserts that "the glossopetrae of Lüneburg do not differ from those of Malta, except that ours are usually smaller and occur in soil rather than in rocks. They do not look like a snake's tongue, as popular opinion would have it, but more like the tongue of a woodpecker, as Agricola already noticed. There is hardly a reason to doubt that they are teeth of sharks, which belong to the genus of whales, or of dogfish [...]. These teeth differ from one another as they do in living animals; often they appear to be serrated at the margin, others are smooth. The colour also varies, which is due, I believe, to the surrounding soil. Ours [sc. glossopetrae] are blackish or nearly grey. The pointed part is like horn, shiny, polished, and hard, whereas the opposite part closer to the root is softer, spongious, rough, and darker. [...] I shall add pictures of our glossopetrae and of the Maltese ones, so that those who have already

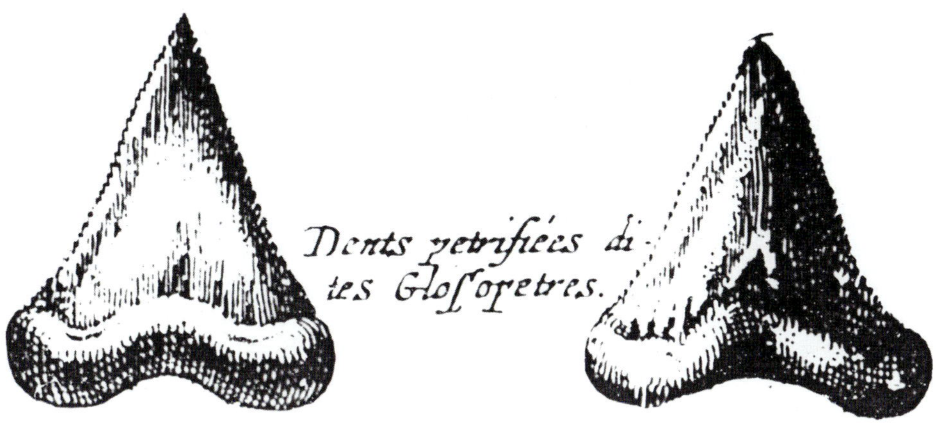

Figure 4
Pictures of *glossopetrae*. From Boccone: *Recherches* (Paris 1671).

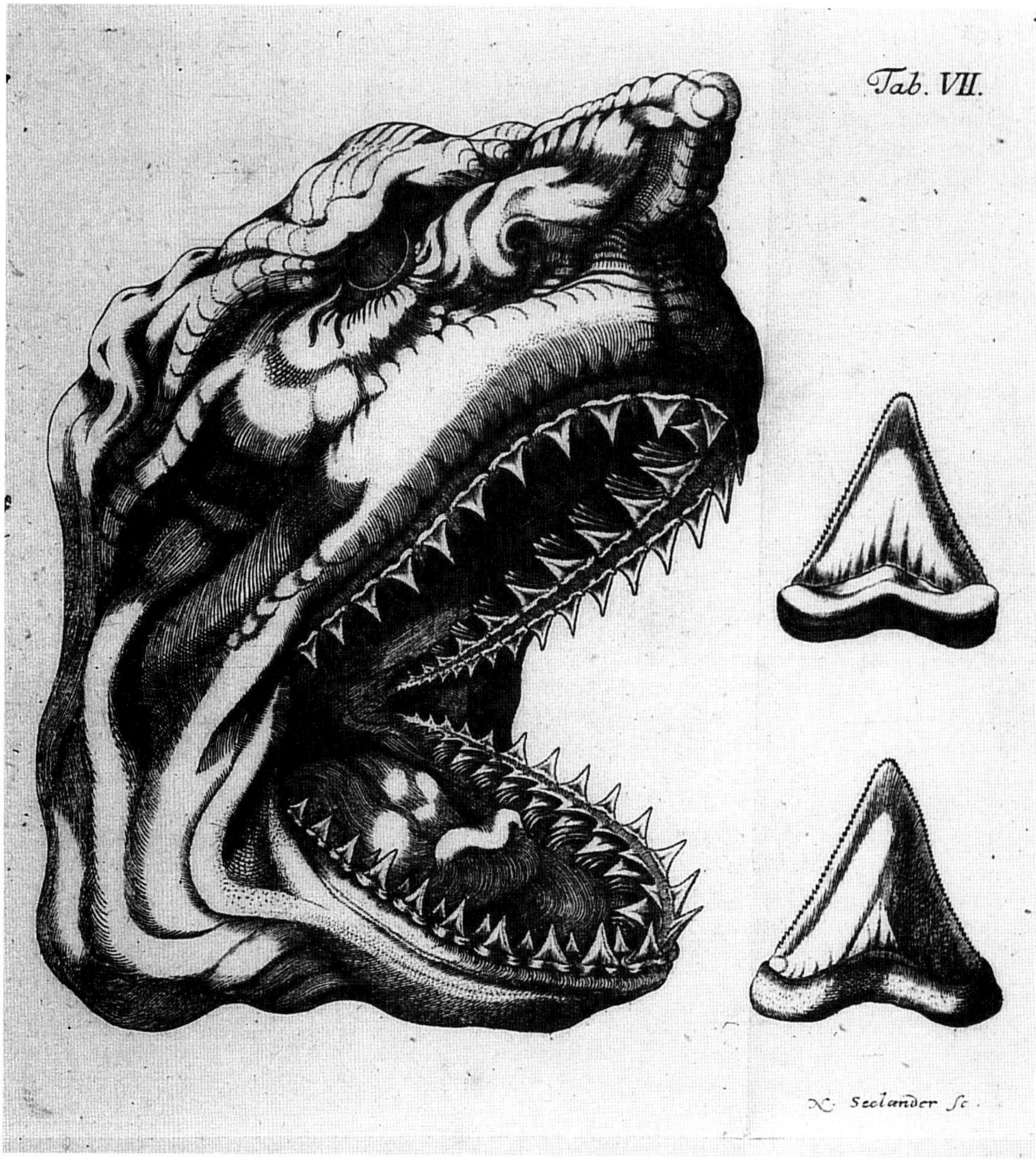

Figure 5
Image of Steno's image of the white shark. From Leibniz: *Protogaea* (Leipzig 1749) plate 7.

seen such teeth of sharks may judge by their own eyes that there is no difference."[19]

Instead of having Seelander design specimens of *glossopetrae* (see fig. 2) as announced in the *Protogaea*, Leibniz might of course have asked his engraver to provide copies of representations stemming from other sources as well, such as Robert Hooke's *Posthumous Works* (cf. Hooke [1705]), (see fig. 3) or Paolo Boccone's Recherches (see fig. 4), (cf. Boccone [1671]).These representations, too, show that no significant differences between these *glossopetrae* and those of Lüneburg are noticeable.

In addition to the pictures of regular *glossopetrae*, Leibniz provides another engraving which reveals itself to be a direct iconic quotation (by way of faithful reconfiguration of an already published picture): "Fur the purpose of comparison, we also want to include the head of a white shark and of its teeth following Steno's plate."[20] This iconic quotation (see fig. 5) is supposed to demonstrate that the teeth of sharks vary in size, shape, and thickness in normal individuals of this species and to visualise the typicality shared by *glossopetrae* on the one hand, and their twin specimens in modern sharks, on the other. The illustration is also meant to provide means by which natural historians and curious readers might bridge the gap between the encounter of single *glossopetrae* (without trace of the fish's skull) and the anatomy of sharks. "One should not be surprised that the jaws do not occur themselves [...]. For scholars have long ago found that the shark's teeth are not firmly fixed into the bones [as in other animals], but simply hang in the flesh. The teeth have thus easily been removed from the jaws by the water and transported far away by currents."[21]

Another way of handling the iconic information to which Leibniz resorts is that of quoting from captions which accompany plates published elsewhere. Indeed, several pages of the Protogaea are filled with captions extracted *verbatim* from Friedrich Lachmund's *Oryktographia Hildesheimensis* by means of which the philosopher refers (without further comments) to a great number of fossils already depicted (see figs. 6–7).[22]

In conclusion, Leibniz uses iconic information in at least four different, yet complementary ways:

(1) iconic quotations in the strict sense of the word (redesigned pictures of pictures),
(2) references to pictures by way of quoted captions of pictures,
(3) textual redescriptions of pictures contained in published or unpublished sources, and
(4) iconic representations of single fossil specimens.

But what happens when most of the iconic information fails, or when no such information is available? Leibniz' attempted reconstruction of an unknown species of unicorn from fossil bones illustrates how *mental* representations take over (or can take over) and inform the design of a picture in the absence of conventional, stable, normalised iconic patterns.

In 1663, several large bones said to belong to the skeleton of a unicorn were found near Quedlinburg. Leibniz described these bones in the following terms, which are largely inspired by a passage in Otto von Guericke's treatise on the void: "[...] the posterior part of the body was inclined, as is usual in animals; the head was elevated and bore an elongated horn on its front, approximately five cubits long, in width like a human leg, but of decreasing proportion. Due to the ignorance of the diggers, the skeleton was broken and extracted piece by piece; the horn with the head, some ribs, the spinal column, and bones were finally handed to the sovereign abbess of the place. This has been reported to me; a picture is added, which is not unsuited to render it [sc. the skeleton]."[23]

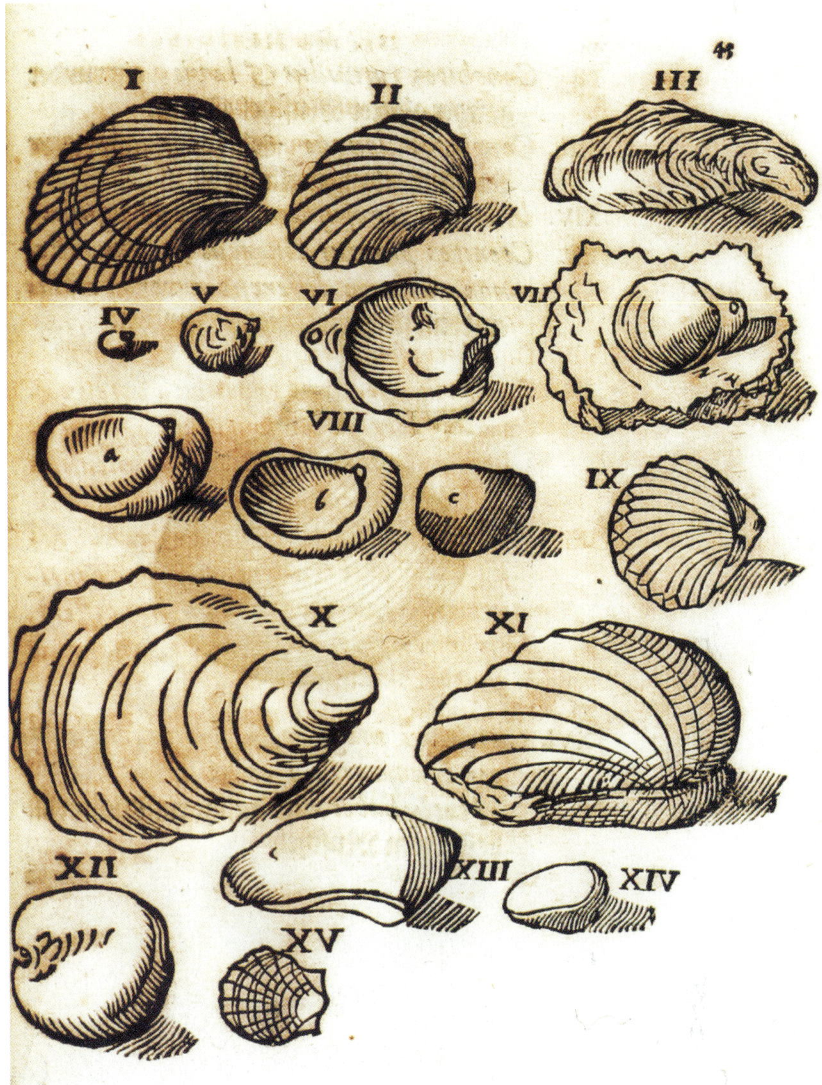

Figure 6
Fossil shells. From Lachmund: *Oryktographia* (Hildesheim 1669) 43. Leibniz refers to the captions of these figures.

One doesn't know for sure what the words 'This has been reported to me' were meant to say. We may surmise that Leibniz did not rely exclusively on Guericke's monograph, and that he had sketches or drawings of some or all parts of the skeleton at hand (he might have made the sketches or drawings himself when he saw the relics, or someone could have made them for him). The visualisation of the unicorn (see fig. 8), however, stems from the philosopher's *mental* representation only.

The animal thus created nearly from scratch did not resemble its forerunners from the classical bestiaries. In this respect, the picture designed for the purpose of publication of the *Protogaea* departed from the medieval and Renaissance imagery of

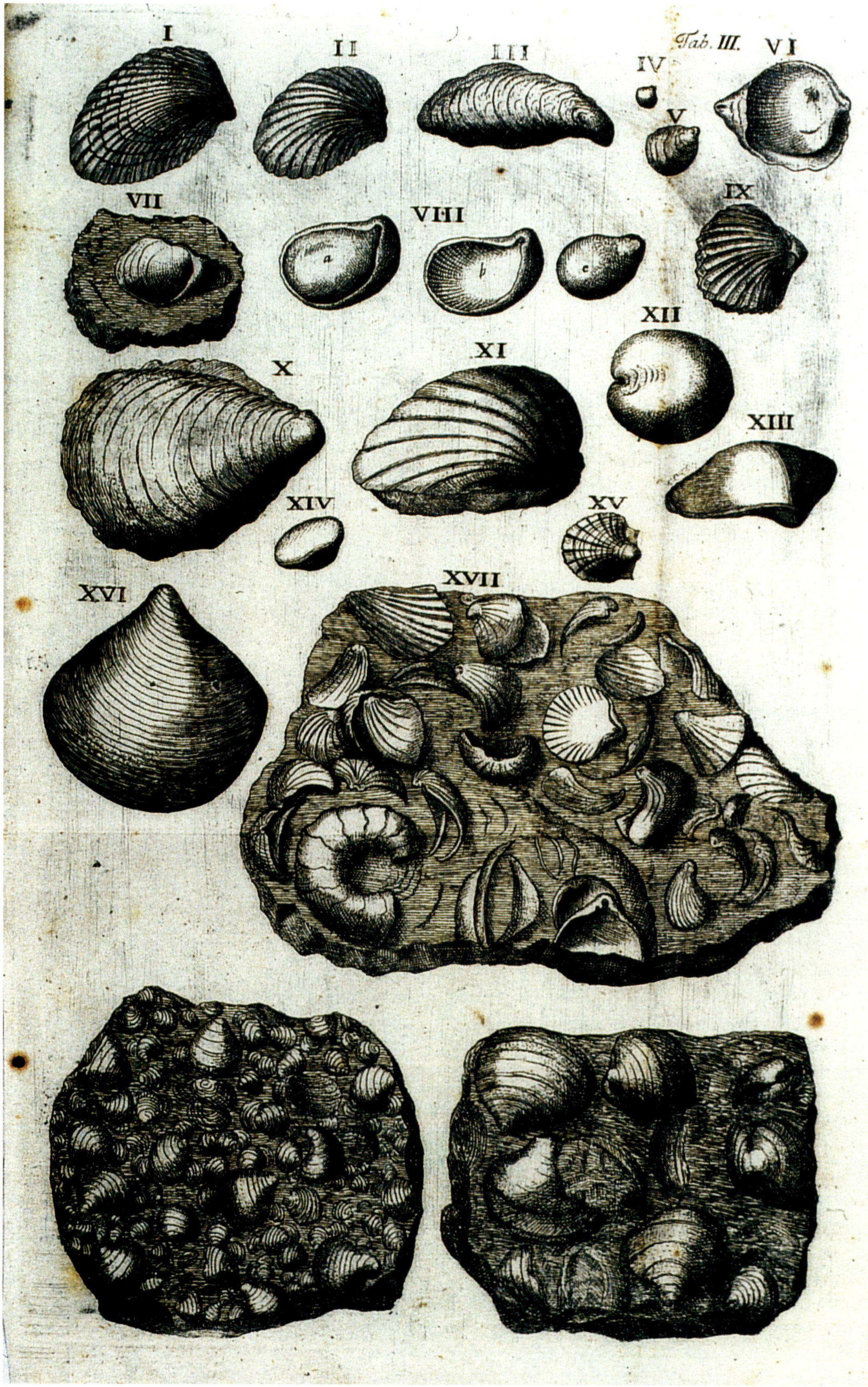

Figure 7
Pictures of fossil shells.
From Leibniz: *Protogaea* (Leipzig 1749) plate 3.

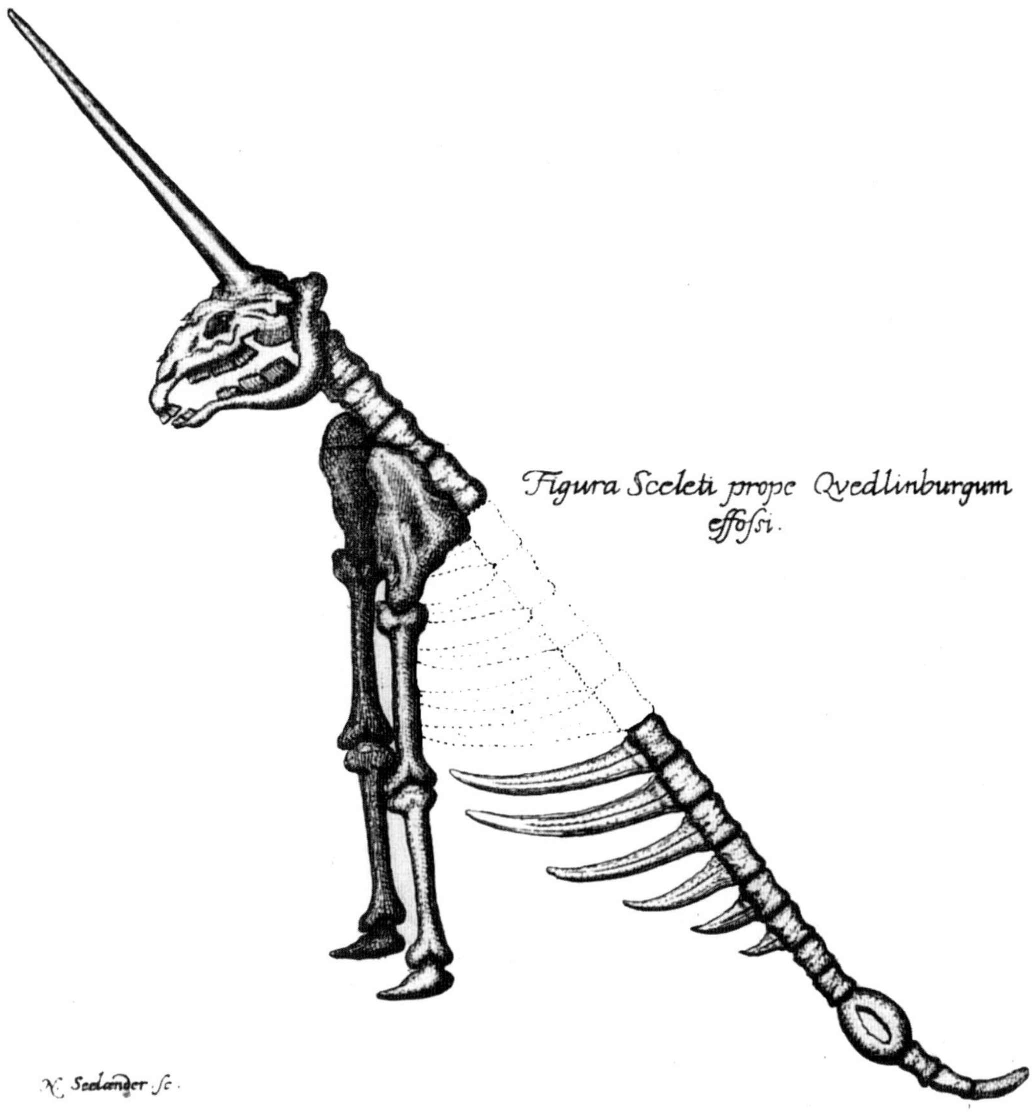

Figure 8
Leibniz' visualisation of the unicorn of Quedlinburg. From Leibniz: *Protogaea* (Leipzig 1749) plate 12.

unicorns: the quadruped, horse-shaped, mythical animal was dead for the polymath who had carefully studied the controversial literature on the topic (cf. e.g. Bacci [1573] and Bartholin [1628]). But other animals endowed with horns (narwhals, elephants, mammoths) did not seem to fit either, nor, for various reasons, did the petrified relics found at Quedlinburg. The occurrence of elephants in northern countries caused some theoretical troubles about which Leibniz had not yet made up his mind. The narwhal's anatomy did not correspond at all to the fossil document as described (and seen? or more or less faithfully depicted?). And mammoths, in turn, were far from constituting a theoretically and iconically tamed species to which natural historians could refer as easily as they did to sharks, snails, or barbels. Thus, Leibniz created the picture of a chimera by putting the bony parts of one or several petrified organisms together according to *his* imagery – an imagery by and large shaped by the iconic standards of the experts of the time, yet still harbouring some elements of previous epochs. As Claudine Cohen puts it: "The fact that Leibniz wanted to have among the pictures of his treatise the figure of this improbable creature [...] shows that the fabulous bestiary remains, at the dawn of the Enlightenment, a likely source of zoology" (Cohen [1994] 86).

Otto Selz Institute,
University of Mannheim

Notes

[1] The full title is: *Protogaea, sive de prima facie telluris et anitquissimae historiae vestigiis in ipsis naturae monumentis dissertatio.*

[2] Leibniz to Duke Ernst August, mid-January 1691 (N. 21): "[...] so werde ich erstlich von denen entfernsten antiqvitäten dieser Lande handeln müßen, wie nehmlich solche allem ansehen nach vom Meer biß wenigstens am Harz unter waßer gestanden; wie sich zu Lüneburg die so genanten Natterzungen wie in der Insul Malta finden, so nichts anders seyn, als Zähne von gewißen Meerwundern, wie sich in der Bumanshöhle und Scharzfeldischen loch knochen unbekannter thiere ereignen [,] ich auch selbst etwas dergleichen aus der Bumanshöhle heraus gebracht; wie ganze seen verschüttet, und zu steinen worden, die gestalt der fische in dem stein geblieben, wie der fliegen dem Agtstein, und sich in die löcher hernach Metallische materi geleget, und dergleichen curiose dinge so dem werck nicht wenig Zierde geben können." (Leibniz 1970a, p. 23).

[3] Leibniz to A. G. v. Bernstorff, early September 1715: "I forgot to tell Your Excellency that the preliminary discourse on the tranformations of the land [...] is finished [J'ay oublié de dire à V. E. que le disours préliminaire sur les changemens du pays [...] est fait]" (Leibniz [1886] 333f.).

[4] The biographical information is drawn mainly from Müller et al. [1969].

[5] Leibniz to Hernri Justel, September 6, 1692 (N. 242): "I am told that Mr Ray has published a book on the transformations of the earth, which I have not seen yet. But I myself have meditated extensively on this subject matter, and I intend to write a discourse, entitled *Protogaea* [...] (J'ay appris que Mons. Ray a publié un livre, des changemens de la terre, que je n'ay pas encore veu. Cependant, j'ay fort medité moy même sur cette matiere, et j'avois projetté un discours intitulé: *Protogaea* [...])" (Leibniz [1970b] 413).

[6] Another document, related to the presentation of the *Protogaea*, has been found and published recently; Leibniz [1997].

[7] On the "new science of natural geography", cf. Leibniz [1949] 18.

[8] Leibniz [1949] 62: "[...] visuntur piscium formae, exacte et a fabre delineatae, quasi artifex nigro lapidi sectilem metalli materiam inseruisset, ichthyomophum aliqui indigitant. [...] Talis lapidis unum mihi fragmentum est ab utraque parte piscis imagine, sed diversa, signatum."

[9] These effigies were, ontologically speaking, the opposite of ideomorphous stones which represent something else without being a physical trace of the thing represented; for the at that time conventional definition of ideomorphous stones (*lapida ιδιομορφοι*), cf. Mercati [1717] 215–220.

[10] Cf. Leibniz [1949] 92 on "plastica facultas."

[11] Cf. Leibniz [1949] 96: "Sic enim profecto se res habet, ut nunquam perfecte veritatem casus imitetur."

[12] Leibniz [1949] 40: "Operae pretium autem facturum arbitror, qui naturae effect ex subterraneis erute diligentius conferet cum foetibus laboratorium [...] quando mira persaepe in natis et factis similitudo apparet."

[13] Cf. Leibniz [1949] 46: "Nemo autem ignorat, Cinnabarin etiam ex sulphure et hydrargo per artem parari [...]."

[14] Cf. Leibniz [1949] 48: "Ex auripigmento vi ignis attollitur Rubini quodam [...] genus [...]."

[15] Leibniz [1949] 64: "[...] nam plerumque genus piscis agnoscas prima obtura, neque unquam a symmetria abit animal, aut magnitudinem not habet suam."

[16] Leibniz [1949] 64: "Habui ipse ibi in manibus mugilem, percam, alburnam, petrae insculptos."

[17] Cf. Leibniz [1949] 64: "Visi et marini generis, ut raja, halex et lampreta [...]. There also appeared marine fish, such as the ray, the herring, and the lamprey [...]."

[18] Such iconic prototypes then may guide the researcher's act of visual recognition in the field or in the cabinet.

[19] Leibniz [1949] 100 and 102: "Glossopetrae autem Luneburgenses nihil a Melitensibus differunt, nisi quod minores nostrae esse solent, nec ut illae in saxo, sed in terra. Linguam repraesentant non serpentis, ut volgo volunt, sed pici potius, ut jam Agricola noster observavit. Lamiarum, cetacei generis piscium, aut canum marinorum dentes esse [...]. Variant forma, ut in ipsis animalibus, nam serratae persaepe comparent in margine, interdum nudae. Color diversus; credo ab ambiente. Nostris fere nigricans aut subcineritius. Pars acuminata, quasi cornea, laevis et polita et dura est; contra postica et crassior ad radices, mollior est, spongiosa, scabra, obscuriorque [...]. Placet figuras sujicere, et nostrarum et Melitensium, ut, qui Lamiarum dentes videre, ipso oculorum judicio, quam nihil intersit, arbitrentur."

[20] Leibniz [1949] 102: "Caput etiam canis carchariae ex Stenonis delineatione cum dentibus subjiciemus in comparationem."

[21] Leibniz [1949] 102: "Nec mirum esse debet, quod maxillae ipsae non comparent [...]. Nam dudum observatum est a curiosis, Laminarum dentes non aeque in ore firmos esse, sed membranae tentum haerere. Itaque evulsi motu aquarum, longuisque provecti, maxillas suas facile deseruere."

[22] Pages 108–116 and 116–122 in Leibniz [1949], correspond to pages 40–42, 44, 46, 48, 50, 52–54, and 56, respectively, in Lachmund (1669).

[23] Leibniz [1949] 128: "Gerikius [...] narrat, repertum sceleton unicornis in posteriore corporis parte, ut bruta solent, reclinatum, capite vero sursum levato, ante frontem gerens longe extensum cornu quinque fere ulnarum, crassitie cruris humani, sed proportione quadam decrescens. Ignorantia fossorum contritum perticulatimque extractum est, postremo cornu cum capite et aliquibus costis, et spina dorsi atque ossibus Principi Abbatissae loci allata fuere. Eadem ad me perscripta sunt; additaque est figura, quadam subjicere non alienum erit." In chapter 3, book 5 of the treatise on the void, Guericke reports: "Accidit quoque hoc ipso anno 1663 *Quedlingurgi,* quod in monte, quem volgo, *den Zeunickenberg* vocant, ubi materia calcis effoditur, et quidens in ejus quadam rupe, repertum est Sceleton Unicornis, in posteriore corporis parte, ut bruta solent, reclinatum, capite vero sursum elevato, ante frontem gerens longe extensum cornu crassitie cruris humai, atque ita secundum proportionem longitudine quinque fere ulnarum. Animalis quidem hujus Sceleton primum ex ignorantia fuit contritum, et particulatim extractum, donec caput una cum cornu et aliquibus costis, spina dorsi, at ossibus, *Reverendissimae Principi Abbatissae,* ibidem degenti, fuerit traditum"" (Guericke [1672] 155).

References

Agricola, G.
1955 *De natura fossilium* (Textbook of mineralogy) (1564). New York.

Bacci, A.
1573 *L'alicorno. Discorso [...] nel quale si tratta della natura dell'alicorno, e delle sue virtù eccellentissime*. Florence.

Bartholin, C.
1628 *Opuscula quatuor singularia: I. De unicornu ejusque affinibus et succedaneis opusculum* [etc.]. Strasbourg.

Boccone, P.
1671 *Recherches et observations naturelle sur la production de plusieurs pierres, principalement de celles qui sont de figure coquille, & de celles qu'un nomme corne d'ammon; sur la petrification de quelques parties d'animaux; sur les principes des glossopetres; sur la pierre etoilée, et sur l'embrasement du Mont Gibel ou Etna arrivé en l'an 1669*. Paris.

Cohen, C.
1994 *Le destin du mammouth*. Paris.

Guericke, Otto von
1672 *Experimenta nova (ut vocantur) magdeburgia de vacuo spatio* [etc.]. Amsterdam.

Gurwitsch, A.
1973 *Leibniz. Philosophie des Panlogismus*. Berlin and New York.

Hooke, R.
1705 *The posthumous works* [etc.]. London

Lachmund, F.
1669 *Oryktographia* [romanized form] *Hildesheimensis, sive, Admirandorum fossilium: que in tracta Hildesheimensi reperiuntur, descriptio iconibus illustrata, cui addita sunt alia de calculis de fontibus, &c.* Hildesheim.

Leibniz, G. W. F.
1693 "Protogaea autore G. G. L." In: *Acta eruditorum* anno MDCXCIII: 40–42.

1749 *Protogaea, oder Abhandlung von der ersten Gestalt der Erde und der Spuren der Historie in den Denkmaalen der Natur aus seinen Papieren herausgegeben von Christian Ludwig Scheid*. Leipzig and Hof.

1886 "Briefwechsel mit dem Minister von Bernstorff und andere Leibniz betreffende Briefe und Aktenstücke aus den Jahren 1705–1716," mit einer Einleitung herausgegeben von Archivar Dr. Doebner. In: *Zeitschrift des historischen Vereins für Niedersachsen* 188: 205–339.

1949 "*Protogaea*." In: *Werke*, ed. W. E. Peuckert, vol. I. Stuttgart.

1970a "Letter to Duke Ernst August, mid-January 1691 (N. 21)." In: *Allgemeiner politischer und historischer Briefwechsel*, vol. VI, 22–31. Berlin.

1970b "Letter to Henri Justel, September 6, 1692 (N. 242)." In: *Allgemeiner politischer und historischer Briefwechsel*, vol. VIII, 412–414. Berlin.

1989a "Mémoire sur les pierres, qui renferment des plantes et des poissons dessechés." In: *Opera omnia nunc primum collecta* [etc.] (1768), vol. II/2, 178–179. Hildesheim, Zürich and New York.

1989b "Epistola ad autorem dissertationis de figuris animalium quae in lapidibus observantur, et lithozoorum nomine venire possent." In: *Opera omnia nunc primum collecta* [etc.] (1768), vol. II/2, 176f. Hildesheim, Zurich and New York.

1997 "Leibniz on geology: A newly discovered text." In: *Studia Leibnitiana* 29: 6–11.

Mercati, M.
1717 *Metallotheca*. Rome.

Müller, K. and G. Krönert
1969 *Leben und Werk von Gottfried Wilhelm Leibniz*. Frankfurt am Main.

IV.
Depicting the World at Large and the Hidden Potential of Images

Planetary Diagrams – Descriptions, Models, Theories

From Carolingian Deployments to Copernican Debates

Bruce Eastwood and Gerd Graßhoff

Introduction

Through three distinct periods from ca. 800 to ca. 1600 we find that European astronomers were concerned with questions about the planets that involved the discussion and invention of models without quantitative expression. This qualitative tradition was first developed in the ninth century in the course of studying ancient Latin texts on cosmology and astronomy. The diagrams, used to represent different phenomena and aspects of planetary motion, continued to be employed as long as they were found useful for teaching, for exposing questions, or for proposing theoretical positions. The history of this tradition of planetary diagrams indicates a constant concern for qualitative theory and the co-existence of both qualitative and quantitative planetary theory after the introduction of the Greco-Arabic mathematical tradition of planetary astronomy in twelfth-century Europe. In the sixteenth century the same qualitative tradition continued as a source for approaches to new phenomena and problems.

The Beginnings in Carolingian Europe

Astronomy and Computus before 800

From the sixth century to the twelfth century in Western Europe there was no direct influence of Greek works in the exact sciences. This was an era in which, despite certain Western exchanges with the Byzantine state and church, no writings in Greek on mathematics or astronomy seem to have crossed the boundaries of these two worlds. Between the Arabic-speaking and Latin worlds, with the exception of an incomplete transfer of knowledge about the astrolabe in the eleventh century, the same sort of barrier appears to have existed for the exact sciences. What had been the western half of the Roman Empire in the fifth century became a mosaic of successor kingdoms in Italy, Spain, the British Isles, and the large area occupied today by France, the Low Countries, Germany, and Switzerland. Within these lands, excepting Spain, which was conquered by Muslims in the early eighth century, Latin was the virtually exclusive language of scholarship through these six, early medieval centuries.

The sixth century saw the decline of astronomical and especially planetary knowledge, witnessed by Cassiodorus in Italy and by Gregory, bishop of Tours, in the kingdom of the Franks. Away from the Mediterranean, the seventh century, both on the continent and in the British Isles, was an era when the classical tradition of "liberal arts" was deemed of little use. The actual curriculum of studies in the only organised schools of the time, the ecclesiastical schools, focused on grammar, computus, and chant. Grammar was increasingly devoted to rules for Latin writing, with the examples being chosen much less from classical sources and much more from the Bible, stories about holy men and women, and Christian poetry. Chant was a practical study, focused on the needs of Christian ritual. Computus, the sole "scientific" discipline by modern lights, attended to counting, basic arithmetical operations, and the knowledge needed to find the proper date for Easter and thereby the Sundays and all other Christian feast days throughout the year. This knowledge did not include precise astronomical observation nor any theory of the motions of the sun or the moon, but it did require calculation with lunar and solar time intervals given by definition. In the early eighth century the situation began to change with the example set by Bede (ca. 673 – 735), who included observations of the sun and moon at crucial points in their cycles, at the monastery of Jarrow-Wearmouth in northern England.[1] Bede, however, did not depart from the goals of computus, nor did he integrate planetary study into the discipline.

The reappearance of astronomy as a discipline, recognisably related to the astronomy of the fourth and fifth centuries in the Roman world, occurred about the turn of the eighth century in the kingdom of the Franks under King Charles (768 – 814), known today as Karl der Große or Charlemagne. As part of his search for greater stability in the kingdom, in 789 Charles sent out the *Admonitio generalis*, which ordered a reform of the clergy and required the establishment of schools at all monasteries and cathedrals and further prescribed that the Psalms, proper chant, writing, grammar, and computus be taught. The king also pointed out in a separate letter that the liberal arts were ill-known by the religious and should be cultivated by those who were able (MGH/L 59f., 79f.). His close advisor, Alcuin (ca. 730 – 804), in a letter sent from the abbey of Saint Martin of Tours in 797, mentioned that studies there followed the pattern of the royal court, which included not only the Bible but also the arts, especially grammar and astronomy. Elsewhere Alcuin wrote as part of his own grammar text a preface, commonly referred to as "De vera philosophia," in which he quoted from the biblical book of Proverbs the statement that Wisdom built her house with seven pillars, and he proceeded to connect the seven gifts of the Holy Spirit with the seven liberal arts, which, Alcuin said, would lead up the steps to perfect knowledge (*perfecta scientia*). These steps he identified explicitly as the arts, ending with astronomy (*astrologia*), and emphasised that they were needed for the enlightenment of the soul as well as for the defense of the faith against heresiarchs (MGH/E 176f.; PL, vol. cI, coll. 849–854). Along with Alcuin, the king not only supported study of the arts but also had a personal interest in both computus and astronomy.[2]

Schools and Texts

The study of astronomy in and outside the schools from the time of King Charles onwards required the recovery and dissemination of texts that seem to have lain relatively unknown during the preceding two hundred years. Beyond the works of Cassiodorus, Isidore, and Bede, none of which gave adequate guidance for a coherent understanding of planetary astronomy and none of which presented a useful picture of the celestial sphere without supplementary oral instruction, there were six Roman Latin texts that provided essential knowledge. The descriptions of the celestial sphere and the constellations appeared in Aratus' *Phaenomena*, as translated by Germanicus (and Cicero and Avienus), and in Hyginus' *Astronomia*. The foundations of planetary astronomy came to Carolingian scholars in Pliny's *Historia naturalis*, Macrobius' *Commentarii in somnium Scipionis*, Martianus Capella's *De nuptiis Philologiae et Mercurii* (Book VIII), and Calcidius' *In Timaeum commentarius*. This second group of four works is where we must turn for planetary diagrams.

The study of these Roman works and the development from them of a noteworthy foundation for planetary astronomy occupied more than half a century, beginning in the late 700s. For none of the four is there a continuous prior history of a manuscript tradition. Between Boethius in the early sixth century and Isidore in the first third of the seventh century we find the last references to these astronomical sources until the time of Bede or later. Even with Bede only limited acquaintance with Pliny's *Historia naturalis* is certain. He made use of one work by Macrobius but made no mention of the commentary on Cicero's *Somnium*.

The earliest surviving manuscripts of the four Roman works are from the ninth century, a testimony both to their earlier disuse and to their revival in Carolingian times. For the availability of Pliny's work we have as witnesses not only Alcuin's letters to the king but also the survival of three manuscripts from the beginning of the ninth century containing

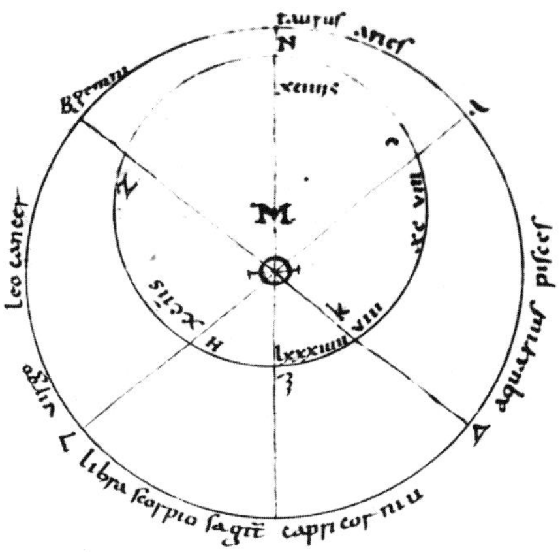

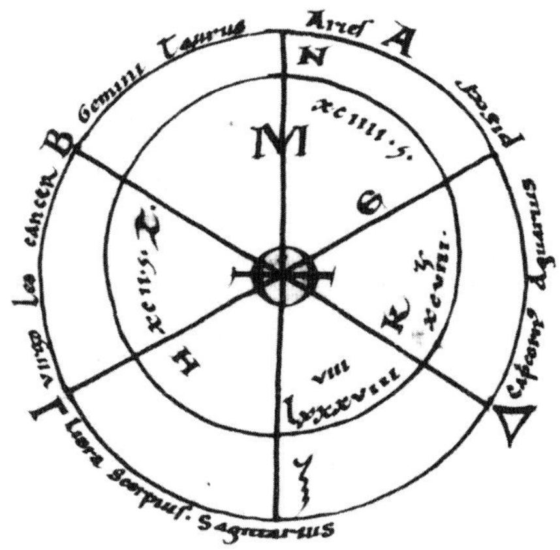

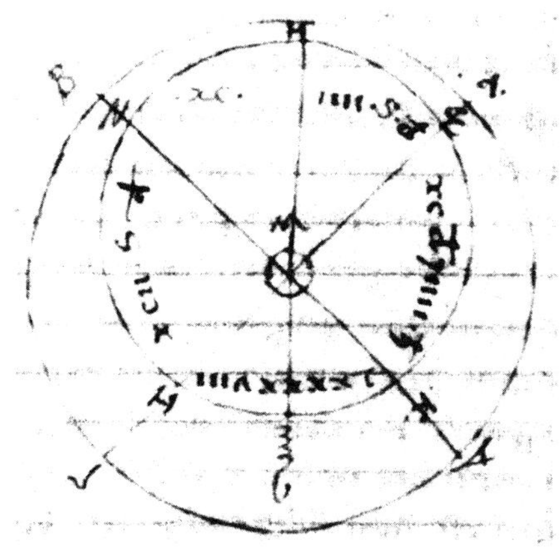

Figure 1 (*left*)
Eccentric model of solar motion. Calcidius' diagram in a Ms of the tenth century (Bamberg Staatsbibliothek ms. Class. 18, f. 35v).

Figure 2 (*centre*)
Eccentric model of solar motion. Calcidius' diagram in a Ms of the twelfth century (London British Library ms. Add. 15293, f. 22r).

Figure 3 (*right*)
Eccentric model of solar motion. Calcidius' diagram in a Ms of the fifteenth century (Vaticano Biblioteca Apostolica Vaticana ms. Vat. lat. 1544, f. 70v).

major portions of the complete work.[3] Macrobius' commentary on the *Somnium Scipionis* was known and being used for its astronomy during Charles the Great's lifetime, as a letter in 811 to the king from the monk Dungal at Saint-Denis makes clear.[4] The text of Calcidius' commentary on Plato's *Timaeus* was used for theological content in the circle of Alcuin ca. 800. Furthermore Calcidius' text was being systematically excerpted before the middle of the ninth century for use in a compendium of astronomical doctrines.[5]

Finally, the astronomical book of Martianus Capella's *De nuptiis philologiae et mercurii* still exists in eighteen different manuscripts surviving from the ninth century, and we know that at least the Capellan book on grammar was already being used in the late eighth century.[6] With undated extended glosses from the first half of the century, we have the first approximately dated and attributable commentary, by John the Scot, during the decade of the 850s.[7] Irish scholars seem to have had a special interest in Capellan astronomy during the era of Charles the Bald (843 – 877). The surviving manuscripts were copied or studied as far east as Fulda and Freising, although the focus of ninth-century production was in northern and northeastern France.

The schools which came to teach these works were both cathedral and monastic schools, which admitted both secular and clerical students. The religious reformers of the early ninth century succeeded in 817 in legislating, at a council in Aachen, the exclusion of all but oblates from the monastic schools. The consequent burden upon the episcopal schools appears to have been too much. There is mention of so-called external schools, where monastic teachers may have continued to teach students other than those planning to become monks. Certainly some monastic schools opened their doors to persons not professing a monastic vocation. And monastic teachers were occasionally sought to assist in or to revive the education of students at cathedral schools. For the ninth century we know of seventy schools, those which left any record of activity.[8]

The variety of training in astronomy in the schools was huge. Monastic and cathedral schools could be depended upon to cover computus. Not only the king's *Admonitio generalis* but also a subsequent series of ecclesiastical statements repeated the requirement that all parish priests know computus.[9] In small schools a single teacher gave instruction in all disciplines. Both the capabilities of the students and the inclination of the master influenced the extent of astronomical study beyond that required for computistical competency. For a monastery like Fulda we know that the student population was at different times between 26% and 40% of the total population of the abbey, and this must have provided opportunities for some students to study astronomy beyond computus. Certainly the interests of a scholar like Raban Maur (ca. 780 – 856) would have encouraged students at Fulda to further study. Beyond the basics of computus, students could find excerpts from authors like Pliny and Macrobius, added to exemplify the wider use of numbers and arithmetical reason in the order of the cosmos.[10]

Ambitious masters and ambitious students forged ahead to the fuller study of these and other authors, especially from the second quarter of the century onwards.[11] They also made direct observations of constellations and of planetary locations in the zodiac.[12] The sundial was known from Antiquity. Use

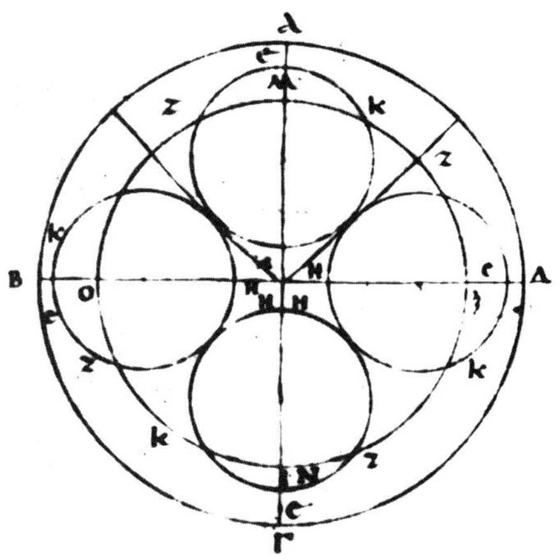

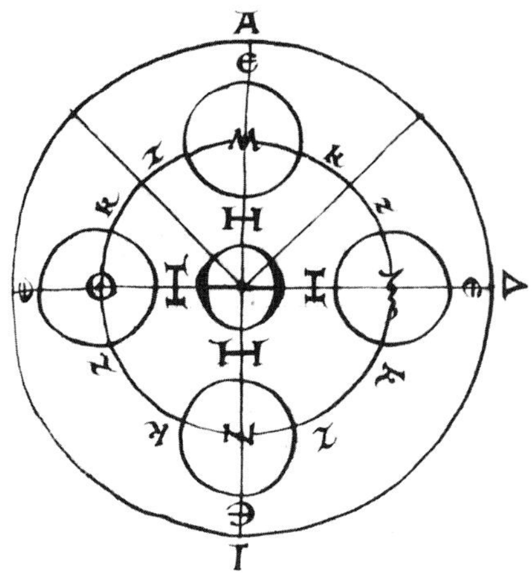

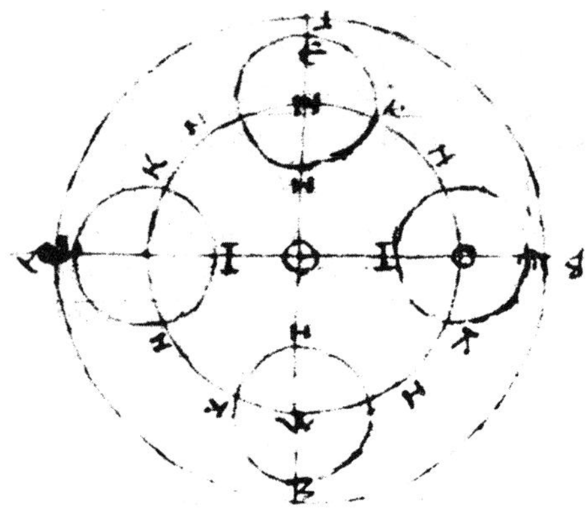

Figure 4 (*left*)
Epicyclic model of solar motion. Calcidius' diagram in a Ms of the tenth century (Bamberg Staatsbibliothek ms. Class. 18, f. 36v)

Figure 5 (*centre*)
Epicyclic model of solar motion. Calcidius' diagram in a Ms of the twelfth century (London British Library ms. Add. 15293, f. 23r).

Figure 6 (*right*)
Epicyclic model of solar motion. Calcidius' diagram in a Ms of the fifteenth century (Vaticano Bibliotheca Apostolica Vaticana ms. Vat. lat. 1544, f. 71r).

of a gnomon or of other empirical tools for discovering the times of equinox and solstice and other data was practiced.[13] And when they found texts conceptually troublesome they reflected upon the theoretical problems they found.

Diagrams and the Study of Texts

Beginning in the ninth century, consequent to the revival of astronomy as one of the *artes*, planetary diagrams provided striking evidence of philological reflection, observational practice, and theoretical elaboration. These diagrams, invented by ninth-century scholars, appeared primarily in manuscripts of the Roman texts listed above. It is important for us to notice both the spatial placement of a diagram in relation to the text and also the different ways that a diagram can connect with a text. The most common addition to a text by a medieval scholar-scribe was the gloss. Usually a gloss was a word or more, inserted above the text line, providing a synonym or other form of definition of a word in the text. Sometimes a gloss appeared in the margin, when there was need to explain more fully or to provide further information. A diagram could function exactly like a verbal gloss. Infrequently we find mini-diagrams between textual lines. Commonly we find diagrams in the margins near the relevant texts. Sometimes diagrams appear at the end of the text, as if an appendix, either at the end of the chapter or even at the end of the book. In all this we might see that such placements of diagrams give the same visual impression as do the scholarly notes used in books today - as footnotes or as endnotes.

A diagram is connected to the proper text in two ways. The simplest is physical proximity. If a diagram is in the margin, it usually stands parallel to the related text with the assumption that a reader can easily make the connection. And when diagrams appear at the end of the chapter or the book, the diagram will often have a reference word, or words, placed beside it. The word, or phrase, comes directly from the text, identifying both by vocabulary and by its grammatical inflection the exact point in the text to which it relates. To all this we must add the diagrams at the ends of chapters where there appear no explicit clues for finding the related texts. When such diagrams occur, they are normally canonical additions to the work and are so recognised; we expect this only in texts widely read or used as school texts. And there are the diagrams without connecting references, which are of such a nature that their specific relevance is obvious. This is the case for diagrams of the planets Mercury ad Venus circling the sun, when placed at the end of Capella's book. No reader could fail to make the connection between text and diagram here, unless he has not read the text at all.

Dynamics of Diagrams: Calcidius and Pliny

Why and how did medieval planetary diagrams develop? Each text offers its own situation and a different account. In considering the dynamics we look for answers to four questions. Why were diagrams produced? Why were changes made to diagrams? Why were new features added? Why were diagrams discontinued?

While pictorial images have more complicated interactions with their texts, we usually expect geometrical diagrams to follow and clarify the texts they illustrate. A diagram should present in an unambiguous way

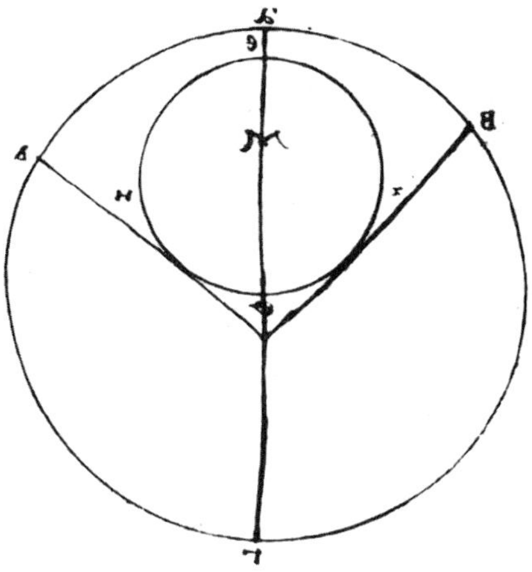

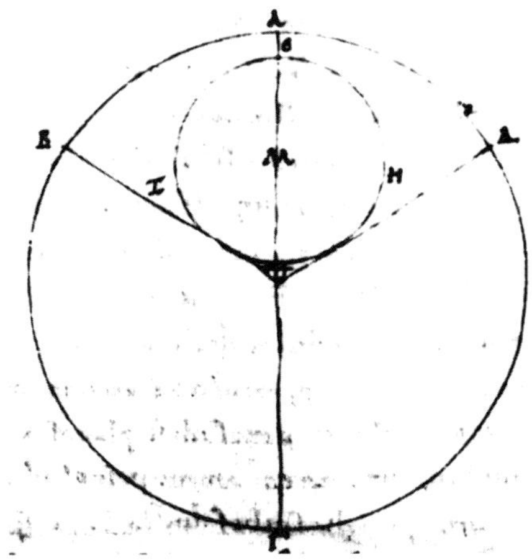

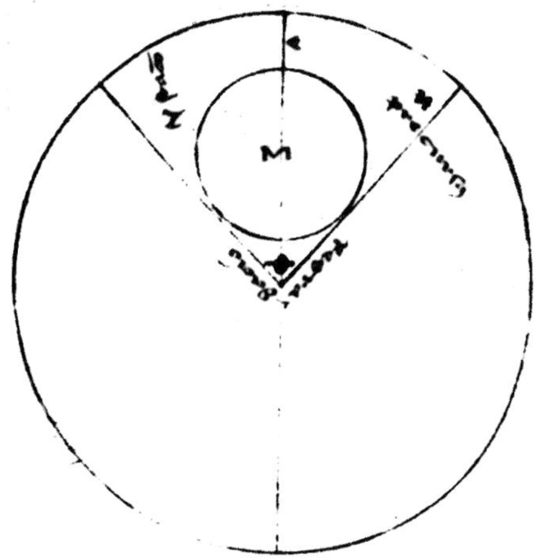

Figure 7 (*left*)
Epicyclic model of planetary motion. Calcidius' diagram in a Ms of the ninth century (Paris Bibliothèque Nationale de France ms. lat. 2164, f. 38v)

Figure 8 (*centre*)
Epicyclic model of planetary motion. Calcidius' diagram in a Ms of the eleventh century (Paris Bibliothèque Nationale de France ms. lat. 6282, f. 34r).

Figure 9 (*right*)
Epicyclic model of planetary motion. Calcidius' diagram in a Ms of the fifteenth century (Cambridge Fitzwilliam Museum ms. McClean 169, f. 116v).

the elements of the text. Of the Roman astronomical texts it is Calcidius' commentary on the *Timaeus* that shows us this situation. Calcidius took words and phrases of Plato and spelled out in elaborate commentary with accompanying diagrams how the words of Plato should be understood according to late Hellenistic astronomical theory. Although lacking mathematical arguments, these Calcidian elaborations stated how the planetary arrangements should be drawn, using the vocabulary of geometrical astronomy to construct the models intended by the text. Since we have no examples of manuscripts or diagrams before the Carolingian era, our basis for evaluating the history of Calcidian diagrams is the group of extant ninth-century copies. And when we look at the images in this work over the succeeding centuries, we find different traditions for different concepts or models.

Stability is the strongest trait of the Calcidian diagrams describing the varying lengths of the seasons of the year, according to both eccentric and epicyclic models of solar motion. While some individual manuscripts may show modifications, the tradition is remarkably stable (figs. 1–3, 4–6).[14] The same can be said for the general diagram for epicyclic planetary motion from the ninth to the fifteenth century (figs. 7–9).[15] Furthermore we can see that these diagrams survived the fourth to eighth centuries in understandable and useful form.

Corruption and correction, however, appear in the manuscript tradition of the Calcidian diagrams for the bounded elongation of the planets Mercury and Venus. The text presents two different ways of understanding the apparent retrograde and progressive motion of the inner planets, but the diagrams in all manuscripts surviving from before the eleventh century show corrupt and incomprehensible versions of the original forms of these two diagrams. In a few manuscripts of the eleventh century there appear clear inventions, corrections made to the corrupt traditions and traceable to a scholar's work on one particular manuscript of the time. The corrector reformed the traditional diagrams so that they would be in accord with the text. Notably, no attempt was made to change the text to fit the traditional diagrams (figs. 10–12).[16]

While the answer to the first of our dynamical questions is obvious for the work of Calcidius – the diagrams in the original version were described by the text and completed the text – the answer to the second question is an interesting example of both recognition of and solution to a problem of diagrammatic representation. The diagrams were changed because prior corruption had made them useless as completions of the text and as visualised models.

Diagrams for planetary motion do not appear in Pliny the Elder's *Historia naturalis*, nor were they added to the margins of the medieval manuscripts, yet we do find medieval diagrams invented to represent lists of data taken from Pliny. Astronomical excerpts from Pliny's *Historia naturalis*, incorporated in computistical collections during the ninth century, described four elements of planetary orbits, giving the simplest information about each. Soon after the selection of these excerpts, there appeared an elaborated graphic representation for each of the four: (a) relative order from the centre to the periphery, (b) harmonic intervals between planetary

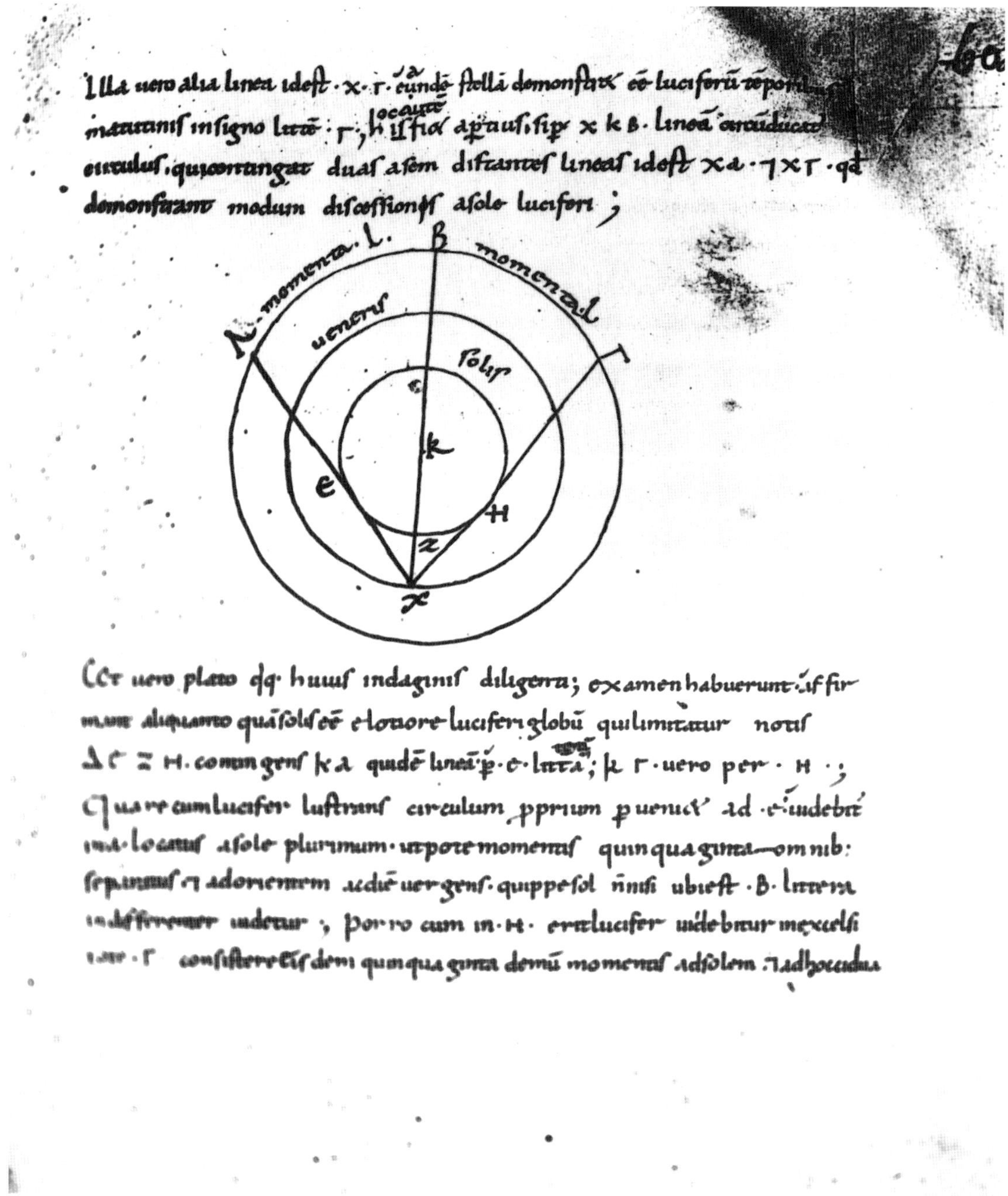

Figure 10
Non-epicyclic model of the bounded elongation of Venus. Calcidius' diagram in a Ms of the eleventh century (Valenciennes Bibliothèque municipale ms. lat. 293, f. 60r).

orbits, (c) the zodiacal locations of apogees and perigees, and (d) the inclinations, or latitudes, of the orbits to the ecliptic.[17]

Probably the most interesting of these diagrams for moderns is the last, regarding latitudes. In Ptolemaic planetary theory, latitudes were among the more difficult topics. However, Pliny and other Roman textbook writers simplified this topic almost unbelievably. Latitudes which were on rotating planes became fixed, and the complex Ptolemaic geometrical structure for planetary latitude was transformed into a simple number. This number represented the fixed number of degrees of inclination of the planet's orbital plane to the reference plane of the ecliptic, or the central line through the zodiac. If the zodiac is a band of constellations around the earth as centre and is tilted at 24 degrees to the plane of the earth's equator, each planet moves on an orbit that is contained within the width of the zodiacal band but is also inclined at a unique and specific angle to the ecliptic plane, which passes through the centre of the zodiacal band. Ptolemy described this in sophisticated geometrical language. Pliny

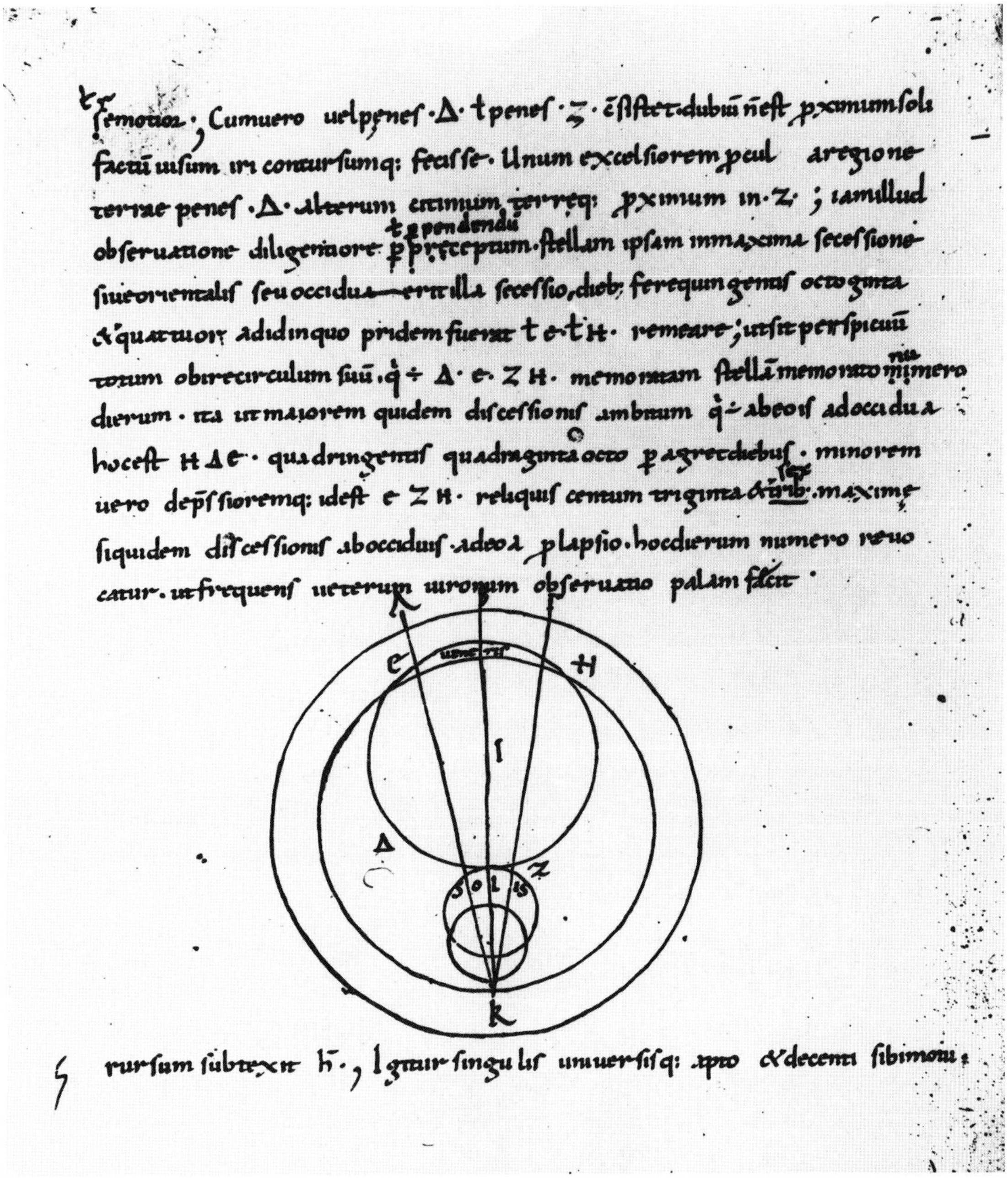

Figure 11
Epicyclic model of the bounded elongation of Venus. Calcidius' diagram in a Ms of the eleventh century (Valenciennes Bibliothèque municipale ms. lat. 293, f. 60v).

simply gave numbers for the sizes of the angles. The Plinian excerpt for planetary latitudes said:

> *Why do the sizes and colours of planets change, and why do they move toward the north and then away to the south? There is latitude and obliquity to the zodiac, whereby these appearances occur. And only those parts of the earth under the zodiac are inhabited, while the remainder languish under the influence of the poles. Venus runs beyond the zodiacal band by two degrees. The moon traverses the full width of the band but never exceeds it. Aside from these, Mercury is the most variable, but of the twelve degrees in the band, this star does not traverse more than eight, of which two are in the middle, and two are below. The sun travels in the middle, between two degrees, unequal, on a serpentine path. Mars travels on the four middle degrees; Jupiter on the middle and the two above; Saturn on two like the sun.*

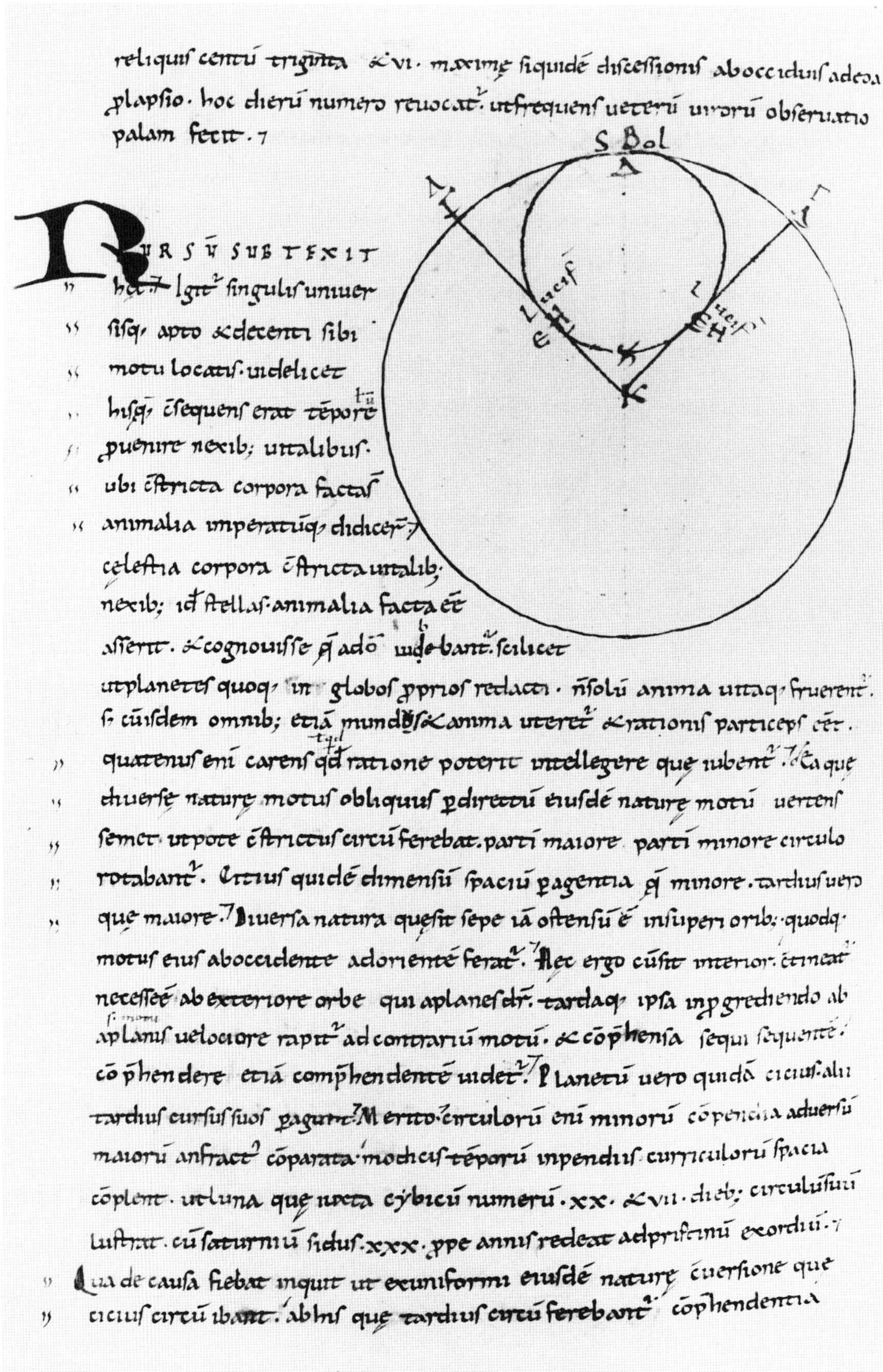

Figure 12
Epicyclic model of the bounded elongation of Venus. Eleventh-century reconstruction of Calcidius' diagram. (Wien Nationalbibliothek ms. lat. 443, f. 183v).

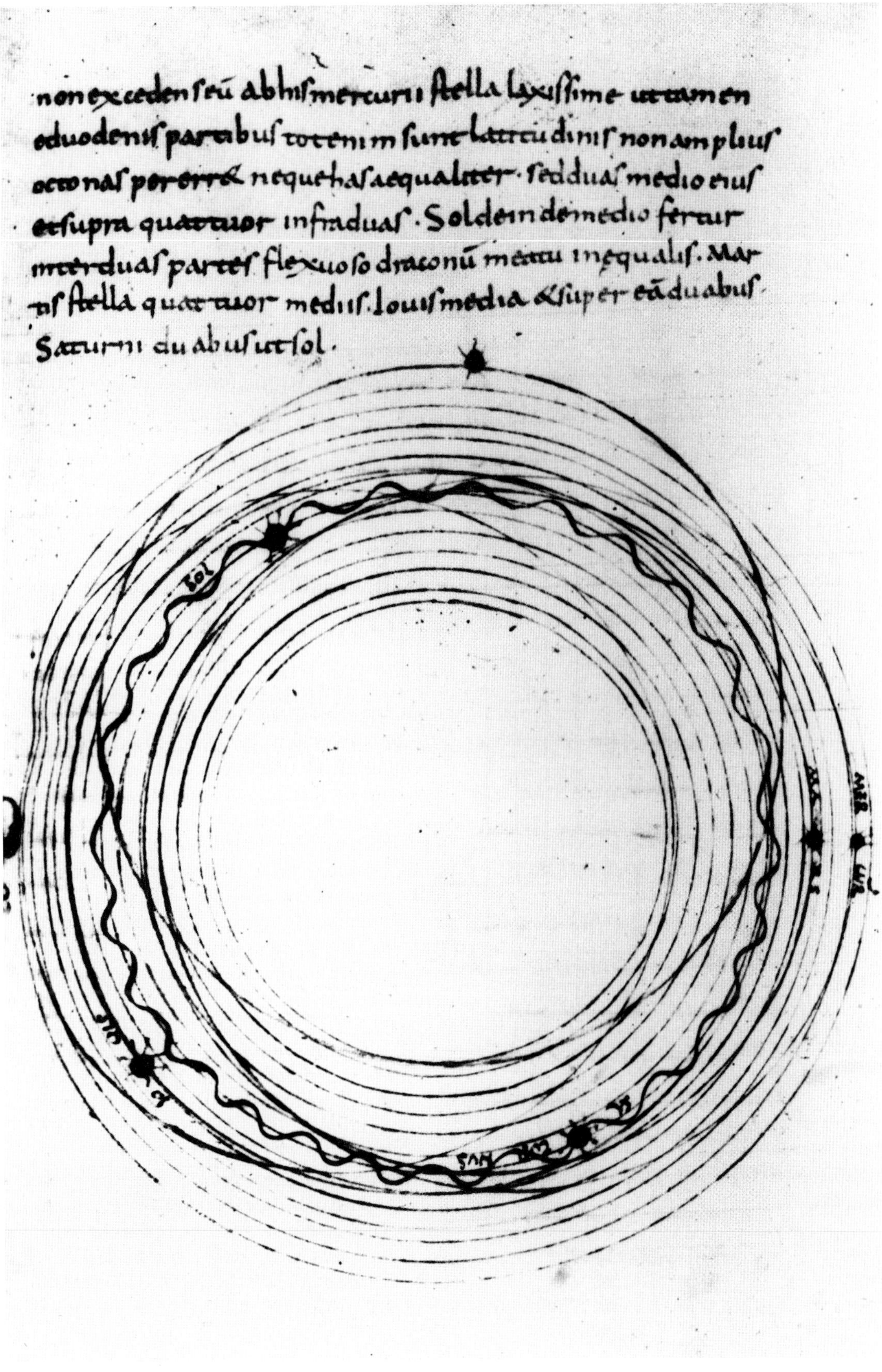

Figure 13
Pliny's planetary latitudes. Circular diagram in a Ms of the ninth century (Monza Bibliotheca capitolare ms. lat. F.9.176, f. 73r).

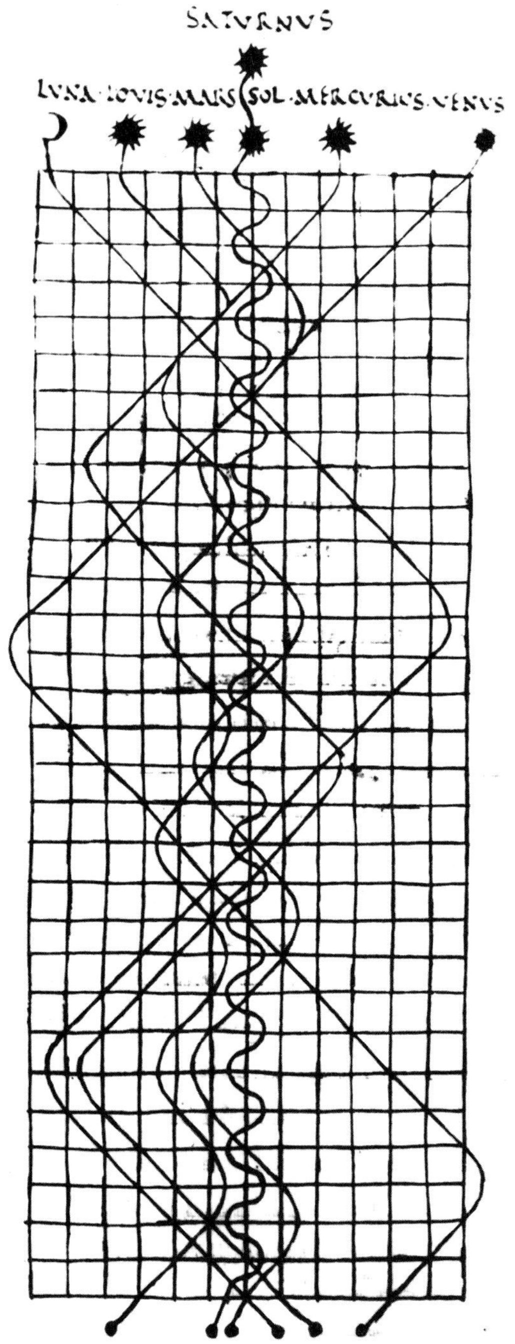

Figure 14
Pliny's planetary latitudes. Rectangular diagram in a Ms of the tenth century (Strasbourg Bibliothèque Nationale et Universitaire ms. lat. 326, f. 123r).

Figure 15 (*opposite, above*)
Pliny's harmonic planetary intervals. Circular diagram in a Ms of the tenth century (Strasbourg Bibliothèque Nationale et Universitaire ms. lat. 326, f. 122v).

Figure 16 (*opposite, below*)
Pliny's harmonic planetary intervals. Vertical list in a Ms of the tenth century (Paris Bibliothèque Nationale de France ms. lat. 5239, f. 38v).

Despite the length of Pliny's statement, it tells the reader essentially only the following numbers of degrees for the latitudes of the planets:

Venus	–	14
Moon	–	12
Mercury	–	8 (5 above and 3 below the ecliptic)
Jupiter	–	4
Sun	–	2 (on a wavy line centred on the ecliptic)
Saturn	–	2

The Plinian diagram shows us the theoretical planetary framework within which the simple list of numbers for planetary latitudes was to be understood (fig. 13). This diagram reiterates the circularity of the zodiac, the regularity of the imagined circles for the twelve degrees of the band of the zodiac around us, the circularity of the planetary orbits, and how much of the zodiacal band each planetary circle transits. It is possible, although no textual evidence exists in support, that this diagram was conceived as a stereographic projection from the pole of the ecliptic.[18] In any case, the invention of the diagram solved the problem of presenting to students the spatial meaning of the Plinian text.

This Plinian diagram answers the implied question of how to present certain planetary data stated in a text that lacked immediate reference to the appropriate theoretical framework. However, within a few decades after its creation the circular latitude diagram was replaced by another, a rectangular diagram, which reduced the amount of theoretical content added to the relevant Plinian text and also offered a more easily produced and more quickly read image (fig. 14).

This two-stage process of taking a simple recitation of planetary data and then diagramming it within a theoretical framework, which came subsequently to be reduced by subtracting the more easily assumed elements, occurred with each of the four planetary excerpts from Pliny.

To exemplify the process again, let us look at the excerpt for planetary harmonic intervals (figs. 15–16). The Plinian text runs as follows.

> *While some persons have argued that the sun-moon distance is nineteen times the earth-moon distance, Pythagoras computed the distance from earth to moon as 125,000 stadia, with the sun-moon distance double that and the sun-zodiac distance as triple the earth-moon. And at times Pythagoras used musical intervals with one tone* (tonus) *for the earth-moon distance, for Moon-Mercury 1/2, Mercury-Venus 1/2, Venus-Sun 1-1/2, Sun-Mars 1, Mars-Jupiter 1/2, Jupiter-Saturn 1/2, Saturn-fixed stars 1-1/2. Thus by the seven tones the harmony is made which they call the octave* (diapason).

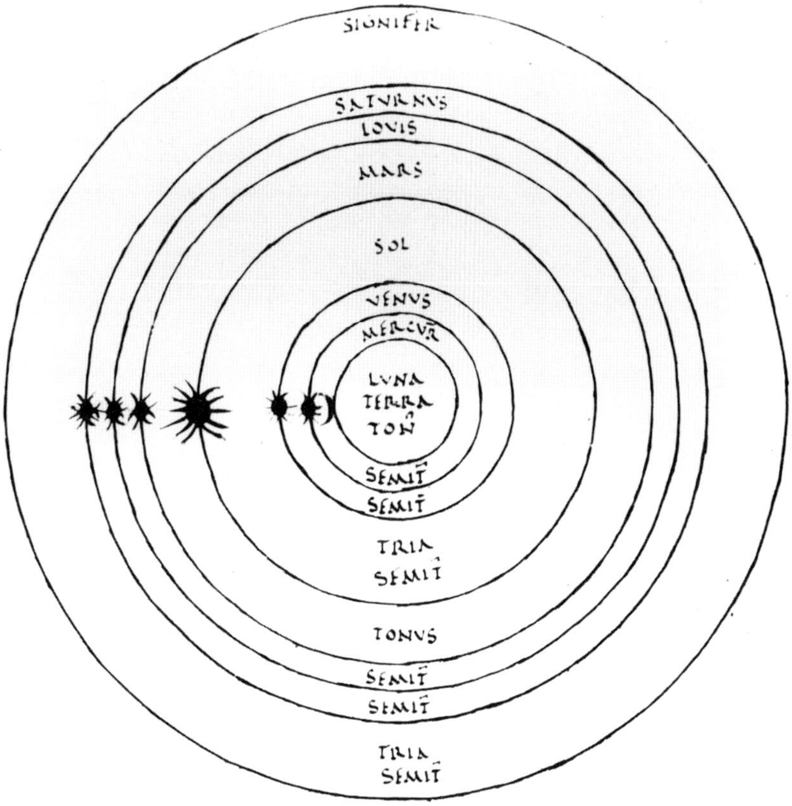

Here, the original conception for an illustrative diagram included concentric circles for all the planets. This element was dropped after a few decades, leaving only a vertical list of planetary names alternating with the names of appropriate harmonic intervals to show the sequence of planets and intervals.[19] With the second stage of development in the Plinian diagrams we observe a revision to the answer to the implied question of how much theoretical context to provide for the Plinian data. Rather than assuming that the first response to the question was in error, we can instead assume that the first response was what was needed at the time. The second step, or stage, which reduced the amount of theoretical framework in the diagram, was taken only after a few decades and represented the achievement of a new and higher level of theoretical background on the part of the students using these excerpts and the diagrams. The newer diagrams – the same two-step process occurred in all four types of Plinian diagram – assumed more prior knowledge and presented less elementary theory in the images. However, one theoretical property of planetary paths that the newer form of latitude diagram made much clearer than the initial form was symmetry. And we can find this attribute given even more emphasis in examples from the later tenth century and beyond. But the basic pattern of the Plinian rectangular latitude diagrams remained stable over the ensuing centuries.

Dynamics of Diagrams: Martianus Capella

Having briefly reviewed (a) diagrams that correspond directly to texts by completing them (Calcidius) and (b) diagrams that give needed theoretical framework for texts in order to introduce new data from those texts to novices (Pliny), we turn now to (c) diagrams which show reflection about the theory contained or implied in the texts. In the prior cases we have seen diagrams that presented models of planetary theoretical matter, but they did not raise questions about these models. In the medieval manuscripts of Martianus Capella we find many diagrams of clarification, some of which did raise questions about planetary models. The Capellan diagrams were of two general types. One type followed the more usual function of the verbal gloss, which was to define a term or a phrase. In the ninth century a group of ten astronomical diagrams was assembled and

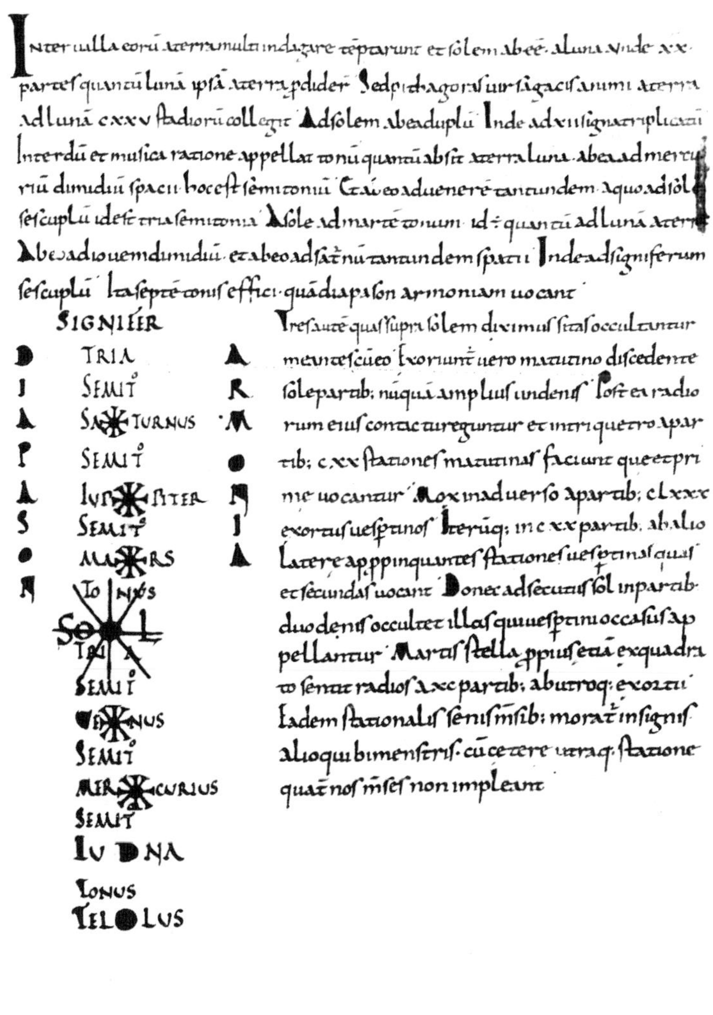

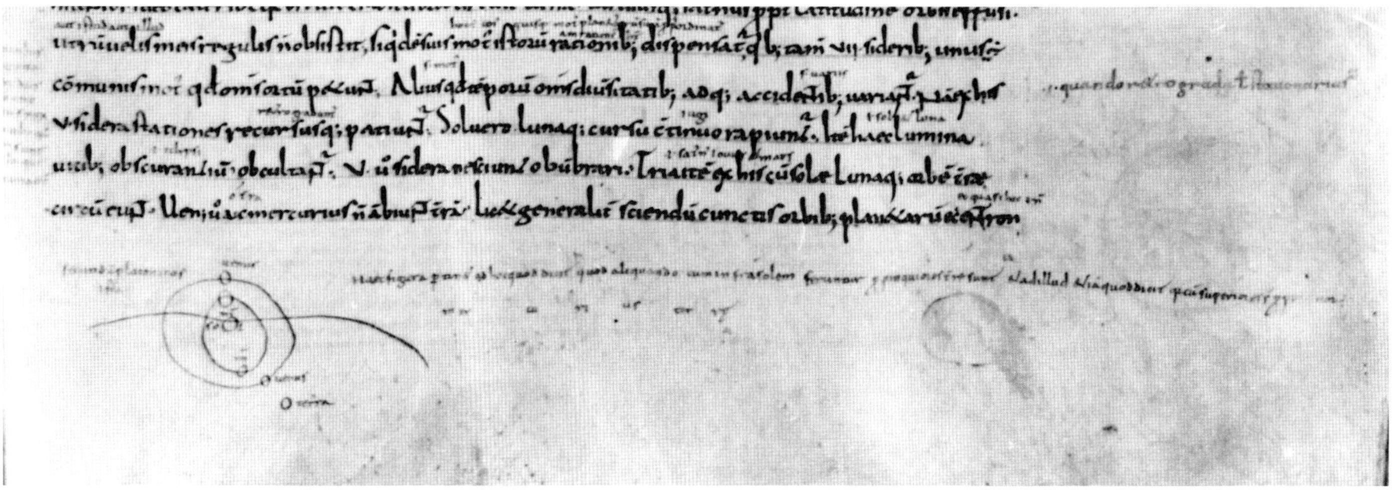

Figure 17
Concentric circles around the sun for Mercury and Venus (Leiden Universiteitsbibliotheek ms. Voss. lat. F.48, f. 79r).

travelled as an appendix to the astronomical book of Capella's *De nuptiis*, usually appearing at the very end of the final book, along with similar groups of diagrams for each of the other Capellan books on the quadrivium.[20] Of these ten diagrams one has special significance, because it evolved from a series of questions raised about the text of Martianus at two points where he described the paths of the planets Mercury and Venus.

Giving no clue about his source for the idea, Martianus Capella went against both Pliny and Macrobius in asserting that the two inner planets circled around the sun as their centre rather than around the earth as did the moon, the sun, and the three outer planets. His first statement of this pattern of planetary motion said, "Along with the Sun and the Moon three other planets circle around the orb of Earth, while Venus and Mercury do not go around Earth" (Martianus Capella [1983] 323.21-22 (VIII, 854)). This description was quite imprecise, but it was made more specific some lines further on by the following.

> *Now Venus and Mercury, although they have daily risings and settings, do not travel about the earth at all; rather they circle the sun in wider revolutions. The centre* (centron) *of their orbits* (circuli) *is set in the sun. As a result, sometimes beyond the sun, more often beneath it, they are closer to the earth. The greatest elongation of Mercury and Venus from the sun is one and one-half signs, or 46 degrees.* When both planets have a position above the sun, Mercury is closer to the earth; when they are below the sun, Venus is closer, inasmuch as it has both a "chaster" and a more open orbit* (orbe castiore diffusioreque).[21]

This description of the planetary paths elicited extensive comment from ninth-century readers. Of the sixteen extant manuscripts with the text of this description, six make no change in the text and add no gloss or diagram in the margin. Of the other ten manuscripts, five have both changes to the text by way of gloss or direct emendation and added diagrams to elaborate and clarify textual ambiguity. Five other manuscripts have verbal glosses or emendations that follow the first five in meaning, but they do not contain diagrams. The five manuscripts with diagrams show us a careful line of reasoning developed and transmitted by ninth-century readers in response to challenges posed by the text.

The first step in response to the text given above was a diagram for the first part of the text, as far as the mention of the elongation from the sun. To this point Capella had offered a model adequately represented by two concentric circles for Venus and Mercury with their common centre in the sun, and this picture was drawn in the margin near the text (fig. 17).

Here the marginal pictorial gloss simply specified concentricity for the two planets, since this property was not stated by Martianus. But the reader faced a troubling problem with the latter part of the text given above. Beginning with "When both the planets have a position ...," certain readers determined that the text required a change, which in turn produced a different diagram for the two circumsolar planets. This image showed the two planetary circles intersecting (fig. 18, lower diagram).

Interpreting the word *"castiore"* at the end to mean narrower, tighter, or closer, these scholars then decided that the word *"terris,"* for "earth," should be changed. A look at the Latin text here is useful. This last part reads, "Sed cum supra solem sunt, propinquior est terris Mercurius, cum intra solem Venus, utpote quae orbe castiore diffusioreque curvetur."[22] Initially as an interlinear gloss for *"terris,"* later as a direct replacement, the word *"ei"* (occasionally *"soli,"* which produces the same change of meaning) was inserted. If we now refer back to the translation given above and, firstly, replace the word "chaster" with the ninth-century understanding of *castior* and, secondly, make the indicated change to *terris*, we have the following reading of the last part of that text. "When they are above the sun, Mercury is closer to it (*"ei"*); when below the sun, Venus, inasmuch as it circles more closely [when below] and more openly [when above]." Hence the resulting intersecting circles, quite different from the previous concentric circles.

What we have here is a complex development in the exploration of a model, which the designer of the diagram has neatly synthesised. Encountering a text which, with increasing specification, did not become clearer but rather more ambiguous, the earliest scholar we can observe attacked this ambiguity from more than one direction. First he found that the apparent meaning of *castior* required an emendation in the preceding text in order to allow the planet Venus to be closer to the sun when below it in the circumsolar path Venus followed. The revised text then clearly required intersecting circles, and this was made explicit by the marginal diagram. We find these steps taken by the scholar who produced the glosses and diagrams to the Capellan astronomy in one particular manuscript in the middle of the ninth century, a Leiden University manuscript catalogued as Vossius latinus F.48. Furthermore this scholar continued to explore the topic, not being satisfied with the Capellan text alone. He introduced as well the alternative model which was attributed to Pliny the Elder in a marginal gloss on the same page of this manuscript, and he produced a diagram for the Plinian model. This model, as he understood it, required both planets to travel only within the circle of the sun, that is, on pendant loops intersecting each other – they also had to fit the revised Capellan text – and always between the earth and the sun's path, never crossing the circle followed by the sun (fig. 18, upper diagram).[23] The scholar of this Leiden manuscript made the text of Capella more precise by revision, explored its two different meanings, and sought further insight

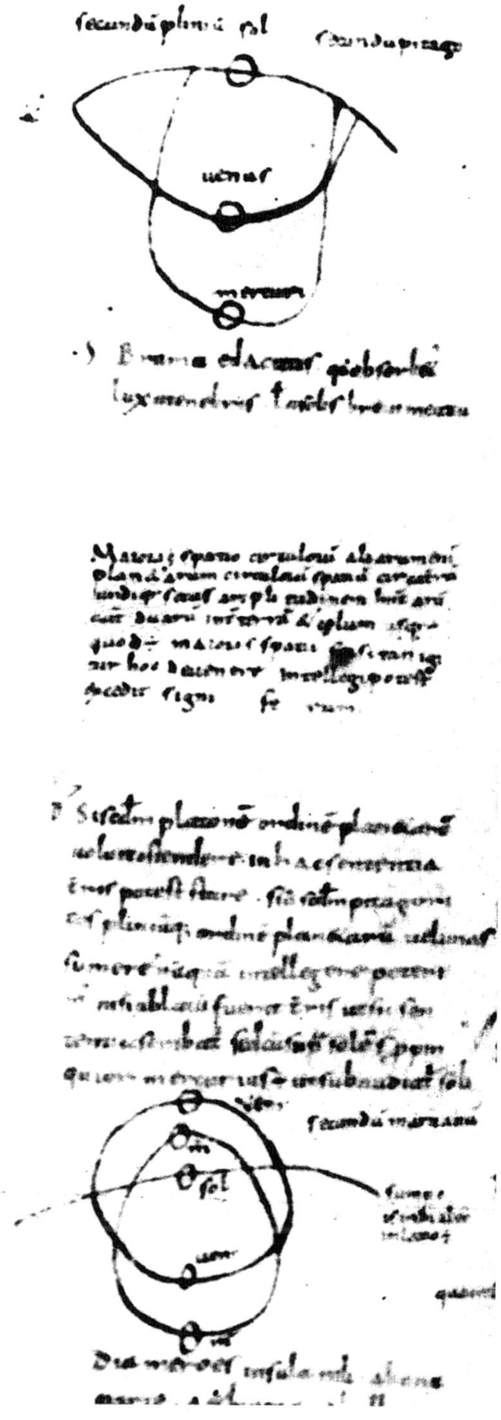

Figure 18
Lower diagram: intersecting circles around the sun for Mercury and Venus. Upper diagram: Plinian model for circumsolar Mercury and Venus (Leiden Universiteitsbibliotheek ms. Voss. lat. F.48, f. 79v).

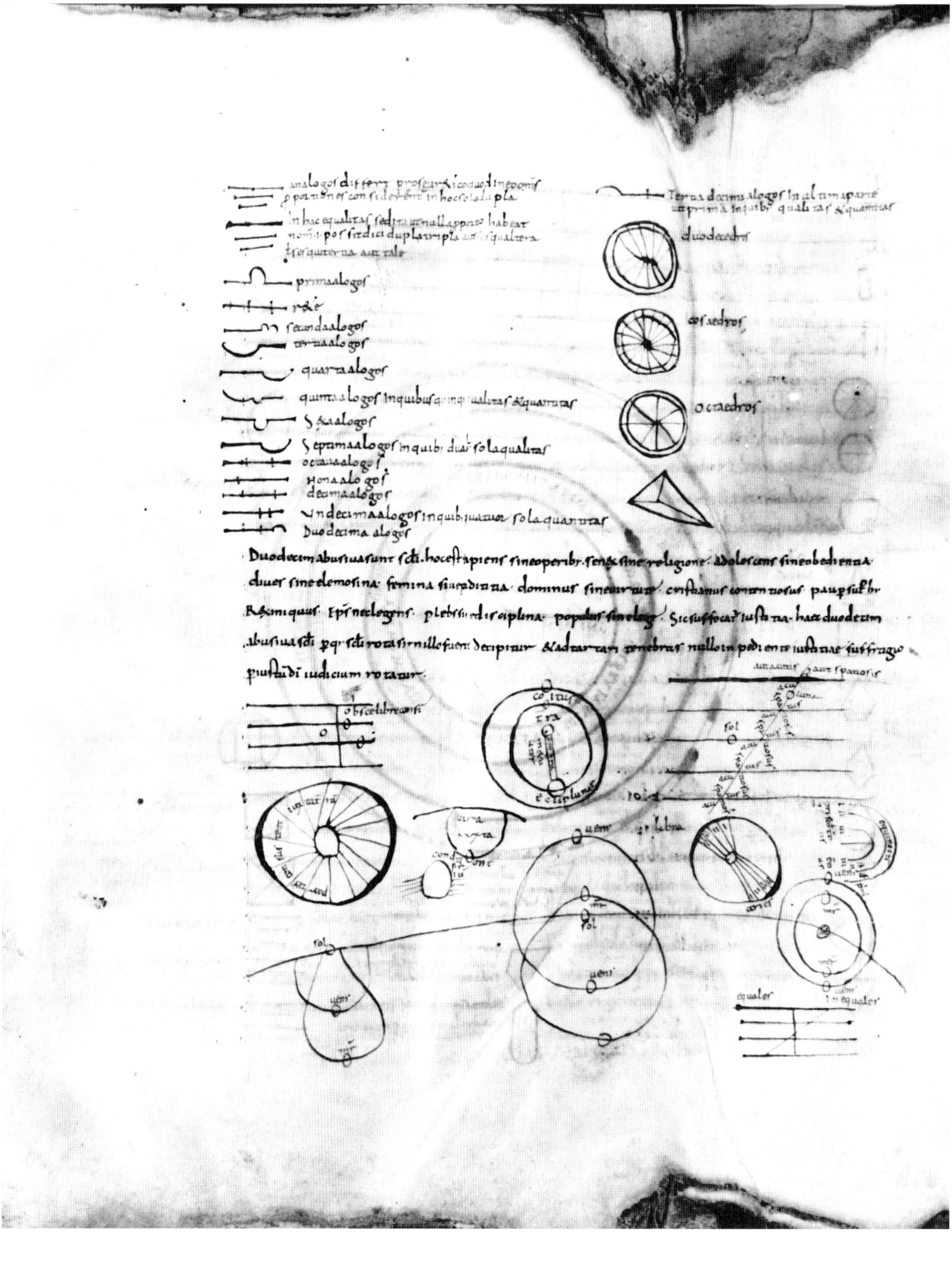

by looking at another text of a recognised authority, Pliny the Elder. And he did all this with the help of and by means of diagrams of his own devising. He resolved questions about the text of Capella by a series of diagrams, or planetary models, that made quite explicit the different possibilities of the Capellan text.

The same ninth-century scholar, perhaps working somewhere in the Loire valley, determined that the text proposed a set of challenges that could not be satisfied by one diagram, and he drew together his three alternatives, shown here in figures 17–18, and placed them in a continuous set of three diagrams on a single arc for the sun's path. He placed this triple-version diagram at the end of the Capellan handbooks, in the company of nine other astronomical diagrams, of which he seems also to have been the author (fig. 19).[24]

By constructing this composite diagram, he concluded that the Capellan text could not be reduced to a single and unambiguous meaning. Three variants of circumsolar path by the two planets were contained in the text of Martianus Capella. Only diagrams could make this completely clear.

The compound diagram with three variants, first devised by the scholar writing in the above mentioned Leiden manuscript, was copied in at least three more ninth-century manuscripts of the Capellan handbooks.[25] The copying of this diagram signified the acceptance of this solution to the problem posed by Capella's text on the paths of the inner planets around the sun. None of the three known copies, clearly made from the original Leiden manuscript scholar's design or a faithful intermediate, bothered to include the marginal diagrams that were designed and used by the original scholar as he worked out his solution to the meaning of the text. His step-by-step progress was no longer necessary for subsequent readers; the final result in the compound diagram was quite sufficient.

How do we explain the other ninth-century manuscripts that did not adopt this very interesting solution to the Capellan textual problem? First, we should remember that six manuscripts show no gloss or emendation to indicate that the meaning of "*castiore*" was troubling to the copyist or users. Presumably, their readers either ignored any uncertainty, or else they understood that Capella meant only concentric circles for the paths of Venus and Mercury around the sun. Five more manuscripts show that the same sort of revising gloss or emendation was made to the text as that produced by our Leiden manuscript scholar, but these five contain no diagram for the paths of the inner planets around the sun. The conclusion we draw about these manuscripts is that they represent followers of the same line of reasoning, actually deriving their glosses or emendations to the text from one or another manuscript indebted to that Leiden manuscript.

Finally, we have a small group of three manuscripts that show partial agreement with our innovator but unwillingness to go as far as him in the third model, attributed to Pliny. One of these three manuscripts shows a changed version of the text itself with a confirming gloss added, a rather sure sign that the manuscript text derived from the line of development we have already traced.[26] In the margin beside the modified text is a diagram of the two planets on intersecting circles around the sun, with no further diagram for these planets anywhere in the manuscript. The other two recalcitrants, if we may be excused this colourful label, present the first two diagrams, showing first concentric and then intersecting circumsolar circles. They also include the marginal gloss giving a description of Pliny's model, but they do not add the third diagram proposed by our innovator of Leiden Vossius F.48, nor do they, of course, include his compound diagram.[27] The copyists or the directing scholars for these two manuscripts appear to have determined that Pliny's model was the same as that with intersecting circles. Hence only the two diagrams copied were needed to solve the problem in the Capellan text.

Whether we look at the main line of interpretation following directly from Vossius F.48 or at divergent lines of interpretation, we find that the composition of diagrams to make individual planetary models clear and precise and also to explore the relevance of models in other texts was an essential activity in the Carolingian study of the astronomy of Martianus Capella. The Capellan astronomy does not seem to have been fully studied before the second quarter of the ninth century, at which time glossing and diagramming began. Diagrams were the means for stating, exploring, and resolving questions about Capellan planetary models.

Figure 19 (*opposite*) Quadrivial diagrams, including a set of ten astronomical diagrams with the three versions of circumsolar planetary motion placed together on an arc (Leiden Universiteitsbibliotheek ms. Voss. lat. F.48, f. 92v).

Qualitative Theory in the High and Later Middle Ages

At the beginning of the eleventh century the planetary diagrams for Plinian astronomical excerpts from the preceding two centuries witnessed a significant level of doctrinal and conceptual knowledge among monastic and secular clergy at major centres. The images for planetary apogees and for planetary latitudes had become common knowledge, sets of basic astronomical data with conceptual frameworks that students could be expected to understand. Less widespread than the Plinian texts and diagrams but considered important at certain centres was the commentary on Plato's *Timaeus* by Calcidius, which had already been excerpted by the middle of the ninth century for its explanations of variation in solar speed (by eccentrics) and of planetary retrograde motion (by epicycles). At the end of the tenth or beginning of the eleventh century Abbo of Fleury (ca. 940–1004) chose a set of Calcidian astronomical diagrams that emphasised eccentrical and epicyclical interpretations of solar and planetary motions for hiscomputistical collection (in Berlin Staatsbibliothek ms. Philipps 1833). Very soon a set of appropriate explanatory excerpts from Calcidius was added to these images in Abbo's computus. And Martianus Capella's astronomy with its introduction of circumsolar orbits for Mercury and Venus became the single most widely used source for planetary astronomy through the ninth and tenth centuries. On the basis of these sources, available and widely read by the eleventh century, a body of sophisticated qualitative planetary theory set the stage for further or more complete model-building.

Dynamics of Diagrams: Construction of a Planetary Theory

At the beginning of the eleventh-century the text of Martianus Capella's astronomy stimulated a new stage in planetary model building by way of diagrams. In a manuscript by a French hand, written at the end of the tenth or beginning of the eleventh century, the scholar who copied, or directed the copying of, the text made many additions in the textual space rather than marginally or at the end of the book, the more customary place for such additions.[28] An outstanding example was his drawing of a model for the paths of all the planets (fig. 20).

He also gave headings to sections of the astronomy book, and this diagram follows the section which he headed, "The earth is not the centre for all the planets."[29] In this section, comprising a short part of Capella's text dealing with the arrangement, speeds, and sizes of the planetary orbits,[30] our scholar isolated a text that began with the statement that no planets centre upon the earth, because all move eccentrically, and then went on to identify the two circumsolar planets as more extreme examples of the non-centrality of the earth. Next, very briefly, Capella's text mentioned the differences of opinion among unnamed authorities with regard to the order of the planets immediately above the moon. It closed this section with an introduction to the dimensions of the planetary orbits, a topic directly related to the order of the planets. Figure 20 occupies the full page following, with the new planetary model taking the top two-thirds of the page. The three-version diagram for Mercury and Venus is in the bottom third of this page. For the designer of the planetary model here, the correct choice among the three was obvious, which he showed by using the Capellan model of intersecting circles for the inner planets.

The planetary theory described in this elaborate diagram has many striking properties. Most notable is the distance of the earth from the centre of any planetary orbit, including the moon. Second is the location of the sun closer to the centre of the model (momentarily) than the earth is. Third, the circumsolar and intersecting orbits of Mercury and Venus stand out sharply. Another striking property is the extra arc, or tumor (two for Mars), that is inserted in the orbital circle of each of the outer planets. Each of these elements of the theory deserves discussion.

The displacement of the earth from the centre was noted in other Latin writers on astronomy. However, this was contrary to the view of the widely known *Commentarii in somnium Scipionis* by Macrobius. One of Macrobius' canonical diagrams, found in most of the medieval manuscripts, located each of the seven planets in the sign of the zodiac in which it had originally been created. The constant character among these diagrams was the concentric circles for the

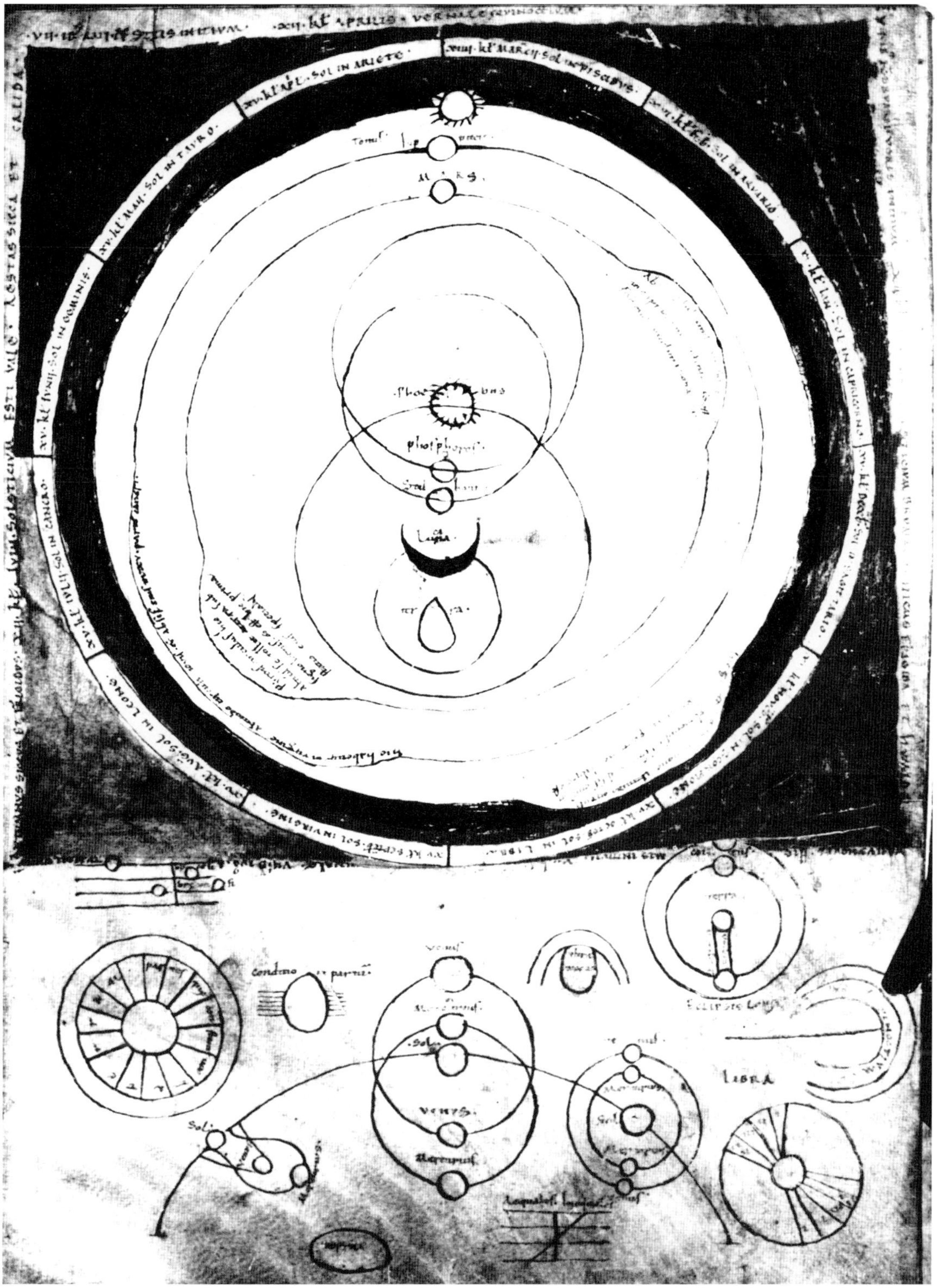

Figure 20
Composite Capellan planetary theory. Diagram in a Ms of the eleventh century (Firenze Bibliotheca Laurenziana ms. San Marco 190, f. 102r).

planets. These circles held the earth at the centre and the zodiac as the outermost circle, or band.[31] But Capella joined with both Pliny and Calcidius in making the planets follow circles centred far from the earth. We can make more sense of this apparent disagreement between Macrobius and the other writers, at least as they were seen by eleventh-century scholars, by recognising that Macrobius offered a more generalised and cosmological position. Conceptual order was more important to him than details of planetary position and motion. Pliny and Calcidius, well-known authorities in astronomy by the eleventh century, described the bounded elongation of Mercury and Venus, the eccentricity of planetary orbits, and the varying speeds of different planets. They were outlining models, or parts of models, for astronomical theory. Capellan astronomy developed in the eleventh century much more in association with this body of thought.

Location of the sun very near the centre of the picture of the model simply emphasised the circumsolar element of the model. In fact, the major visual emphases are two, the circumsolar planets and the protrusions for locating planetary apsides, since these are the two most obvious structural elements of the Capellan planetary theory. In the image these elements are over-emphasised; they do not appear as described verbally. The centrality of the sun is actually the centrality of the circumsolar pattern, a focal theme that makes Capella's account clearly different from others and which had aroused significant comment and development before this eleventh-century image appeared. The sun and its two satellites were not physically central, but they were central to the understanding of Capellan planetary astronomy.

The circles of Mercury and Venus are large. Their orbital circles around the sun are far larger than that of the lunar orbit around the earth. Indeed, the size of Venus' circle around the sun is close to the size of the sun's circle around the earth. Here again we find an emphasis – physically it is an over-emphasis – in order to identify unmistakably the leading element of Capellan planetary astronomy. The image points the viewer to the outstanding conceptual component by exaggerating its size and shifting its position. In a qualitative model this procedure is both appropriate and effective.

The protrusion from each of the three outer planetary circles represents the location of that planet's apogee and does so in a novel way. The standard Plinian diagram for this phenomenon simply contains six eccentric circles for the six planets beyond the moon and locates the centre, and therefore the apogee as well, of each planetary circle under the designated zodiacal sign for that planet. Unlike such a standard apsidal diagram, this Capellan image places each outer planet on a concentric circle, the centre of which is not the sun or the earth but the geometrical centre of the diagram, and then gives to each of the three outer planets a sort of bubble, or stretched arc. This bubble lies under the sign designated by Capella's text for the planet's apogee. Within the bubble a brief text gives the data not only for the apogee but also for the astrological exaltation according to Capella's text. The circle for Mars is a special case, for Capella tells us that this planet has two stations located 180 degrees apart, and the image shows us two bubbles, on opposite sides of the circle, for Mars. Just as with other elements of this diagram, the outer planetary circles with their bubbles present us with a way to recognise apsidal locations that goes beyond a geometrical representation. It is extravagant in going beyond, but it is an effective device that immediately directs the viewer's eye. A qualitative model does this in seeking to emphasise the various parts of a theory without presenting a quantitatively designed diagram.

The Capellan Tradition through the Fifteenth Century

From the beginning of the eleventh to the end of the thirteenth century the Latin tradition in astronomy was profoundly affected by its interaction with the Arabic and Greek astrological and astronomical materials that were translated into Latin during this era. There is a long-standing argument that Latin practical and instrumental developments, especially those concerned with the computus and the astrolabe, were the core and stimulus of achievements in Western astronomy during these centuries.[32] What has not been noted is the continued study of qualitative theory, such as Capella's model for the inner planets.

When we look at the development of astronomical study at the schools of the twelfth and thirteenth centuries, many peculiarities appear. While the translations of Ptolemy's *Almagest* from the Greek in 1160 and from the Arabic by 1175 would seem to have set a completely new standard for planetary theory, it becomes clear that this standard was not soon used in schools. At the University of Paris in the first half of the thirteenth century the standard text required for student preparation in the liberal art of astronomy was that of Martianus Capella. An anonymous introduction to the arts in the early 1240s, the *Compendium circa quadrivium*, divided the "demonstrative" science of astronomy into that which establishes effects from causes as taught by Ptolemy and that which presents the content of the discipline as taught by Martianus. This *Compendium* then proceeded to outline for many paragraphs the content of the Capellan book and never returned to Ptolemy.[33] As the thirteenth century progressed a succession of texts came to the fore in the Parisian arts curriculum. In John of Sacrobosco's *Sphaera* (ca. 1230), which offered a more systematic and more explicitly geometrical approach than Capella, the student found the simplest of introductions to the stellar sphere, which could be combined with the study of computus for the motions of the sun and the moon. For the planets, Sacrobosco's work gave only a hint of what the contemporary student of astronomy should know. The *Sphaera* became the standard required, basic work by the 1250s, but Capella's text continued to be widely consulted, at times in preference to Sacrobosco's. An anonymous thirteenth-century commentator on the *Sphaera* named six authorities on mathematical astronomy, three of whose texts he determined to be demonstrative – Ptolemy, Geber, Thabit – and three he categorized as narrative – Alfraganus, Martianus, Sacrobosco. This commentator clearly distinguished the works of these six writers as mathematical astronomy. The mid-thirteenth-century master, Lambert, at the Dominican convent of Auxerre, supported the astronomy of Capella as the most suitable introduction for students.[34]

Because Sacrobosco's *Sphaera* offered no notable treatment of the planets, there quickly appeared a series of anonymous *Theorica planetarum* texts, the earliest of which was used by Michael Scot before 1235.[35] The somewhat later *Theorica planetarum* by Campanus of Novara brought into the astronomy curriculum of the fourteenth century a more complete insight into Ptolemaic models.[36] Yet students continued to consult Sacrobosco's *Sphaera*, and, in turn, glosses to this work continued to mention Capella, as at least three fourteenth-century manuscripts show (Thorndike [1949] 88, n. 53). In the latter half of the thirteenth century a fairly lengthy summary of astronomy, apparently intended as an introduction for students, included a number of simple diagrams of epicyclic planetary motions.[37] An outstanding image among these is a large design of the planets, all on epicycles moving on deferents around the earth and within an encompassing zodiac. The two planets Mercury and Venus are on intersecting circles that are both circumsolar (fig. 21).

In other words, these two planets have a pattern that we have learned to associate with Martianus Capella since the invention and dissemination of this model in the ninth century. Furthermore the associated chapter (on the subsequent folio) on the paths for Mercury and Venus has language showing its significant dependence on the tradition of Capellan circumsolarity developed in the ninth century. The large design in figure 21 contains various astronomical and astrological doctrines inscribed on the diagram in brief texts that derive partly from Arabic but mostly from traditional Latin sources. The diagram itself offers a combination of newer and older astronomical models. On one hand we see the outer planets all on epicycles moving on deferents around the earth. This part of the image calls to mind the models of the newer Greco-Arabic astronomy, and it may be noted as well that such explanations were proposed, even if not fully worked out, in the commentary of Calcidius on Plato's *Timaeus*.[38] On the other hand, we have the older Capellan model for the inner planets, which does not partake of the newer models; the sun carries the inner planets on their intersecting circles around on its own eccentric circle. This image shows us an example of the continuity of qualitative models well into the era when Ptolemaic quantitative models were known and used. The lesson here is the co-existence of newer along with traditional models rather than the replacement of the earlier by the later.

The evidence of surviving manuscripts attests the unbroken study of Martianus Capella's astronomy beyond the eleventh century. In a lengthy Oxford manuscript of the twelfth century, the *Liber Yparchi*, a revised form of Capella's astronomy, is one of fifteen items, all of which are mathematical, chronological, computistical, astronomical, or cosmological. In a later twelfth-century codex of Aristotelian natural philosophical works there was, still in the thirteenth century but no longer today, a copy of *Martianus de astrologia*. Again, in a codex from ca. 1200 Book VIII of Capella was one of eleven items in a collection of cosmological works. A col-

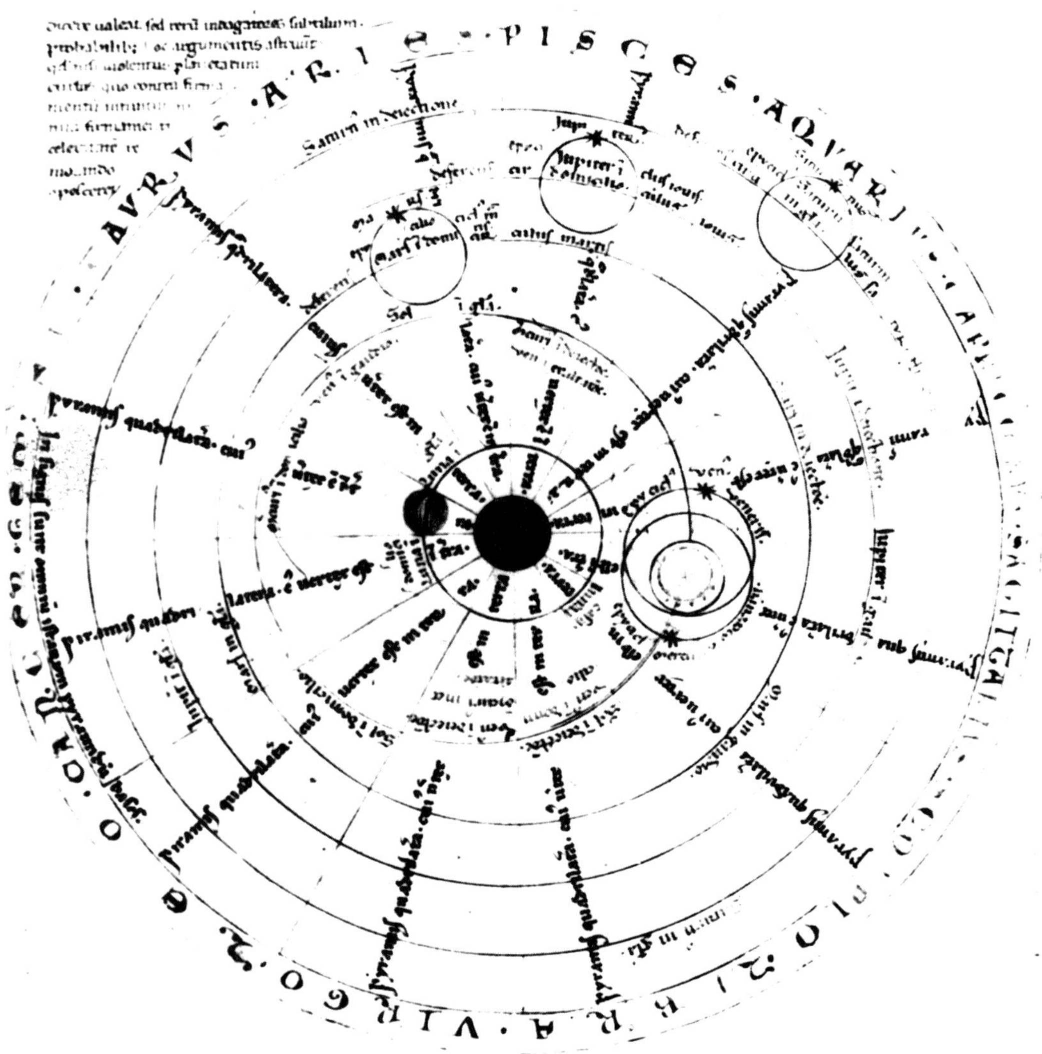

Figure 21
Capellan model incorporated into a Greco-Arabic-inspired epicyclic planetary theory (København Kongelige Bibliotek ms. G. K. S. 277. fol., f. 49r (1254?)).

lection of scientific works – astronomy, cosmology, physics, optics, weights, medicine, et al. – of which all the others came from the twelfth century or later, still saw fit to include Capellan astronomy in the thirteenth to fourteenth century. And from the thirteenth to fourteenth century we have a compilation of mostly Greco-Arabic astrological works in which Book VIII of Martianus is the second of twenty-one items.[39] These manuscripts do not all give Capella's work the same status, but they all witness its continued importance in interrelated categories of texts: technical, natural-philosophical, cosmological, astrological, and astronomical.

The number of surviving manuscripts of Capella's astronomy suggests a continuing healthy interest in its doctrines. The twelfth century has left to us seven copies of all nine books of Capella's work as well as fourteen copies of Book VIII (on astronomy) alone. From the thirteenth century we have five copies of the whole work and eleven of the astronomical book by itself. The fourteenth century saw a reduced copying, for we have two copies of all nine books and eight copies of the astronomy. With the fifteenth century we can see greater attention, for thirteen copies of De nuptiis survive along with three copies of Book VIII alone. From these centuries there remain forty-two separate diagrams of Capellan circumsolarity in Capellan manuscripts: twenty-five from the twelfth century, ten from the thirteenth, and ten from the fourteenth. Of the fifteenth-century copies at least five contain astronomical diagrams, including the diagram for three versions of circumsolar Mercury and Venus as it had been drawn in the ninth century, and three of these five include as well the large drawing (fig. 20) of Capellan theory as created in the early eleventh century. Among the three copies in the fifteenth century of Book VIII alone we should note that one appears in a collection of various writings on the stars; the Capellan text took its place in this collection as the only source on the planets, and only the planetary astronomy of Capella was excerpted for this astronomical compilation.[40] The number of copies of Capella's work, the number of these copies that repeated one or more of the diagrams for Capellan circumsolarity, the excerpting of Capella's description of Venus' and Mercury's paths around the sun in various collections of astronomical materials, the continued association of Capellan astronomy with other scientific texts, and the tacit and explicit recommendations of Capellan astronomy through the high Middle Ages – all these facts make it clear to us that the Capellan qualitative model for the inner planets was known and given attention to the end of the Middle Ages. Along with Ptolemaic and other, quantitative models for planetary motion this qualitative model remained current in the study of astronomy.

Merging Two Traditions: The Sixteenth Century

Medieval Latin qualitative astronomy and classical Greek mathematical astronomy did not simply coexist independently from each other in the world of scholarly learning throughout the fifteenth and sixteenth century. Moderns often describe the development of theoretical astronomy as a linear succession from Ptolemaic astronomy, with intermediaries in Arabic astronomy up to Copernicus' heliocentric revolution and Kepler's final abolition of epicyclic models. Contrary to this view we find that only the synthesis of both qualitative and quantitative astronomical traditions set for Copernicus the questions leading to a heliocentric cosmology. Three theoretical elements come together at the basis of the new mathematical astronomy.

(a) *Geometry as deductive means.* Ptolemaic astronomy is geometrical at its core, perfect for computational purposes. The qualitative medieval tradition lacked any computational possibilities, which made it impossible to derive even the simplest positional information about the celestial bodies. Yet this tradition was rich enough to add two further essential elements for Ptolemaic astronomy.

(b) *Explanatory requirement.* Pliny required that astronomy should causally explain the qualitative astronomical phenomena.

(c) *Specific limits of a circumsolar model.* Capella restricts only the two inner planets to circumsolar motion.

The combination of these three theoretical elements requires three significant changes for both traditions.

(1) The inclusion of element (a) into qualitative astronomy requires a strict geometrical reinterpretation of qualitative astronomical models. Only geometrically well-defined circles are superimposed to describe planetary motion by their combined rotations about a geometrically defined centre. Capellan diagrams were seen from a stricter geometrical viewpoint, which led to a significant change in the understanding of Capellan astronomy in the sixteenth century, as we show below. The intersecting-circle model, canonically attributed to Capella, no longer suffices as a possible model, since its centre is geometrically ill-defined.

(2) The geometrical interpretation of element (b) specifies that explanations of qualitative astronomical phenomena are provided only if qualitative phenomena can be deduced geometrically from a model.

(3) Capellan circumsolar motion of the inner planets provides a model for the explanation of a number of qualitative phenomena, most notably the fact that the inner planets never exceed a certain angular distance from the sun (bounded elongation).

These three elements are the basis from which Copernicus undertook his new theoretical enterprise. They appear frequently in *De Revolutionibus*.[41]

As a mathematical astronomer Copernicus had to base the geometry of his astronomy on Ptolemy's models. Yet, one could transfer those models into empirically equivalent heliocentric models. For doing so, Copernicus referred to one and only one reason; he expressed it in different ways and metaphors. In a very prominent passage, in Book One, Chapter X of *De Revolutionibus*, Copernicus gives credit to Martianus Capella for having shown him the right way to understand the motion of the planets. Capella and other Latin authors posited Mercury and Venus to circle the sun. Their reason for doing so was, according to Copernicus (Copernicus [1984] 19):

> *Existimant enim, quod Venus et Mercurius circumcurrant Solem in medio existentem, et eam causam illo non ulterius digredi putant, quam suorum convexitas orbium patiatur, quoniam terram non ambiunt ut ceteri, sed absidas conversas habent.*

Rosen translates this as:

> *This is the reason, in their opinion, why these planets diverge no farther from the sun than is permitted by the curvature of their revolutions. For they do not encircle the earth, like the other planets, but "have opposite circles"* (Copernicus [1992] 20).

Rosen's translation is unintelligible. There is nothing like a "curvature of ... revolutions." He omits the historical background to which Copernicus alludes. Copernicus' explanation why the inner planets cannot deviate from the sun more than the bounded elongation makes use of three unusual phrases.

(i) Contrary to Rosen's translation, "*suorum convexitas orbium*" states "the convexity of their orbits," an expression that was specifically used by Pliny (Pliny, *Historia Naturalis* II). It is an expression describing the geometrical properties of the orbits as drawn *in a diagram*. The closed orbit of each of the inner planets appears convex from the position of the earth, outside their planetary circles about the sun. On the other hand, the orbits of the outer planets appear concave from the earth, which is inside their circles. Such geometrical premises lead to a simple geometrical proof for the bounded elongation of the inner planets (that they cannot exceed a certain angular distance from the sun): convex orbits exclude the planet's movement in opposition to the sun, since "convex orbits" always circle about the sun but cannot ever circle about the earth.

(ii) The phrase "*terram non ambiunt*" is a specific expression from Martianus Capella saying that the inner planets do not move around the earth.

(iii) "*Absidas conversas*," finally, is an expression of special importance. Rosen's translation as "opposite circles" obscures a complicated background. The expression "*absidas conversas*" does not occur in the text of Capella but very prominently in Pliny's *Natural History*. Yet there it is used unclearly so that it elicits a number of unhelpful glosses from medieval commentators. Yet Pliny does not say anything about a heliocentric motion of the inner planets. He simply requires repeatedly an explanation for their maximum elongation from the sun, but he does not provide a geometrical model for such an explanation. Capellan diagrams in their geometrical reinterpretation do add such models.

In order to describe the requirement for a heliocentric theory and assign it to ancient authorities, Copernicus needs the combination of the Plinian demand for qualitative explanation and a geometry as its solution. He can only find it in Capellan *diagrams*. In the diagram of the three versions for the motion of the inner planets the name of Pliny is attached to a circumsolar motion. Through the diagram Copernicus could identify Pliny's requirement for a causal explanation with a *geometrical* argument that produces the fact of bounded elongation of the inner planets. Pliny then could be cited authoritatively for circumsolar motion. In an argumentative style frequent at the beginning of *De Revolutionibus*, Copernicus uses the rhetorical means of citing known phrases from Latin authorities to synthesise the requirement for explanation with the model of circumsolar motion. The sequence of subjects in *De Revolutionibus* follows that of Ptolemy's *Almagest*. He opens the book with a discussion of whether the heaven is spherical. While in the *Almagest* Ptolemy carefully develops geometrical arguments for a *rotating spherical* universe and tests his conclusions on a whole range of observational phenomena, Copernicus does nothing of the sort.[42] Instead of recounting observational data and providing geometrical descriptions, he rephrases Pliny as a ground for the view that the universe is spherical (Copernicus [1984] I.1 7. Pliny II.2, II.65, II.70. Ptolemy [1984] 38). Copernicus could have seen the Capellan diagram during his studies of astronomy in Italy, in the Florentine manuscript image in figure 20 or one of the many triple-version diagrams in circulation. Here the connection between Pliny and circumsolar motion is explicit.

Pliny's particular role in the heliocentric transformation is visible in Rheticus' account of the Copernican achievements (Rheticus [1982]). Rheticus was professor at the University of Wittenberg and travelled in 1539 to Frauenburg to learn from Copernicus about his new astronomy. The *Narratio prima* was the first published account of Copernicus' views. Here Rheticus summarises the reasons for preferring the Copernican model in one section. When citing ancient authorities supporting Copernicus' innovation, Rheticus names Pliny and not Martianus Capella in the context of the motion of the inner planets (Rheticus [1982] 55). Rheticus can only have connected Pliny with Copernicus through direct conversation with him, because Pliny's name is not mentioned at the corresponding passage in *De Revolutionibus*, and Pliny does not talk about the subject. Pliny does not describe the heliocentric motion of Mercury and Venus as an explanation for their bounded elongation at all. It is his request for such a causal explanation that intrigued both Copernicus and Rheticus in referring to Pliny's work. The methodological preference in the new theory for additional explanatory content is the sole reason that Copernicus introduced a heliocentric mathematical astronomy. The Copernican theory has no other advantage over its Ptolemaic competitor. It is neither empirically more accurate, nor does it use fewer circles for the construction of the models.

In the Ptolemaic system the planets move on epicycles about the earth O (fig. 22). The earth O is eccentric to the centre C of the large circle, the deferent, on which the epicycle moves with the planet P uniformly in relation to the equant point E. Such models successfully describe the motion of the planets in the zodiac as follows:

(i) The non-uniform velocity of the planets in their orbit about the sun (first anomaly). From a modern perspective this is due to the elliptic shape of the orbit. The epicyclic model accounts for this variation of speed largely by (a) the eccentric position of the earth at O and (b) the introduction of the equant in respect to which the planet moves uniformly on its epicycle.

(ii) The retrograde motion is a perspective effect of the terrestrial motion (second anomaly) at opposition to and conjunction with the sun. It can be modelled by adding the motion of the planet on an epicycle.

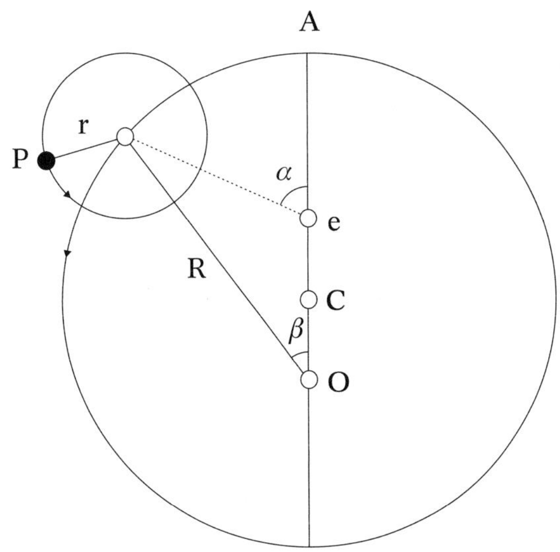

Figure 22
Ptolemaic planetary model of longitudinal motion.

Copernicus began his astronomical work by studying Regiomontanus' *Epitome*, where the geometrical properties of Ptolemy's models are explored fully. For a mathematically trained astronomer it was standard procedure to transform models into equivalent geometries which describe the same apparent motions of the planets. They could be transformed in such a fashion that the planets would move around the sun instead of the earth. This was well known to the astronomers of Copernicus's time. But why should anyone do so, especially since Aristotelian physics would pose severe difficulties in explaining spatial motion on a moving earth? Several often-cited reasons do not explain Copernicus' preference for the heliocentric theory. Foremost among such reasons are those referring to the equant and to empirical adequacy.

(i) It is true that Copernicus was uncomfortable with Ptolemy's use of the equant to model the non-uniform motion of the planets, but this did not provide a reason for a heliocentric transformation. His substitution of additional epicycles for the equant was indifferent to whether the arrangement was heliocentric or geocentric.

(ii) The demonstrated equivalence of heliocentric and geocentric models with respect to the apparent planetary motion also excludes empirical adequacy as a criterion for deciding between the two cosmologies. For any heliocentric model Copernicus might compose, there would always be a geocentric version that produces the same apparent motions of the planets.

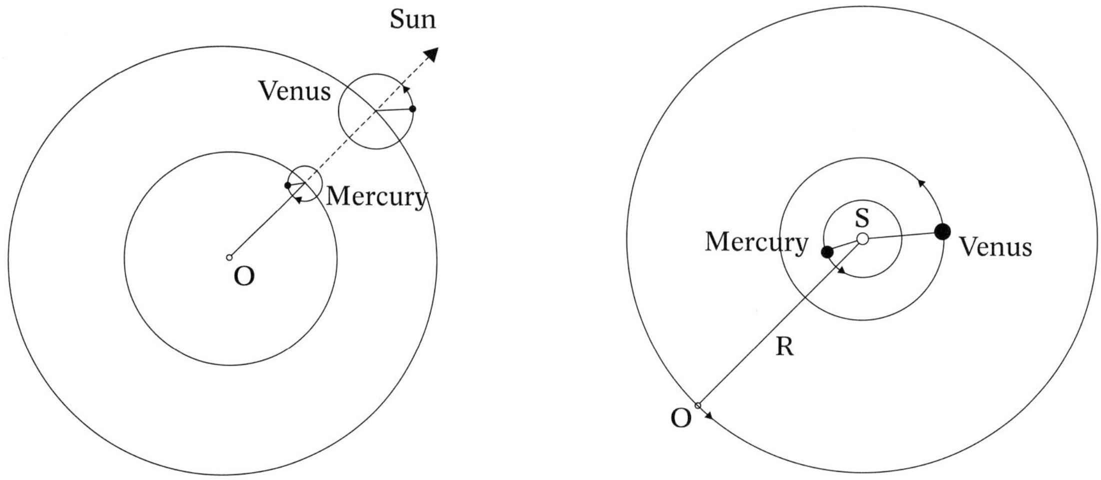

Figure 23
Simplified Ptolemaic models for the inner planets (left). Equivalent model with circumsolar motion (right).

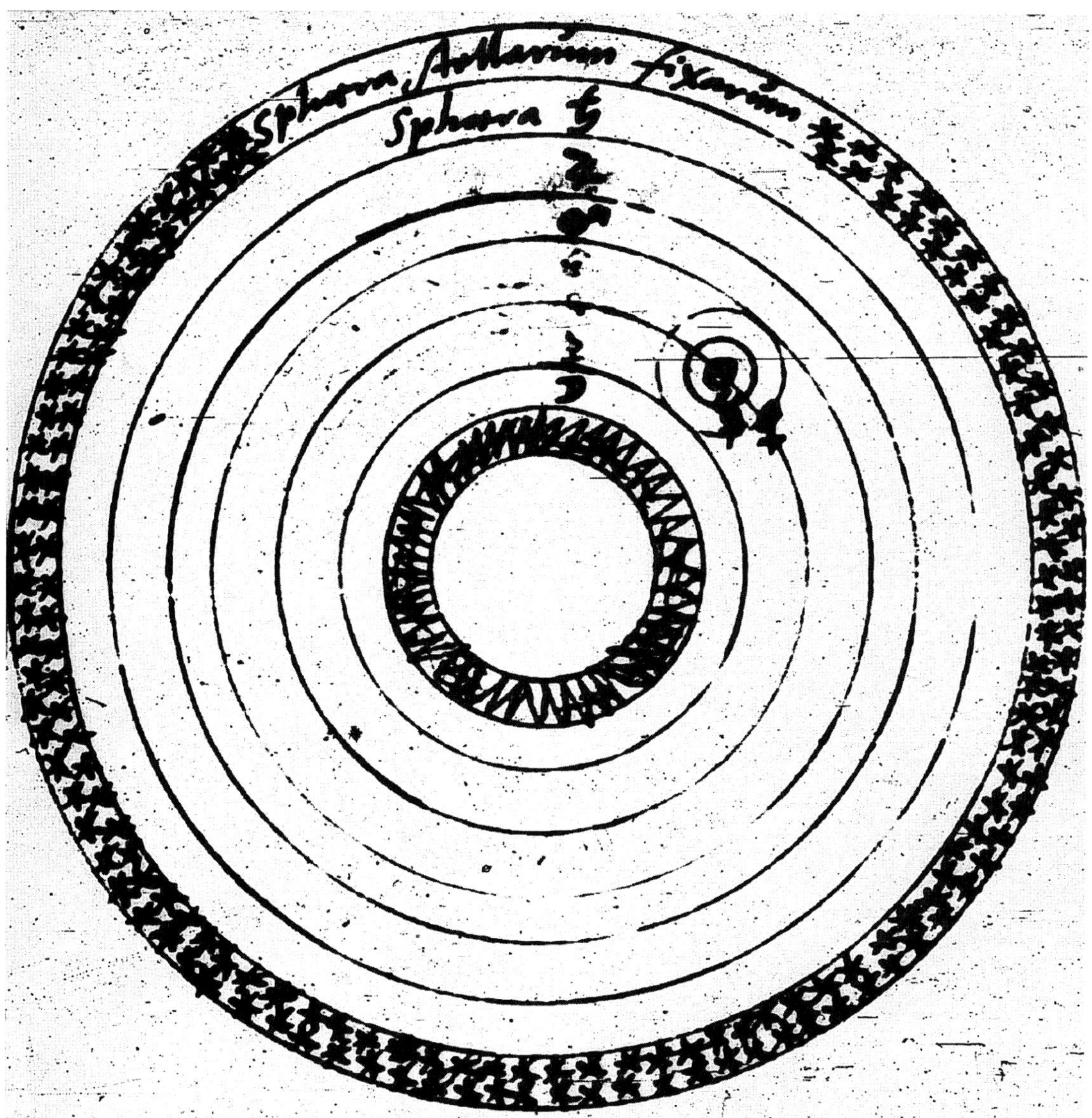

Figure 24
Capellan model. Sixteenth century version by Christoph Rothmann (Murhardsche Bibliothek der Stadt Kassel und Landesbibliothek 4° Abs. astron. 11 H. 37/72, f. 2r).

If neither empirical adequacy nor an apparent violation, via the equant, of Aristotle's principle of uniform motion led Copernicus to design a heliocentric planetary theory, what else could have guided him? Early documents confirm that Copernicus initially worked on the transformation of Ptolemaic models for the inner planets into their equivalent heliocentric counterparts (cf. Swerdlow et al. [1984] 54ff.). Capellan diagrams set the pattern. Their sole advantage was to provide explanations, as required by Pliny, for a number of qualitative astronomical properties. Both models are compared in figure 23, which is simplified in so far as the geometrical features required for the first anomaly are omitted.

The left arrangement is Ptolemaic with the orbits of Mercury and Venus arranged within the circles of the sun. The empirically equivalent model of heliocentric motion is shown on the right. It can be proven geometrically that both models produce the same observational data. Yet only the heliocentric model allows a geometrical proof for the bounded elongation of the inner planets. In Ptolemy's model the maximal elongation could be different. His model must require, in addition to the geometrical arrangement, that the centres of the epicycles of Mercury, Venus and the sun lie on one line. The additional requirement is not a product of the geometry. This difference in construction is the sole reason for Copernicus' new cosmology.

Although the medieval tradition visualised Capellan circumsolar motion of the inner planets with a triad of orbital arrangements and identified Capella with the intersecting-circles model, Copernicus mentions neither of the other two arrangements for the Latin authors nor does he attribute intersecting circles to Capella. He tacitly attributes concentric circumsolar motion to Capella and the other Latin authors together. The medieval alternatives disappear. Strikingly, this is the case without exception for sixteenth century astronomers. Yet like Copernicus, they not only analysed Ptolemaic and Copernican models but also considered medieval authors and their qualitative models as serious alternatives. In the middle of the sixteenth century Wilhelm IV of Hesse-Kassel called upon astronomers and instrument makers to work in his observatory. One of them was Christoph Rothmann, who wrote a lengthy manuscript surveying astronomy at a basic level. In the beginning of this work Rothmann laid out the order of the subject and describes possible cosmological arrangements, among them the cosmological order of spheres of the Ancients. Drawn on a separate page (fig. 24) is Rothmann's presentation of ancient cosmological thought, with Mercury and Venus on circumsolar paths about the sun. For Rothmann, this model was the best to display the ancient views, although Ptolemy's different order could not have escaped him.[43]

There is no mention of intersecting circles in connection with Capella. The same thing happens with the first printed diagram of a Capellan model by Valentin Naibod (fig. 25). And when Kepler's teacher Michael Maestlin annotated his copy of *De Revolutionibus*, he glossed Copernicus' reference to Martianus Capella by the remark "Egyptian order". Even later, G. B. Riccioli in the mid-seventeenth century subsumed the Latin astronomers, including Capella, under one "*Systema Aegyptiorum*," without intersecting circles (fig. 26).

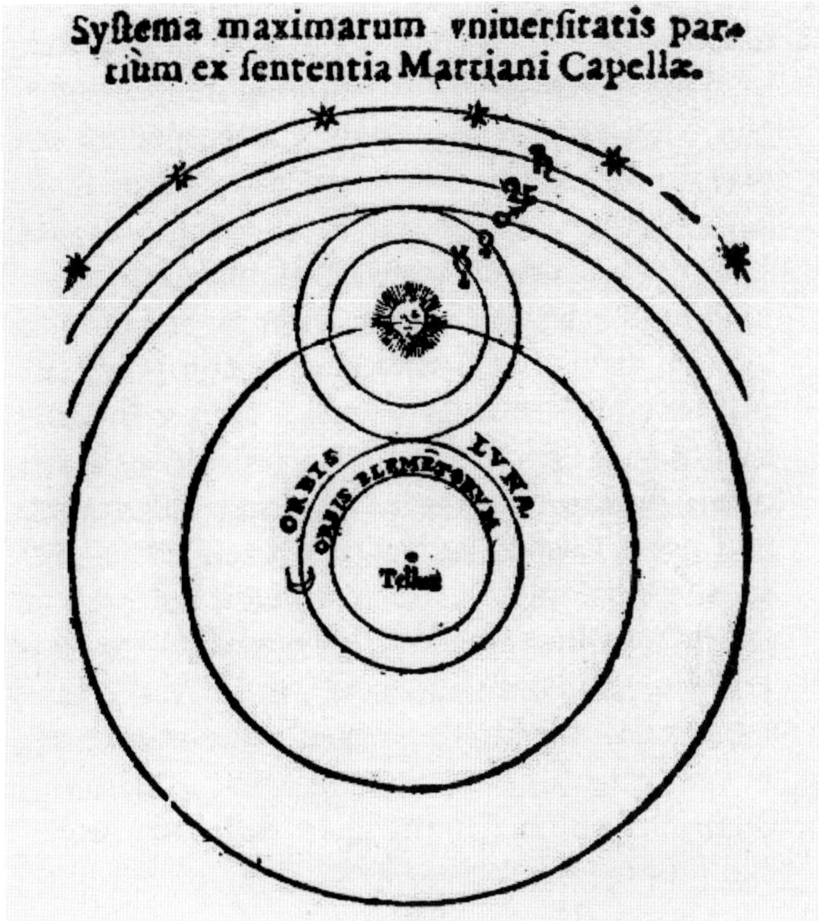

Figure 25
Capellan model as printed in Naibod: *Primarum de coelo et terra institutionum* (Venice 1573) folio 41r. Cited from Westman [1975] 323.

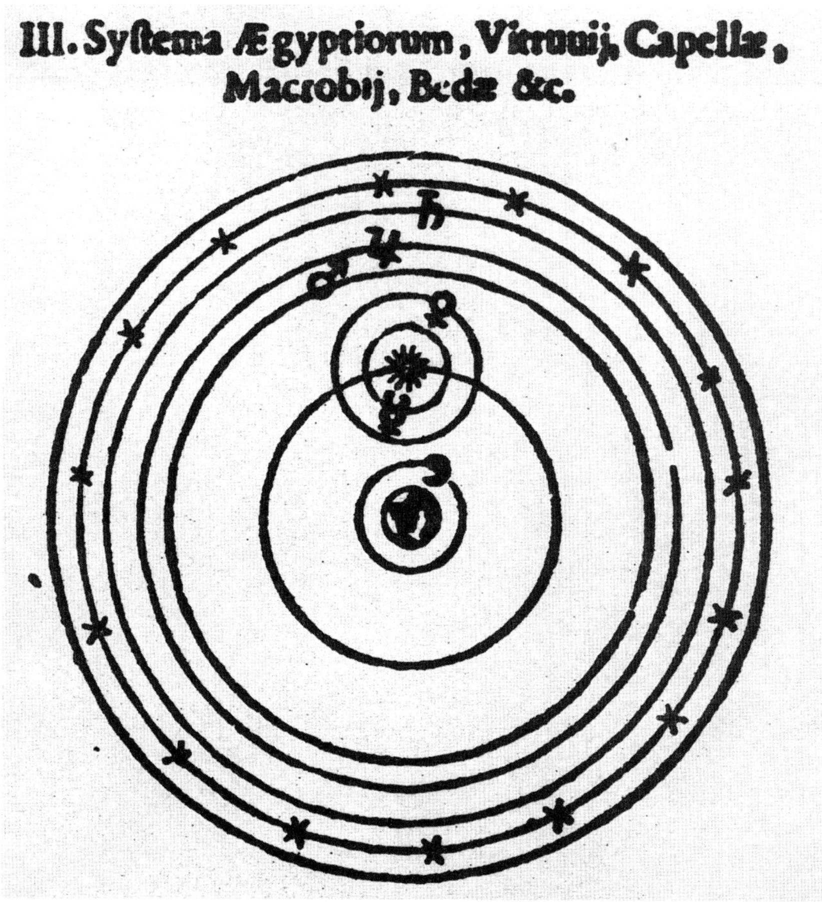

Figure 26
Capellan model as printed in Riccioli: *Almagestum novum astronomiam* (Bologna 1651) 101.

When Capellan diagrams for the inner planets merged with mathematical astronomy, their meaning changed and only concentric orbits remained possible in planetary arrangements. These Capellan models contributed important explanatory elements to the dominant Ptolemaic astronomy and eventually were a crucial stimulus for a quickly developing heliocentric astronomy. The process of constructing this new astronomy took place almost exclusively by means of diagrams.

University of Kentucky
and Universität Bern

Notes

1 General history of education in this era appears in Riché [1962]; translated as Riché [1976]. An excellent background to the early medieval computus up to the seventh century appears in O'Connell [1936]. For description of the seventh-century study of computus see Stevens [1995] ch. 4. Chapter 2 (ibid.) depicts the reappearance of limited astronomical observation within the framework of computus in the work of the most remarkable scholar of the early eighth century, the Englishman Bede.

2 In addition to Charles' convening of the computistical conference that led to the Computus of 809, various letters in his correspondence with Alcuin show the king's active interest. See the examples in MGH/E 185–187 (nr. 126), 228–235 (nrs. 144–145), 237-245 (nrs. 148–149), 278–283 (nrs. 170–171).

3 In a letter of early September 798, Alcuin not only referred to the planetary astronomy of Pliny but also suggested that the king had available to send to him a copy of the first books of Pliny's 37-book opus. See MGH/E 250 (nr. 155). The three early manuscripts are ms. Vat. lat. 3861, ff. 173 (Books II - XIX, the first fragmentary, the last incomplete); Paris BN ms. lat. 6796, ff. 81 (Books XIV - XIX); Leiden UB ms. Voss. lat. F.61, ff. 152 (Books XX, 186 - XXXVI, 97). For the first and last of these see Munk Olsen [1985] 249; for the Paris ms. see Munk Olsen [1989] 110. Bischoff [1961] 54, gives the dating and claims an origin at Corbie for the Paris ms. The Vatican and Leiden mss. appear to be from northeastern France, possibly Corbie, as well.

4 Barker-Benfield [1983]. The most comprehensive published list of the mss. appears in Eastwood [1994a] 138–155. Regarding the use for astronomy see Eastwood [1994b].

5 Ineichen-Eder [1978] esp. 193, 196 (Nr. 14), 199. On uses of Calcidius' translation of and commentary on Plato's Timaeus, see McKitterick [1992]. In Paris BN ms. lat. 13955, ff. 46v–60r, a composition possibly made at Fulda before mid-century, we find (a) an extensively glossed copy of Martianus Capella's astronomical book, (b) a copy of the Aratean text "Duo sunt extremi vertices mundi quos appellant polos ...," (c) an astronomical compilation derived from carefully excerpted and intermingled texts from Pliny, Macrobius, and Calcidius. The Calcidian material comes from nine chapters of the commentary and is described by Eastwood [1999] 172–178.

6 Leonardi [1960] 57: Karlsruhe Badische LB. ms. Reichenau Fragm. 136, s. VIII ex., contains Bk. III, 313–319.

7 The manuscripts are listed in Leonardi [1960] 1–99, 411–524. From Leonardi's list I include numbers 8, 20, 26, 28, 73, 81, 82, 83, 84, 89, 101, 144, 160, 161, 162, 171, 208, 210. The dating for John the Scot's commentary, despite various attempts at more precision on the basis of the text, cannot be better determined than this, as proposed by Liebeschütz [1973] esp. 53.

8 On the schools in Carolingian Europe see Contreni [1995a]; and [1995b].

9 For example, Bishop Hayto, or Hatto, of Basel, who was also abbot of Reichenau, in 821 required that all priests know computus as part of their essential training for proper parish leadership. Mansi [1902] col. 395. The same set of requirements for parish priests was repeated by Burchard, bishop of Worms, (d. 1025) in his *Decretum*, II, 2 (Migne, PL 140, col. 625), and later by Ivo of Chartres and Gratian.

10 Consequent to the computistical conference of 809, two large computi that were compiled soon thereafter contained excerpts from Pliny and Macrobius. For the conference of 809, see the highly interpretive account by Borst [1993]. On the contents of the so-called Three Book Computus and the Seven Book Computus, see King [1969] 3–22, 28–44. Both compilations used Plin-

ian excerpts; only the later collection, the Seven Book Computus, had material from Macrobius.

11 In 811 the Irish monk Dungal composed a letter to Charlemagne in which he set forth a general introduction to planetary astronomy, based primarily on Macrobius, with some information included from Pliny. This letter subsequently appeared in manuscripts with further excerpts from Macrobius and excerpts from Isidore of Seville as well, all on cosmological and astronomical topics. It seems clear that Dungal's letter became a distinct source itself on planetary astronomy and was copied as such in the manuscripts. See Eastwood [1994b] esp. 132–134. For information on the tradition of a set of astronomical excerpts from Pliny through the ninth century and beyond, see King [1969] passim; also Eastwood [1987].

12 The *Computus* of Rabanus Maurus witnesses both of these. See Rabanus Maurus [1979] 252, 259. In the first of these places (ch. 40) Rabanus says that a student will learn the signs of the zodiac better by having them pointed out to him by someone who describes their appearance in the sky than by reading a description in a book. In the second place (ch. 48) Rabanus lists the placement of the visible planets in the signs of the zodiac on a given date.

13 Stevens [1995] ch. II, 18–20 for Bede; Helperic of Auxerre, Liber de computo, c. 31 (PL, vol. CXXXVII, coll. 40–43).

14 Examples are Bamberg Staatsbibliothek ms. Class. 18, s. X; London British Library ms. Add. 15293, s. XII in.; and Vaticano ms. Vat. lat. 1544, s. XV. In these mss. the diagram for the eccentric model (figs. 1–3) appears respectively on ff. 35v, 22r, and 70v; the epicyclic model (figs. 4–6) appears respectively on ff. 36v, 23r, and 71r. The texts are found in Calcidius [1962] 128–130 (eccentric model) and 131–134 (epicyclic model).

15 The text in Calcidius [1962] is on pp. 136f. See the ms. diagrams (figs. 7-9) in Paris Bibliothèque Nationale ms. lat. 2164, f. 38v (ca. A.D. 800); Paris BN lat. 6282, f. 34r (s. XI m.); Cambridge Fitzwilliam Museum ms. McClean 169, f. 116v (s. XV).

16 Here we present only specimen diagrams. The full details of the history of these diagrams and the eleventh-century corrections appear, along with references to all pertinent manuscripts, in the article by Eastwood [1992]. The key manuscript in the process of correction is Wien Nationalbibliothek ms. lat. 443 (s. XI–1/2). The innovations in the various planetary diagrams in this manuscript are described Eastwood [1999] 186–193.

17 These Plinian astronomical excerpts have been identified and discussed in terms of their manuscript traditions most thoroughly by two scholars: see Rück [1888]; and King [1969]. Study of the excerpts with special concern for the development of the diagrams that were invented to accompany them appears in Eastwood [1987]; also, Eastwood [1989] chs. 5–6.

18 This possibility and its limitations are discussed in some detail Eastwood [1995].

19 In addition to the last two items in n. 17 above, there are further examples of this process of simplification of the Plinian diagrams shown in Eastwood [1986] 197–251.

20 We can see this group of astronomical diagrams in its earliest form in Leiden UB ms. Voss. F.48, f. 92v, perhaps produced at Auxerre. This ms. is of special importance for Capellan diagrams, as we discuss below. The layout on the page was improved in the copy in Leiden UB ms. B. P. L. 36, folio 129r, perhaps written at Lorsch; it appears in the same form in Paris BN ms. lat. 8671, 84r. An incomplete copy appears in Paris BN ms. lat. 8669, folio 122v. Finally, from either the ninth or the tenth century there is the excessively crowded example in Munich SB clm 14729, folio 221v. This group of four or five ninth-century examples shows the early establishment of this tradition of a definite group of diagrams, although the tradition did not become universal in Capellan manuscripts.

21 At the asterisk in this translated quotation, I follow the emendation to the text given in the edition of James Willis (Martianus Capella [1983]) 324.15; the full text is at 324.10–17 (VIII, 857). The final line in Willis' text has the word vastiore rather than the castiore in my version. As Willis' critical apparatus notes almost cryptically, his choice of word at this point comes from an editorial emendation by A. Dick in an earlier edition, who in turn made the emendation on the basis of a simple suggestion by the nineteenth-century editor U. F. Kopp, who had, however, preserved the manuscript readings with castiore. The choice of reading here by Dick and Willis is supported by no ms. and contradicted by all. Virtually every ms. has the castiore. For detailed discussion of this text and editorial choices made in the past, see Eastwood [1989] ch. 2, esp. 146-148; the medieval interpretation of this text is discussed at pp. 149-155 and here below in a different context.

22 Above n. 21; the Latin given here is found at 324.16-17 in Martianus Capella [1983], the last section of the text translated above.

23 The diagram in the outer margin of Leiden UB ms. Voss. lat. F.48, f. 79v, has labels that not only identify the intersecting planetary loops but also specify that this image accords with Pliny and the Pythagoreans (*secundum plinium, secundum pitago[ricos]*). In addition an explanation is given in an outer marginal gloss farther down the page that says, in part, "If in fact we want to assume the order of the planets according to the Pythagoreans and Pliny, we shall never be able to comprehend it [in this text of Capella] unless the word *'terris'* has been dropped out so that the sentence reads, 'when they are above the sun Mercury is closer' and it is understood to mean 'to the sun.'" This gloss was copied from an earlier manuscript. The peculiarities of the diagram constructed here and assigned to Pliny by the scholar of this ms. are the result of his reading of Pliny's *Historia naturalis* II 73 (Pliny I 150.11–12), a text which is quite unclear.

24 Leiden UB ms. Voss. lat. F.48, folio 92v. The three separate models brought together here were laid out initially in close proximity to the texts they illustrated on folio 79r-v. The development of the ten astronomical diagrams presented on folio 92v as a group is described by Eastwood [2000] esp. 9–17.

25 The compound diagram appears also in Leiden UB ms. B. P. L. 36, folio 129r; Paris BN ms. lat. 8669, folio 122v; Paris BN ms. lat. 8671, folio 84r.

26 This manuscript is Leiden UB ms. B. P. L. 87, folio 124v. The original text reported here has already been changed to read, "Sed cum supra solem sunt, propinquior est ei mercurius ...," with an interlinear gloss to the "ei" that reads, "soli".

27 These two manuscripts are Leiden UB ms. B. P. L. 88, folio 162r-v, and a manuscript apparently copied from it, Vat. Regin. lat. 1987, folios 127v–128r.

[28] This manuscript, Firenze BL ms. San Marco 190 (folio 102r for the diagram concerned), was discussed primarily for the Capellan book on geometry/geography by Leonardi [1956–57]. It is manuscript no. 60 in Leonardi's catalogue of Capellan mss. See Leonardi [1960], where some further bibliography appears.

[29] This heading appears in boldface script: "Quod tellus non sit centrum omnibus planetis" (folio 101r, line 36).

[30] This separately labelled part of the text covers sections 855 to 857 and most of 858 in Book VIII: Martianus Capella [1983] 323.23 - 324.24.

[31] While these manuscript diagrams varied in details, the concentric circles were universal. The standard diagram usually appeared in physical proximity to the text on the subject, *Commentarii in somnium Scipionis*, I.xxi.24–26 (Macrobius [1963] 89).

[32] Pedersen [1973/1975], outlined this interpretation of the eleventh through thirteenth centuries clearly. More recently McCluskey [1998] chs. 9–10, has followed and provided much more detail for the same line of interpretation.

[33] The *Compendium circa quadrivium* is edited by Lafleur [1988] 362–371 for astronomy; see ibid. 131f. for Lafleur's conclusion that Capella was replaced by Sacrobosco's *De sphaera* in the 1240s as the standard text for arts students.

[34] On the rise of Sacrobosco's astronomical text in the curriculum, see Beaujouan [1997]; also Lafleur and Carrier [1997a]; see also Lafleur and Carrier [1997b]. See Thorndike [1949] 413, for the anonymous commentator's discussion. Lambert of Auxerre [1971] 4.

[35] Pedersen [1962], [1981]. Edwards [1985] esp. 339f. for the dating.

[36] A useful summary of the introduction of Ptolemaic models, both for stellar and for planetary motion, appears in McCluskey [1998] 188–206.

[37] The tract, entitled *Septima liberalium artium scientia, scilicet, astrologia*, appears in Copenhagen, Kongelige Bibliotek, ms. G. K. S. 277. Fol., ff. 42v–53v. At f. 50ra–b is a chapter headed "De circulis veneris et mercurii." The large diagram shown in our figure 21 appears on f. 49r. An acephalic copy of Book VIII of *De nuptiis* appears at ff. 56r–58v (VIII, 833–887); a folio containing the initial part of the astronomical book was lost between the medieval and the modern page numbering. Dating is based on two different points in the manuscript. Within the tract on astronomy is found a reference (f. 47rb, 28–29) to the *Anticlaudianus* of Alan of Lille, thus placing the tract later. At f. 77r in the ms. there appears a horoscope for a date in 1164, clearly an example from a previous time (and before Alan of Lille's work), and a set of planetary positions for the year 1254, which may or may not have been the current year of writing.

[38] See Calcidius [1962] 134–138 (cc. 83–86), including a figure for epicyclic planetary motion. In the Copenhagen ms., diagrams and discussion of planetary epicyclic motion follow immediately upon this large composite in figure 21.

[39] The manuscripts described in this paragraph are the following, listed with the number in Leonardi's manuscript catalogue for Martianus Capella (Leonardi [1960]: Oxford, Bodleian Library, ms. Auct. F.1.9 (Leonardi no. 139); Oxford, Bodleian Library, ms. Selden Supra 24; Paris, Bibliothèque Nationale, ms. lat. 6415 (Leonardi no. 152); Munich, Staatsbibliothek, clm 534 (Leonardi no. 117); Vatican City, Bibliotheca Apostolica Vaticana, ms. Regin. Lat. 1452 (Leonardi no. 207).

[40] The five fifteenth-century manuscripts with diagrams for three versions of circumsolarity are: Basel, Universitätsbibliothek, ms. F. V.40, folio 154r; Florence, Bibliotheca Laurenziana, ms. Pl.51.13, folio 128v; Naples, Bibliotheca Nazionale, ms. V. A.16, folio 228v; Vatican City, Bibliotheca Apostolica Vaticana, ms. Urb. Lat. 329, folio 139v; Venice, Bibliotheca Nazionale, ms. XIV.35, folio 143r. The three that include the larger conception of Capellan theory of all the planets are the Florentine, Vatican, and Venetian manuscripts in this group. The astronomical collection with Capellan planetary astronomy is in Bibliotheca Apostolica Vaticana, ms. Urb. Lat. 1358, folios 152r–v, 161r–163v.

[41] Especially in Book One there are numerous implicit references to Pliny. Copernicus' own annotated copy of Pliny's *Natural History* is still preserved and can be consulted in the library of Uppsala University. The apparatus of the critical edition identifies some but not all references to Pliny by Copernicus.

[42] He cannot copy Ptolemy's argument, since he views the stars as spatially static and all apparent motion as a result of the earth's rotation. Hence, Copernicus has no convincing argument for assuming a spherical stellar sphere.

[43] Goldstein and Barker (Goldstein et al. [1995] 388) print a facsimile of another diagram (Ms. astron. 11, folio 9) with circumsolar motion that they misleadingly call "Inverted Copernican hypothesis". It shows a heliocentric arrangement of the planets. But they fail to see the connection with diagram shown in fig. 24 which establishes the Capellan link.

References

Barker-Benfield, B.
1983 "Macrobius." In: *Texts and Transmission. A Survey of the Latin Classics*, ed. L. D. Reynolds, 222–232. Oxford.

Beaujouan, G.
1997 "Le quadrivium et la Faculté des arts." In: *L'enseignement des disciplines à la Faculté des arts (Paris et Oxford, XIIIe - XVe siècles)*, ed. O. Weijers and L. Holtz, 185–194. Turnhout.

Bischoff, B.
1961 "Hadoardus and the Manuscripts of Classical Authors from Corbie." In: *Didascaliae. Studies in Honor of Anselm M. Albareda*, ed. S. Prete. New York.

Borst, A.
1993 "Alcuin und die Enzyklopädie von 809." In: *Science in Western and Eastern Civilization in Carolingian Times*, ed. P. L. Butzer and D. Lohrmann, 53–78. Basel.

[Calcidius]
1962 *Timaeus, a Calciddio translatus commentarioque instructus*, ed. J. H. Waszink (= Plato latinus, ed. R. Klibansky, vol. IV). London.

Contreni, J.
1995a "The Pursuit of Knowledge in Carolingian Europe." In: *'The Gentle Voices of Teachers.' Aspects of Learning in the Carolingian Age*, ed. R. E. Sullivan, 106–141. Columbus, OH.

1995b "The Carolingian Renaissance: Education and Literary Culture." In: *The New Cambridge Medieval History*, ed. R. McKitterick, 709–757. Cambridge.

Copernicus, N.
1984 *De Revolutionibus* (= Copernicus Gesamtausgabe, II). Hildesheim.

1992 *On the Revolutions*, trans. and comm. E. Rosen. Baltimore.

Eastwood, B. S.
1986 "Plinian Astronomy in the Middle Ages and Renaissance." In: *Science in the Early Roman Empire: Pliny the Elder, his Sources and Influence*, ed. R. French and F. Greenaway, 197–251. London.

1987 "Plinian astronomical diagrams in the early Middle Ages." In: *Mathematics and its applications to science and natural philosophy in the Middle Ages*, ed. E. Grant and J. E. Murdoch, 141–172. Cambridge.

1989 *Astronomy and Optics from Pliny to Descartes*. London.

1992 "Heraclides and heliocentrism: texts, diagrams, and interpretations." In: *Journal for the History of Astronomy* 23: 233–260.

1994a "Manuscripts of Macrobius' *Commentarii in Somnium Scipionis*, before 1500." In: *Manuscripta* 38: 138–155.

1994b "The astronomy of Macrobius in Carolingian Europe: Dungal's letter of 811 to Charles the Great." In: *Early Medieval Europe* 3: 117–134.

1995 "Latin Planetary Studies in the IXth and Xth Centuries." In: *Physis. Rivista internazionale di storia della scienza*, n.s. 32: 217-226.

1999 "Calcidius's Commentary on Plato's Timaeus in Latin Astronomy of the Ninth to Eleventh Centuries." In: *Between Demonstration and Imagination. Essays in the History of Science and Philosophy*, ed. L. Nauta and A. Vanderjagt, 171–209. Leiden.

2000 "Astronomical Images and Planetary Theory in Carolingian Studies of Martianus Capella." In: *Journal for the History of Astronomy* 31: 1-28.

Edwards, G.
1985 "The Two Redactions of Michael Scot's Liber Introductorius." In: *Traditio* 41: 329–340.

Goldstein, B. and P. Barker
1995 "The Role of Rothmann in the Dissolution of the Celestial Spheres." In: *British Journal for the History of Science* 28: 385–403.

Ineichen-Eder, C.
1978 "Theologisches und philosophisches Lehrmaterial aus dem Alkuin-Kreise." In: *Deutsches Archiv* 34: 192–201.

King, V. H.
1969 *An Investigation of Some Astronomical Excerpts from Pliny's Natural History found in Manuscripts of the Earlier Middle Ages* (B. Litt. thesis, Oxford University; Bodl. Libr. Ms. B. Litt. d. 1465).

Lafleur, C.
1988 *Quatre introductions á la philosophie aux XIIIe siécle*. Montréal.

Lafleur, C. and J. Carrier
1997a "Les Accessus philosophorum, le recueil Primo queritur utrum philosophia et l'origine parisienne du Guide de l'étudiant du ms. Ripoll 109." In: *L'enseignement de la philosophie au XIIIe siècle. Autour du ‚Guide de l'étudiant' du ms. Ripoll 109*, ed. C. Lafleur, 589-642. Turnhout.

1997b "Le réglementation curriculaire (de forma) dans les introductions à la philosophie et les guides de l étudiant de la Faculté des arts de Paris au XIIIe siècle." In: Lafleur and Carrier [1997a] 521–548.

Lambert of Auxerre, L.
1971 *Logica (Summa Lamberti)*, ed. F. Alessio. Florence.

Leonardi, C.
1956–57 "Illustrazioni e glosse in un codice di Marziano Capella." In: *Archivio paleografico italiano, Bulletino*, n.s. 2-3/2: 39–60 and 2 plates.

1960 "I Codici di Marziano Capella." In: *Aevum* 34.

Liebeschütz, H.
1973 "The place of the Martianus Glossae in the development of Eriugena's thought." In: *The Mind of Eriugena*, ed. J. O'Meara and L. Bieler, 49–58. Dublin.

Macrobius, A. T.
1963 *Commentarii in Somnium Scipionis*, ed. J. Willis. Leipzig.

Mansi, J. D.
1902 *Sacrorum conciliorum nova et amplissima collectio*, vol. 14. Paris.

Martianus Capella
1983 *[De nuptiis philologiae et mercurii]*, ed. James Willis. Leipzig.

McCluskey, S.
1998 *Astronomies and Cultures in Early Medieval Europe*. Cambridge.

McKitterick, R.
1992 "Knowledge of Plato's Timaeus in the Ninth Century: the implications of Valenciennes, Bibliothèque Municipale MS 293." In: *From Athens to Chartres*, ed. H. J. Westra, 85–95. Leiden.

MGH/E
1895 *Monumenta Germaniae Historica. Epistolae IV: Karolini aevi II*, ed. E. Duemmler. Berlin.

MGH/L
1883 *Monumenta Germaniae Historica. Legum Sectio II: Capitularia I*, ed. A. Boretius. Hannover.

Munk Olsen, B.
1985 *L'Étude des auteurs classiques latins aux XIe et XIIe siècles*, vol. II. Paris.

1989 *L'Étude des auteurs classiques latins aux XIe et XIIe siècles*, vol. III/2. Paris.

Naibod, V.
1573 *Primarum de coelo et terra institutionum quotidianarumque mundi revolutionum libri tres*. Venice.

O'Connell, D. J.
1936 "Easter Cycles in the Early Irish Church." In: *Journal of the Royal Society of Antiquaries of Ireland* 66: 67–106.

Pedersen, O.
1962 "The *Theorica planetarum* Literature of the Middle Ages." In: *Classica et Mediaevalia* 23: 225–232.

1973/1975 "The Corpus Astronomicum and the Traditions of Mediaeval Latin Astronomy. A Tentative Interpretation." In: *Colloquia Copernicana* III/1973. Torun, and in: *Studia Copernicana* III/1975 67–78. Wroclaw.

1981 "The Origins of the 'Theorica planetarum.'" In: *Journal for the History of Astronomy*, 12: 113–123.

PL
Patrologiae cursus completus/Series Latina, ed. J. P. Migne. Paris

Pliny (Plinius Secundus, Gaius)
Historia Naturalis, eds. L. Jan and K. von Mayhoff. Leipzig.

Ptolemy, C.
1984 *Almagest*, trans. G. J. Toomer. London.

Rabanus Maurus (Hrabanus Maurus)
1979 *Martyrologium. De computo* (Corpus Christianorum: Continuatio Mediaevalis 44), ed. J. McCulloh and W. Stevens. Turnhout.

Rheticus, G. J.
1982 *Narratio prima*. Wroclaw.

Riccioli, G. B.
1651 *Almagestum novum astronomiam veterem novamque complectens*. Bologna.

Riché, P.
1962 *Education et culture dans l'occident barbare, VIe - VIIIe siècles*. Paris.

1976 *Education and Culture in the Barbarian West*. Columbia, SC.

Rück, K.
1888 *Auszüge aus der Naturgeschichte des C. Plinius Secundus in einem astronomisch-komputistischen Sammelwerke des achten Jahrhunderts. Programm des Königlichen Ludwigs-Gymnasiums ... 1887/88*. Munich.

Stevens, W. M.
1995 *Cycles of Time and Scientific Learning in Medieval Europe*. Aldershot.

Swerdlow, N. and O. Neugebauer
1984 *Mathematical Astronomy in Copernicus' De Revolutionibus*. New York.

Thorndike, L.
1949 *The 'Sphere' of Sacrobosco and Its Commentators*. Chicago.

Westman, R.
1975 "Three responses to the Copernican theory: Johannes Praetorius, Tycho Brahe, and Michael Maestlin." In: *The Copernican Achievement*, ed. R. Westman. Berkeley.

Images, Models and Symbols in Copernican Propaganda

Giancarlo Nonnoi

Pictures of the Moon and the "Supreme Patron"

When Galileo Galilei's Sidereus Nuncius appeared on the European intellectual scene in March 1610, the thin volume in quarto – composed of just 30 sheets – stood out in the panorama of the scientific literature of the times owing to a series of components of different kinds that took their place alongside the numerous and extraordinary celestial novelties contained in the text.[1]

Written hurriedly between January and March of the year in which it was published, Galileo, with the publication of the *Sidereus*, wished to reach at least three other closely interdependent goals which were partly outside the observational content of the text. Counting on the absolute scientific originality of the *Starry Messenger*, the Pisan scientist expected on the one hand to become a permanent member of the elite of the international scientific community and, on the other hand, finally to be able to return to his homeland with the highest intellectual and economic recognition after almost eighteen years spent in the service of the *Serenissima* Republic of Venice (Westfall [1985]; Biagioli [1990], [1993]). And from there, on the strength of the further prestige that came to him from the protection of the Medicis, he intended to launch his personal campaign in favour of the Copernican hypothesis to which he asserted he had privately adhered starting from 1597 (Galilei [1890-1909] II 198, X 68).

Because, and a consequence, of these objectives which were collateral to the cosmological themes discussed in it, the first of the Galilean masterpieces was also marked by an important change and innovation in the canons and communicative strategies in use in scientific treatises in general, and in astronomical ones in particular (Moss [1993] 76).

Galileo's adoption of novel and more sophisticated communications instruments did not stop at the rhetorical level, in the prose felt so expressively, in the flat and discursive style in which the discoveries were announced and the dissertation approached; nor did he limit himself to the use of certain narrative styles which diverged from the assertive and apodictic forms used in the phrasing of much of traditional natural philosophy.[2] The difference in style between Galileo's writing and the other astronomical literature of the epoch is also marked by the presence in Galileo's works of a unique and graphically appreciable iconographic display which almost lends concreteness to verbs such as *videre*, *conspicere* and *observare* and to other synonyms that return at crucial moments in the text (Battistini [1989] 34–36).

With the *Sidereus Nuncius*, besides the classic geometrical diagrams referring to the *ratio* of the instrument and to the traditional punctiform and linear representation of celestial bodies and their motions (Heninger [1977]; Kemp [1996]; Roche [1998]), for the first time in a work on astronomy some figures were used as an attempt to represent the phases of the moon and the arrangement of the stars as they were observed by Galileo with the telescope.[3] Conceived to make clearly perceptible the "terrestrial" conformation of the orography of our satellite, especially the drawings of the moon, thanks to a wise blend of light, shadow and halftones, are still today the bearers of a strong impressionistic message, and from this peculiarity draw a communicative power that is more eloquent than any words (see fig. 1). These are such as to produce in the reader the illusion of looking through the eyepiece of the *perspicillum* in the nights between December of 1609 and January of 1610.[4]

Among the numerous aspects of Galileo's drawings of the moon there is also the ambition to supply the reader with a "realistic" documentation of the observations;[5] these are not dissimilar from what we today succeed in achieving with far more effective photographic techniques. Naturally, Galileo's plates cannot hope to rival a photograph or to show a fidelity comparable to that means.[6] Galileo's images of the moon are rather the idealised synthesis of several consecutive observations in the same visual conditions.[7] They are the same ones that the *Serenissimo* Prince could see through the telescope given to him by Galileo, together with the dedication of the just-discovered satellites of Jupiter.

Undoubtedly there is an inseparable epistemological coherence between the iconographic elements and the content of the announcement. For the telescopic astronomy inaugurated by Galileo, compared to the mathematical astronomy of Copernicus' *De revolutionibus*, the documentation of conditions and observational data constitute a prime methodological factor in relation to the theoretical explanation itself. However, it is just as indubitable that, thanks to this epistemological specificity, the *Starry Messenger* acquires a pronounced uniqueness, making it the right instrument to use in reaching the author's personal goals and starting a campaign of veritable proselytism in favour of the new astronomy (Galilei [1890–1909] X 299–302), but at the same time making it particularly attractive and interesting from the aesthetic standpoint. This epistemological coherence was but one peculiarity, certainly more superficial than the others, but for that very reason capable of attracting not only mathematicians, but also men of culture in general, on whom the *Sidereus Nuncius* exerted an irresistible fascination that few non-Galilean scientific texts can boast.[8]

Astronomy, on the other hand, had become a question of public interest also because of the exegetic questions immediately implied in it and the importance that these questions had in the confrontation between the Protestant world and that of the Counter-reformation. It can thus come as no surprise to find that the campaign in favour of Copernican astronomy could resort to the aid of images as effective subsidiary rhetorical instruments capable of spreading good will and consensus in favour of the hypothesis, as well as facilitating comparison with the far more difficult mathematical demonstrations (Moss [1993] vii).

This was evidently an effect subordinated to the scientific content to be communicated. In any case, there is no doubt that Galileo did not overlook the amount of added value that the realistic illustrations of the moon gave to his work.[9] He was in fact perfectly aware of the evocative force of images and symbols.[10]

Confirmation of this can be found in the intention, in the light of the planned second edition of the *Sidereus*, which was never published, of enhancing the visual evocativeness of the work "*con molte et bellissime figure* [...] *le quali sono cosa mirabile da vedersi*."[11] On the other hand, it must be kept in mind that those to whom the ambitious propagandistic operation in favour of heliocentric astronomy undertaken with the *Starry Messenger* was addressed were not only the narrow circle of astronomers. It was also destined for the vaster public of the courts, the curiae, the nunciatures, the chancelleries, the universities and the *collegi studiorum*. This was a rather diversified public as far as interests, competence, taste and sensitivity were concerned: the common denominator was their knowledge of Latin.

The white-hot disputes that characterised Galileo's return to Florence and the "*salutifero editto*" of 1616, which imposed silence concerning the mobility of the Earth, caused a brusque cutback to, if not the total stoppage, of the plan to involve the prince and spread consensus for the Copernican hypothesis. Many of Galileo's initiatives in this period can be considered defensive actions from several viewpoints. And this imposed behaviour

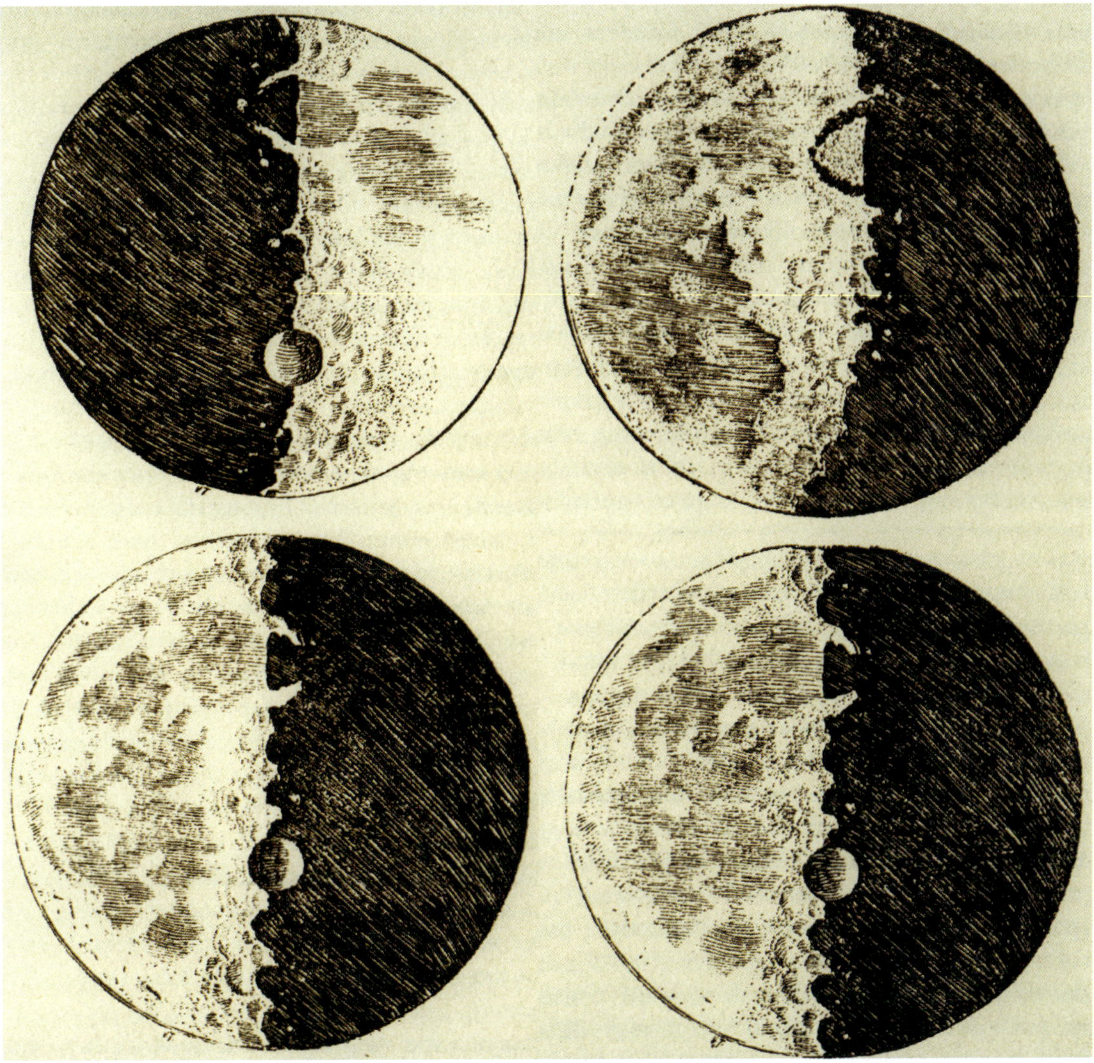

Figure 1
Galileo's pictures of the moon in the *Sidereus Nuncius* (Venice 1610).

also reflected on the expressive and communicative means used, as well as on the typographical aspect of Galileo's works which were published after 1616, such as the *Letters on Sunspots*,[12] the *Discourse on Bodies in Water* and *The Assayer*. These three works were in fact more in line with the standards of traditional treatise writing, with a strong prevalence of linguistic elements over iconographic ones which, the few times they appear, are reduced to the essential and are drastically schematised.

In August of 1623, the Florentine Maffeo Barberini, a cultured humanist and lover of the arts and sciences, a member of the *Accademia dei Lincei* and an admirer of Galileo's, was elected pope under the name of Urban VIII. This event was greeted with exultancy among progressive intellectuals, who saw in him the "supreme patron."[13] Galileo in particular interpreted this as a turn in favour of Copernicus. He believed the time had finally come to render public that oft-announced and oft-postponed "*Sistema Massimo*" (Galilei [1890–1909] X 350–352, 422).

However, implementation of the old plan went far beyond the dissemination of the results of several decades of research work or the publication of a volume, no matter how important this would have been. In Galileo and his friends, the decision to conclude the *Dialogue on the Two Chief World Systems* was linked to the hope of turning to the advantage of the Copernican system the "*mirabil congiuntura*" which had come about with the election of Maffeo Barberini (Galilei [1890–1909] XIII 134f.), and of beginning to acquire support and consensus equal to what had been gained through the publication of the *Starry Messenger*.[14] With the not negligible difference that the person to whom it is dedicated, the prince whose benevolent patronage is invoked for the success of the new enterprise, is not the lord of a noble family of continental fame and prestige, but the "*Mecenate supremo*" (Galilei [1890–1909] XIII 121), he who had inherited from the prince of the Apostles the highest see and throne.

It has often been pointed out that through the *Dialogue on the Two Chief World Systems* Galileo was confident that he would be able to swing the Church over to positions which, although not official, were at least not openly hostile to Copernicus' thesis. But, more pragmatically, Galileo and his friends probably would have been satisfied if through their initiative they had succeeded even in obtaining a less restrictive interpretation, and consequently a more lenient application of the edict of 1616 promulgated by Paul V with which the profession of the heliocentric thesis had been interdicted. This would have represented a loosening of the tethers of censorship that was necessary from the political viewpoint as well, or at least this was the opinion of the "Galileans," in order to confute concretely the denigratory campaign fed by Protestant circles in which it was insinuated that the condemnation of propositions favourable to the Earth's motion was due to the ignorance and intellectual and scientific weakness of the Church hierarchy and Catholic men of science.[15] It was thus a confrontation, with the Italian Copernicans having to fight on different fronts, from which Galileo counted on emerging as the champion of the renewed challenge to the heretics "*on the other side of the mountains* [...] *all of the same opinion as Copernicus* [...] *to show them that the Catholics*" were not inferior and far from being "*ignorant of the human disciplines*."[16]

As concerns this ambitious design, if the naming of Jupiter's satellites in honour of the Medici family had been commensurate with the position occupied by the noble Florentine family in the Catholic political and religious universe, in the same way the "*ingenious fantasy*," which is to say the ebb and flow of the sea – on the basis of which Galileo thought he would be able to prove the "*vertigine della Terra (the earth's turning)*" –, would have acquired an advantage and an incomparable echo if it had been announced under the aegis and with the moral support of the holder of the highest spiritual power on earth, whose see and power centre were in the Eternal City.[17] It goes without saying that in this well-planned strategy for the acquisition of papal consent, even the sole authorisation to publish the *Dialogue* in Rome was considered a sure sign of benevolence (Galilei [1890–1909] XIV 87f., 77f., and 92). This was a tactical choice that had been highly successful in the past on the occasion of the publication of *The Assayer*,[18] but this second time, contrary to the expectations of Galileo and his friends of the *Accademia dei Lincei*, it did not produce the same positive results.

With *The Assayer* still a rough draft, Galileo set in motion his vast circle of relations. Between April and June of 1624, encouraged by the Roman "Galileans" the Tuscan mathematician and philosopher travelled to Rome to render homage to the new Pope.[19] He was greeted most cordially by Urban VIII and was received as many as six times by the Pope. But, as we know, this was not sufficient to make the creators of the ambitious design successful in reaching the objectives they had set for themselves.[20] In fact, even when the much desired *imprimatur* for publication of the *Dialogue* in Rome was issued, this was not followed up by the actual printing of the volume in the city of the Popes for reasons that still have not been determined exactly. Despite this, more than a few traces of the initiative aiming at the legitimisation of Copernicanism at the highest level remain in the text of the *Dialogue* and these are also of a certain consistency, some of them the direct consequence of adaptations and the playing down of the text as imposed by the revisers at the personal disposal of the Pope.[21]

Among the most characteristic aspects of the *Dialogue* that refer to Galileo's far-sighted plan are certainly the several rhetorical instruments he used in his masterpiece. The force of conviction of the *Two Chief World Systems* is no longer entrusted to the admirable discoveries and the original illustrations of the moon contained in the *Sidereus Nuncius*. The images of the mind, the conceptual associations evoked through a refined use of dialogue take the place of the telescopic images.[22] In Galileo's hands the ancient Socratic art acquires a formidable argumentative and representative force,[23] capable of entertaining and persuading even, and perhaps above all, the educated non-scientific reader.[24] Indeed, in the *Dialogue*, even lacking incontrovertible empirical proof and absolutely new observations of the kind presented in the *Sidereus Nuncius*, Galileo did not refrain from theatricality. The phenomenon of the tides (which Galileo considered the product of the combination of the rotation and revolution of the Earth) tries, in verbal form, to replace and reproduce the sensation aroused by the images of the moon and the first vision of Jupiter's satellites.[25]

The illustration preceding the title-page of the *Dialogue* can also be considered a further, although still partial, sedimentation of the ambitious plan for bringing the Church and its universe closer to Copernicanism. This is an engraving by Stefano Della Bella, a Florentine artist who at the time he placed his fine hand at the service of the most illustrious Italian scientist of the time was little more than twenty years old, and later went on to have a successful career in the service of the most important European courts.[26]

Many important and stimulating pages have been written about this *"antiporta,"* the engraved plate facing the title-page, which through the many subjects to be found in the illustration announces the central themes discussed in the work (Ciardi et al. [1985]; Tongiorgi Tomasi [1986]; Galilei [1998] II 108–110). With great perspicacity, the correspondence between the different subjects shown in the plate and the theoretical questions of which they are emblematic have been identified (see fig. 2). The arrow at the feet of Aristotle, for example, symbolises the peripatetic theory of movement with special emphasis on the central function assigned to the medium in Aristotle's dynamics of natural motions as well as of violent motions (cf. Aristotle, *Physics*, 215 a1–216 b20; idem, *De coelo*, 301 b17–31. Galilei [1890–1909] VII 175ff.).

The armillary sphere held by Ptolemy and around which the discussion of the three characters appears to revolve is a model of the geostationary system. While Copernicus, slightly to one side, as if awaiting the moment to enter the scene definitively once the inadequacy of the ancient astronomy had emerged, holds in his left hand along his side a *tellurium*, a schematisation of the heliocentric system reduced to sun, earth and moon. We can also see how the shell at Copernicus' feet, the harbour constructions and the fleet aground that can be seen on the right side of the scene presage the theory of the ebb and flow of the sea as a "confirmation" of the Earth's motion. The heavy vessel that can be glimpsed between the figures of Copernicus and Ptolemy, lying at anchor with its sail half unfurled, appears to suggest the idea of a science in anguish awaiting the result of the dispute between the two systems so that once freed of prejudices of the senses and the intellect it can finally sail away in safety.[27] And not even the drape held by the two putti at the top and framing the scene is a mere ornamental accessory. The composition of the ample drapery in fact gives architectural emphasis to the space rendering homage to the ancient protector, whose name is thus found at the top of the scene in an effective

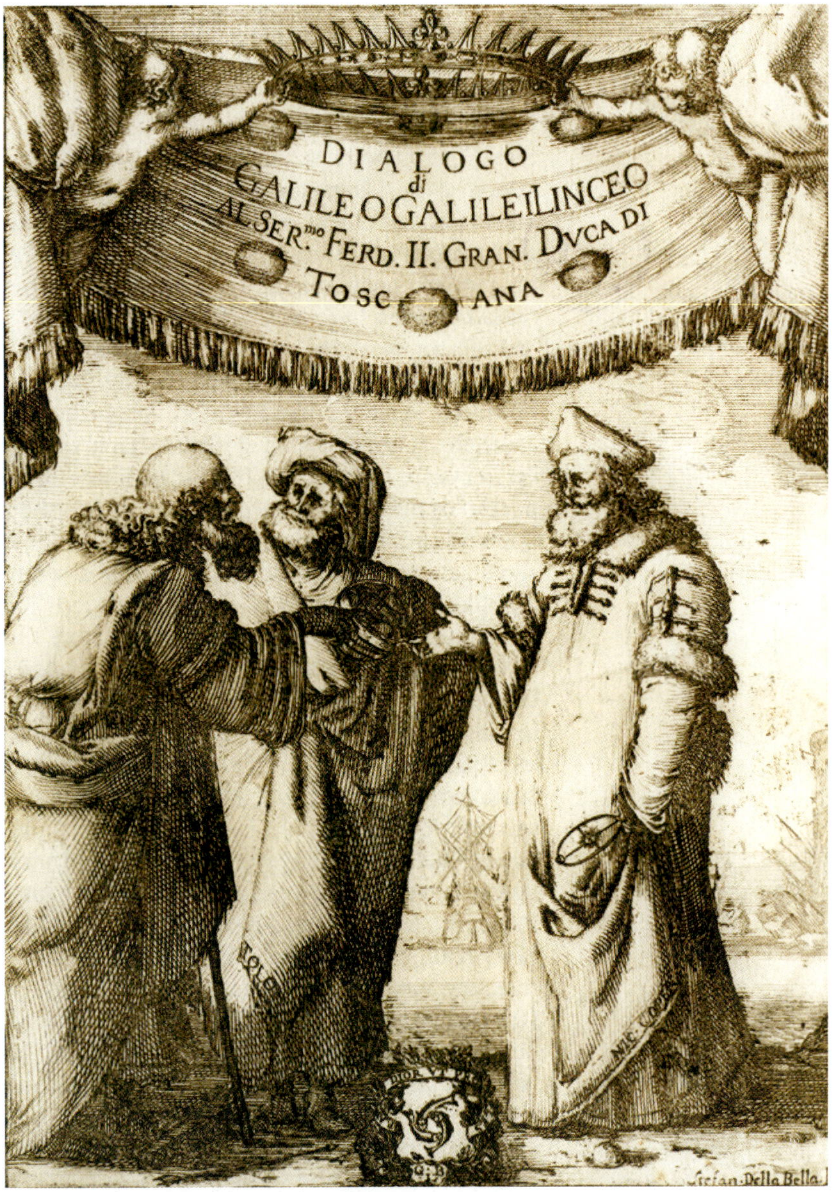

Figure 2
"Antiporta" of Galileo: *Dialogo sopra i due massimi sitemi del mondo* (Florence 1632), engraved by Stefano Della Bella. Courtesy of the Biblioteca Universitaria di Cagliari.

and elegant synthesis in which the title of the work and the heraldic frieze of the Medici family, with the semispheres composing it recalling the court of Jupiter's moons named after the members of the Grand Duke's family. The design also allows a glimpse of some atmospheres and sketches some lines referring to places in which "congresses" are held or the protagonists in the course of the "ragionamenti" are mentioned.[28]

Perhaps less attention has been paid to the characters and the abundant symbology of which they are the bearers. Aristotle, for example, is represented with his back to the viewer and is the only figure with his face partly hidden.[29] Of the three, the philosopher from Stagira is clearly the one who bears the signs of age: his tired body leans on a staff, just like his limping philosophy that turns its back on the present.

Like actors on an ideal philosophical stage, the figures shown in the *"antiporta"* are arranged in chronological order along an ideal time trajectory simulating the passage from west to east of the Earth's motion around its axis. Aristotle is the only one whose head is not covered; his vast baldness contrasting with the gaudy turban covering Ptolemy's head. A head-dress which, because of its evident Middle East origin, visualises the distance separating Alexandrine astronomy from the Latin-Christian tradition. Copernicus wears the narrow three-cornered hat of the ecclesiastics. The astronomer from Thorn, as is known, had been a canonical at the Cathedral of Frauenburg, and of the three he is the only son of the Church of Christ but, at the same time, a devote observer of the teachings and precepts of the See of Peter.

The Forerunner of the Manifesto of the Copernicans

Among the many external references that may have influenced the compositional structure and the elements included in the frontispiece (*antiporta*) of the *Dialogue on the Two Chief World Systems*, a connection has been noted between the latter and the imposing fresco painted by Raphael some time around 1510 in the Stanza della Segnatura in the Vatican and known as the *School of Athens*.[30]

However, this certainly pertinent observation deserves further, more detailed attention. I feel that the hypothesis that sees in Raphael's masterpiece the model that inspired the frontispiece of Galileo's *Dialogue* can be supported by other evidence. This is especially true when we leave aside the analogies of the details, which are certainly there, and concentrate our attention on the general intellectual atmosphere that inspired the two works, and in particular on the philosophy surrounding the principal nucleus that keeps them in close communication. I refer to the archetypal value contained in the dialogue between Plato and Aristotle, upon which the Renaissance masterpiece makes the eye converge (see fig. 3).

It is to be remembered that in the intentions of Julius II, the Pope who was the patron of Bramante and Michelangelo, besides Raphael, the complex system of frescoes that decorate the large room were to celebrate the apotheosis of divine science and its supremacy over human knowledge. On the wall opposite the one on which the *School of Athens* is painted, the Pope had Raphael paint the just-as-renowned *Disputa del Sacramento*, before which every theologian, after much painstaking searching, bows down before the mystery of the Incarnation and gives up all ambition of reaching autonomous knowledge of the ultimate aim of man and the universe.[31] The spirit of the complex ideological framework that confers compositional and philosophical continuity on the symbolic nuclei present in the different pictorial elements found between the walls and ceiling of the *Stanza* was a common notion of late Renaissance and early seventeenth-century culture, owing to the *Lives* of Vasari (as well as to others), in which he had devoted several soon-to-be-considered memorable pages to the *School of Athens*.[32]

In continuity with the reading of Vasari, of great interest with respect to the thematic content of Galileo's *Dialogue* is the large number of direct and indirect references to the study of the stars present in Raphael's allegory. The cosmological issue was, in fact, one of the prime topics of the universal philosophical dialogue shown in the painter's masterpiece. Among the numerous sages in the dialogues in the airy spaces of Bramante's architecture, we can recognise, among others,

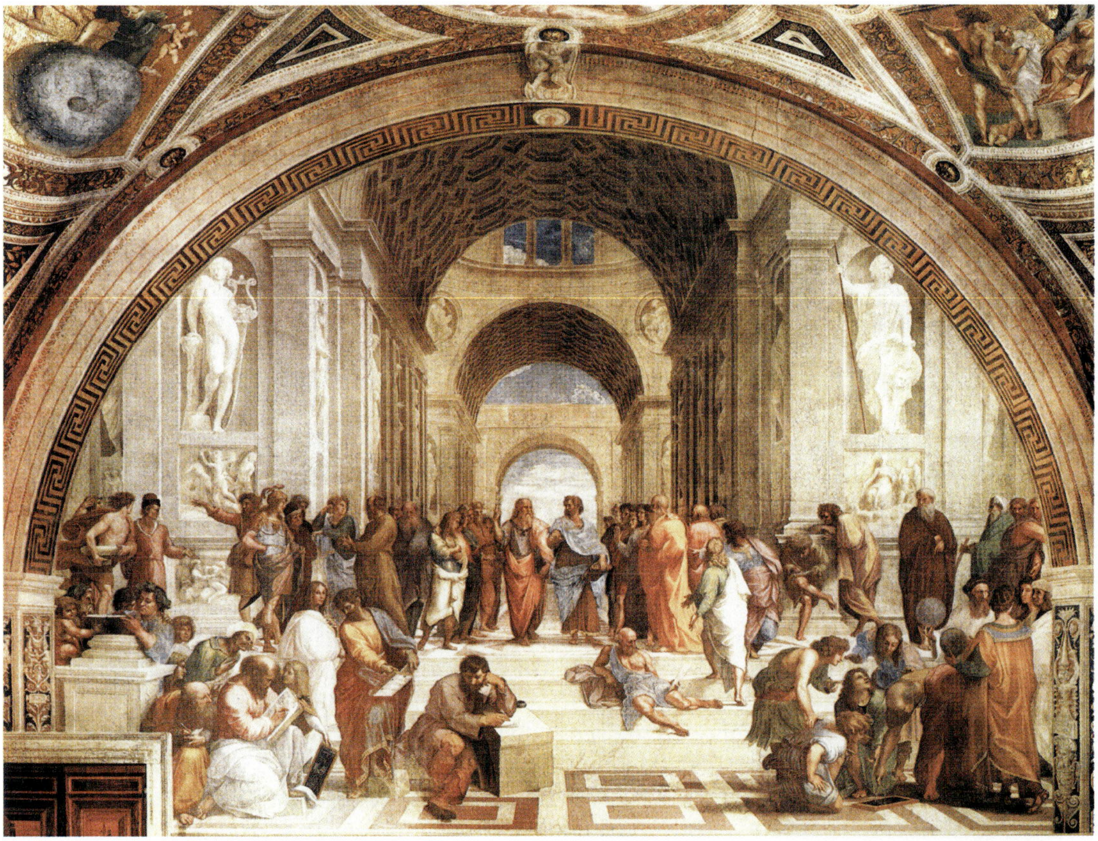

Figure 3
Raphael: *The School of Athens.*

Epicurus, Zeno, Parmenides, Pythagoras, Averroes, Euclid (or perhaps Archimedes), Zoroaster and Ptolemy. As concerns the latter two in particular, they are part of a group in which one of the many disputes framing the supreme confrontation between the philosophy of the Academy and that of the Peripatos is in progress. The dispute, which is being listened to at a distance by a figure with Raphael's face, may be about the Renaissance question of the intimate symmetry between the different parts of the cosmos. Zoroaster is in fact holding in his right hand a diaphanous globe of the sphere of the stars while Ptolemy displays the more opaque and palpable terrestrial sphere (see fig. 4). The analogy to the philosophical confrontation seen in the frontispiece of the *Dialogue on the Two Chief World Systems* is too strong to be considered casual. Among other things, besides the situational recurrence just mentioned, it is to be recalled that the Ptolemy present in Della Bella's print holds an armillary sphere on which the intellectual gaze of the characters in the scene converges.

The cosmological problem is thus found to be one of the main topics of the universal

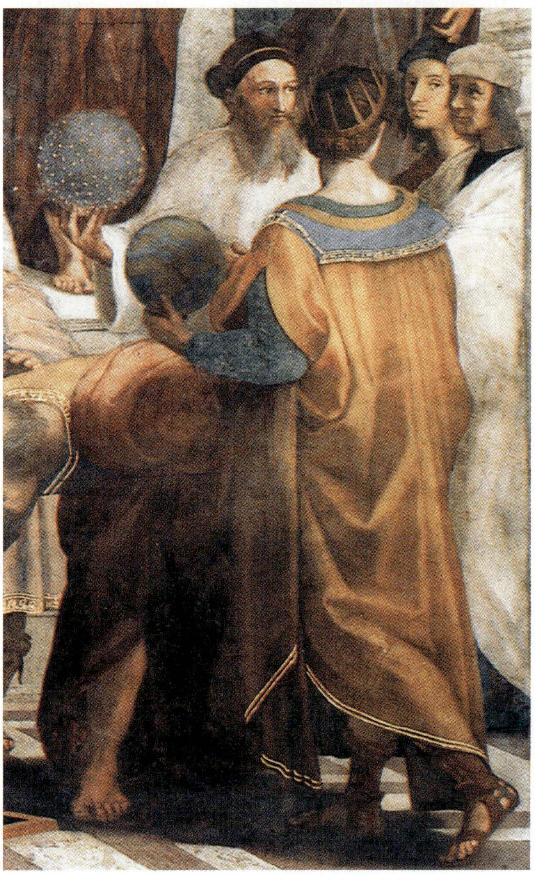

Figure 4
Raphael: *Zoroaster and Ptolemy.* Detail of *The School of Athens.*

philosophical dialogue portrayed in Raphael's masterpiece. But beyond the specific areas of the painting devoted to it, from an overall observation of the fresco, among the several personages and the different allegories emerges a wider-ranging weft of latent relationships which manifest themselves through the potential concatenation of the numerous geometrical and astrological symbologies disseminated over several points of the work. On the ceiling of the room, for example, the different pictorial modules are connected with one another by means of the different itineraries of a pathway made of allegories and emblematic forms that act as junctions between the large scenes painted on the vertical planes of the room. Within this labyrinth of paths, in a frame placed at the zenith of the arch framing the large fresco of the *School of Athens*, we find throned, as the sole master of the cognisance of cause, Philosophy, at whose right sits Ethics, while in a frame placed symmetrically opposite we see Astronomy, magnificently portrayed by the crystalline sphere of the Ptolemaic world, in the act of receiving the starting motion from an angelic power placed outside of it.[33] This closely woven network of references and the hierarchical system that follows illustrate how astronomy can create moments of philosophical tension with theology when human cognisance of the ways of the heavens is not in harmony with dogmatic and revealed wisdom.

If there is any truth in the supposition that the point of convergence in Raphael's fresco, together with some of the thematic modules connected to it, constituted a model for the structure of the scene depicted by Stefano Della Bella, then the *"antiporta"* of the *Dialogue* assumes meanings beyond those recognised up to now. These meanings emerge by reflecting on the consolidated semantic paradigms of the model to which it refers. A matrix that cannot be neglected when we attempt to clarify the limits and ambitions of the work for which the protection of the most "Catholic" of all patrons was being solicited.

In consideration of this perspective it is to be pointed out that the *School of Athens* is only one of the elements in a more articulated system of symbolisations that come together to exalt and celebrate the spirit that animates and qualifies the function of the rooms in which the painting is found. The Stanza della Segnatura was one of the most solemn and prestigious places in the Pope's apartments. Used as the Pope's private library, this large room takes its name from the *Signatura Gratiae*, one of the sections of the supreme tribunal of the Curia to which the faithful addressed petitions for intercessions and supplications and through which Christ's Vicar, with total and paternal discretion, dispensed favours and pardons. Thus, every reference, even implicit, to the *School of Athens*, as an emblematic structural element of the place in which expression was given to the illuminated wisdom of the Pope, the generous magnanimousness of the Church and its unchallenged spiritual hegemony, undoubtedly constituted a reference to an image that was quite well known to believers in Catholicism. In the first place, for the members of the Roman Curia and, more in general, for the vast world of itinerant and influential European Christianity which, with its numerous followers, dwelled at length and assiduously in that unique *theatrum mundi* which was the Rome of the Counter-reformation (Haskell [1980] part. I).

Galileo was well aware that the search for and protection of the highest spiritual guide in the Latin West would necessarily involve more than one concession in the autonomy of research and in the free expression and dissemination of its results. However, it was a price he was willing to pay, as can be seen clearly in the words he wrote to Andrea Cioli during the delicate negotiations in view of the publication of the *Dialogue* in Rome.[34] Galileo was in fact aware that an authorisation to print his work, albeit *sub condicione*, would signify, thanks to the infinite variety of Baroque ambiguities, a point in favour of Copernicus' system.

As is known, even what appeared in the eyes of Galileo and his friends to be the wisest tactical moves wound up being turned against those who had devised them and fell far from reaching the mark. In any case, apart from the conclusion of the issue, it is difficult to think that Galileo did not know Raphael's frescoes and their symbolic value, at least through the afore-mentioned description by Vasari. On the other hand, it is not improbable that Galileo had even had the opportunity to admire them *de visu* during one of his sojourns in Rome and on the several occasions he had to meet the Pope in his apartments, just at the time the strategy for legitimisation of Copernicanism and the *captatio benevolentiae* of Barberini was taking shape.[35] In the same way, it appears difficult to believe that the admirable allegory of the comparison of the two chief philosophies of the past produced by Raphael did not belong to Galileo's cultural background and iconographic memory. A similar eventuality would in fact be difficult to reconcile with that page of Viviani's *Racconto Istorico* in which he reports that Galileo used to say that if in his youth he had been able to choose a profession he would have chosen painting.[36] On the other hand, an important reference to Raphael is also found in the *Dialogue*, when in the final lines of the First Day, on bringing the conversation on the "acuteness of human ingeniousness" and man's capacity for understanding *intensively* (Galilei [1997] 113), the painter from Urbino, together with Michelangelo and Titian, is cited as an example of the highest expression of art in "representing all visible objects" (Galilei [1890–1909] VII 130). Moreover, when we turn our attention to theoretical background factors, we must consider how deeply, starting from the middle of the fifteenth century, the tensions connected with the confrontation between Plato and Aristotle on the one hand, and between Mannerism and Classicism on the other, had marked Florentine philosophical and artistic culture, penetrating into its most intimate fibres, and how greatly this confrontation forged the formation and development of Galileo's thought.[37]

For the same reasons, no less motivated would be the hypothesis of Stefano Della Bella's lack of knowledge of Raphael's work. Della Bella, who grew up in a family of artists, was an painter and engraver who had been an apprentice in the school of Cesare Dandini (Vesme [1906] 68f.; Forlani Tempesti [1983] 6f.; Negri Arnoldi et al. [1986] 129, 162f.). He was an artist with a fervid imagination who between the 1630s and 1660s became well known as a master engraver of celebrative and commemorative works and who was greatly appreciated in Europe.[38]

As is known, in the course of the sixteenth and seventeenth centuries, print-making in Italy and the rest of the continent went through a period of extraordinary development. And along the lines of what took place on a larger scale for the book, prints from etched plates, thanks to their natural reproducibility, became an instrument for the dissemination and divulgation of figurative subjects which otherwise would not have reached the general public (Ivins [1953] 21–50). It is to be kept in mind that in this work of diffusion and divulgation of the image, an important part of the work of master print-makers was devoted to the reproduction of the works of the great masters. These printed reproductions were those of the most famous works and those that were more immediately appreciated by the authorities and the public. And from the free play of this mixture of intellectual and mercantile elements, the *corpus* of the *"a fresco"* paintings of the great Raphael were for the artists of the burin one of the most exploited and fecund sources of inspiration. One of the aspects of most interest to us in Raphael's popularity among sixteenth- and seventeenth-century engravers is represented by the considerable reproduction and divulgation of the scenes painted in the loggias and rooms of the Vatican.[39] This was a phenomenon that remained fairly constant and enduring and reached its apex in the Baroque Age, when veneration for Raphael ([Raphael] [1984]) induced aristocrats and kings to order prints of his works, especially those with subjects and allegories which in some way had reached the level of reassuring emblems of power in its sumptuous and solemn official guise, a stylistic cipher that especially distinguished the artist's Vatican frescoes.[40] It is intuitive that in the operation of transposition from the painting to the print many of the motifs that had inspired the artist were lost, together with the function the work was destined for. However, for the very reason of the objective impossibility of remaining faithful to the original, the finest graphic transpositions are those which at a glance, as in a didactic projection of the artist's thought, succeed in bringing into focus the main points of the composition and the situational components enclosed within the pictorial archetype.

As concerns the amalgamation of the iconographic and hierarchical reference codes, we can certainly say that Della Bella's *"antiporta"* of Galileo's *Dialogue* was in fairly strong agreement with the canons of the artistic genre to which it belongs, but most of all it showed itself to be perfectly suited to the new purpose for which it had been conceived. In some way, this suitability is confirmed by the meticulous scrupulousness with which Nicolò Riccardi, Master of the Sacred Palace, in light of the authorisation to publish in Rome, supervised the revision and "adaptation" of the *Dialogue*. Riccardi, "*the Father mostro*," was also careful to have the Pope's approval of each of the elements of the book, even discussing with him the title and the "title-page of the work" (Galilei [1890–1909] XIV 120, XIX 327). For this reason, with the knowledge we have at our disposal, we cannot even exclude the fact that the idea of the allegorical scene in the frontispiece may have come to light during one of those meetings and in a place which, if not the same room, at least one that was not too far away from the one containing Raphael's masterpiece.[41]

Without going into any further speculation on the transposition of ideas from the great fresco to the more modest engraved plate, what the visible citation of Raphael's *School of Athens* contained on the engraved page on which the *Dialogue* is announced suggested to those of Galileo's numerous contemporaries capable of understanding it fully was the intimate symmetry and equivalency of the problematic nature of the philosophy intercurrent between the ancient dispute illustrated in the fresco in the Stanza della Segnatura, and the controversy between the two chief world systems.

On the basis of the correlations that have been pointed out, we can thus affirm that the "*antiporta*" is an integral part of Galileo's work and not simply an elegant ornament. From a certain viewpoint, the frontispiece in Galileo's *Dialogue* can actually be seen as a true illustrated synopsis of the theses and problems to be discussed during the four days. From another standpoint, the "scene" that precedes the real title-page appears in allegorical form as the philosophical manifesto of Galileo's science. A manifesto in which the remembrances of Raphael, which are especially evident in the classical taste of the sumptuous clothing worn by the characters, have the ring of an act of deference and supplication addressed to the Church of Rome for a philosophical choice, if not actually in favour, at least not hostile, to the Copernican hypothesis.

The important function of promoting Copernicanism that had evidently been assigned to the print appearing at the beginning of the *Dialogue* was clearly understood by contemporaries of the Copernican school, who exploited its immediate communicative force even later in the battle for heliocentrism by reproposing it in some of the editions of the *Dialogue* published during the seventeenth century. The same allegory is to be found in the first Latin edition of the *Dialogue* edited by Elia Diodati[42] and Matthias Bernegger which, with the financial support of the Elzeviris was issued in Strasbourg in the spring of 1635 (Galilei [1635]). The same illustration is also found in the subsequent Latin edition that appeared in Lyon in 1641 (idem [1641]), and in the one published in London in 1663, which is practically a reprint of the Strasbourg edition (idem [1663]). One important aspect to underscore is that all of these publishing initiatives taken as a whole, which were instrumental in ensuring diffusion of the *Dialogue* throughout Europe (Drake [1967]; Westman [1984]; Beaulieu [1984]), were conceived and implemented in Protestant circles and on the initiative of persons who through this *also* intended to spread anti-Roman propaganda.[43]

On examining it carefully, the frontispiece of the Latin editions mentioned above were not, however, totally faithful reproductions of the engraving that "decorated" the first Florentine edition in 1632, but was rather a reworking, not without intelligence, of the original. In this new version, although the original *ratio* of the scene and its visual impact are not impaired, some details had been drastically changed (see fig. 5). These changes clarify the intention of better calibrating and updating the message that the images were to convey in light of the different function assigned to the Latin edition of the *Dialogue*: the consequence of the new stage in the struggle for Copernicanism caused by Galileo's condemnation and abjuration.

Entrusted to the steady hand of printmaker Jacob van der Heyden from Strasbourg,[44] the revision of the scene of the astronomical dialogue is quite effective in delivering its message. In some respects, the new images, thanks to the greater sharpness of the contrasts and accentuation of the halftones, are more legible than the Italian specimen. By means of fairly simple touches and additions, the Galilean elements present in the 1632 prototype were not only conserved in the new composition, but even strengthened, thus amplifying the comparison and the opposition to Aristotle's physics.

Having removed the emblem of the Florentine printer, which was originally at the feet of Ptolemy and Copernicus, the space thus made available was occupied by a stone which appears in contraposition to the arrow symbolising Aristotle's theory of motion. This was clearly a reference, now made even more explicit, to the laws of motion discovered by Galileo and to the principle of the relativity of motion which, especially on the Second Day of the *Dialogue*, are discussed and illustrated through various experiments.

In the new print the Vatican and Renaissance atmospheres which, as we have seen, characterised Della Bella's work, were completely eliminated. The three characters conversing, although they maintain the same postures and relative positions, no longer have the inspired and stereotyped physiognomy that they have in Raphael's portrayal of the ancient sages.[45] Each person has a countenance of his own. In particular, Copernicus is freed of the head-dress that had in the meantime become characteristic of the Catholic clergy and reacquires the lineaments of the face in the portraits that circulated at the time and which have been passed down to us.[46] From these same images come the details of dress, which are redesigned with more austere and *Central-European* taste which act as counterpoint to the voluminous tunics of vaguely classical style that cover the Greek philosopher and the Alexandrine astronomer.

The printed words on the inside of the drapery above the scene deserve special mention. In this 1635 print, the body of the letters making up the word *Dialogus* is sharply reduced in size in comparison to the far greater emphasis given to the key words of the title *De Systemate Mundi*. A title[47] that is deliberately assertive, not problematic nor purposefully toned down, as had been imposed by the Pope through the Master of the Sacred Palace in the first Italian edition (Galilei [1890–1909] XIX 327). Beside the name of Galileo the appellative *Linceo*, which appeared in the Italian edition of the *Dialogue* was maintained.[48] A title that signified little in a European context but which, in the light of the adverse fortune that tormented the Academy and some of its members both before and after the Galileo *affaire*, was placed there not only as the well-deserved recognition of the support that the prestigious Roman association founded by Federico Cesi had offered in the battle for the Copernican system, but I should say, above all, as a renewed challenge to the Jesuits launched by those "Galileans" who had raised *The Assayer* to the level of manifesto and symbol of their cultural battle.

The *Dialogue on the Two Chief World Systems*, which on the initiative of Galileo's friends on the other side of the Alps had become *Systema Cosmicum*, began circulating among the intellectuals of Europe. The first thing the readers came in contact with was a redesigned and corrected version of the print that had appeared in the Diodati edition

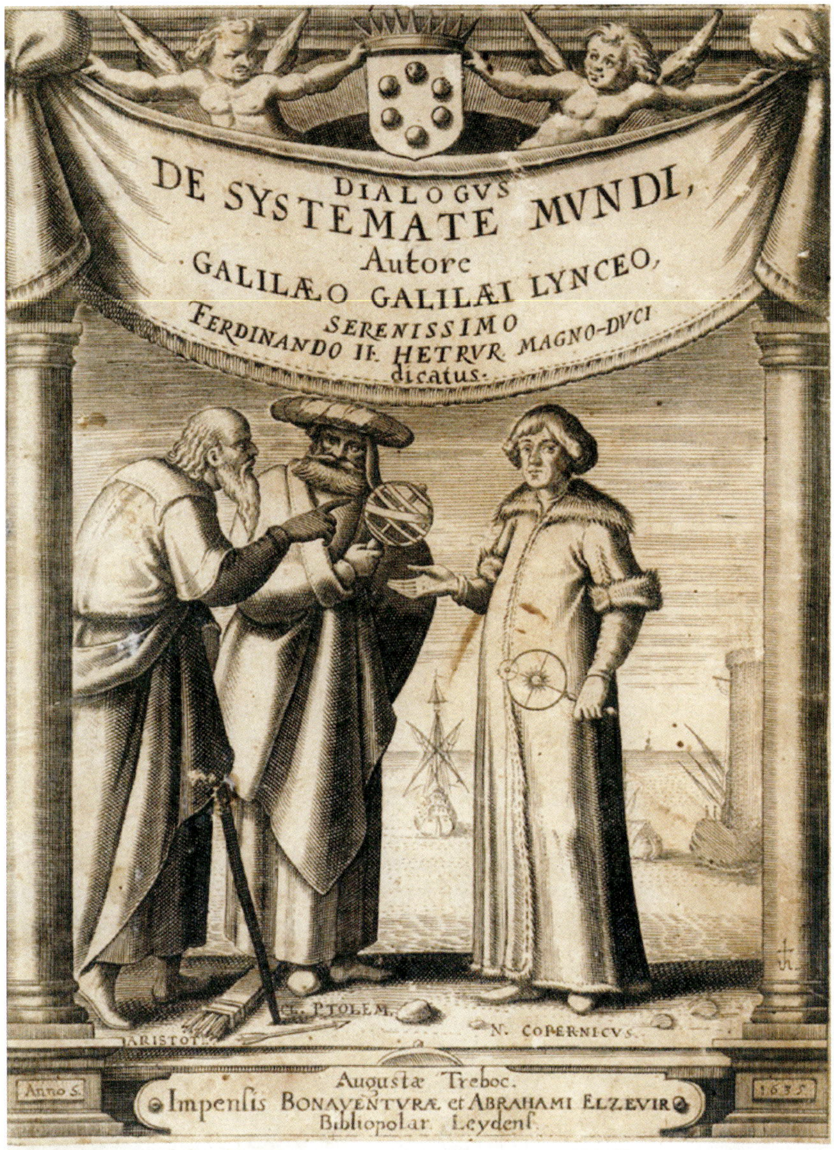

Figure 5
"Antiporta" of Galileo: *Systema Cosmicum* (Strasbourg 1635). Courtesy of the Biblioteca Universitaria di Cagliari.

of Galileo's masterpiece. The fact that the same engraving accompanied the three Latin editions of the *Dialogue* that appeared in the seventeenth century leads us to believe that the content of the engraving, although it had changed at the hand of the artist who etched it and in its reference points, was a confirmation of its role as the vehicle of propagation of Copernicus' ideas. In this role, the portraits of Aristotle, Ptolemy and Copernicus, in the last and decisive challenge, become the icon of the *Dialogue* and the manifesto of heliocentrism, with both represented symbolically. When we then consider these elements of content together with the vast echo and the great impression made by Galileo's condemnation and the inclusion of the *Chief Systems* in the *Index Romanum Librorum prohibitorum*, one can see how the *Systema Cosmicum*, and consequently its iconographic synopsis, could so quickly become an effective way of exposing the Galileo *affaire* and keeping alive the Protestant campaign against the Catholic Church (Eisenstein [1979] II 649, 666). Thus it was that thanks to the multiplicity of meanings and the extra-scientific designs behind the Latin edition of the *Dialogue*, especially in Protestant Europe, curiosity concerning Galileo's work was at a high pitch, and this interest evidently made for the rapid passing from hand to hand of the volume. This explains why, not lastly thanks to this component, the image that represents its synthesis, for the very reason that it is placed on the opening page, gained further notoriety and became fixed in the memory even of those who had held the Dialogue in their hands but had read it only in part or perhaps not at all.

Beyond the 'Dialogue,' Imagining Other Worlds

In England, more than in any country on the continent, Copernicanism had found a favourable intellectual climate and had immediately found supporters, in particular among Elizabethan physiologists and later among the Puritan virtuosi. The *Nuncius Sidereus* with its discoveries and astronomical observations which had awakened fantasy and imagination, had made Galileo quite well known (Galilei [1890–1909] III-1 188–190; Nicholson [1956] ch. I and II; Boas Hall [1967]; Drake [1967]; Feingold [1984]). Later, interest in the *Systema Cosmicum* was no less widespread and in just thirty years two editions were published, one in Latin and the other in English (Salusbury et al. [1661]).

In this particular environment, which was undoubtedly less hostile to Copernicanism than elsewhere (Westman [1986]), just three years after the first edition of the *Dialogue*, a book by an anonymous author appeared in 1638 under the ingenuous and engaging title: *The Discovery of a World in the moon, or a Discourse tending to Prove that 'tis probable there may be another habitable World in that Planet.*[49] The real author was John Wilkins, a young Puritan clergyman just twenty-four years of age who was destined shortly to become one of the major animators on the English intellectual scene (Shapiro [1969]; Stimson [1931]). While still a student at Magdalen Hall in Oxford, under the influence of the *Lectures* delivered by Henry Briggs, Wilkins accepted the Copernican hypothesis, and was particularly taken with the question of the plurality of worlds and the moon's being inhabitable.

As is known, the idea that there may be worlds other than ours is one that goes far back in time. In modern times it had been given new life by Nicholas of Cusa, but most of all by Giordano Bruno during his sojourn in England, where the sensation raised remained in the memory for quite some time (McMullin [1986]; Massa [1977]; Ciliberto et al. [1997]). An ample debate on the subject began only within the more general framework of the dispute on the world system and following publication of Galileo's *Nuncius Sidereus* and later that of Kepler's *Somnium sive de astronomia lunari* (McColley [1936b]).

In particular, the thesis that the moon might or might not be inhabited, even at some future time, by living beings similar to humans, had its supporters and critics. Ben Jonson, for example, considered it a fool's fantasy and used his satire in dealing with the ever-growing amount of news coming from the moon,[50] while Francis Godwin, almost at the same time as the publication of Wilkin's *Discovery*, tried his hand at a literary genre that was to become popular in the decades and centuries to come, with the publication of a novel under the rather direct title: *The man in the moone: or a discourse of a voyager thither by Domingo Gonsales. The Speedy Messenger* (Godwin [1638]). Although in paraphrastic form, the reference to Galileo's Sidereus *Nuncius* is perfectly clear. It is to be noted, however, that Ben Jonson's masque and Godwin's work of fiction, taking into account the linguistic register, the genre and the authors themselves, were addressed to a diversified public, but one that was outside the circle of specialists in astronomy, mathematics and the geography of the heavens. In contrast, from this viewpoint Wilkins' *Discovery,* thanks to a wise blend of physical, astronomical and exegetic arguments, evidently wished to interest both the specialist and the layman. In any case, the historiographically important aspect is the question of whether the moon is inhabitable or not presupposes a world structure along Copernican lines, even though the hypothesis of a Selenite population or lunar colonisers appeared by no means to be a necessary consequence of the heliocentric hypothesis.[51]

The success of the young Puritan clergyman's book was immediate. The copies of the first edition were sold out in a few months so that a reprint was called for in a short time and met with the same reception as had the first edition (McColley [1936a]). Together with success, and for that very reason, came the inevitable attacks, especially from conservative astronomers and traditionalist theologians (McColley [1938]; Dillenberger [1960] 104–112).

To satisfy the unmet demand for the volume and at the same time to counter the severe criticism that had been levelled against the new astronomical hypotheses as being in disagreement with the sacred texts, Wilkins published a second edition of the work in 1640. Once again London, a "market" that was always thirsty for and open to the new, was chosen for publication along with a printer, John Norton, who enjoyed greater commercial prestige than the publisher of the first edition, and whose sales outlets were in Fleet Street at the heart of the city (McKerrow [1968]; Plomer [1907]).

Together with the new edition of the *Discovery,* which contained one more chapter than the first, for the first time Wilkins published another voluminous work with a slightly misleading title: *A discourse concerning a New Planet tending to prove, that 'tis probable our Earth is one of the Planets.* This was in reality a full defence of the Copernican system, conducted for the most part on the basis of scriptural exegetics oriented in an historicistic direction, as concerned topics that were not a question of faith. The *Discovery* and the *Discourse* are thus not treatises on astronomy in the strict sense. They did not contain scientific novelties, neither theoretical nor observational. Their purpose was, as Wilkins himself declared, to free Copernican astronomy from the prejudices that surrounded it (Wilkins [1970] I 254). Thus they were works of propaganda and proselytism which, when limited to this aspect, are related to similar attempts such as the *Lettera sopra l'opinione de' Pittagorici e del Copernico* by Paolo Antonio Foscarini, the *Apologia pro Galileo* by Tommaso Campanella (McColley [1939]), the *Lettere copernicane* by Galileo, but, above all, as we shall see shortly, to some of the communicative strategies inaugurated by Galileo with *Sidereus Nuncius* and the *Dialogue.*

The importance of Wilkins' two works in the dissemination of Copernicanism in the British Isles and in continental Europe has often been emphasised.[52] However, this fitting recognition was not always followed by an adequate evaluation of the subsidiary instruments, especially the iconographic ones, which Wilkins in his struggle in favour of Copernicanism placed side by side, and sometimes superimposed, on traditional rhetorical and argumentative means of a linguistic nature.

From this viewpoint, the first thing that deserves mention is that Wilkins adopts Galileo's iconography as presented in *Sidereus* and the *Dialogue* almost *tout court.* In Wilkins' two treatises there is a reproduction of the images of the moon seen through the telescope and numerous diagrams and drawings which in the *Chief Systems* give intuitive and visual form to some of the passages and mathematical and mechanical explanations. But what makes Wilkins' work even more singular and far more effective on the communicative plane is the frontispiece under which the two treatises are grouped in the 1640 cumulative edition and which, like a modern blurb, was placed on the bookseller's tables to attract the attention of the prospective buyer. The peculiarity of the first page of the volume, a mixture between the title-page and the real frontispiece, is not only in its semantic richness, which is evident even to a fleeting glance, but perhaps above all in two separate and complementary aspects together. Within the same frame, on the one hand Wilkins attempts to create a kind of synopsis, an *abstract*, in images illustrating his conceptions – in some cases going beyond the contents included in the verbal part of the book –, and on the other hand he succeeds in enclosing within the limited space of a single sheet an extraordinary number of symbols, models, formulas, suggestions and allegories, many of which had previously appeared in some of the most important works on Copernican astronomy.

Here we can neither point out nor discuss the many references included in the engraving by William Marshall[53] under the evident dictation of Wilkins. On the other hand, we cannot expect to catch all the messages that were clear in a cultural environment so far removed from ours. We shall therefore limit ourselves to naming and commenting on some of the elements the interpretation of which is less uncertain. The complex print, which is divided into four frames, one of which containing information that with a certain approximation we may consider of the commercial kind, is structured around a scene based on three figures which, from right to left, represent Copernicus, Galileo and Kepler (see fig. 6). One receives a strong impression of resemblance between this frame and the frontispiece of Galileo's *Systema Cosmicum* in the 1640 Strasbourg edition. On closer perusal, however, we realise that it is an effect of the Gestalt type, since in detail the two engravings present innumerable differences. Despite this, there is no doubt that the William Marshall print is a reinterpretation of the model represented in the Latin edition of the *Dialogue*: the figure of Copernicus in the *New World* is indeed practically identical to that of the Strasbourg edition of Galileo's work. If we observe them specularly, we can see that the two figures correspond almost point by point. The only difference that deserves consideration, except for the position of the left arm of the Polish astronomer, is the quality of the two works, with the English one arriving a poor second. The undoubted iconographic derivation from the Galilean volume highlights some characteristics of the instruments of diffusion and Copernican propaganda adopted in the new phase which had its beginnings in England.

First of all, on the general plane it is clear that Wilkins explicitly intended his work to be a continuation of the battle against the pigheadedness of conservative theologians which had been started by Galileo.[54] Secondly, on the strictly iconographic plane, it is just as clear that the renewed utilisation of the paradigm of the three personages goes in the direction of an exploitation of the immediate capacity of an image to bring concepts into focus, which had become consolidated in the European intellectual memory. This was an operation that took advantage of the same register of associations that had been used to such good effect by Galileo and Della Bella with Raphael's *School of Athens*.

However, Wilkins had not passively taken an illustration that had to some extent become a classic. Of the original image, as had occurred with the Strasbourg edition of the *Dialogue on the Two Chief World Systems*, a rearrangement had been made so as to make it respond more closely to the new public to which it was addressed and to the specific intellectual reasons which had prompted the new work, among which there are even some cultural designs deriving from Bacon, with whom Wilkins, as was true of most Puritan intellectuals, was strongly sympathetic (Jones [1961]; Hill [1965]). To tell the truth, compared to the previous manipulation, Marshall and Wilkins' tampering is from several viewpoints an actual alteration of many of the meanings contained in the model of reference. To say it in the manner of Thomas Kuhn, we could speak of a true paradigmatic rearrangement. In fact, in the Latin edition published on Diodati's initiative, the corrections had touched on the elements that gave the figures Roman stylistic overtones: in the case of Wilkins, the changes were made to the personages themselves, two of whom were not present in previous representations. Indeed, Galileo and Kepler take the place that Aristotle and Ptolemy had occupied in the first versions of the same scene. The first indication that this change of characters appears to supply goes in the direction of placing Wilkins' *Discourse* beyond the phase dealing prevalently with the dispute between the Ptolemaic and Copernican models, thus preparing the potential reader to come to grips with the crucial question of the consequences that the heliocentric hypothesis were to have on the factual plane.

To make the message of the engraving less ambiguous and at the same time strengthen its didactic function, Wilkins and Marshall resorted to an expedient, by no means new, but certainly not usual in scientific or philosophical texts. The personages in the plate preceding the *New World* speak by means of ribbons with wide volutes floating above their heads and on which we see, just as in modern cartoons, some very short phrases, all rigorously limited to three words.

Copernicus, the "timid canon," questions himself, while pointing to the diagram of the world system that follows the model bearing his name, on the consequences to reality that may come from experimental confirmation of his hypothesis ("*Quid si sic?*"). The question on the epistemological nature of the Copernican hypothesis was, even for Wilkins, evidently quite far from being answered.[55]

Among the many theoretical unknowns consequent to acceptance of the heliocentric scheme as the real world system and the inevitable down-grading of the Earth to the role of planet, the question of the possible existence of other inhabited or inhabitable worlds was certainly the one with the greatest impact and fascination. In the past, in consideration of the state of astronomical knowledge, the affirmative hypothesis had no tangible evidence worthy of the name to support it, and had remained at the stage of poetic and speculative fantasy, perhaps not lastly because the consequences from the theological standpoint were unprecedented. Only with Galileo's telescopic observations and the publication of *Sidereus* did the extravagant idea receive new lymph, becoming overnight a possibility worth taking into consideration. Kepler in particular, referring to the new empirical evidence, had solicited new philosophical discussions on this point (Kepler [1993] 19f.).

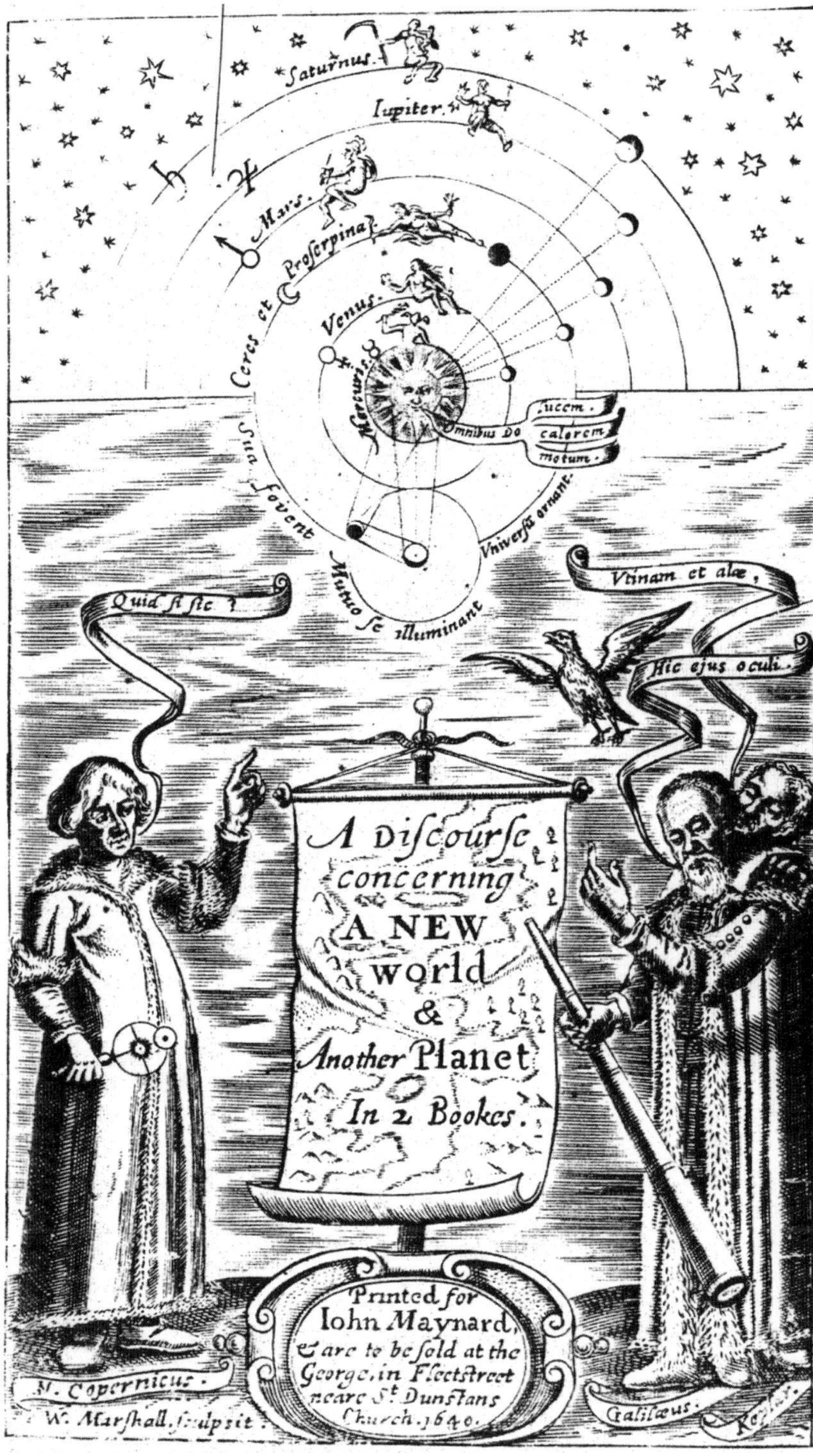

Figure 6
Frontispiece to John Wilkins: *A New World* (London 1640). Courtesy of the British Library, London.

This particular development of the astronomical debate that began with Galileo's observations was underscored by Wilkins, who thus widened the gap between Copernicus and Galileo, placing between them a large, completely unwound scroll. The context and the subsequent content of the treatises hint at the fact that it is a map of the lunar surface, with mountains, valleys, watercourses and vegetation. This is an image that is no different from similar topographical representations of our planet: an intentionally idyllic representation capable of bringing to mind the descriptions of the *new world* being brought back across the Atlantic by pioneers and pilgrims rather than those of the ashen visions and Galilean drawings of the moon. On the same sheepskin we see the title of the work in which the word *new* preceding the word *world* stands out.

On the right-hand side of the scene, wrapped in a rich overcoat similar in fashion to the one worn by Copernicus, we see Galileo, whose lineaments closely resemble those of the fine engraving portraying him in *The Assayer.* Along the lines of a stereotype diffused throughout Europe that considered the Italian astronomer the Christopher Columbus of the heavens,[56] or still more, according to the expression of Wilkins himself, the eyewitness of renewed lunar geography (Wilkins [1970] I 68), Galileo has a telescope in his right hand. With the left hand he is pointing to an ungainly bird, perhaps a bird of prey, flying above him and to which Galileo pronounces the rather sibylline phrase: *"hic eius oculi."* The most immediate sense of the words placed in Galileo's mouth would appear to refer to the human eye's extraordinary gain in visual perspicacity with the use of the instrument, so much so as to rival the proverbial sight of the king of birds.

On closer perusal, however, the overall figure of the Italian scientist and the rest of the scene suggest a more complex and polysemic reading of the metaphor. The position of Galileo's head, bent on his chest, is a singular one. His eyes are closed. His expression is one of submission, of absence. In 1640, England, and Wilkins in particular, were aware of the Pisan's judiciary misadventures caused by his Copernican ideas.[57] It is also probable that the author of the *New World* was informed of the blindness that had struck the elderly scientist, who was forced to live segregated in Arcetri. Between the spring of 1638 and the summer of 1639, John Milton had travelled on the continent and had also been to Italy, where he had visited Galileo in his exile in the hills to the south of Florence (Friedman [1991] 159–174). The impressions of that meeting were later immortalised in the pages of the *Areopagitica* which was published in 1644,[58] but it is certain that the news of Galileo's state had circulated before that within the Puritan intellectual circles that had been most active in this convulsive phase of English history. Indeed, Milton shared the ideals of the supporters of Parliament and was on their side in the struggle for the freedom of expression, in which he participated with great fervour immediately following his return home.

But beyond the elements of Galileo's biography that may have found a place in the representation and attitude in which the Italian scientist is portrayed, even more important to recall is that Galileo nurtured deep reservations concerning the existence of living beings on the moon or in any other place in the cosmos.[59] Thus the head bowed towards the ground, together with the emphasis on the visual nature of Galileo's astronomical experience contained in the words he pronounces, may *also* be the allegorical representation of Galileo's insistent refusal to consider hypotheses that real experiences, owing to their limited gnosiological purport, are not capable of corroborating.

It must not appear singular that Wilkins felt the necessity to document Galileo's methodological position. In the plate preceding the *New World*, even the words pronounced by Copernicus centre on the epistemological problem, as discussed previously. In a certain sense, it would appear that Wilkins' intention was to place the two scientists on two methodologically converging lines. Thus, as Copernicus speculated on the possibility of extrapolating from mathematical astronomy conclusions having validity on the plane of the description of reality, in the same way Galileo was rather sceptical concerning the possibility of taking from observational astronomy empirical elements in favour of the philosophical assumption of another inhabited world on the moon.

The theme of the plurality and inhabitability of worlds as the central theme of the two astronomical treatises tells us that despite the prudence that kept him from accepting the thesis in favour,[60] young Wilkins considered Galileo's scepticism philosophically sterile. Only in this light do the position occupied by Kepler in the scene and the words he pronounces, which otherwise would be enigmatic, acquire full meaning.

I have elsewhere suggested that the singular position of Kepler standing behind Galileo may, within the context of the metaphor of dwarfs and giants so in vogue in English culture of the times, allude to philosophical preeminence of the German scientist (Merton [1985]; Jones [1961] *passim*). I believe that same interpretation lends congruousness and meaning to that *"utinam et alae"* (literally: "perhaps even the wings") that Kepler whispers into the ear of an evidently demoralised Galileo.

In the *Dissertatio cum Nuncio Sidereo*, Kepler, as we mentioned before, had without result asked Galileo the question *si sunt in Luna viventes creaturae* (Kepler [1993] 19). The theme was indeed at the centre of the philosophical interest of the imperial mathematician and represents the fulcrum of the *Sommium sive astronomia lunaris*, a work that describes an imaginary journey to the moon during which the protagonist can observe first-hand the physical geography and the nature of the creatures that inhabit our satellite.[61]

The *Somnium*, together with the parts of Kepler's major works with many speculative digressions, is often mentioned in the course of the *Discovery* and the *Discourse* and constitutes one of the most authoritative and constant references on which Wilkins' theory of the inhabitability of the moon was based.[62] The noteworthy influence of Kepler's *Somnium* on Wilkins' treatises thus conveys the sense of a further removal from the ideas of the Pisan mathematician in Kepler's exhortation to emulate the bird, not only in its acute vision, but likewise in its capacity of flying higher towards the moon.[63] But at the same time, this recommendation represents guidelines for a programme of research in which natural philosophy does not recede in the name of the neutrality of telescopic observations or mathematical astronomy, nor does it relinquish its vocation for formulating cosmological hypotheses.

The important programmatic indication formulated in the mouth of Kepler leads us directly to the upper part of the engraving in which we find the representation of the state of the cosmos arranged according to Copernicus' order. Here too we find many symbolic and didactic elements that unveil the cosmological directrices that are hinted at in the lower part. Also here it is possible to distinguish some Galilean elements which are, however, englobed in a cosmological device of clearly Keplerian design to which some components belonging to the Puritan world vision are not extraneous.

The fulcrum of the representation is in the sun–earth–moon subsystem on which the eye converges so as to render intelligible the phenomenon of the phases of the moon and the other planets, both in relation to their positions with respect to the sun and with reference to the viewpoint of the observer on the Earth's surface. The schematisation and the visual result are no mean achievement, but are not enough to illustrate the complexity and the nature of the phenomenon it was intended to represent. For this reason, Wilkins was once again forced to resort to a series of captions. For example, the words *mutuo se illuminante* that arches below the lunar orbit evidently refers to the phenomenon of the "*secundarius fulgor*" or "ashen light", a phenomenon to which Galileo devotes special attention in the *Sidereus*.[64]

The entire world system is thus seen as enclosed between a pair of propositions of a metaphysical order. On the left we find the principle of the self-sufficiency of the laws of nature, "*sua fovent*," which in relation to the orbital motion of the planets highlights their mechanical autonomy and their independence from any extrinsic cause, including the action of spheres, intelligences or angelic powers (Wilkins [1970] I 243f.). On the right, with the formula "*universum ornant*," the total absence of finality connected with God's creative action is reaffirmed. Together with this is underscored the intrinsic and unconditional magnification of the Divine Maker which spontaneously derives from the harmonious theory of the heavenly bodies that embellish the universe.

In this complex cosmological and theological vision, which is shaded into different forms in the upper part of the plate, the sun, the pivot of the Pythagorean and later of the Copernican system, dominates at the centre of the world from where, according to a vision not without emanatory elements, irradiates around it a "magnetic virtue" that vivifies every body and planet. In this vision the Keplerian origins of the tripartite enunciation "*omnibus do lucem, calorem, motum*," pronounced by the sun are evident (Wilkins [1970] I 244). This is a phrase that represents the development in a cosmological key of the formula "*a me omnes*" of a more eschatological flavour, pronounced previously by the sun in the diagram that represented the cosmic system in the 1638 edition of the *Discovery*. By means of this, Wilkins evidently intended to call attention to some themes emphasised by a certain Puritan reformist current which thought it had recognised in the heliocentric system a living allegory of the power and rule of the Judaic-Christian God.[65]

The attribution to the sun of powers and a nature so exclusive and excellent undoubtedly produces further discontinuity with the rigorously rationalistic model of knowledge postulated by Galileo in the *Dialogue on the Two Chief World Systems* and displayed in the plate preceding the title-page with which Galileo's work was presented. In any case, it was not a traumatic separation but a dialectic one which took place through the assimilation and complete exploitation of the communicative and evocative power of the image which, in a few short years, had become the icon and banner of the Copernicans throughout the European continent and in which the philosophical confrontation between the two astronomical theories was seen to have reached its final act.

Few texts favourable to Copernicanism were as influential and as widely read by the great educated public as Wilkins' two treatises on the *New World*. In the seventeenth century there were four editions, the last of which in 1684, in England alone. Translated into French in 1655,[66] Wilkins' two works were widely circulated in France where, among other things, they represented an important reference point and direct source of inspiration for the science fiction literature of that country (Cyrano de Bergerac [1657]; Fontenelle [1686]). Later, in 1713, there was an edition in German (Wilkins [1713]) of both Wilkins' treatises. The young Puritan clergyman had evidently found the right formula for rendering Copernican astronomy persuasive and captivating, thus making a formidable contribution to the knowledge and affirmation of this world system. It was an action of divulgation and penetration of heliocentrism that met with great success, the effectiveness of which can be measured, for example, by the fact that starting from about the second half of the seventeenth century even the compilers of popular almanacs had in general adopted the Copernican world system (Nicolson [1939] 14ff.). Thus, in 1687, when the first edition of Newton's *Principia* appeared, thanks to the work of propaganda and apologetic accommodation that Wilkins and many others of his generation had conducted for so many years, religious opposition to the work, which decreed the irremediable mathematical and philosophical defeat of the Aristotelian and Ptolemaic system, was quite contained, so much so that in the years immediately following, the new philosophy of the universe became the most powerful instrument in the confutation of scepticism and atheism in the sermons of moderate theologians (Jacob [1990] ch. IV and V).

Università di Cagliari

Notes

1 The many aspects connected with the *Sidereus* were carefully reconstructed by Isabel Pantin in the introduction to Galilei [1992] i–lxxxviii.

2 Italo Calvino wrote that as soon as Galileo "si mette a parlare della luna innalza la sua prosa a un grado di precisione ed evidenza ed insieme di rarefazione lirica prodigiose" (Calvino [1980] 183 and 186f. See also Calvino [1985]).

3 Winkler et al. [1992]. An exclusiveness that the *Sidereus* was to maintain for a long time, up to the *Novae coelestium terrestriumque rerum observationes* (1646) by Fontana and the *Selenographia sive lunae descriptio* (1647) by Hevelius.

4 Concerning the drawings in the *Sidereus* see Tabarroni [1984].

5 On the revolution brought about by Galileo in the representation of objects coming from astronomical observations with the telescope, see Kemp [1992] 93f. On the new "*mentalità visiva*" inaugurated by Galileo, Ezio Raimondi made some interesting observations in his essay *La nuova scienza e la visione degli oggetti*, (Raimondi [1969]); now also in idem [1974] 3–56. More in general on the growing need for illustrations that was emerging from the development of the scientific revolution, the reader is referred to the pages by Paolo Rossi contained in Rossi [1998] I 112ff.; Edgerton [1985].

6 In general terms this point is discussed by Hackmann [1998] 171–173. For an analysis of the many interactions connected with the recording of our visual information of the world on a flat piece of paper, cf. Tufte [1990] ch. I, and Kemp [1992] ch. 2.

7 This aspect was discussed by Righini [1975] 59–76, and Gingerich [1975] 77–88. See also Drake [1978] 144ff.

8 A. Battistini [1989] 38ff.; Nicolson [1962] ch. I and II. According to a recent study by Reeves [1997], the illustrations of the *Sidereus* also led some painters to replace traditional astronomical representations of religious iconography with images of the kind proposed by Galileo.

9 "[...] Voglio disegnare le faccie della Luna di un periodo intero con grandissima diligenza, et imitarle a capello, perché invero è una vista di grandissima meraviglia" (Galilei [1890–1909] X 300).

10 This is what we find in the *Sidereus* in the dedication to Cosimo dei Medici, but also with an eye turned towards the reader: "Eiusmodi est enim humanae mentis conditio, ut nisi assiduis rerum simulacris in eam extrinsecus irrumpertibus pulsetur, omnis ex illa recordatio facile effluat" (ibid., III/1 55. See also X 283–285).

11 "[...] With many and beautiful figures [...] which are wondrous to behold" (ibid., X 299).

12 From this viewpoint, the *Istoria* is in reality a transitional work.

13 On Urban VIII's policy of promoting the arts and letters, see Haskell [1980] part I.

[14] The operation began with *The Assayer* still in the form of a rough draft. On learning of the election of the new pope, Cesarini quickly added a dedication to Urban VIII, whose *"benignissima protezione"* he invoked. Galilei [1890–1909] VI 201.

[15] Drake [1980] 9ff., [1982]; Fantoli [1994] 164, 297. On the Pope's intention to mitigate the effects of the anti-Copernican edict, cf. Galilei [1890–1909] XIII 182 and XIV 87f.

[16] The *Lettera a Francesco Ingoli* contains the principles that had inspired this programme, which was also full of national pride (Galilei [1890–1909] VI 511). Cf. also the preface to the *Dialogue* Al Discreto Lettore (ibid., VII 29 and Galilei [1998] 2/II, Commento, 119–122).

[17] On the structures and rules of the so-called *"system of patronage"* in the case of the publication of the *Dialogue* cf. Westfall [1989] 69–73 (partly already in idem [1987]); Biagioli [1993] 248ff.

[18] The reasons are well summed up in the words of Viginio Cesarini: "[...] le dico che sicuramente vogliamo pubblicar l'opera, e che vogliamo ciò fare a Roma, non ostante la potenza degli avversari, contro i quali ci armeremo dello scudo della verità, ed anco de favori de padroni" (Galilei [1890–1909] XIII 106).

[19] Galileo's intentions were expounded to Prince Cesi in the following words: "Io raggiro per nella mente cose di qualche momento per la repubblica litteraria, le quali se non si effettuano in questa mirabil congiuntura, non occorre, almeno per quel che si aspetta per la parte mia, sperar d'incontrarne mai più una simile" (ibid., XIII 135).

[20] Redondi [1987] sheds light on numerous intellectual and political elements, as well as the context and customs which directly or indirectly conspired towards this end.

[21] On the changes requested and the entire reorganisation of the *Dialogue*, the reader is referred to Altieri Biagi in Galilei [1993], especially 901–916.

[22] Winkler et al. [1992] link the lack of illustrations to Galileo's changed social status and to the more general nature of the topics discussed in the *Dialogue*, which were by their nature and by tradition more suited for a verbal and mathematical description.

[23] The content and rhetorical structure of the *Dialogue* are analysed in depth by Finocchiaro [1980].

[24] "La cerimoniale introduzione e le attaccature dei principii de' dialoghi con le materie seguenti, che son cose più tosto oratorie o poetiche che scientifiche," Galileo confided to his friends and correspondents. Galilei [1890–1909] XIV 60. Cf. Battistini [1989] 114ff.; Vickers [1983]; Altieri Biagi [1983].

[25] "[...] Regarding the present subject – we read in the *Dialogo* – let us be satisfied with what you may have been able to understand with imagination so far." Galilei [1997] 296. Idem [1890–1909] VII 456.

[26] Of the vast bibliography on Stefano Della Bella the print-maker, we mention only a few titles to which to refer. Vesme [1906] 66–332; Forlani Tempesti [1983], idem [1973]; Massar [1971]; Viatte [1974].

[27] Also in the title-page of the *Instauratio Magna* by Francis Bacon there appears a vessel which with full sails courageously points its bow towards the rough and dangerous waters of the *new philosophy*, leaving in its wake the calm waters of ancient knowledge. Cf. Burnett [1998].

[28] The setting of the *Dialogue on the Two Chief World Systems* is Venice, in the *"palazzo dell'illustrissimo Sagredo"* on the Canal Grande. The reason for this is not only to represent a civil and social context and an ideal of research, but also to recall the place in the Mediterranean basin where better than anywhere else it is possible to observe the tides day by day.

[29] On the symbolic meanings contained in the posture of characters see Schapiro [1973] ch. II.

[30] Pantin [1993] 87f., 94f. Also Redondi [1987] 258–262, believed he had found a fundamental link between the imposing doctrinal structure illustrated in the Stanza della Segnatura and Galileo's judiciary vicissitudes. Tongiorgi Tomasi [1986] 166, inclines to consider it, on the contrary, linked to the iconographic scheme as found in *The three philosophers* by Giorgione.

[31] On the infinite number of speculative possibilities offered by Raphael's pictorial synthesis of philosophy in the Stanza della Segnatura see Reale [1997]; Becherucci [1998] 90ff.; and Beck [1976] *passim*.

[32] "Raffaello [...] cominciò nella camera della Segnatura una storia quando i teologi accordano la filosofia e l'astrologia con la teologia, dove sono ritratti tutti i savi del mondo che disputano in vari modi. Sonvi in disparte alcuni astrologi che hanno fatto figure sopra certe tavolette e caratteri in varii modi di geomanzia e d'astrologia [...] Similmente vi è Aristotele e Platone, l'uno col Timeo in mano, l'altro con l'Etica, dove intorno li fanno cerchio una grande scuola di filosofi. Né si può esprimere la bellezza di quelli astrologi e geometri che disegnano con le seste in su le tavole moltissime figure e caratteri" (Vasari [1976] IV 166). On the *Adorazione del Sacramento*, initially most significantly entitled *Il trionfo della Chiesa or Il trionfo dell'Eucarestia*, see ibid. 171–173.

[33] On Raphael's Astronomy see Edgerton [1993] 196–207.

[34] "[...] Ad arbitrio loro potessero i superiori aggiungere elevare e metter protesti a lor piacimento, non recusando io stesso di nominare questi i miei pensieri con il titolo di chimere, sogni, paralogismi e vane fantasie, rimettendo e sottoponendo sempre il tutto all'assoluta sapienza e certa dottrina delle scienze superiori etc." (Galilei [1890–1909] XIV 216).

[35] It is to be recalled that between May and June of 1630, Galileo returned once again to Rome to obtain the *imprimatur* for the printing of the *Dialogue* from the ecclesiastic authorities (ibid., XIV 89, 216).

[36] Viviani [1907] 602, 604. On Galileo's aesthetic conceptions, still of fundamental importance is the study by Panofsky [1954] (in abridged form with some revision it is also to be found in idem [1956a], [1956b]). On Galileo's relationship with the figurative arts see also Edgerton [1984] 225–232; Cavicchi, [1991] 313–315.

[37] On Galileo's stylistic tastes, cf. Panofsky [1956a] 6ff.

[38] Between 1632 and 1639 Della Bella worked in Rome under the protection of don Lorenzo de' Medici before going to Paris in the service of Richelieu and Mazarin. It is interesting to note that during his sojourn in Rome, Della Bella also produced five small prints for Urban VIII's *Poemata*

published in Rome in 1637 (Vesme [1906] 69ff., 240f.).

39 It was a true genre in which some print-makers had specialised. Among these we find Marcantonio Raimondi, Nicolò Nelli and Giorgio Ghisi. Cf. [Raphael] [1985].

40 [Raphael] [1984] 20. This was not only an Italian phenomenon in that it became manifest in France and England as well. Thuillier [1983], Vasselin [1983]; Blunt et al. [1960] 64ff.

41 It can also be said that the *'antiporta'* print received implicit posthumous approval. Indeed, as soon as the volume appeared it was anatomised in minute detail. A question was even raised, for example, on the possible recondite meaning of the different font with which the preface had been printed. The only objection that was raised about the graphic aspects of the work concerned "quei tre delfini che uno tiene in bocca la coda dell'altro." These dolphins were nothing more than the typographic signature of the printer Landini, which appeared both on the frontispiece and on the title-page that followed it (Galilei [1890–1909] XIV 369, 379). Also to be taken into consideration is that in the long list of the initial accusations levelled at Galileo no mention is made of the *"antiporta"*. Ibid. XIX 326.

42 One of Galileo's long-time friends, Diodati continually followed the events connected with the composition and publication of the *Dialogue*. Galilei [1890–1909] XIV 37, 49, 134, 157, 298, 339; Gardair [1984] 391–398. Eisenstein [1979] II 644, underscored how much Diodati did for the diffusion of Galileo's science.

43 Although it followed closely the project of the Strasbourg edition, the second Italian edition of the *Dialogue* (Galilei [1710]) does not contain the print.

44 Zinner [1943] 467; Bénézit [1976] V 530. Very curiously, the mark of the print-maker near the bottom of the column to the left of Copernicus very closely resembles the astronomical symbol for Jupiter, the planet which twenty-five years before had brought Galileo onto the stage of the European astronomical debate.

45 In the Florentine edition the characters have the same face (Panofsky [1956b]).

46 On the portraits of Copernicus, see Zinner [1943] 455–484.

47 The title is, among other things, different from the one that appears on the frontispiece of the work. The second says *Systema Cosmicum*. The difference was pointed out by Besomi in Galilei [1998] II 962 Note 75.

48 In a first stage the idea was to have the Accademia dei Lincei print the volume with costs borne by Prince Federico Cesi, who, however, passed away on 1 August 1630 (Galilei [1890–1909] XIV 67, 216).

49 Printed by E. G[riffin] for Michael Sparke and Edward Forrest, London 1638.

50 *News from the New World Discover'd in the Moone*, 1622. Jonson [1941] VIII 513–525.

51 Galileo, for example, denied the possibility. Among the several sources cf. Galilei [1890–1909] VII 85f., 125f.

52 See among others: Whewell [1857] I 410; Shapiro [1969] 37f.

53 He lived from 1591 to 1649. He was mostly active in the publishing sector, engraving frontispieces and commissioned portraits for several booksellers and publishers (Bénézit [1976] VII 204).

54 Strongly critical opinions of traditionalists, both Catholic and Protestant, are numerous in the two works. Cf. Wilkins [1970] I 21, 136ff.

55 In Wilkins' work the many references to Copernicanism are always prudently expressed with hypothetical locutions. The preface to *De Revolutionibus*, written and encouraged by the Lutheran theologian Andreas Osiander, like the recommendation made to the Carmelite Paolo Antonio Foscarini by Cardinal Bellarmino not to abandon the conjectural mathematical plane when discussing the new astronomical hypothesis, are but two illustrious examples that well represent the dominant position in the world of theologians. Cf. Copernicus [1873]; Galilei [1890–1909] XII 171f.

56 It was Kepler who first used the simile (Galilei [1890–1909] III/1 109).

57 "It is very likely if it had been but a jest, Galileus would never have suffered so much for it as report saith afterward he did" (Wilkins [1970] I 46).

58 Milton [1953] II 538. *The Areopagitica* contains the text of a speech before Parliament against the censoring of books and in favour of freedom of thought.

59 Galilei [1890-1909] VII 85f., 125f. Cf. also *Istoria e dimostrazioni intorno alle macchie solari e loro accidenti*, ibid., III 220; and Galileo's letter to Giacomo Muti, dated 28 February 1616, ibid., XII 240f.

60 "All that hath been said concerning the people of the new world, is but conjectural, and full of uncertainties". (Wilkins [1970] I 109).

61 Kepler [1634]. As we read in the title, the work was published only after the author's death, since Kepler made continuous changes to the manuscript. Besides the metaphysical fantasies, the *Somnium* also contains a long series of notes in which we find the many observations and reflections on lunar astronomy which Kepler conducted and developed over a period of more than twenty years.

62 "It is the opinion of Keplar, that as soon as the art of flying is found out, some of their nation will make one of the first colonies that shall transplant into that other world" (Wilkins [1970] I 111).

63 Of evident literary origin, the image of the wings is fraught with metaphorical meanings. It is interesting to point out that even Galileo, in the preface to the *Dialogue*, had resorted to a similar image ("tarpar l'ale agli intelletti speculativi") in speaking of the Decree with which Copernican philosophy had been prohibited. It must be said, however, that in Bernegger's Latin translation the metaphor does not come through as it should. Galilei [1890–1909] VII 29 and idem [1641] a 3.

64 "Aequa grataque permutazione rependit Tellus parem illuminazionem ipsi Lunae" (Galilei [1890–1909] III/1 74).

65 Already present in the *De Revolutionibus*, l. I, c. X, and the *Narratio prima* by Rheticus, in England it was Thomas Digges who especially insisted on this albeit imperfect analogy. Cf. *Perfit description of the Caelestial Orbes according to the most anciente doctrine of the Pythagoreans, latelye reniued by Copernicus and by Geometricall Demonstrations approved.* It appeared as an appendix of Digges [1576].

66 Wilkins [1655]. The frontispiece copies the English one of 1640.

References

Altieri Biagi, M. L.
1983 "Il dialogo galileiano." In: *Giornate lincee indette in occasione del 350° anniversario della pubblicazione del 'Dialogo sopra i due massimi sistemi,'* 143–166. Rome.

Aristotle
Physics. (Bekker pagination).

De coelo. (Bekker pagination).

Battistini, A.
1989 *Introduzione a Galileo*. Bari.

Beaulieu, A.
1984 "Les réactions des savants français au début du XVII e siècle devant l'eliocentrisme de Galilée." In: *Novità celesti e crisi del sapere*, ed. P. Galluzzi, 373–381. Florence.

Becherucci, L.
1998 "Raffaello e la pittura." In: *Raffaello: la pittura e i disegni*. Novara.

Beck, J. H.
1976 *Raffaello*. New York.

Bénézit, E.
1976 *Dictionnaire critique et documentaire des peintres, sculpteurs, dissinateurs et graveurs*. Nouv. ed. Paris.

Biagioli, M.
1990 "Galileo's System of Patronage." In: *History of Science* 28: 1–62.

1993 *Galileo, Courtier: The Practice of Science and the Culture of Absolutism*. Chicago and London.

Blunt, A. and H. L. Cooke (eds.)
1960 *The Roman Drawing of the XVII & XVIII Centuries in the Collection of H. M. the Queen at Windsor Castle*. London.

Boas Hall, M.
1967 "Galileo's Influence on Seventeenth-century English Scientists." In: *Galileo Man of Science*, ed. E. McMullin, 405–414. New York.

Burnett, A. D.
1998 *The Engraved Title-Page of Bacon's Instauratio Magna*. Durham.

Calvino, I.
1980 *Una pietra sopra*. Turin.

1985 "Le livre de la nature chez Galilée." In: *Exigences et perspectives de la sémiotique. Recueil d'hommage pour A.J. Greimas*, eds. H. Parret and H.-G. Ruprecht, 683–688. Amsterdam.

Cavicchi, E.
1991 "Painting the Moon." In: *Sky and Telescope* 82: 313–315.

Ciardi, R. P. and L. Tongiorgi Tomasi
1985 "La 'scienza' illustrata: osservazioni sui frontespizi delle opere di Athanasius Kirker e Galileo Galilei." In: *Annali dell' Istituto Storico italo-germanico in Trento* A. 11: 69–78.

Ciliberto, M. and N. Mann
1997 *Giordano Bruno. The English Experience*. Florence.

Copernicus, N.
1873 *Nicolai Copernici Thorunensis De Revolutionibus Orbium Coelestium libri sex. Ex autoris autographo recudi curavit Societas Copernicana Thorunensis. Accedit Georgii Ioachimi Rhetici de libris Revolutionum Narratio prima*. Thorn.

Cyrano de Bergerac, S.
1657 *Histoire comique, par Monsieur Cyrano de Bergerac. Contenant les Éstats et Empires de la Lune*. Paris.

De Vesme, A.
1906 *Le peintre graveur italien*. Milan.

Digges, L.
1576 *A prognostication everlastinge of righte good effecte, fruitfully augmented by the autor, containing plaine, briefe, pleasant, chosen rules to judge the weather by the Sunne, Moone, Starres, Comets, Rainebow, Thunder, Cloudes, with other extraordinary tokens* [...]. *Lately corrected and augmented by Thomas Digges his sonne*. London.

Dillenberger, J.
1960 *Protestant Thought and Natural Science: A Historical Interpretation*. Garden City NY.

Drake, S.
1967 "Galileo in the English Literature of the Seventeenth Century." In: *Galileo Man of Science*, ed. E. McMullin, 415–431. New York.

1978 *Galileo at Work. His Scientific Biography*. Chicago.

1980 *Galileo*. Oxford.

1982 "Galileo's 'Dialogue' Al discreto lettore." In: *Scientia* 247–274.

Edgerton, S. Y.
1984 "Galileo, Florentine 'Disegno' and the 'Strange Spotnesse' of the Moon." In: *Art Journal* 44: 225–232.

1985 "The Renaissance Development of the Scientific Illustration." In: *Science and the Arts in the Renaissance*, eds. J. W. Shirley and F. D. Hoeniger, 168–197. Washington, DC.

1993 *The Heritage of Giotto's Geometry Art and Science on the Eve of the Scientific Revolution*. Ithaca and London.

Eisenstein, E. N.
1979 *The Printing Press as an Agent of Change: Communication and Cultural Transformations in Early-modern Europe*. London, New York, and Melbourne.

Fantoli, A.
1994 *Galileo for Copernicanism and the Church*, trans. G. V. Coyne. Vatican City.

Feingold, M.
1984 "Galileo in England: The First Phase." In: *Novità celesti e crisi del sapere*, ed. P. Galluzzi, 411–420. Florence.

Finocchiaro, M. A.
1980 *Galileo and Art of Reasoning: Rhetorical Foundation of Logic and Scientific Method*. Dordrecht, Boston, and London.

Fontenelle, B. LeBovier de
1686 *Entretiens sur la pluralité des mondes. Digression sur les anciens et les modernes.* Paris.

Forlani Tempesti, A.
1973 Mostra di incisioni di Stefano della Bella. Introduction and Catalogue by A. Forlani Tempesti. Florence.

1983 (ed.) *Stefano Della Bella: Incisioni, scelte e annotate da A. Forlani Tempesti.* Florence.

Friedman, D.
1991 "Galileo and the Art of Seeing." In: *Milton in Italy: Contexts, Images, Contradictions*, ed. M. Di Cesare, 159–174. Binghamton NY.

Gardair, J. M.
1984 "Elia Diodati e la diffusione europea del 'Dialogo.'" In: *Novità celesti e crisi del sapere*, ed. P. Galluzzi, 391–398. Florence.

Galilei, G.
1635 *Systema Cosmicum, Authore Galilaeo Galilei Lynceo* [...] *In quo Quatuor Dialogis, De Duobus Maximis Mundi Systematibus, Ptolemaico et Copernicano, Utriusq; rationibus Philosophicis ac Naturalibus indefinite propositis, differitur. Ex Italica Lingua Latine conversum. Accessit Appendix gemina, qua SS. Scripturae dicta cum Terrae mobilitate conciliantur.* Strasbourg.

1641 *Galilae Galilaei Lyncei* [...] *Systema Cosmicum: In quo Dialogis IV. De duobus maximis Mundi Systematibus, Ptolemaico & Copernicano, Rationibus utrique propositis indefinite differatur. Accessit locorum S. Scripturae cum Terrae mobilitate conciliantio.* Lyon.

1663 *Systema Cosmicum, Autore Galileio Galilei Lynceo* [...] *In quo Quatuor Dialogis, De Duobus Maximis Mundi Systematibus, Ptolemaico & Copernicano, Utriusque rationibus Philosophicis ac Naturalibus indefinite propositis, differitur. Ex Italica Lingua Latine conversum, Accessit Appendix gemina, qua SS. Scripturae dictacum Terrae mobilitate conciliantur.* London.

1710 *Dialogo di Galileo Galilei Linceo,* [...] *Dove ne i congressi di quattro giornate si discorre sopra i due massimi Sistemi del Mondo, Tlemaico e Copernicano; Proponendo indeterminatamente le ragioni Filosofiche e Naturali tanto per l'una, quanto per l'altra parte. In questa seconda impressione Accresciuta di una lettera dello stesso autore non più stampata, e di varij trattati di più Autori, i quali si veggono nel fine del Libro.* Florence (actual Naples).

1890–1909 *Le opere di Galileo Galilei*, Edizione Nationale, ed. A. Favaro. Florence.

1992 *Sidereus nuncius, le messager celeste*, trans. I. Pantin. Paris.

1993 *Dialogo sopra i due massimi sistemi mondo di Galileo Galilei*, ed. M. L. Altieri Biagi, (= Letteratura italiana, Le opere, II). Turin.

1997 *Galileo On The World Systems*, trans. M. Finocchiaro. Los Angeles and London.

1998 *Dialogo sopra i due massimi sistemi del mondo tolemaico e copernicano*, Critical and annotated edition, eds. O. Besomi and M. Helbing. Padua.

Gingerich, O.
1975 "Dissertatio cum Professore Righini et Sidereo Nuncio." In: *Reason, Experiments and Mysticism in the Scientific Revolution*, eds. M. L. Righini Bonelli and W. R. Shea, 77–88. New York.

Godwin, F.
1638 *The Man in the Moone: or a Discourse of a Voyager Thither by Domingo Gonsales. The Speedy Messeger.* London.

Hackmann, W. D.
1998 "Natural Philosophy Textbook Illustration, 1600–1800." In: *Non-verbal Communication in Science Prior to 1900*, ed. R. G. Mazzolini. Florence.

Haskell, F.
1980 *Patrons and Painters. A Study in the Relations between Italian Art and Society in the Age of the Baroque.* New Haven and London.

Heninger Jr., S. K.
1977 *The Cosmographical Glass: Renaissance Diagrams of the Universe.* San Marino CA.

Ivins Jr., W. M.
1953 *Prints and Visual Communication.* London.

Hill, C.
1965 *Intellectual Origins of the English Revolution.* Oxford.

Jacob, M. C.
1990 *The Newtonians and the English Revolution.* New York.

Jones, R. F.
1961 *Ancient and Moderns. A Study of the Rise of the Scientific Movement in Seventeenth-century.* Berkeley and Los Angeles.

Jonson, B.
1941 *Ben Jonson, Complete Critical Edition in Eleven Volumes*, eds. P. Herford and E. Simpson. Oxford.

Kemp, M.
1992 *The Science of Art: Optical Themes in Western Art from Brunelleschi to Seurat.* New Haven.

1996 "Temple of the Body and Temple of the Cosmos: Vision and Visualization in the Vesalian and Copernican Revolution." In: *Picturing Knowledge: Historical and Philosophical Problems Concerning the Use of Art in Science*, ed. S. B. Baigrie, 40–85. Toronto and Buffalo.

Kepler, J.
1634 *Ioh. Keppleri Mathematici Olim Imperatorii Somnium sive opus postumum de astronomia lunari. Divulgatum a M. Ludovico Kepplero Filio, Medicinae Candidato.* Frankfurt.

1993 *Dissertatio cum Nuncio Sidereo; Narratio de observatis Jovis satellitibus. Discussion avec le Messager Celeste. Rapport sur l'observation des Satellites de Jupiter,* trans. I. Pantin. Paris.

Massar, P. D.
1971 *Presenting Stefano della Bella, Seventeenth-century Print-maker.* New York.

Massa, D.
1977 "Giordano Bruno's Ideas in Seventeenth-century England." In: *Journal of the History of Ideas* XXXVIII: 227–242.

McColley, G.
1936a "The Second Edition of the 'Discovery of a World in the Moon.'" In: *Annals of Science* 1: 330–334.

1936b "The Seventeenth-Century Doctrine of a Plurality of Worlds." In: *Annals of Science* 1: 385–430.

1938 "The Ross-Wilkins Controversy." In: *Annals of Science* 3: 153–176.

1939 "The Debt of Bishop John Wilkins to the 'Apologia pro Galileo' of Tommaso Campanella." In: *Annals of Science* 4: 150–168.

McKerrow, R. B.
1968 *A Dictionary of Printers and Booksellers in England, Scotland and Ireland, and of Foreign Printers of English Books, 1557–1640.* London.

McMullin, E.
1986 "Giordano Bruno at Oxford." In: *Isis* 77: 85-94.

Merton, R. K.
1985 *On the Shoulders of Giants.* Orlando FL.

Milton, J.
1953 *Complete Prose Works of John Milton*, 4 vols., ed. D. M. Wolfe. New York and London.

Moss, J. D.
1993 *Novelties in the Heavens: Rhetoric and Science in the Copernican Controversy.* Chicago and London.

Negri Arnoldi F. and S. Prosperi Valenti
1986 *Il disegno nella storia dell'arte italiana.* Florence.

Nicolson, M. H.
1939 "English Almanacs and the New Astronomy." In: *Annals of Science* 4.

1956 (1st ed.) *Science and Imagination.* Ithaca NY.

1962 *Science and Imagination.* Ithaca NY.

Panofsky, E.
1954 *Galileo as Critic of the Arts.* The Hague.

1956a "Galileo as Critic of the Arts." In: *Isis* 47: 3–15.

1956b "More on Galileo and the Arts." In: *Isis* 47: 182–185.

Pantin, I.
1993 "Une Ecole d'Athènes des Astronomes? La représentation de l'astronomie antique dans les frontispices de la Renaissance." In: *Images de l'Antiquité. Le Texte et son illustration.* (= Colloque Paris XII, 1991), eds. E. Baumgartner and L. Harf. Paris.

Plomer, H. R.
1907 *A Dictionary of the Booksellers and Printers Who were at Work in England, Scotland and Ireland, 1641–1667.* London.

Raimondi, E.
1969 "La nuova scienza e la visione degli oggetti." In: *Lettere italiane* XXI/3: 265–305.

1974 *Il romanzo senza idillio.* Turin.

[Raphael]
1984 *Raffaello. Elementi di un mito. Le fonti, la letteratura artistica e la pittura di genere storico.* Florence.

1985 *Raphael invenit. Stampe da Raffaello nelle collezioni dell'Istituto Nazionale per la Grafica.* Rome.

Reale, G.
1997 *Raffaello e 'La scuola di Atene.'* Milan.

Redondi, P.
1987 *Galileo Heretic*, trans. R. Rosental. Princeton.

Reeves, E.
1997 *Painting the Heavens. Art and Science in the Age of Galileo.* Princeton.

Righini, G.
1975 "New Light on Galileo's Lunar Observations." In: *Reason, Experiments and Mysticism in the Scientific Revolution*, eds. M. L. Righini Bonelli and W. R. Shea, 59–76. New York.

Roche, J. J.
1998 "The Semantics of Graphics in Mathematical Natural Philosophy." In: *Non-verbal Communication in Science prior to 1900*, ed. R.G. Mazzolini, 197–233. Florence.

Rossi, P. (ed.)
1998 *Storia della scienza moderna e contemporanea*, 3 vols. Turin.

Salusbury, T. (ed.)
1661 *Mathematical Collections and Translations: The first Tome. In two parts. The First Part; Containing, I. Galileus Galileus His System of the World. II. Galileus His Epistle to the Grand Dutchesse Mother, concerning the Authority of Holy Scripture in the Philosophical Controversies* [...]. London.

Schapiro, M.
1973 *Words and Pictures. On the Literal and the Symbolic in the Illustration of a Text.* The Hague and Paris.

Shapiro, B. J.
1969 *John Wilkins, 1614–1672. An Intellectual Biography.* Berkeley and Los Angeles.

Stimson, D.
1931 "Dr. Wilkins and the Royal Society." In: *Journal of Modern History* III: 539–563.

Tabarroni, G.
1984 "I disegni autografi della Luna e altre espressioni figurative dei manoscritti Galileiani." In: *Novità celesti e crisi del sapere*, ed. P. Galluzzi, 51–55. Florence.

Thuillier, J.
1983 "Raphael et la France: Présence d'un peintre." In: *Raphael et l'art français*, eds. J.-P. Boulanger and G. Renisio, 11–36. Paris.

Tongiorgi Tomasi, L.
1986 "Il simbolismo delle immagini: I frontespizi delle opere di Kircher." In: *Enciclopedismo in Roma Barocca*, 165–175. Venice.

Tufte, E. R.
1990 *Envisioning Information.* Cheshire CN.

Vasari, G.
1976 *Le vite*, ed. R. Bettarini. Florence.

Vasselin, M.
1983 "La fortune gravée de Raphael en France, aperçu historique et critique." In: *Raphael et l'art français*, eds. J.-P. Boulanger and G. Renisio, 37–45. Paris.

Viatte, F.
1974 *Les dessins de Stefano della Bella*. Paris.

Vickers, B.
1983 "Epideictic Rhetoric in Galileo's Dialogo." In: *Annali dell' Istituto e Museo di Storia della Scienza di Firenze* VIII/2: 121–128.

Viviani, V.
1907 "Racconto istorico della vita di Galileo." In: G. Galilei [1890–1909] XIX.

Westfall, R. S.
1985 "Science and Patronage: Galileo and the Telescope." In: *Isis* 76: 11–30.

1987 "Patronage and the Publication of the Dialogue." In: *History and Technology* 4: 385–399.

1989 *Essays on the Trial of Galileo*. Vatican City.

Westman, R. S.
1984 "The Reception of Galileo's 'Dialogue.' A Partial World Census of Extant Copies." In: *Novità celesti e crisi del sapere*, ed. P. Galluzzi, 329–371. Florence.

1986 "The Copernicans and the Churches." In: *God and Nature. Historical Essays on the Encounter between Christianity and Science*, eds. D. Lindberger and R. L. Numbers, 76–113. Berkeley, Los Angeles, and London.

Whewell, W.
1857 *History of the Inductive Sciences, from the Earliest to the Present Times*, the third edition with additions, 3 vols. London.

Wilkins, J.
1655 *Le monde dans la lune*, trans. J. de LaMontagne. Rouen.

1713 *Johannis Wilkins Vertheidigter Copernicus, oder curioser und gruendlicher Beweiss der Copernicanischen Grundsaetze*, trans. J. G. Doppelmayr. Leipzig.

1970 *Mathematical and Philosophical Works of the right Rev. John Wilkins, Late lord Bishop of Chester, to which is prefixed the Author's Life, and an Account of his works. In two volumes*, (= facsimile reproduction of the edition London 1802). London.

Winkler, M. G. and A. van Helden
1992 "Representing the Heavens. Galileo and the Visual Astronomy." In: *Isis* 83/2: 195–217.

Zinner, E.
1943 *Entstehung und Ausbreitung der Coppernicanischen Lehre*. Erlangen.

Edmond Halley and Visual Representation in Natural Philosophy

Alan Cook

"I believe it is necessary to adjoyn a Schema, showing at one view all the Tracts and Courses of those Winds; whereby 'tis possible the thing may be better understood, than by any verbal description whatsoever." (Halley [1686])

Introdution

As a very young man of twenty-one Halley obtained the support of very influential members of the court of King Charles II to undertake a mission to the island of St. Helena, newly acquired by the English East Indian Company, to observe stars of the southern hemisphere that could not be seen from Greenwich, and to determine their positions. In all, he established the positions of some 300 stars, some of which could be seen from northern observatories. On his return he published his results in a small book (Halley [1679]) which he accompanied with a planisphere of the southern stars (fig. 1). As was then conventional he delineated the constellations by representations of the beings for which they were named. Because Charles II had supported his project by asking the East India Company to take him in one of their ships to St. Helena and back, he introduced a new constellation, *Robur Carollina*, in honour of the king. Charles had saved himself after defeat at the battle of Worcester, an incident in the English civil wars, by hiding from the forces of Oliver Cromwell in an oak tree.

In representing his results on such a planisphere, Halley was following the practice of all his predecessors and that went back to Ptolemy at the latest. Although it is now apt to be dismissed as fanciful, there are clear advantages to it. The methods that were available for the measurement of positions of stars in the sky were not good enough even in Halley's early years for stars to be identified unambiguously by the angular co-ordinates assigned to them in catalogues. Astronomers therefore identified them by such phrases as "the bright star to the right of the eye of the eagle." Such a phrase not only identified the star but also acted as a mnemonic in reminding astronomers of its existence and position. The practice had its disadvantages, however, the principal being that the descriptions were never wholly free of confusion.

While the constellation as a visual description of the arrangement of stars might with care be adequate, it could not be a useful model for setting out natural phenomena that were more than simple locations.

Trade Winds

Some fifteen years after his observations from St. Helena, Halley assembled from the reports of sea captains a collection of data relating to the trade winds. Trade winds are those winds in the Atlantic that blow from east to west, and are so called because they carried with them ships trading from Europe to America. The return voyages were made in higher latitudes where, to conserve angular momentum in the atmosphere, the winds blow from west to east. Halley made none of the observations himself, with the exception of a few that he took on St. Helena. His account displayed a characteristic that is especially significant in his natural philoso-

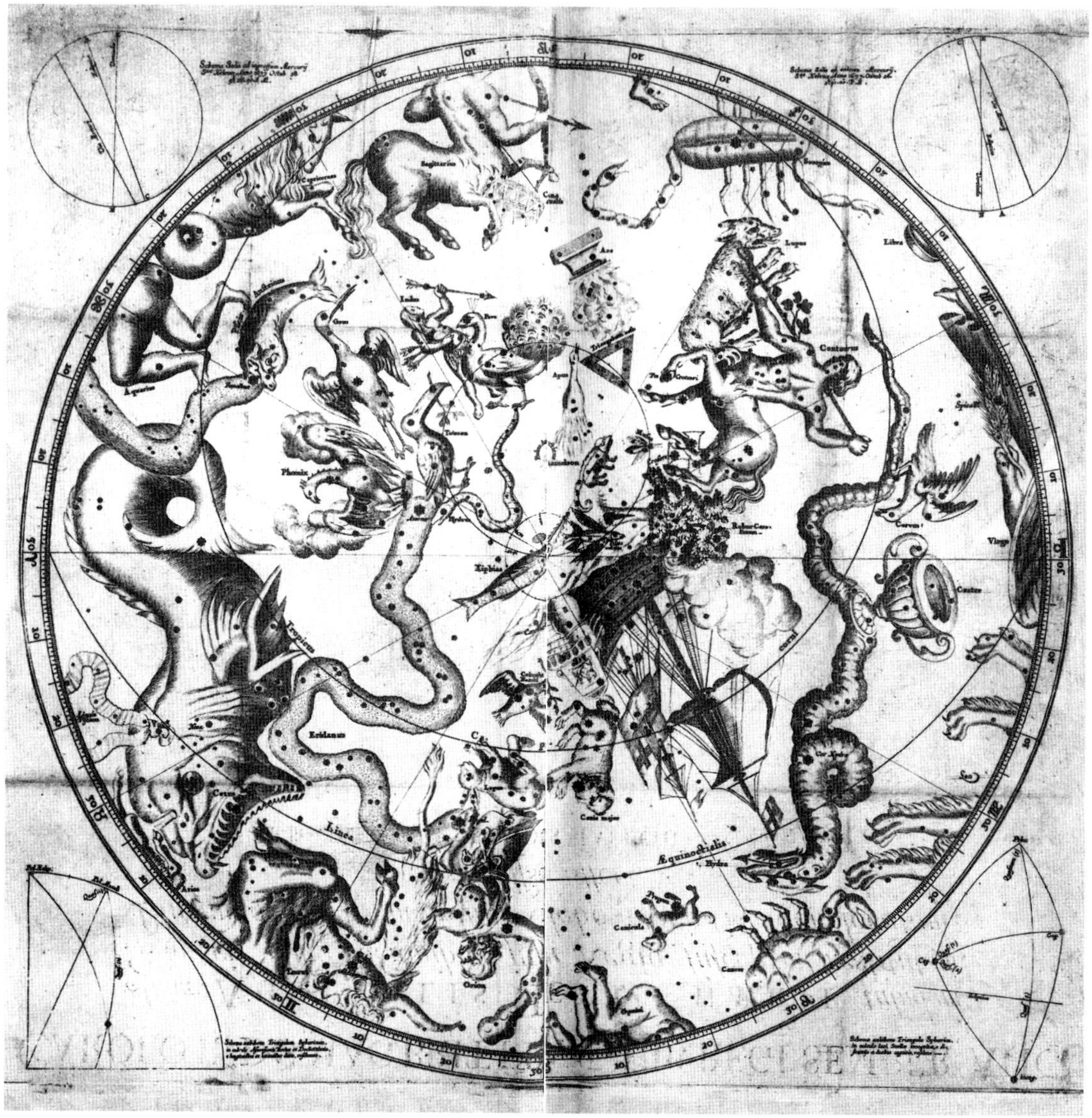

Figure 1
The constellation of *Robur Carollina* with an image of an oak tree imposed upon it. Part of Halley's planisphere of the southern hemisphere.

phy, that is of collecting, representing, and interpreting a large body of observations of some natural phenomenon.

Halley presented his material in a paper in the *Philosophical Transactions* of the Royal Society of London as a global map on the Mercator projection (fig. 2). He imposed upon a chart of the continents an array of short lines drawn parallel to the direction of the winds at any place, in effect a map of their flow lines. Flow lines were later used to describe the flow of liquids in channels and tubes, but no one before Halley had delineated the course of winds in this way. His lines were, however, very effective in showing the worldwide course of the winds, although he apparently did not space them inversely as the velocity of the air, such as would be done now.

Halley did not rest with drawing up the chart. Because he had a worldwide view of the winds, he saw that they were related to the heating of the atmosphere by the sun. He argued that the air would be held by gravity to the surface of the earth and could only move over it as the result of some external force, namely heating by the sun. He envisaged that the heating would reduce the density of the air under the sun so that as the sun moved round the earth, the less dense air would cause denser air to move towards it and thus create a motion of the whole atmosphere following the sun around the earth. Halley was right in seeing that it was the heat of the sun

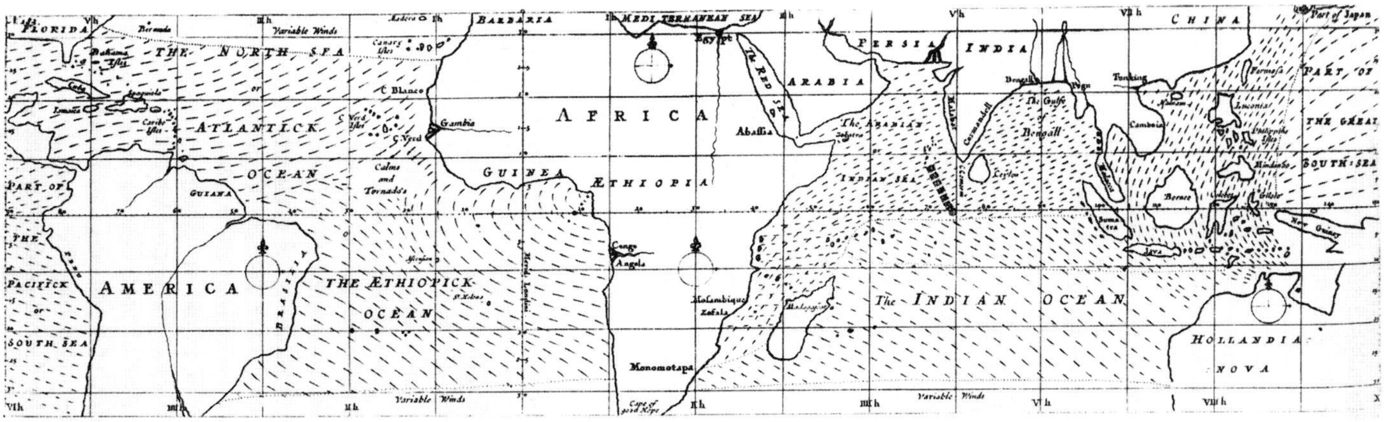

Figure 2
Halley's chart of the trade winds. From *Philos. Trans.* XVI (1686).

that drove the trade winds, but in his day the concept of the Coriolis force had not been formulated, and his mechanism was wrong. The force driving the winds is from the cold dense air at the poles to the hot less dense air at the equator, but the resultant motion on the rotating earth is around parallels of latitude on account of the Coriolis acceleration. As mentioned above, the angular momentum of the whole atmosphere must be conserved, and so the westerly rotation of the trade winds is approximately balanced by the easterly rotation of the winds in higher latitudes.

Halley recognised that the motion of cyclones was imposed on the trade winds, and that cyclones were born in the West Indies; very much later Harold Jeffreys showed that cyclones are an important element in the balance of the angular momentum of the atmosphere.

Halley realised that the trade winds vary with the seasons and that his map did not show this, in particular he called attention to changes in the winds associated with the monsoon season on the Indian Ocean.

Halley's chart was, of course, useful to mariners in giving a systematic account of what in some ways they already knew, and in helping them to plan voyages. Halley found the chart at least as useful in his scientific interpretation, for as a global presentation of the phenomenon it revealed a pattern that suggested an explanation to him. That is undoubtedly one of the outstanding advantages of visual representation in natural philosophy, that by giving a comprehensive prospect of some phenomenon, it calls attention to patterns that demand and suggest explanations.

The Magnetic Field over the Atlantic

In the two years from 1699 to 1701 Halley had a commission as a captain in the Royal Navy to command the pink *Paramore* in cruises around the Atlantic Ocean to measure the magnetic variation (declination in modern terms). Pinks were small storeships; Halley had a crew of twenty. His logs survive and were edited and published by Professor N. W. J. Thrower about fifteen years ago. Halley not only observed the magnetic variation on most days at sea, but also determined the longitudes of numerous places on land – only on rare occasions, when a favourable eclipse occurred, could he find longitude at sea.

Halley published the results of his cruise as a chart of the Atlantic on which he drew lines along which the variation had the same value – isogones as they are now known (fig. 3). This was the first time that anyone had published a map with any sort of isoline – a line on which some quantity is constant. Nowadays we are so familiar with all manner of isolines showing the values of some quantity over a large or small area that they pass without comment as the obvious way of representing geographical, meteorological, demographic and many other observations, but no one had printed and published such a map before Halley. The value of his innovation was immediately and widely recognised and isogones soon became known as *Halleyan lines*. A year or so after his first Atlantic chart appeared, Halley published another of the whole world on which he supplemented his own Atlantic observations with data from other people; many were reports from sea captains that came to him at the Royal Society.

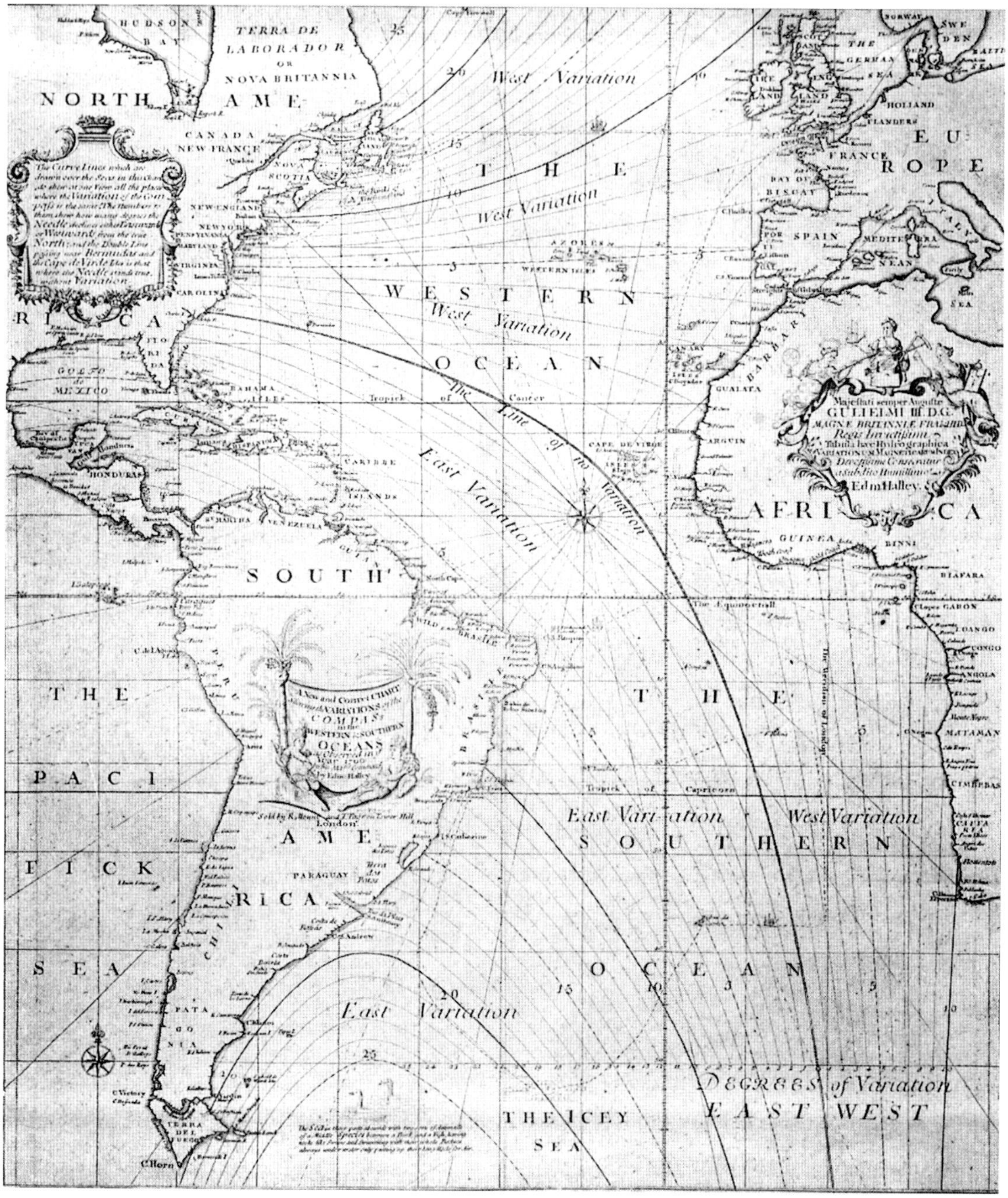

Figure 3
Halley's Atlantic chart. From Thrower: *The three voyages of Edmond Halley* (London 1981).

Halley claimed that the idea of isogones was new and was his own. Thrower discussed Halley's claim in his edition of the logs, and believed it to be justified. A manuscript chart with contours of the depths of water had been produced before, and Athanasius Kircher, in his book on magnetism (1643), appears to refer to a manuscript chart with isogones. Halley referred elsewhere to some magnetic apparatus that Kircher describes, but it is not clear if he knew of either the isobath manuscript or the magnetic chart. In any event no one had previously engraved, nor printed nor published any chart such as Halley now presented.

It is hardly possible to exaggerate the importance of Halley's innovation. It led to the use of isolines in all manner of other fields. In geomagnetism it provided a firm base for the study of the changes in the earth's field from that day to the present. From now on natural philosophers had at hand convenient means of studying how phenomena varied from place to place over the earth or over the sky; and by comparing maps corresponding to different times they could readily see the dominant features of changes in the phenomena. Two of Halley's abiding concerns were just those, the study of phenomena over the whole earth (and sky), and the study of the history of natural phenomena. The fact that the deviations of the earth's magnetic field from a simple dipole field drift westward with time had been known for about half a century before Halley's cruises; Halley had assembled data about the drift, and in two papers had suggested that it arose from different magnetisations of a core and of a shell of the earth, the core and the shell being in relative rotation. As with the trade winds, Halley was looking for the physics behind the pattern. Physics then was again inadequate for his purpose, but, again, Halley established for the first time the phenomena that had to be explained. When he set out his account of the westerly drift, he had not yet produced his Atlantic chart but, when he did produce it, that defined more sharply those phenomena.

Today we see Halley's Atlantic chart as a fundamental contribution to scientific method and to the science of geomagnetism, but these were not the reasons that his cruises were made at the public charge and that he was commissioned as a captain of the Royal Navy. The cruises had a strictly practical purpose: to provide new more reliable measurements of the magnetic field for use by navigators on the open seas.

Charts such as Halley's are based explicitly or implicitly upon the profound metaphysical assumption, that nature behaves regularly. Although he drew isogones uniformly over the oceans, Halley had observed for the most part only around the coasts. His lines over the unsurveyed seas were ways of interpolating between the observations on the coasts; they would have been invalid and misleading if values over the unsurveyed areas could not be inferred from those that had been surveyed. Halley does not comment on his assumption, but more than twenty years before, in the *Philosophiae naturalis principia mathematica* (1687), Newton had set out his finite difference methods for interpolating between series of numerical data. Halley in drawing isolines effectively was employing a finite difference graphical procedure in two dimensions, a procedure that remained effective until modern computers could handle the large quantities of data that result from observations in two or more dimensions. The formal justification for all such methods is that data in one, two or more dimensions may be represented by series of orthogonal functions. Whether by algebraic analysis, by finite difference methods or by drawing isolines, a certain number of such functions is determined, up to the point that the uncertainties entailed by the paucity and inaccuracies of the data are greater than the apparent values of the functions. The geomagnetic field can be analysed into a set of spherical harmonics, the amplitudes of which decrease sufficiently rapidly with order that Halley's relatively crude interpolation accords reasonably well with the actual field.

Tides in the English Channel

Almost immediately after he had ended his second Atlantic cruise Halley set forth upon a third cruise in command of the *Paramore* to observe the tides in the English Channel, principally the times of high water and the direction and magnitude of the flow. He did not in general measure the height of the tide, although it is plain in other surveys of his that he measured depths in the Thames approaches from an effective low water datum. Yet again he devised an appropriate new method to present his results (fig. 4). Halley supposed that the time of high water at any place was related just to the phase of the moon, the relation varying from place to place. This is not quite sufficient, for the phase of the tide depends on the position of the sun as well as on that of the moon. Friction on the sea bed as the tidal wave travels up the Channel also leads to deviations. Halley seems to have assumed that the tidal flow which he observed vanished at high and low water, which he did not observe; he appears to have known it was not exactly so. Nonetheless, his chart shows the ages of the moon at which high water, defined as the end of the flood, occurred. He also showed by arrows the direction of the flood and some indication of the maximum speed.

Halley's Channel survey was not repeated for a century and a half, and comparisons with a recent Admiralty chart have shown

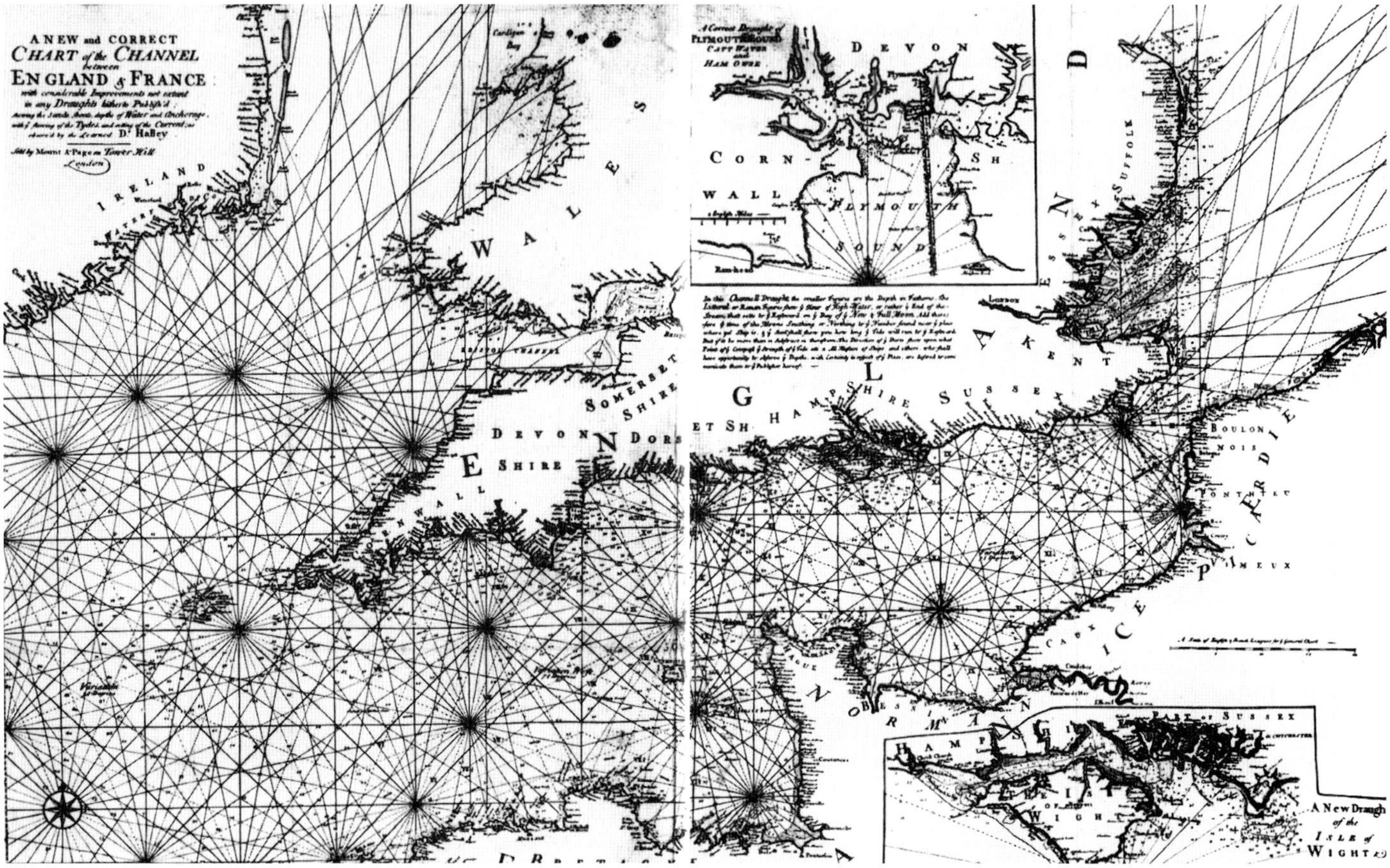

Figure 4
Halley's tidal chart of the English Channel. From Thrower: *The three voyages of Edmond Halley* (London 1981).

the chart to be remarkably accurate. The original publishers, Page and Mount of London, kept the original plates and used them in many books even after Halley's death and without acknowledgement. Thus it is Halley's chart that appears at the beginning of an oriental pilot from Page and Mount which consists of charts and sailing directions for voyages between London and Indian ports.

Aurorae

Halley observed some striking aurorae himself from London and in addition collected data from other observers in different latitudes. He published drawings of the bright bands as he observed them, and drew some conclusions about their height from comparisons between observers in different places. He likened the appearance of the bands to the rays round the Star of the Garter or to haloes round the heads of saints in paintings. He concluded from the form of his sketches that the coloured bands followed the lines of magnetic field emerging from the poles of the earth. Halley did suggest an explanation of why that should be so, but a comment by Bernoulli indicates why Halley could not give the correct explanation. Bernoulli pointed out that Halley had not explained the colours of the auroral rays. Halley could not do so because the concept of ionisation did not then exist, nor the idea that ions would emit radiation of characteristic energy on recombination. This is closely related to the form of the rays, for as the ions are charged they circulate in spirals around the magnetic field lines and so are effectively attached to them. As with Halley's other studies, patterns themselves cannot lead to an explanation of the phenomena if the physics behind the patterns is lacking.

Pattern Recognition in Natural Philosophy

The particular value of maps and charts in natural philosophy is that it is easy to see from them how some quantity varies from place to place, both on the large scale and on the small scale. Furthermore, if maps can be constructed for successive epochs, then the way in which the pattern changes over time can be seen. In addition, correlations between quantities can be detected from the relation between maps of those quantities. Visual inspection of maps is thus a powerful way of appreciating patterns of variation, how they change in time and how the patterns of different quantities may be related. Pattern recognition has thus become a popular technique in recent years in geophysical exploration and geological prospecting, and sophisticated mathematical methods have replaced simple visual inspection for detecting patterns and their correlations.

If a pattern over a surface is analysed into a series of orthonormal functions, large coefficients of functions of low order correspond to a large-scale feature of the pattern, while the functions of high order correspond to the local variations imposed on the large-scale pattern. The analysis into orthonormal functions is a purely mathematical operation quite independent of the physical origin of the pattern, so that for any interpretation of the pattern to be possible, the mathematical analysis must be supplemented by some hypothesis about the physical origin. If, as is often the case, the observed quantities are derived from a potential, then the sources of the local variations must be relatively close to the surface, while those of the large-scale pattern may lie much deeper. Halley had some intuitive understanding of that conclusion when he discussed the worldwide variation of the magnetic field of the earth; it is a conclusion from which important information about the structure of the earth may be inferred. It is, however, important to emphasise that because the same pattern with the same analysis into orthonormal functions may be generated by very different physical processes, the identification of a cause requires information beyond that provided by the pattern itself.

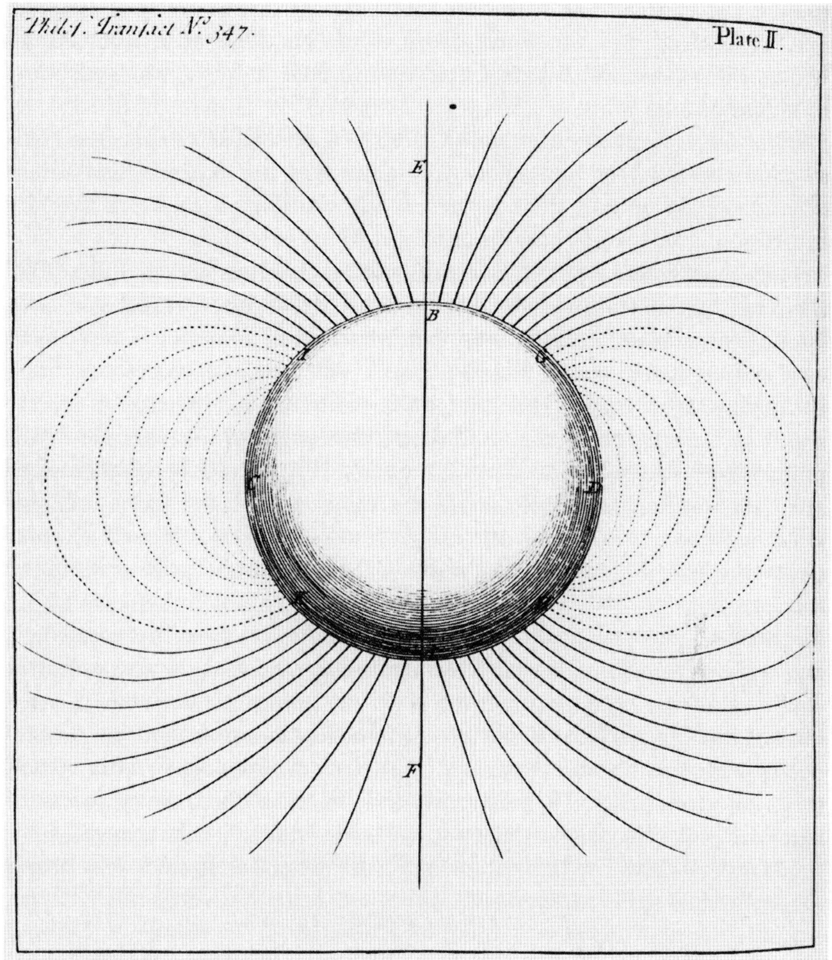

Figure 5
Magnetic field lines of the earth; Halley showed that the luminous bands of the aurorae followed the field lines. From Cook: *Edmond Halley* (Oxford 1998).

The Solar Eclipse of 1715

A total solar eclipse was predicted for London and the south-eastern part of England, and Halley prepared for it by drawing up and publishing a map of the predicted course of the shadow across England (fig. 6). He had two purposes. One was to show people that an eclipse was perfectly predictable and should occasion no panic. The second purpose was to arrange for observers, both practising astronomers and casual viewers, to make observations of the passage of totality and send them to him. This was the first time that any such co-ordinated programme of observations in natural philosophy had been organised; nowadays it is common practice in astronomy, geophysics, oceanography and other fields.

Halley, with his own interest in the history of the natural world, would no doubt have been gratified, had he lived so long, to see how his collected data have recently been put to use. They enable the edge of totality to be defined closely, so much so that it is possible to show that the diameter of the sun has not changed by more than one part in 900 from Halley's day until now (Morrison et al. [1988]).

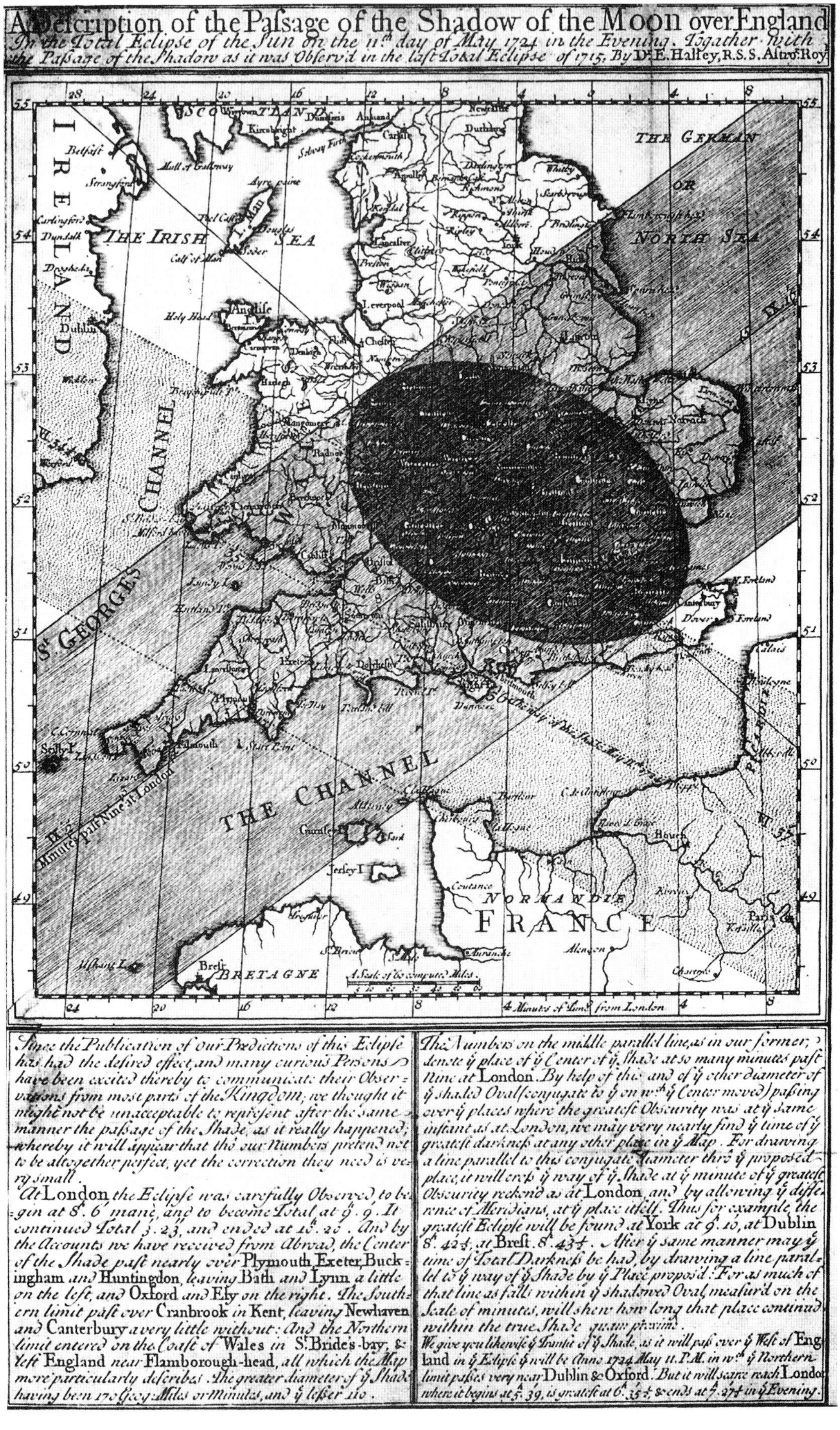

Figure 6
Halley's eclipse map for 1715. From Cook: *Edmond Halley* (Oxford 1998).

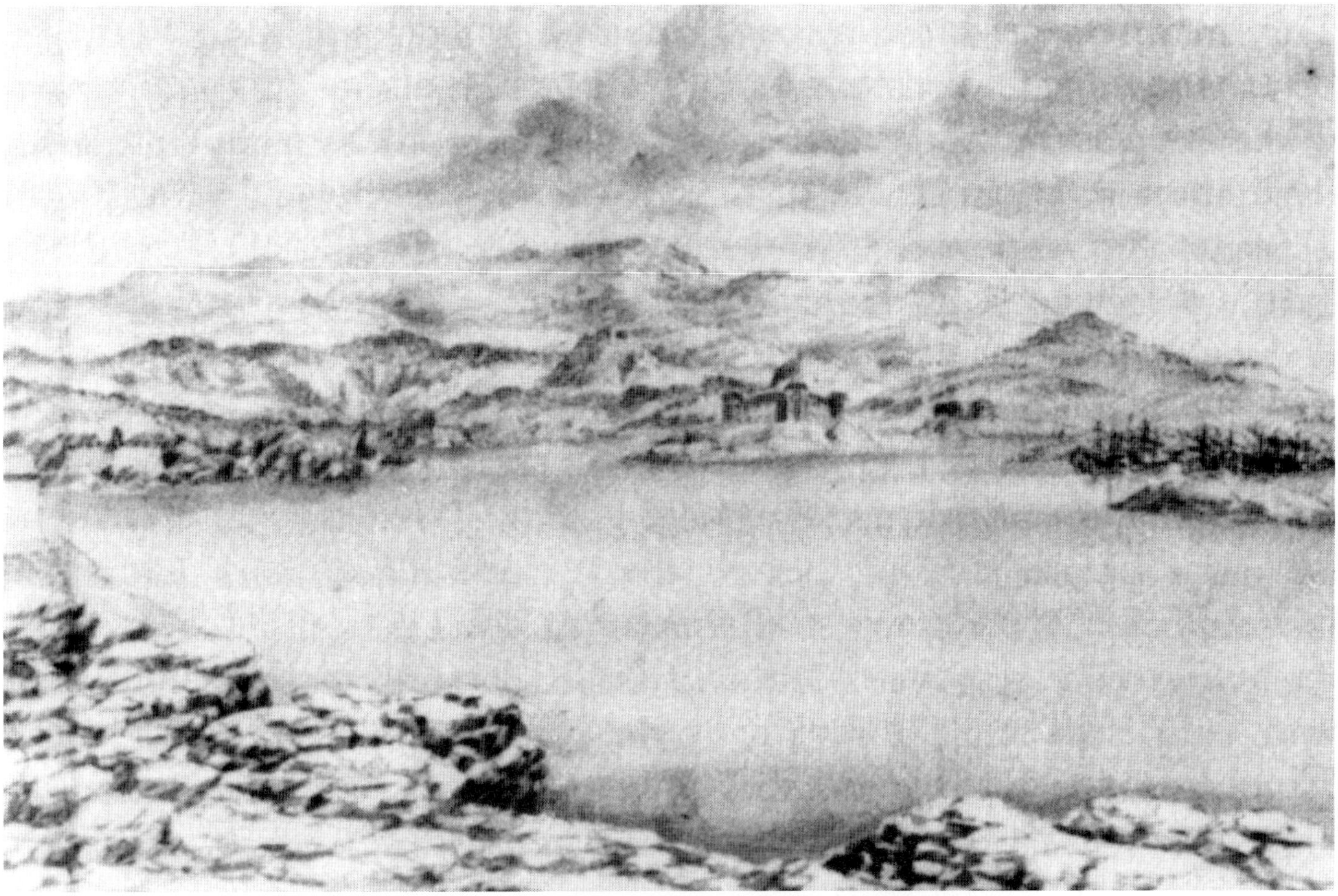

Figure 7 (*above*)
Sketch by Matthias Weiss of the entrance to the harbour of Buccari, showing batteries possibly constructed by Halley. From Cook: *Edmond Halley* (Oxford 1998).

In preparation for a later eclipse that was seen in France, Mme. Lepaute, wife of the clockmaker to the king of France drew up a similar map. It was she who had taken a major part in the calculations for the return of Halley's comet.

The Practice of Natural Philosophy

In 1703 Halley was sent by Anne, the queen of England, to survey ports in the lands of the Austrian Empire at the head of the Adriatic and select one that might serve as a winter base for an English squadron in the War of Spanish Succession. Unfortunately the charts that he prepared seem not to have survived, but the Venetian ambassador to the Imperial Court managed to obtain a report of his surveys and from this it is possible to locate the batteries that Halley constructed to defend his chosen harbour of Buccari (Bakar) to the south of Fiumi (Rijeka). Some thirty years

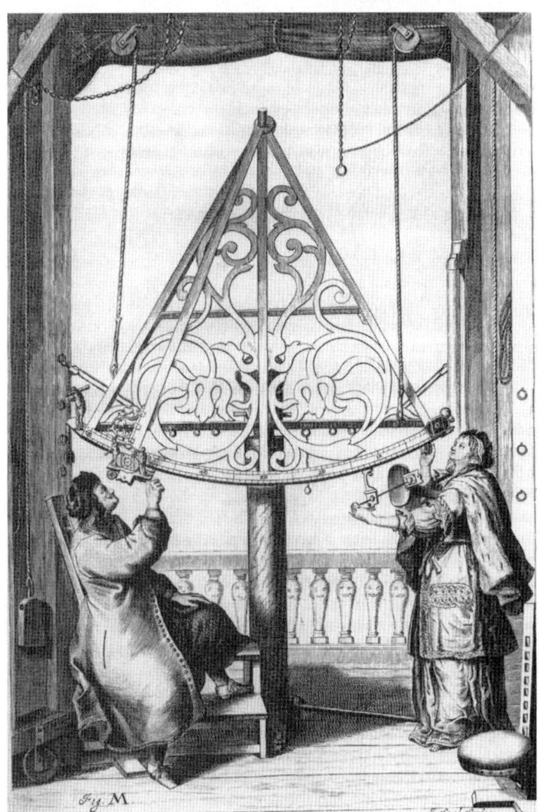

Figure 8 (*left*)
Hevelius and Elizabeth Hevelius observing with the great sextant. From Hevelius *Machina coelesti* (Danzig 1673).

Figure 9
Cassini and his Observatory. From Cook: *Edmond Halley* (Oxford 1998).

later an Imperial engineer, Matthias Weiss, made water-colour sketches which show batteries on the sites that Halley chose and that may well be those he saw built (fig. 7). Here is an example of how pictures made for different purposes may contain important information for the history of science.

Natural philosophers in earlier times often went about their studies in very different ways from those that scientists follow today, and it is very important, and very difficult at times, to appreciate the preconceptions they had, the apparatus and methods they used, and the social circumstances in which they worked. Verbal accounts may be inadequate, assuming in their readers knowledge that we no longer have. Pictures of natural philosophers at work can remedy these omissions.

In 1679 Halley visited the venerable astronomer Johann Hevelius in Danzig, and observed with him using his instruments. Hevelius's principal instrument was a large sextant of about six feet radius, much like that used by Flamsteed and Halley at about the same time, although they, unlike Hevelius, had telescopic sights. Hevelius described all his instruments in Part I of his *Machina Coelestis* (1673), from which the picture in figure 8 comes. This shows Hevelius and his young wife Elizabeth observing with the great sextant. It shows how cumbersome the instrument was and how difficult it must have been to use it. The picture is certainly the first to show a woman actually engaged in a scientific pursuit. Elizabeth observed from time to time with Hevelius and his colleagues and visitors (including Halley) and after his death she brought out his last book posthumously.

Not long after his visit to Hevelius, Halley spent five months in Paris where in the first weeks of 1681 he observed the great comet of November 1680 with G. D. Cassini at the Observatoire de Paris. The Observatoire is to be seen in two engravings, one of Cassini (fig. 9) and the other of a visit of Louis XIV to the Académie des Sciences (the Observatoire could not in fact be seen from the seat of the Académie). The pictures show that the Observatoire was indeed a grand building, but do not reveal that, as a result, it was not suitable for fundamental observations of positions of celestial objects.

Originating still within Halley's lifetime, in the Vatican Pinacoteca there is an interesting set of pictures showing observations of the sun, the moon, the planets and the comets. They were painted by Donato Creti about 1712 at the behest of Count Marsigli as a gift to the Pope. Their purpose was to enlist his aid in the construction of a public observatory in Bologna, in which they were successful. They remain of interest for their depiction of the very long telescopes which were necessary then to reduce aberrations, and for what they show of contemporary knowledge of the satellites of Jupiter (fig. 10), markings on the planets, the rings of Saturn and the phases of Venus, knowledge that can be gained at a glance from the pictures rather than by laboriously reading through many papers.

Halley is probably best known for his prediction of the return of ‚his' comet, which when it was succesfully observed, was taken to show that the cosmos behaved in a way determined by mechanistic laws, and that comets were not, as previously feared, harbingers of great events such as the Norman victory and the English defeat in the Battle of Hastings of 1066 (fig. 11) and the birth of our Lord (fig. 12).

Conclusion

Edmond Halley was very conscious of the value of images in recording natural phenomena and in suggesting their interpretation. Of all that he did in his long life, his stimulus and support for Isaac Newton in the genesis, printing, and publication of the *Philosophiae naturalis principia mathematica*, must ever be pre-eminent, but his introduction of isolines surely ranks closely behind it in the development of natural philosophy, geography and technology. His own uses of images show both the power of pictures and the limitations of pattern recognition in our understanding of the natural world.

Selwyn College,
University of Cambridge, UK

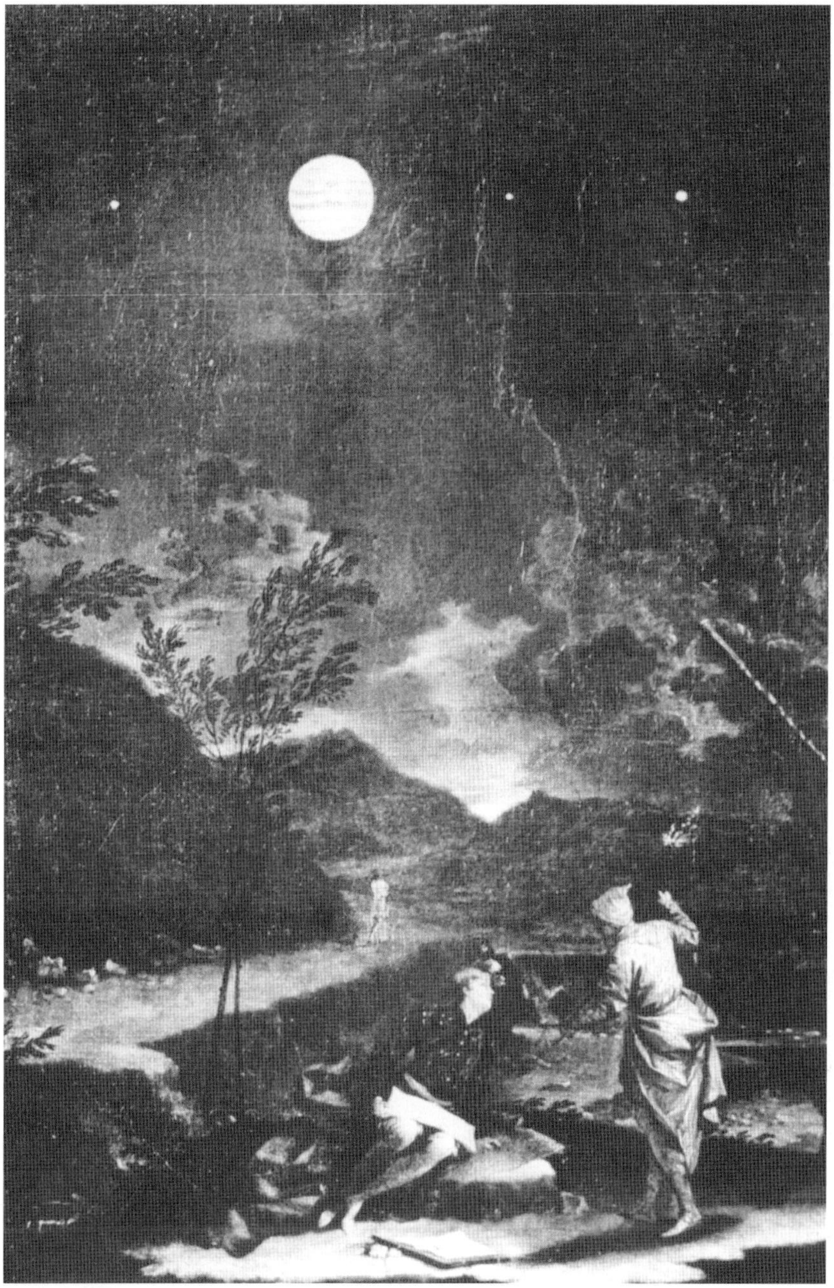

Figure 10
Observations of Jupiter. Painting by Donato Creti (1671–1749).

References

Cook, A.
1998 *Edmond Halley: Charting the heavens and the seas.* Oxford.

Halley, E.
1679 *Catalogus stellarum australium supplementum catalogi Tychonici.* London.

1686 "An historical account of the trade winds and monsoons observable in the sea between and near the tropicks, with an attempt to assign the cause of the said winds." In: *Philosophical Transactions of the Royal Society* XVI: 153–168.

Hevelius, J.
1673 *Machina coelestis*, Pt 1. Danzig.

Kircher, A.
1643 *Magnes, sive de arte magnetica, opus tripartium.* Cologne.

Morrison, L. V., F. R. Stephenson, and J. Parkinson
1988 "Diameter of the Sun in AD 1715." In: *Nature* 331: 412f.

Thrower, N. J. W. (ed.)
1981 *The three voyages of Edmond Halley in the Paramore*, 1698–1701, London.

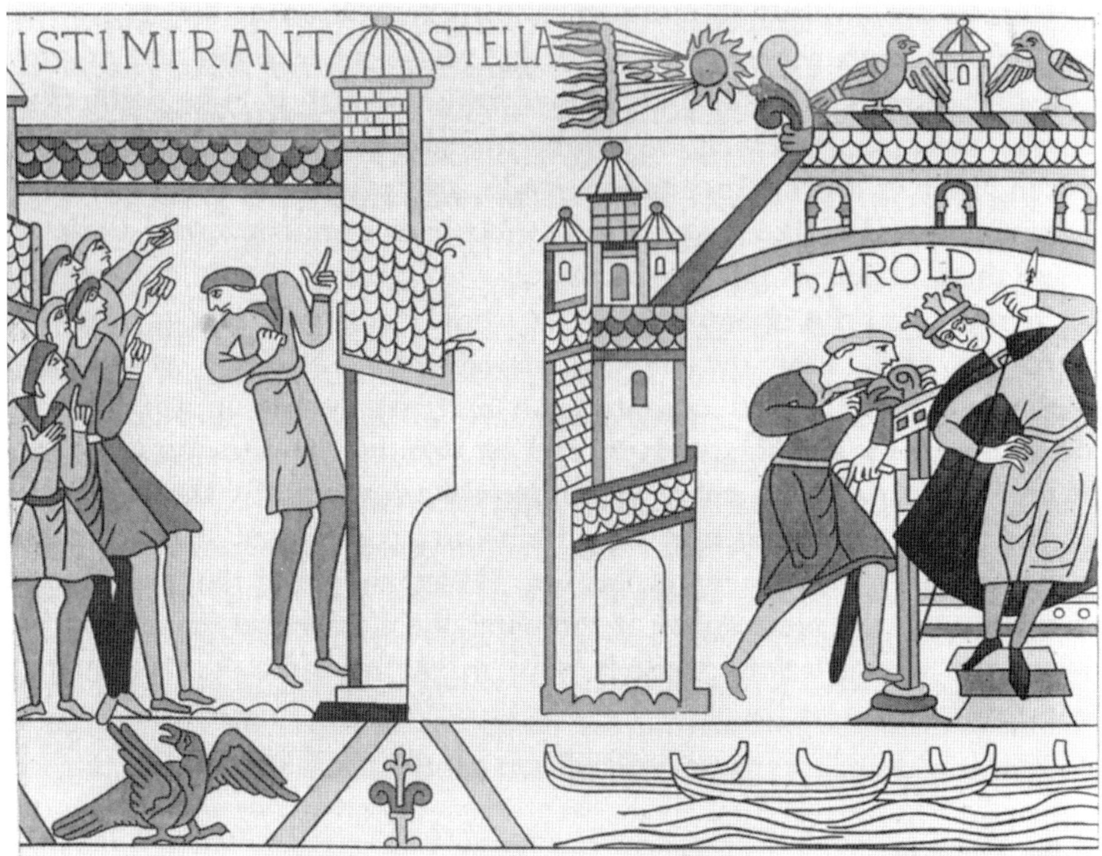

Figure 11
Halley's comet in the Bayeux Tapestry (AD 1066). From Cook: *Edmond Halley* (Oxford 1998).

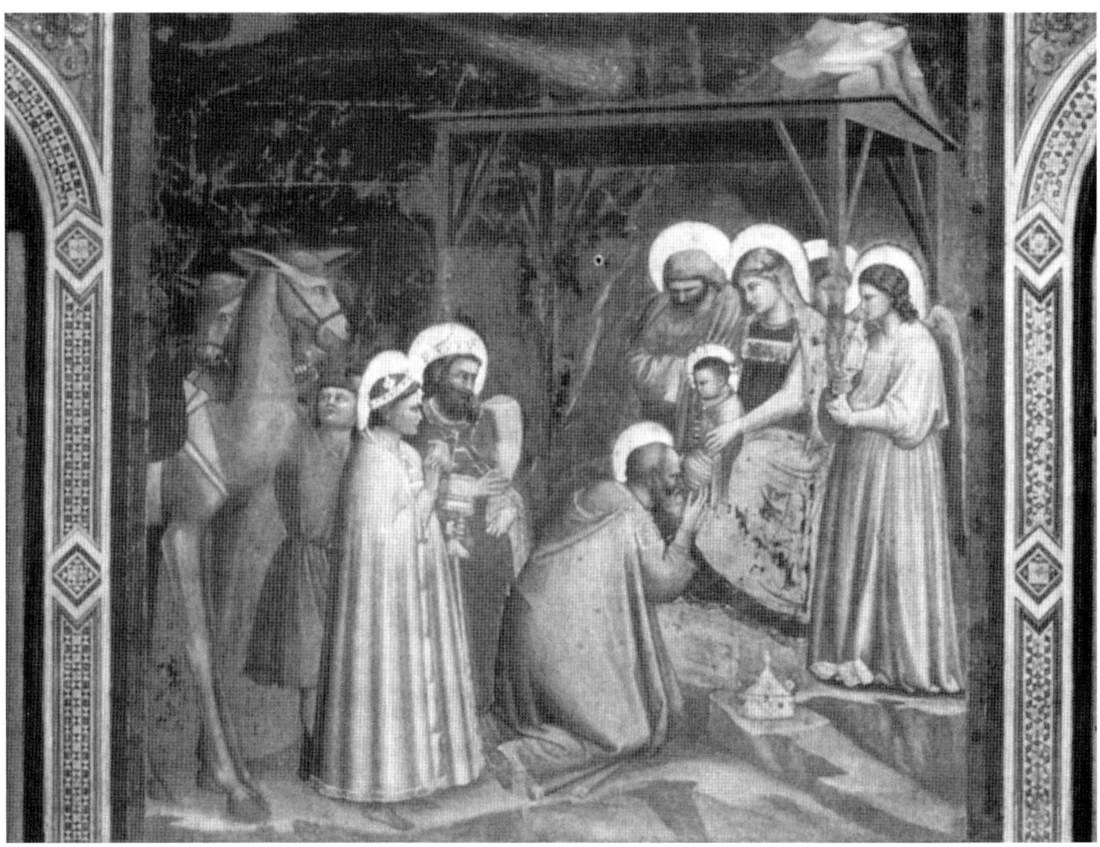

Figure 12
Halley's comet in the fresco by Giotto at Padova. From Cook: *Edmond Halley* (Oxford 1998).

V.
Systems of Knowledge and their Representation by Images

Encyclopaedias and Architecture in the Sixteenth Century

Annarita Angelini

Rather than *mirrors of the world* or, rarely, *trees*, many of the *summae* of knowledge composed between the mid-sixteenth century and the early years of the seventeenth century were intended as buildings or imaginary constructions within which portions of reality, fragments of learning, symbols or elements of knowledge were given form, unity, order and completeness. Aware of this, their authors often declared themselves, to be not specialists, but rather *architects* and *perspectivists*, capable of transferring a paradigm, a *design* which was primarily in their minds, to pre-existing materials or elements, thus making the resulting work, the *building* of knowledge, conform to this design. The titles they gave their works are the first confirmation of the "architectural" or "scenographic" role assumed by the authors.[1]

There are also cases in which an actual building – perhaps non-existent, but described as though real – is used to give knowledge the architectural framework required to support and unify it. This is the case, for example, of *Phantoteque – bati* [...]. *afin qu'il contient tout l'univers et ses parties* – where Jean Bodin sets his *Colloquium Heptaplomeres* (Bodin [1984] 3; see Vasoli [1990]), or the Palazzo of Accademia della Fama in Venice where the rooms bear the names of disciplines and the layout of the building is itself indicative of the way academics divided up knowledge (see Badoer [1560] c. 1v).

Less immediate in its evocative capacity and more recent chronologically is the image of the building to allude to the book as to the most complete and perfect creation – ordered, harmonious, measured and symmetrical in all its parts[2] – of which human intelligence is capable. An example: in the introduction to the French edition of *De Harmonia Mundi* by Francesco Giorgi, Guy Le Fevre de la Boderie presents the book as a *building* (the book is like the world; the world is like a building; the book is like a building) and its author as the architect responsible for its design even more than its production. The author, he wrote,

> *ne s'est pas contenté de comprendre et embrasser au rond de son entendement ce qui est contenu et comprins en la Sferique rondeur de ceste grande machine mondaine,* [...]*: mais a bien osé penetrer outre la derniere enciente de ce grand Cirque et Theatre iusque dans l'Archetype, monde intelligible* [...]. *Donc nostre autheur imitant au plus pres qu'il a peu l'ordre du monde, et l'Ouvriere et l'Architecte d'iceluy, a premierement divisé et comparty ceste harmonie du monde en trois Cantiques, chasque Cantique en 8 Tons ou Livres, et chasque Ton en plusieurs chapitres, si bien conioncts et ailliez, et si cointement distinguez et ordonnez tous ensemble: que la Methode et clarté, voire le mesme ordre du monde univers, s'y est reluyt manifestement* (Le Fevre de la Boderie [1578a] cc. aijv-aiijr).

At the beginning of the following century, Alsted drew on the same *volumen/fabrica* analogy to describe the encyclopaedia as "*systema omnium systematum*."[3] Sciences and arts are represented in the form of a building like in Robert Record's *Castle of*

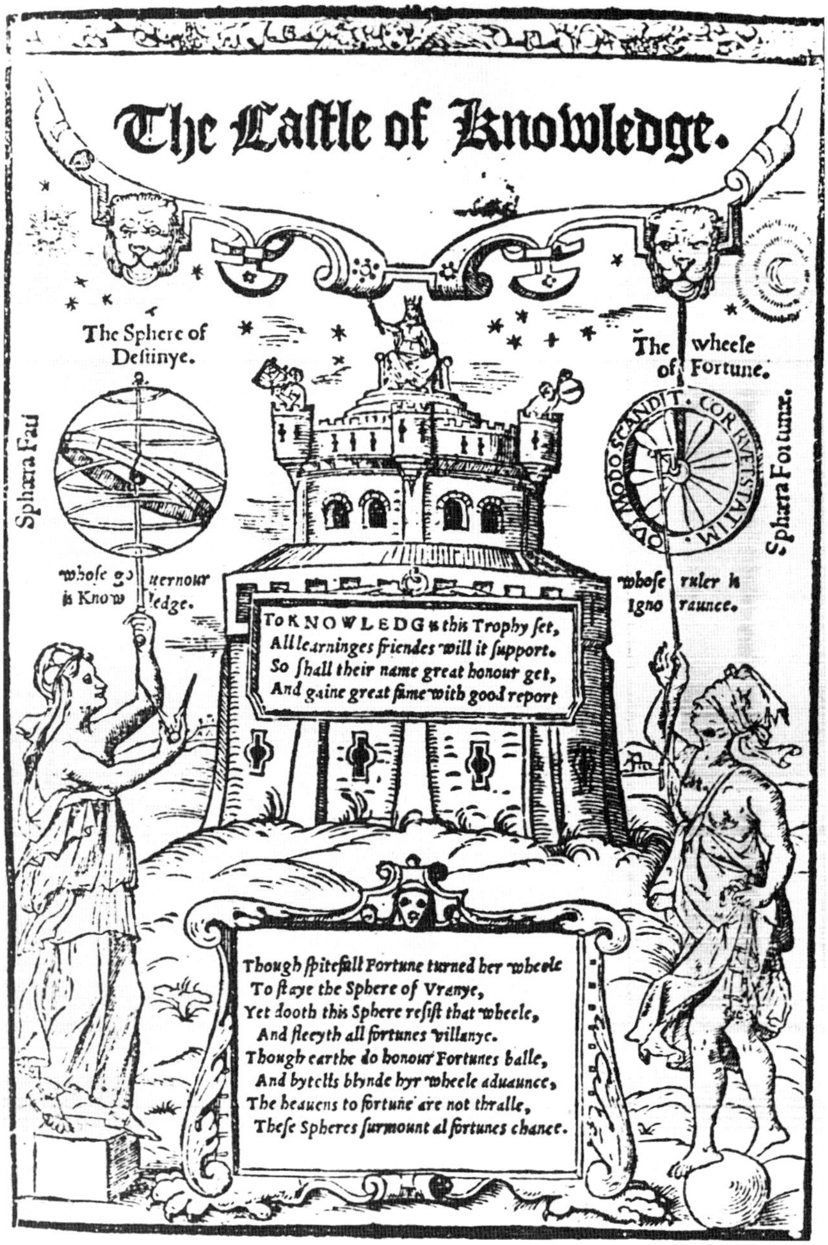

Figure 1
The building of mathematical sciences. Title-page of Record: *The Castle of Knowledge* (London 1556).

Knowledge (fig. 1), or inside a building like in the frontispiece of Alsted's *Encyclopaedia* (fig. 2), or, and this is the case in Andreae (fig. 3) and Campanella among many others, like quarters of a city.

Moreover, the whole world conceived as the most general object of knowledge is represented like a building between sun and moon, as in the *Theatre of world* engraved in 1588 by N. Bonifazio (fig. 4).

There is no doubt that to many authors from the fifteenth and sixteenth centuries, the architectural metaphor, evoking a Platonic image of the divine architect of the universe, served to underline the authority of man, created in the image and likeness of God, in an undertaking such as the ordering of the universe of knowledge. This in itself is significant and largely extraneous, at least in its possible consequences, to the cultural climate of fifteenth-century humanism. It is true that the scholar, like the Demiurge, works on pre-existing material, the objects and contents of knowledge, but in ordering this material and giving it form and a unitary framework, he enjoys the same freedom which Plato attributed to the Architect of the Universe. During the sixteenth century, the image of God with his compasses marking out and measuring the world appeared as a fitting evocation of the first act in the foundation of knowledge, that which precedes identification of specialisation and represents learning as a whole.[4]

But does a relationship actually exist, other than that which links the two terms of a metaphor, between the *ars aedificatoria* and encyclopaedic learning, between an "architectural" science and the capability of the human mind to embrace the infinite variety of objects of knowledge?

In 1556, a much annotated edition of Vitruvius' *De Architectura* was published in Venice; its author, the Veneto scholar, Daniele Barbaro. This was not the first time Vitruvius' text had been printed and annotated, but Barbaro's notes were unusual in that they presented architecture as the synthesis and maximum expression of all human knowledge. According to Barbaro, architecture should not be identified with a particular science or technique, but with knowledge and heroic virtue, in other words, with the universal knowledge of everything the human mind can contain and put into practice (Barbaro [1556] 21).

He who understood the real significance of architecture, wrote Barbaro, will also understand the circle and concatenation of the sciences ("il giro, et la raccomunanza delle scienze"), in other words, what in Greek would be known as *εγκυκλοπαιδεια*. This in short was Barbaro's argument: all sciences and all arts, potentially infinite in their number and immeasurable in their advancement, share a limited number of common terms and principles. This is not all. All can be considered as nothing more than the various consequences and developments of those original principles and characteristics. Unlike

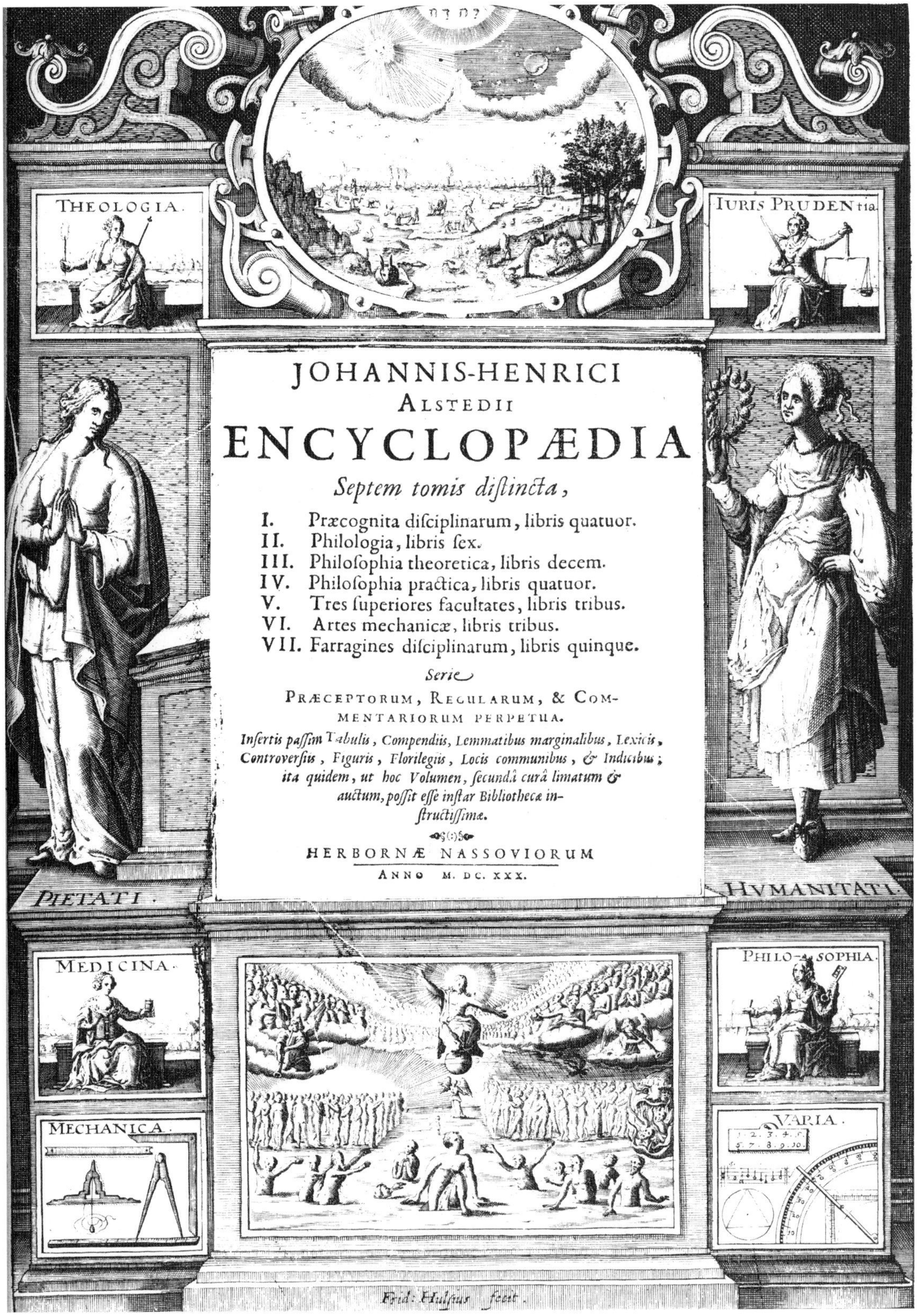

Figure 2
The disciplines theology, jurisprudence, medicine, philosophy, mechanics, *varia*. Title-page of Alsted: *Encyclopaedia* (Herborn1630).

all other disciplines, architecture defines and brings together the elements common to each individual part of knowledge at the height of their universality. These common elements are the dissecting and analytical habitus of the mind and its subsequent re-assembling attitude, called by Barbaro *discorso* and *fabrica*, as well as the six categories of classical *ars aedificatoria – ordinatio, dispositio, symmetria, eurithmia, decor, distributio* – which

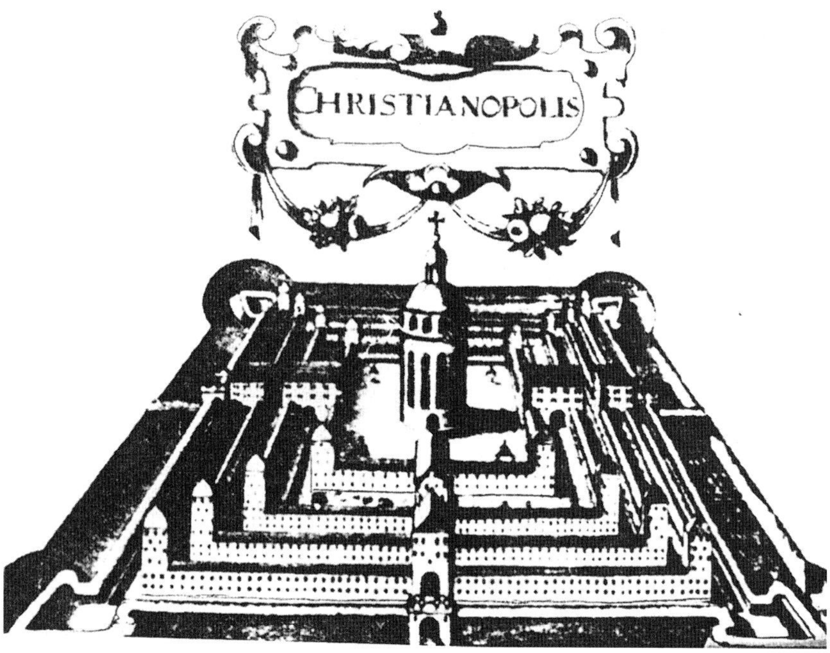

Figure 3
The ideal city. From Andreae: *Respublica Christianopolitana* (Strasbourg 1619).

represent the organisational criteria for each discipline. Thus encyclopaedia, in the sense of universal knowledge, and architecture, signifying the generality of knowledge and the universality of the disciplinary principles, seem to represent two aspects of the same object (Barbaro [1556] 23f.). This knowledge is very different from the mediocre competence in the eleven disciplines Vitruvius' aspiring architect was expected to have.

Neither the extension of architecture beyond the borders of *ars aedificatoria* nor the idea of learning as an active virtue was new to Western culture in the sixteenth century. What was new, however, was their identification with "encyclopaedia," a term obtained from classical texts, but which in the classical world and even during the years in which Barbaro annotated Vitruvius, had no connection with the notions of either "universal knowledge" or "wisdom."

The quotation most common among late sixteenth-century systematists of knowledge comes from book I of Quintilian's *Institutio Oratoria*, where the Greek *"encyclion paedian,"* literally *ordinary education*, is periphrased in Latin as *"orbis doctrinae"* meaning cycle of studies. A series of disciplines, seven according to Quintilian, represented the basic training before acquisition of a specialisation, in this case represented by oratory.[5]

Among the first to propose anew Quintilian's affirmation in the sixteenth century was the Humanist Filippo Beroaldo. This was the first in a chain of quotations which was to last throughout the following century.[6] In 1609, ninety years after Beroaldo's oration, a correspondence between *enciclopedia* and *orbis omnium doctrinarum* could still be found, in Rudolfus Goclenius's *Conciliator* and in his *Lexicon Philosophicum*, with the meaning of *absolute* knowledge of *all* the sciences ("absoluta cognitio omnium scientiarum")[7] and, with the same meaning, in J. H. Alsted's *Philosophia digne restituta*.[8] The

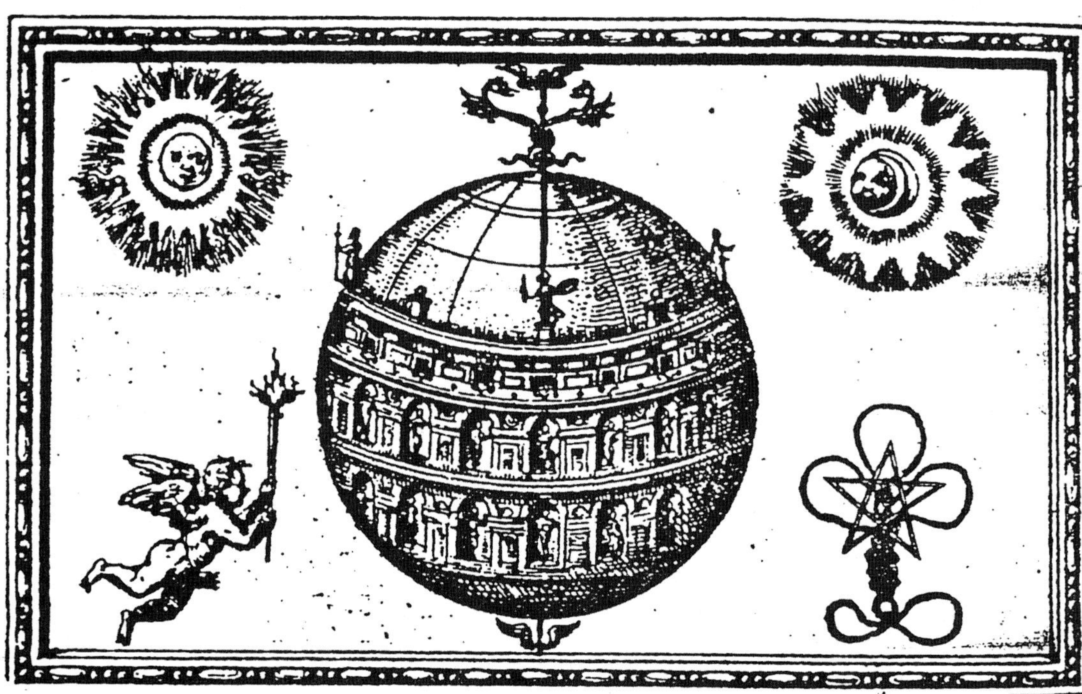

Figure 4
The *Theatre of world*, this is the world in the shape of a building. Engraving by N. Bonifazio (1588).

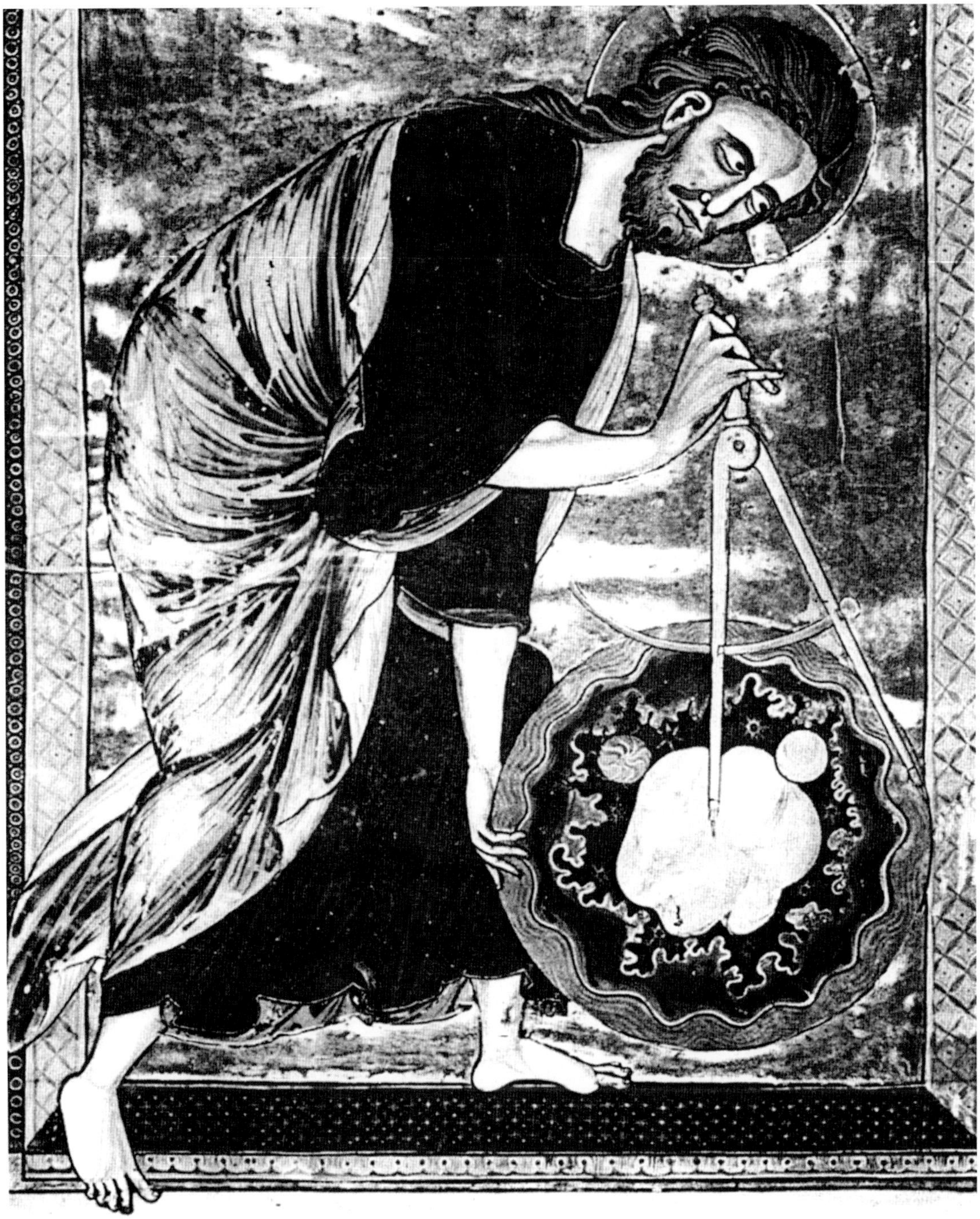

Figure 5
God the *Pantocrator* marking out and measuring the world. Miniature from *Bible moralisée* (13th century).

only, but important, difference with respect to Quintilian and Beroaldo was the addition of the adjective *"omnium"* agreeing with the plurals *"doctrinarum"* and *"scientiarum"* with which they gave the encyclopaedia the character of completeness.

Beroaldo was writing at the beginning of the sixteenth century; Goclenius and Alsted in the first twenty years of the seventeenth. In the meantime, the Quintilian quotation had considerable fortune. Among the many, Pierre de la Ramée, at the height of his polemic against the teaching imparted at the university, had taken up the quotation from the *Institutio Oratoria* again when presenting his own *philosophica disciplina* as the *entire universe* which some have called *εγκυκλοπαιδεια*.[9]

Three years later, in *De veris principiis*, Mario Nizolio, professor at Parma and champion of Ciceronian oratory with an anti-metaphysical, anti-scholastic function,

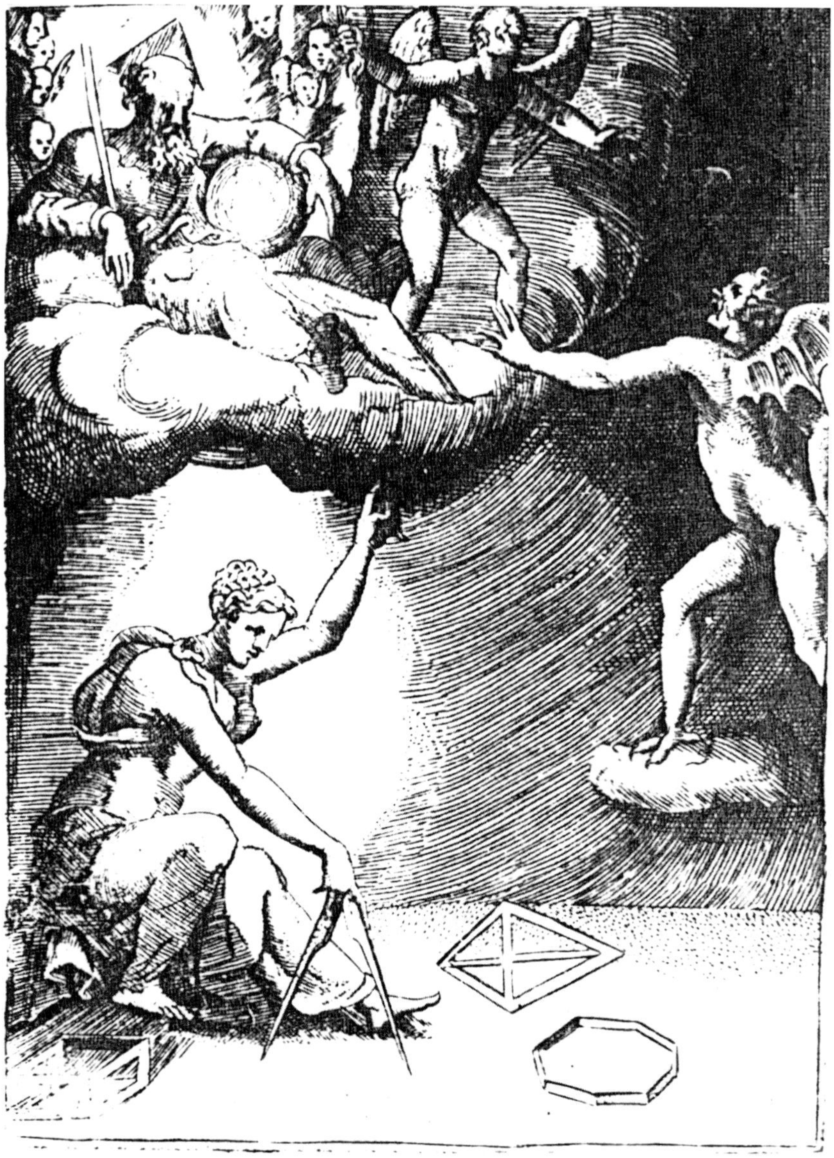

Figure 6
The human knowledge. Symbol CXXIIII of Bocchi: *Symbolicarum Quaestionum.* (Bologna 1555). Compare the attitude of the female figure with the compasses to that of God measuring the world (fig. 5).

had drawn an analogy between the totality of the sciences and that which the Greeks called encyclopaedia and the Romans the circle of all disciplines ("*omnia* literaria studia, et *universam* rem literariam quam Graeci *εγκυκλοπαιδιαν* Latini vero circulum *omnium* doctrinarum vocant") (Nizolio [1952] II, 34).

Another master of eloquence, Guillaume Budé – like la Ramée, linked to the experience of the Collège Royal and, like him, promoter of an educational programme different from that proposed at the Sorbonne – remained true to that definition, considering the encyclopaedia as "*universa* literarum disciplina *orbisculasrisque* eruditionis curriculum" (Budé [1966] I, 16).

Less known but no less omni-comprehensive and interesting was the definition proposed by a curious individual, the Hungarian Paulus Scalichius, promoter of a symbolic philosophy: a mixture of Lullism, cabbala and forms of hermetic Platonism. In 1559, Scalichius gave the name *Encyclopaedia* not to a collection of precepts and disciplines, but to the ideal container in which, he wrote, there would be space for all sciences, sacred and profane, for all philosophies, whether active or contemplative, and even agreement between all schools and sects.[10]

Thus halfway through the sixteenth century, the adjective *completus* in Ramo, *omnius* and universus in Nizolio and Budé, was already distorting the classical notion of encyclopaedia in the direction which was to lead it towards Alsted: from basic training or a cycle of preliminary studies, to a total and general system of knowledge.

But we need to go even further back in time to find – as far as I am aware, for the first time – the adjective *absolutus* with which Goclenius gave the encyclopaedia a character of completeness found in neither the Greek nor the Latin tradition and certainly not in that of the Humanists. To be precise, we need to go back to Joachim von Ringelberg's *Ratio Studii*, published in Paris in 1530 and again in 1538 with the title *Lucubrationes, vel potius absolutissima κυκλοπαεδια*. Like Budé, friend of Erasmus and part of that dialectic tradition which was to lead from Rudolphus Agricola to de la Ramée, when introducing the chapters of his *cyclopedia*, Ringelberg had preferred to set aside the criterion of order, too closely tied to the Aristotelian tradition from which, he made it clear, he intended to keep his distance, in favour of the completeness of a course of studies going way beyond the traditional arts of the Trivium and Quadrivium, from grammar to medicine and from geomancy to agriculture and physiognomy.[11]

Earlier, Rabelais had disparaged the term "*encyclopédie*" associating it with useless scholastic erudition, emblematically shut away in the library of the St. Victor Abbey, in contrast to the total and reformed knowledge (*toutes disciplines restitutées*) which Gargantua did not just extend to sciences and techniques, but also wanted to see renewed in terms of its didactic criteria (Rabelais [1979] I 255 and 206).

In the authors so far considered, not the Greek word *encyclopaedia* but the meaning now attributed to it of *universe of all*

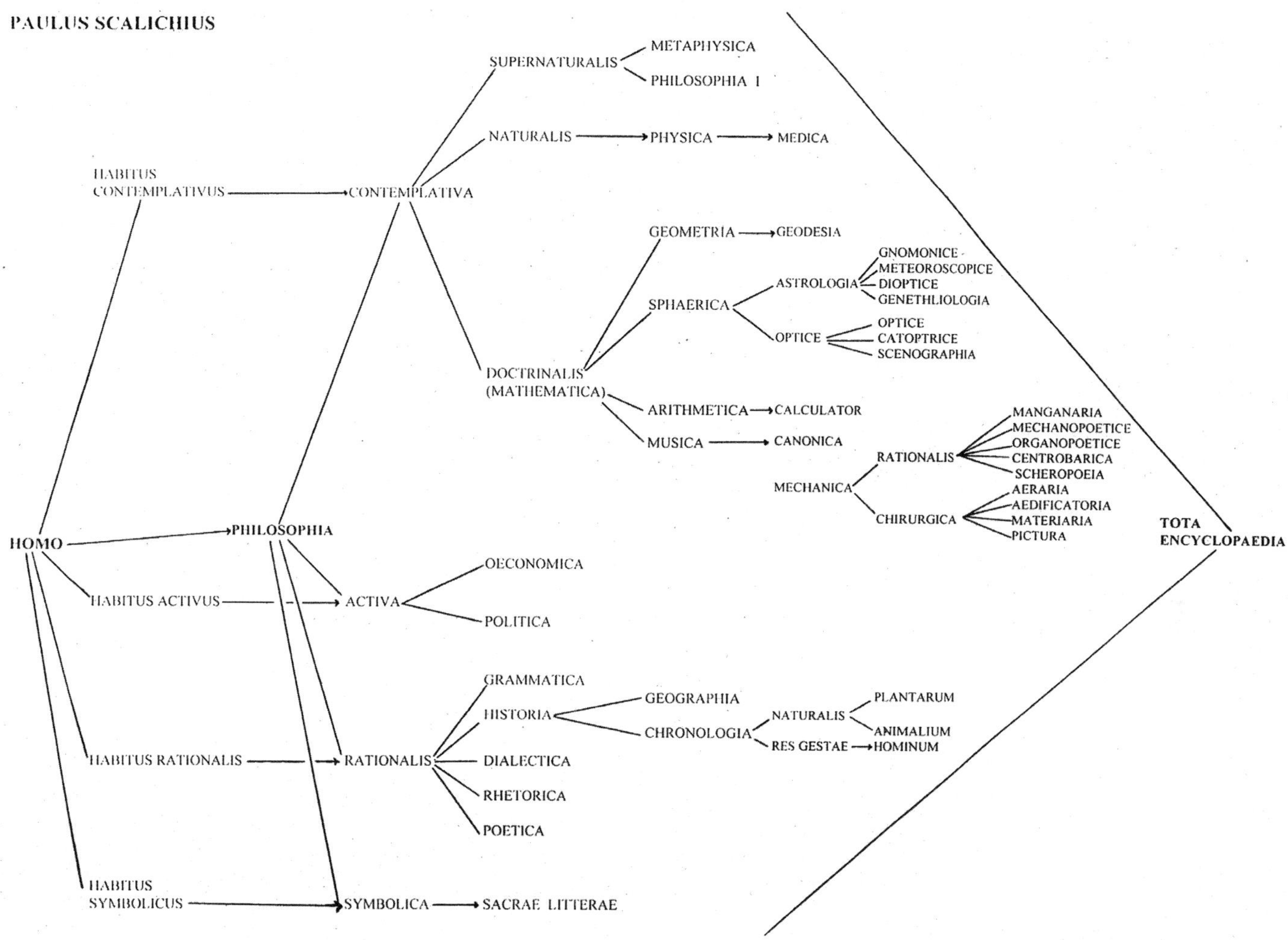

Figure 7
The *orbis disciplinarum.* From Scalichius de Lika: *Encyclopaediae* (Basel 1559).

knowledge was accompanied by battles no less determined than that of Rabelais against the "pseudophilosophy" and "contentious knowledge" of the schools. In fact, it is precisely this pressing need for reform, the urgency for knowledge founded on a solid base but susceptible to growth and capable of guiding man in the "operations" of earthly life, that appears as the most evident common trait accompanying sixteenth-century encyclopaedic enthusiasm. Almost as though the encyclopaedia *omnium doctrinarum* – its shape and outline rather than the contents it transmits – were indissolubly linked to the need to extend the training of students beyond the seven liberal arts to the limits of "practicable" knowledge and, above all, to the intention of eliminating every rigid demarcation which had the aim or effect of isolating the theoretical levels of science from the applicative and practical levels. Almost as though, hidden behind an ancient term, knowledge-in-a-circle was precisely *per sè* the alternative and almost the antidote to what appeared as the limits of the hierarchical knowledge of the schools: the excessive formalism, the contentious character, the incapacity to respond to the urgencies of trade with the world, the inability to describe a universe which was not just revealing a number of its *arcana* and therefore different from that handed down from classical times, but which was apparently also susceptible to modification by the mind, hand and discourse of man. The knowledge under attack was a pyramidal knowledge constructed through the progressive exclusion and subordination of disciplines: the knowledge which was trying to assert itself, a knowledge capable of including and reconciling languages, disciplines, attitudes and sects; the latter, circulating, mobile and flexible; the former, largely static in its overall structure and unmodifiable in its parts. This knowledge constituted a cognitive framework subordinated to ontology and limited by theology, as opposed to

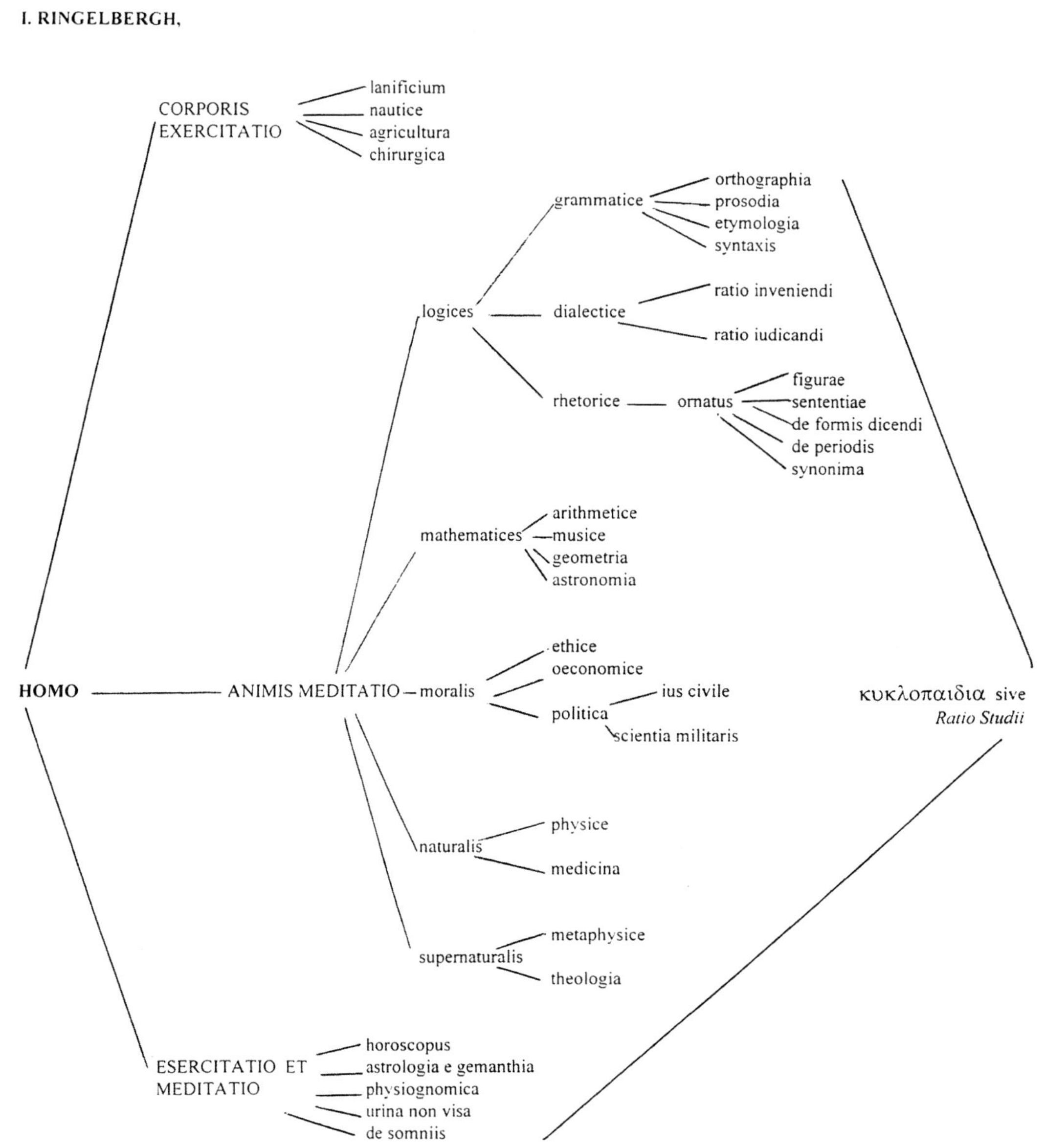

Figure 8
The scheme of human knowledge. From Ringelberg' *Lucubrationes* (Basel 1538).

"buildings" of knowledge constructed by the art of man, largely "indifferent" to ontologies and "neutral" with respect to theologies, yet able to modify the *ordo rerum* and bridge the distance between God and man, between eternity and time, because, again in the words of Daniele Barbaro, to practice architecture and thus possess true knowledge means to satisfy in part the desire for eternity (Barbaro [1567] 21).

It is precisely the semantic shift undergone by the term "encyclopaedia," from the classical meaning of "ordinary education" and "basic educational cycle" to that of general and universal knowledge, which reveals the new awareness of procedures for acquiring and ordering knowledge and, above all, the awareness of a new identity for the arts as a whole, characterised not by their being a "mirror of nature," but by their being the complete and perfect work of man, closer to the irreprehensible regularity of the divine design than to the tangible world.[12]

For Barbaro and many other of his contemporaries, the model of the total and exhaustive organisation of knowledge is supported by an analogy to the image of the universe, conceptualised and symbolised by astronomers and cosmographers. Completeness and perfection, wrote Barbaro, characterise, although in two different spheres, the cosmos created by God and the universe of all the arts and sciences in which man, even before investigating the world which surrounds him, discovers and expresses his own intelligence and ability to intervene and modify the order of tangible reality.

Guy le Fevre de la Boderie was of the same opinion when in his *encyclie* entitled *La Galliade ou de la Révolution des arts et des sciences* he presented his knowledge as divided into spheres.[13] According to Pierio Valeriano's *Hieroglyphica,* such a cosmographic scheme was very close to the Egyptian notion of encyclopaedia (fig. 10).

The main difference between Valeriano's symbol and the notion of encyclopaedia we are describing consists in his conceiving theology as the external sphere which includes all other disciplines and parts of knowledge and wisdom. Another cosmos represents Bocchi's idea of *sapientia* (fig. 11).

A chain links that universe to God's hand and inside it there is a light which symbolised the almost divine power of the human mind (Bocchi [1555] symb. CXXXVI).

Quintilian's definition must then have seemed extremely apt and appropriate, identifying the "encyclopaedia" as the *"orbis doctrinae."* It should however be emphasised that *"orbis"* in Quintilian's Latin, as in Beroaldo's, had no connection with the "orbit" or "orb" of the astronomers, but simply translated the Greek *εγκυκλιον* with the meaning of "cycle" rather than "circle." Despite this, as early as the second half of the sixteenth century, Quintilian's expression, precisely as a result of the scientific and cosmological connotations it had acquired, lent itself to more evocative analogies than other terms which could, however, boast a longer and more apt tradition as to meaning. More than the medieval *summa* or the *polimathia* of Heraclius, debased in the Middle Ages to mere *curiositas*, or the *panepistemon* recently reinstated by Poliziano, the sixteenth-century *orbis disciplinarum* could express and bring together two notions extraneous to the *εγκυκλοπαιδια* of the Greeks and the "orbis disciplinae" of the Romans, but of considerable importance for the "encyclopaedists" of the early modern age, namely that of a *universal system* (complete, ordered, cohesive and centralised), comparable to the image of the cosmos on the one hand and on the other that of the *circulation* and *interweaving* of disciplines which was to become established primarily starting in the following century, to the extent that it became the dominant char-

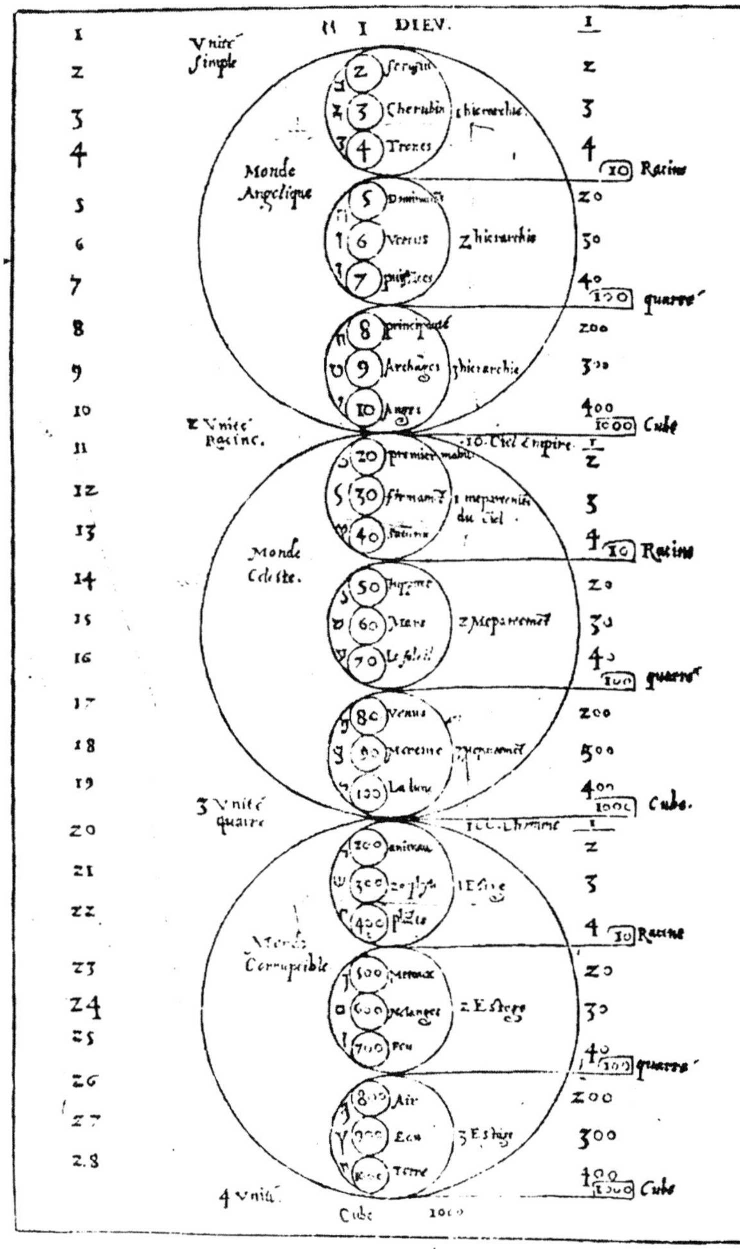

Figure 9 (*above*)
The three worlds and their numerological correlations regarded as a model for organising sciences. From Le Fevre de la Boderie [1578a].

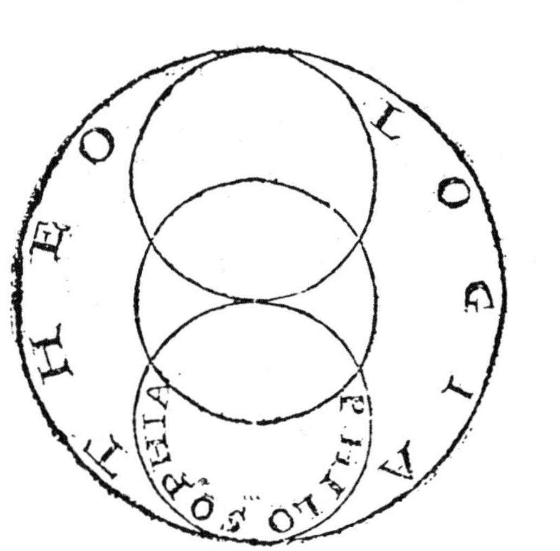

Figure 10 (*left*)
The symbol of the encyclopaedia. From Valeriano: *Hieroglyphica* Basel 1556).

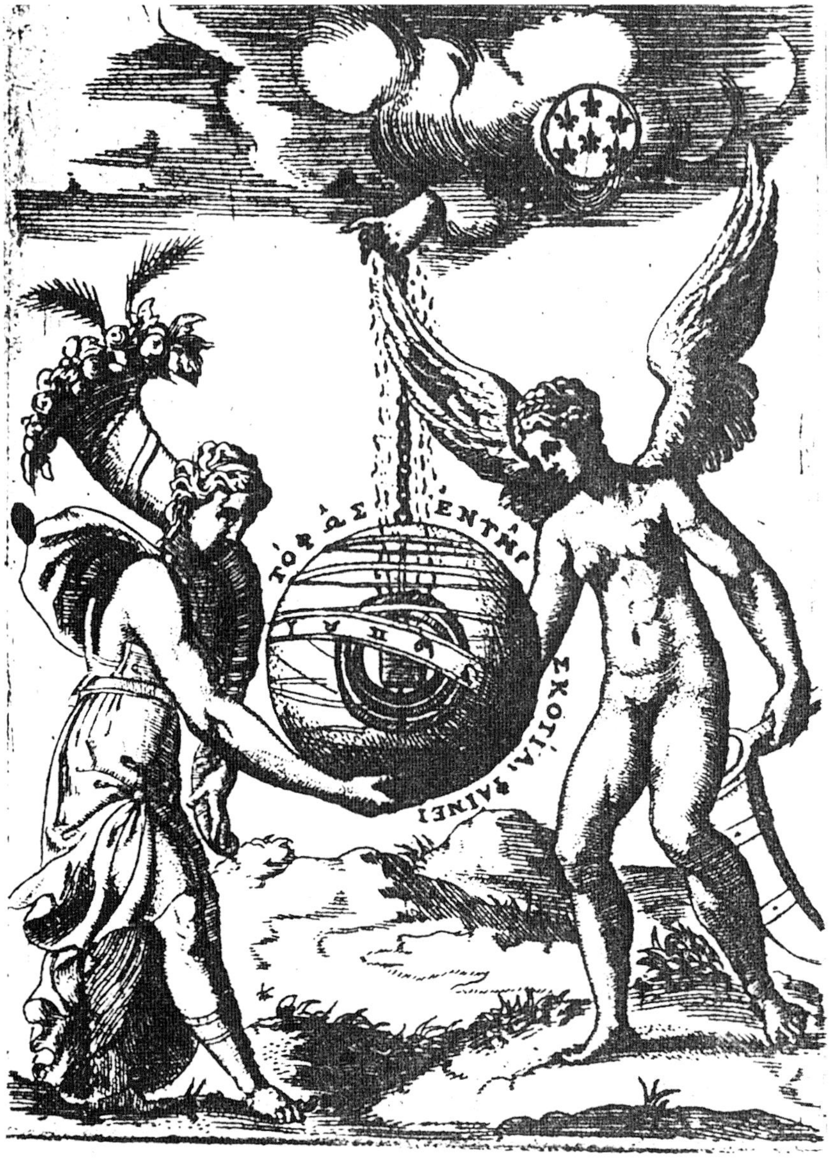

Figure 11
Symbol representing the almost divine power of human mind. Bocchi, *Symbolicarum Quaestionum*. (Bologna 1555), symbol CXXX.

acteristic of the "modern" notion of encyclopaedia. It is on these two aspects of *universal system* and of *interweaving of disciplines* that the sixteenth-century encyclopaedia, notwithstanding the continuity of the past suggested by the name, broke its links with tradition, including that most recent and conceptually closest, described in texts such as Poliziano's *Panepistemon* or Reisch's *Margarita Philosophica*[14] (figs. 13–14).

It was precisely the idea of the interweaving and superimposition of disciplines which allowed Quintilian's "didactic cycle" and the "ordinary education" of the Greeks to be transformed into "spheres" and even "cosmos." It thus enabled the "cycle of knowledge" to be made total and even universal, in that merely by interweaving and constructing levels of knowledge which could be superimposed one on top of the other, it could seem possible to contain not just *some arts* (*caeteres artes*) as Quintilian and the Humanists wanted, but the *absolute* knowledge of all the sciences in the same, single *"orbis."*

Either the idea of *"orbis"* had to be discarded, and with it the idea of the continuity of knowledge, in order to propose the inclusion of all the disciplines, but separately as Rabelais had done, or the number of sciences inevitably had to be reduced to the detriment of the universality of the whole. This had been the road most often taken by the encyclopaedic tradition which had numbered seven, nine or, at maximum with Vitruvius, 12 *artes*.

Another possibility, and this was the completely new road, consisted in limiting the space reserved for each particular sphere. The encyclopaedia no longer conceived as a classification of the *artes* placed side by side, each representing the boundaries of the next, but instead became a "system" of disciplines whose paths continually interweave.

This second aspect – indirectly implied by exacting adjectives such as *omnius, completus, universus* and *catholicus*, absent from the classical tradition – and explicitly acknowledged by encyclopaedic schemes such as those of Nizolio, Ramo and Scalichius – could not, at least for the moment, be borrowed from astronomy. This was not possible for simple chronological reasons. The Copernican image of the universe – the only innovation circulating in the mid-sixteenth century – was no more help to a view of the encyclopaedia as a perfectly contained circle and the inextricable interweaving of disciplines than was the geocentric Aristotelian vision, both perfectly spherical, but delimited by real and impenetrable internal boundaries (see Ong [1956]). These boundaries, had they been the spheres of the elements or the crystalline orbs, would nevertheless have obstructed the "circling" of the disciplines. Again in the words of Daniele Barbaro, "il giro delle scienze." No *"orbis"* without at least the same diameter as the real universe would have been capable of containing them *all*; no mind, not even the "false bottomed" mind of Pantagruel, no book, no library, no theatre of memory would have been capable of containing the innumerable elements of knowledge in succession.

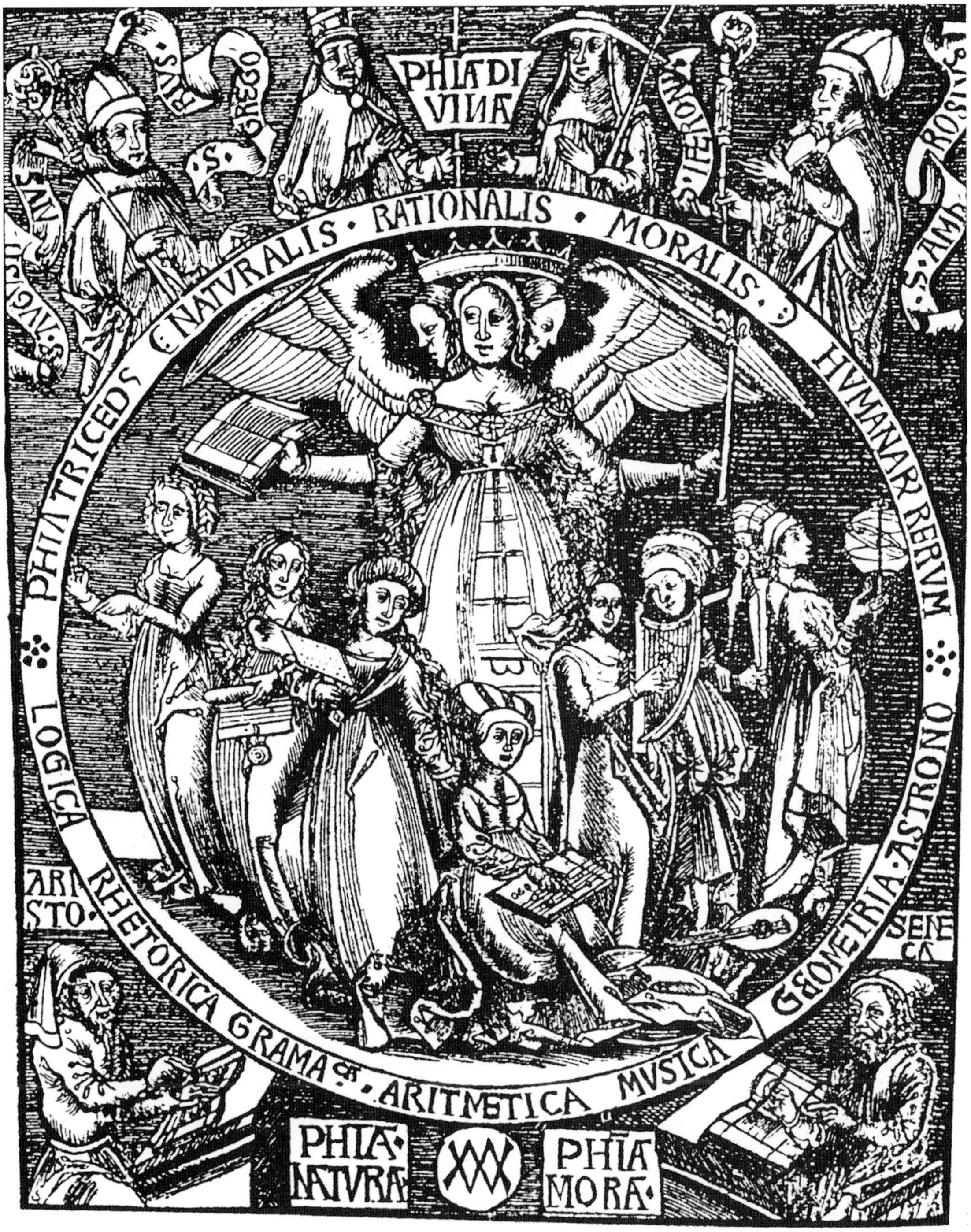

Figure 12
The allegory of Philosophy. From Reisch (Heidelberg 1496).

While waiting for post-Copernican astronomy to authorise a conception of "*orbis*" as a system of mutually intercepting paths, the sixteenth-century "cosmographers of knowledge" could obtain this requisite of an encyclopaedia as the composition and mutual copenetration of disciplines and levels of knowledge from another "place" in the "topic" of tradition, more precisely, from a statement by Cicero from *Pro Archia* according to which:

> *omnes artes, quae ad humanitatem pertinent, habent quoddam commune vinculum et quasi cognatione quadem inter se continentur* (Cicero I, 2).

Cicero, however, had never associated this affirmation with the term "encyclopaedia" as Vitruvius had done in chapter I of *De Architectura*. After reviewing all the competencies required by the architect, Vitruvius declared himself understanding of those who

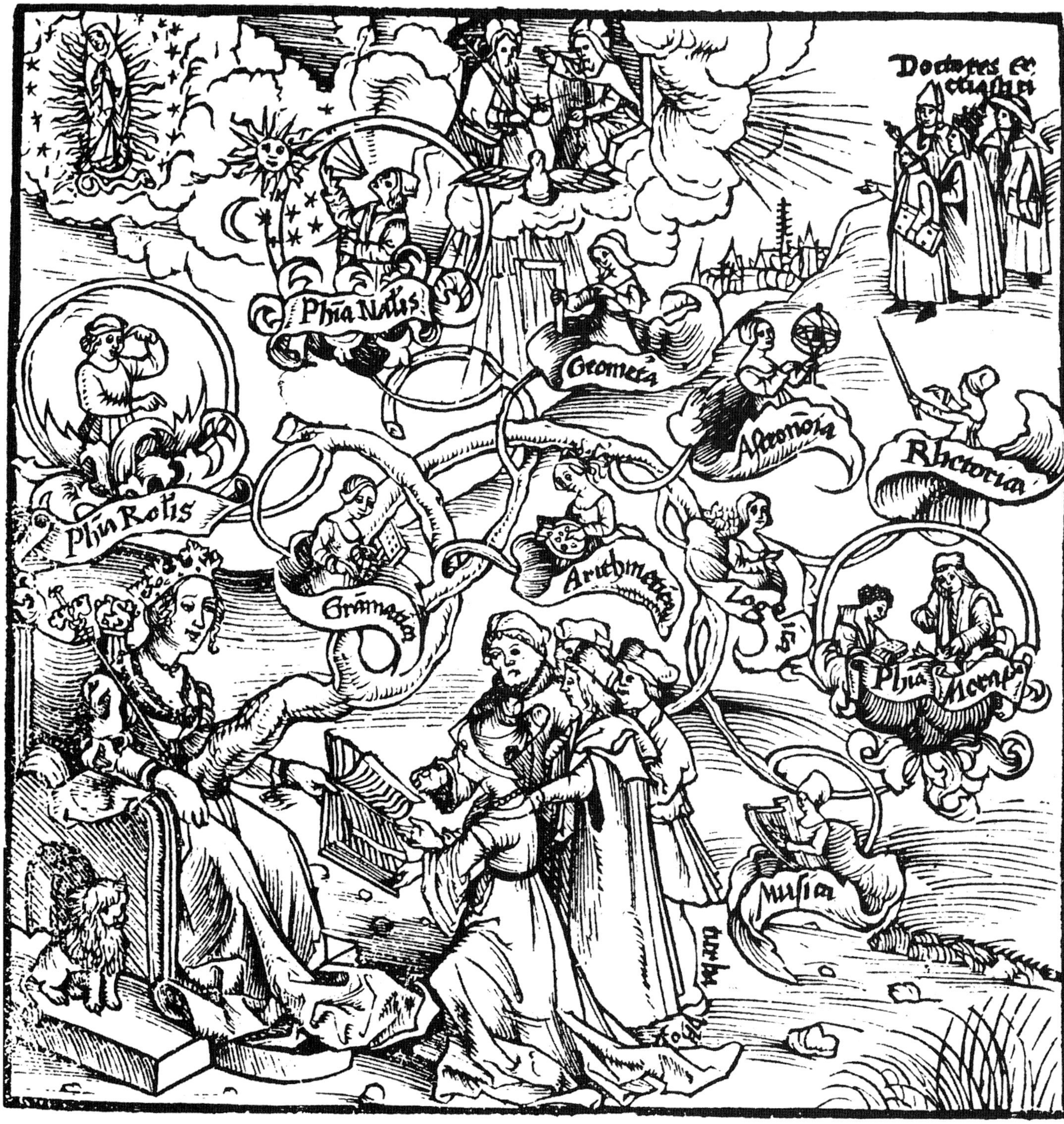

Figure 13
Philosophy is represented as the trunk of the tree of liberal arts whose branches are interlaced with one another. From a mid-sixteenth-century edition of Reisch: *Margarita Philosophica.*

expressed doubts as to the capacity of human intelligence to assimilate such a quantity of doctrines. He realised that any such doubt would be easily eliminated only if it were possible to understand that:

> *omnes disciplinas inter se coniunctionem rerum et communicationem habere* [...] *encyclios enim disciplina uti corpus unum ex his membris est composita* (Vitruvius I, 1, 12).

From the phases of a *curriculum* which was even longer and more complex than that followed by the Quintilian's orator, attention had to be turned to the points of contact

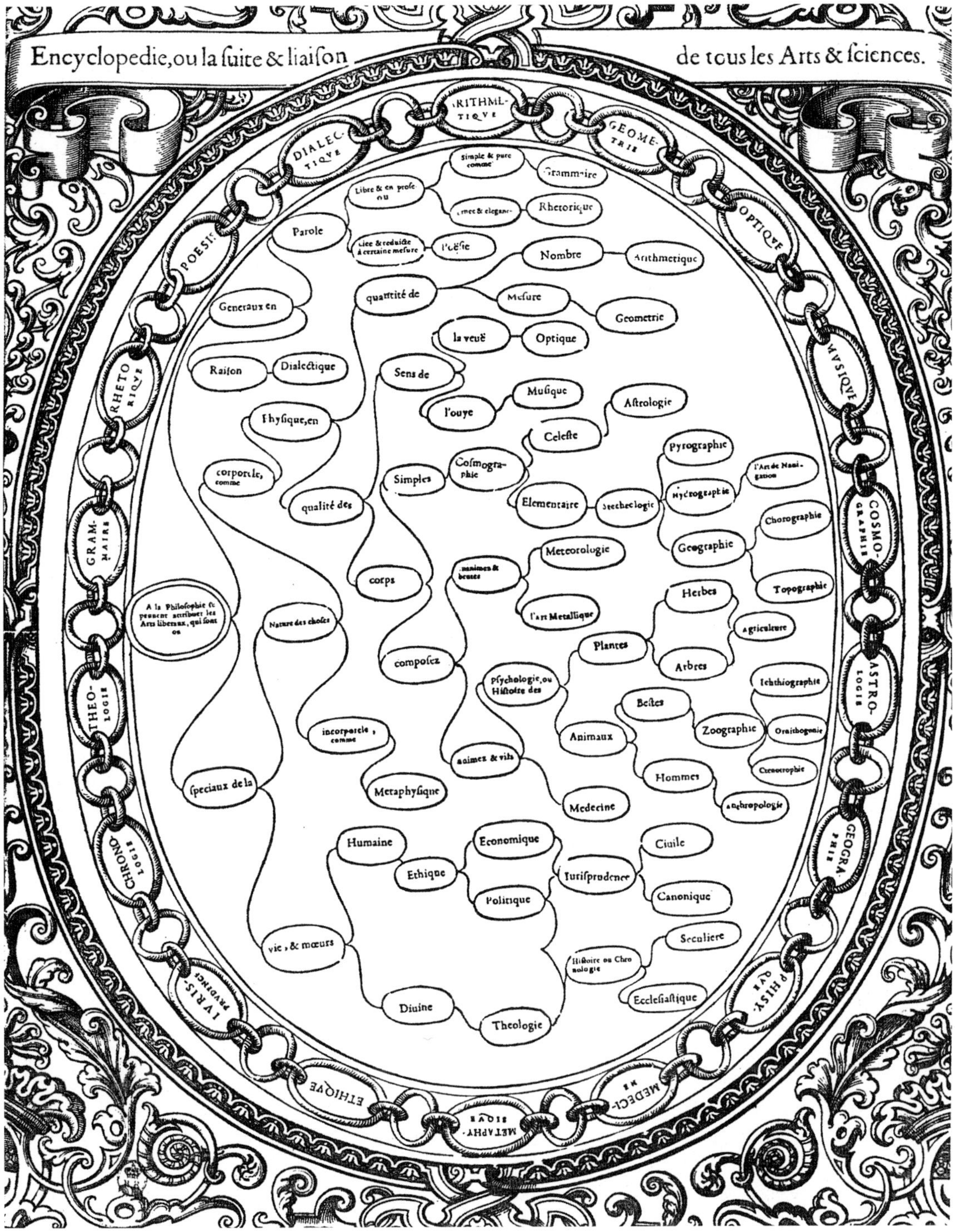

Figure 14
The general plate of Savigny: *Tableaux accomplis de tous les arts libéraux* (Paris 1587).

between the disciplinary segments. Thus the universality of knowledge, the encyclopaedia, had to be related not to "all disciplines" – or to the "*caeteres artes*" of Quintilian's encyclopaedia – but rather to the "*coniunctio*" and the "*communicatio*" which turned all these into a "*corpus unum.*"

In short, Vitruvius suggested the encyclopaedia should be defined as the "*orbis doctrinae*" of the *Institutio Oratoria*, within a conception of knowledge as *corpus unum*, complete, cohesive and coherent, with all the constituent elements interacting simultaneously. This was a concept extraneous to Quintilian and in general to Greek and Roman "encyclopaedism." Two hitherto independent aspects of classical thought on knowledge were therefore superimposed, one which considered the encyclopaedia as a didactic programme oriented towards giving

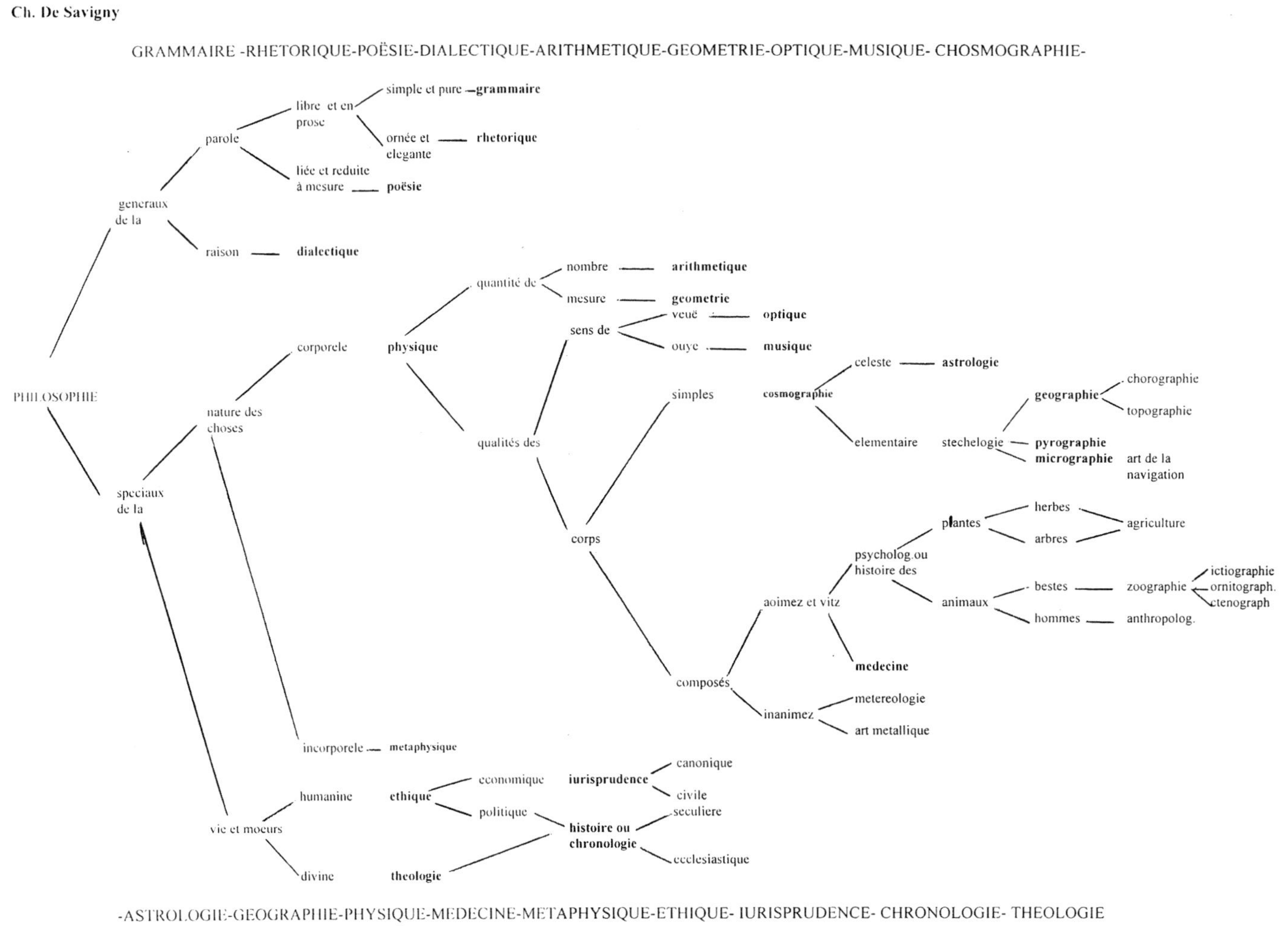

Figure 15
The encyclopedical scheme of Savigny: *Tableaux*.

the student global training before moving on to specialist knowledge and the other an organicistic metaphor of knowledge as a self-sufficient and self-contained body. A metaphor which in the sixteenth century could also count on harmony with the "divine inspired being" as the universe had seemed to many currents of Renaissance thought.

And so, with respect to the classical pedagogical tradition and way beyond the intentions expressed in *De Architectura*, the fusion between the affirmations of Quintilian and Cicero ended up by distorting the very meaning of the term *εγκυκλιον* and the Latin "*orbis*" which translates it. Initially without any ambitions of completeness, it was now used to indicate a unitary and omni-comprehensive structure which rotates on itself around a universe in which the disciplines – *omnes* and *absolute* – were not side by side, but co-penetrating and integrated.

The encyclopaedia thus became the ideal place imagined by Scalichius or conceived by Budé in *De l'institution du Prince*, where:

> *toutes les sciences s'entretiennent comme font les parties d'un cercle qui n'a ny commencement ny fin* [...] *ayans les dictes sciences et disciplines connexité mutuelle et coherence de doctrine et affinité d'estude qui ne se doibt ny peult bonnement separer ny destruire par distinction de faculté ou profession* (Budé [1966] 88).

The role of the encyclopaedia was not to be *summa*, to put the various disciplinary domains in succession, but to restore the link, the *vinculum* which makes the organism capable of moving, of developing limitlessly, but without renouncing its own unity as a result. Before any other aspect of knowledge, it expressed the *connexité mutuelle* of the disciplines declared by Budé, the *entresuite des arts* of Le Fevre de la Boderie, the *dialogus* between sciences and traditions exemplified by Schalichius in the dialogue between Epistemon and Philomusus which introduced his encyclopaedia; even the *chaos*

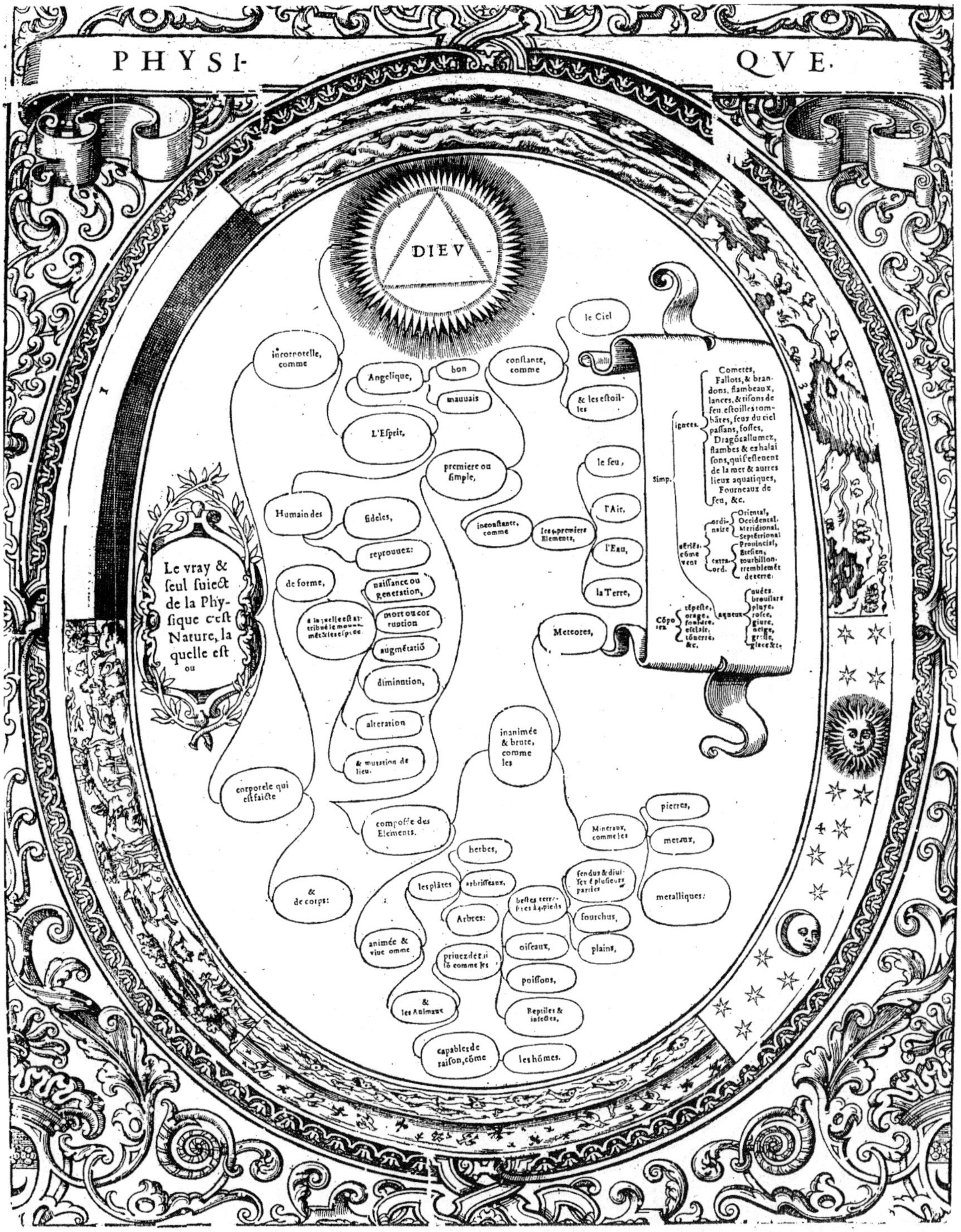

Figure 16
The plate of the Physics of Savigny: *Tableaux.*

of individual disciplinary spheres, as Ringelberg called it, as long as it were contained within a general design.

The apparent rejection of order by many of the sixteenth-century "encyclopaedists" was rather the assumption of a criterion of order different from that of previous ages.

The order of time which had defined the cadence of the formative *curriculum* from *prius* to *posterius* and given the classical and medieval encyclopaedia a linear organisation, despite its very name, was replaced with a criterion which, as wrote Budé, enabled knowledge with neither a beginning nor an end to be organised simultaneously in the space available. A physical and physiological space, finite and visualisable, which therefore had to be economised on and exploited to the full, to the very limit of its capacity.

The intersecting of disciplines, the simultaneous movement of elements of knowledge within the *"orbis"* encircling them all,

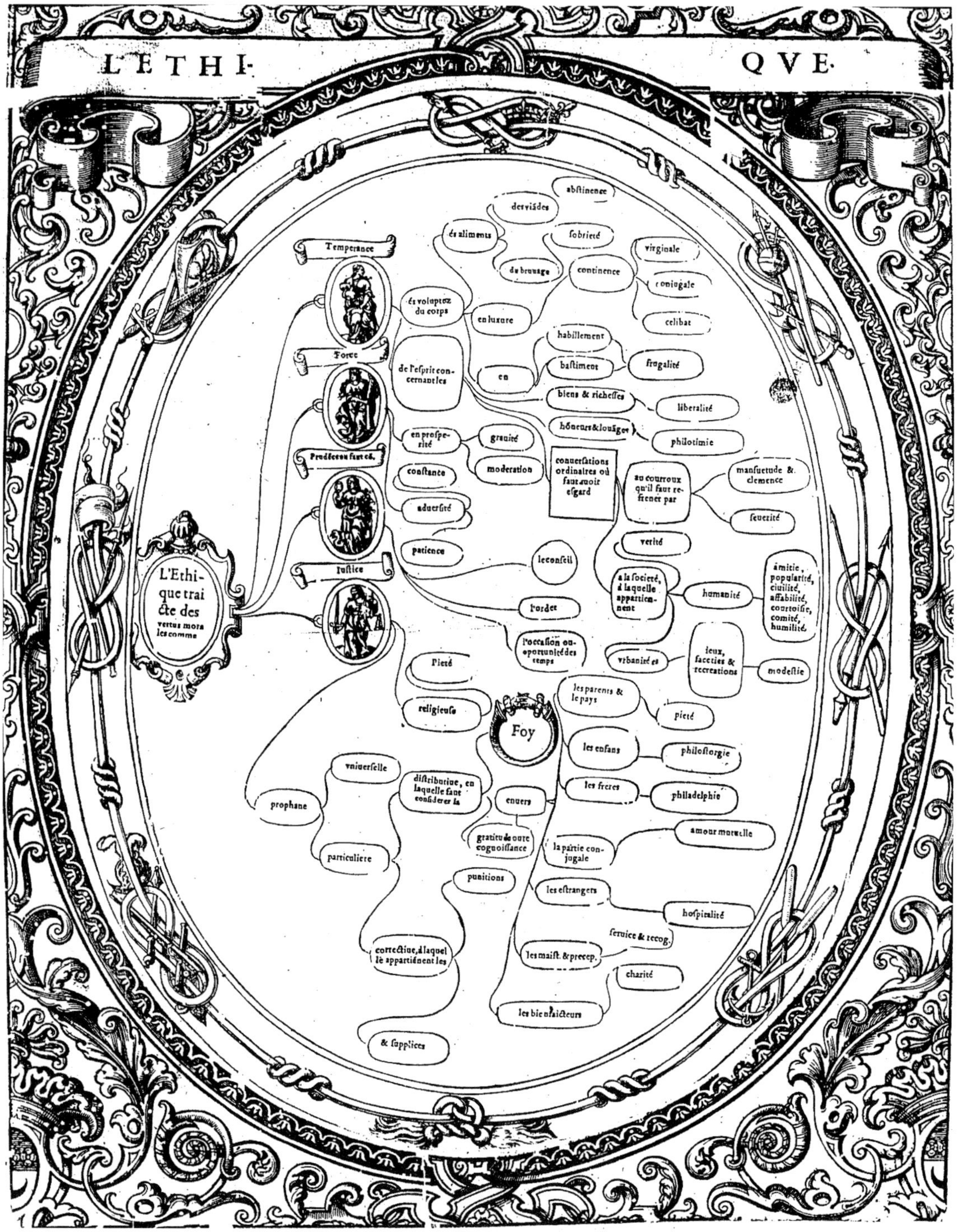

Figure 17
The plate of the Ethics of Savigny: *Tableaux.*

is perhaps the most significant aspect of "encyclopaedism" in the second half of the sixteenth century. An aspect acknowledged by Daniele Barbaro when he states that encyclopaedic universality and unity derive from the "superposition" and "circulation" of the *artes*; recognised by Nizolio when he suggested the universe of knowledge should be considered as made up of various levels of knowledge, shared "*simul et semel*" by all the sciences; not extraneous to the cabbalistic tradition of Paul Scalichius who proposed the unity of knowledge in a sort of topic of "encyclopaedic terms [...] common to the entire encyclopaedia."

An aspect which could not fail to be antithetical to the one which had led to the disciplines being organised in pyramidal form and linked via a relationship of subordination, one-way and irreversible from the top downwards. An aspect which is immediately and perceptibly evident in the illustrated plates of Christophle de Savigny's *enciclopédie ou liaison de toutes les arts et sciences,* drawn up within Ramist circles and published for the first time in 1587 (figs. 14–15).

The circular character of knowledge is here reinstated by the first of the plates which accompany the individual treatises (Savigny [1587]). Grammar, rhetoric, poetry, dialectics, arithmetic, geometry, optics, music, cosmography, astrology, geography, physics, medicine, metaphysics, ethics, jurisprudence, chronology and theology are the rings in the uninterrupted chain of knowledge; a chain closed in on itself to form a circle, this too with neither beginning or end, containing the corpus *unum* and *universum* of knowledge. Outside this sphere, nothing, not even theology, which in the immediately preceding years had prudently been kept outside the circuit of human knowledge. Inside, the confusion of all the branches of knowledge, sacred and profane, speculative and practical. But what at first sight appears to be a chaotic labyrinth was simply the effect produced by the interweaving of a great many disciplines which *simultaneously* occupied the ideal space of knowledge to the point of using it up. And this because, according to Savigny, the encyclopaedia is above all *liaison de touts les arts et sciences.* The same disciplines which had once represented the rings of the chain bounding the circle of knowledge figure as the same number of branches forking off the common trunk of philosophy and dividing into smaller branches according to the Ramist criterion of decreasing generality. As though obliged by the need to economise on space on the page, in the mind and in the memory, the disciplines communicate with one another, they intersect in a number of points, sometimes dividing into numerous lesser orbs corresponding to the main divisions of the encyclopaedia, sometimes stopping suddenly. But none of them escapes the chain which encloses them and each considered individually, demonstrates the same circularity, the same interweaving as the "system" which contains it (figs. 15–17).

Slightly more than half a century separates Savigny from Beroaldo, less than a century from Poliziano. In the meantime, the skies had changed motion and form, the earth had gained a new continent but lost its central role in the universe, two reforms had split Europe in half while the inhabitants of the Old Continent, divided by faith, law and language, resembled ever more closely the island dwellers of a great archipelago.

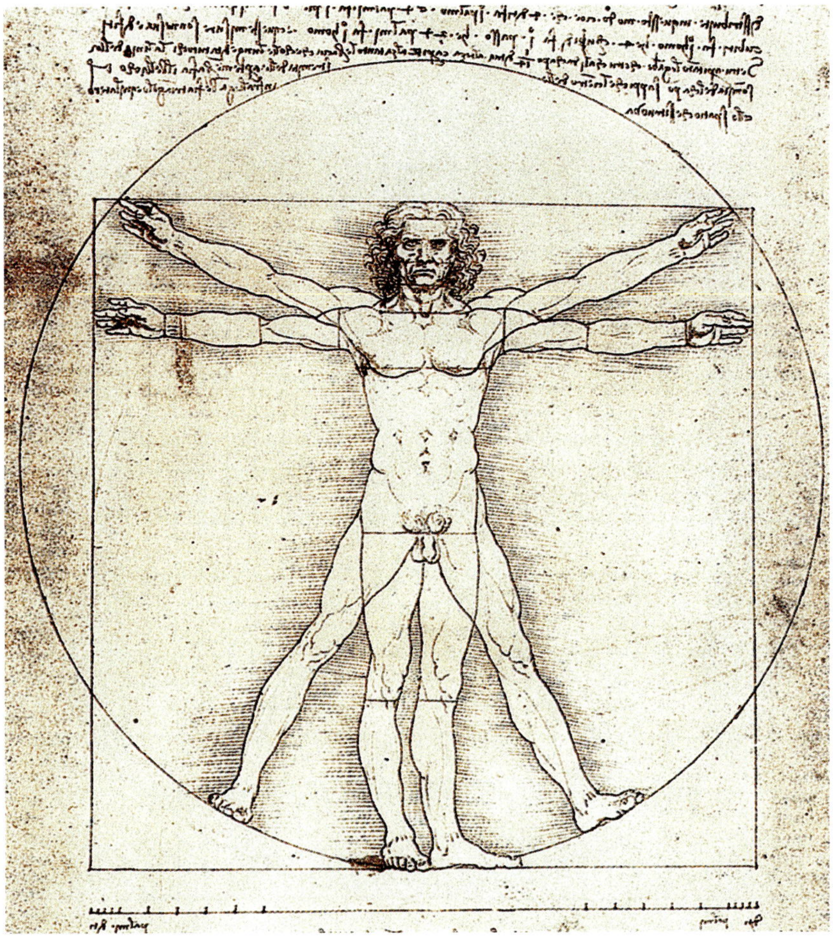

Figure 18
The Vitruvian *homo ad circulum et ad quadratum.* Drawing by Leonardo da Vinci (Gallerie dell'Accademia Venice).

It would have been a miracle if knowledge had not also changed its structure, if the notion of encyclopaedia had remained outside the changes and revolutions.

The educational programme of the Roman orator and together with it the very name of encyclopaedia had come to the support of the antischolastic reaction of the fifteenth and early sixteenth centuries. But it was already clear to Poliziano that the doctrines necessary to enable the modern scholar to debate "*in omni materia et per omnes disciplinas*" were considerably more numerous and more varied than those which appeared sufficient to Cicero or Quintilian. The didactic programme was amplified by practical and operative branches of knowledge, traditionally excluded from the formation of free man, the seven arts branched out, they combined with each other and gave rise to a quantity of sub-

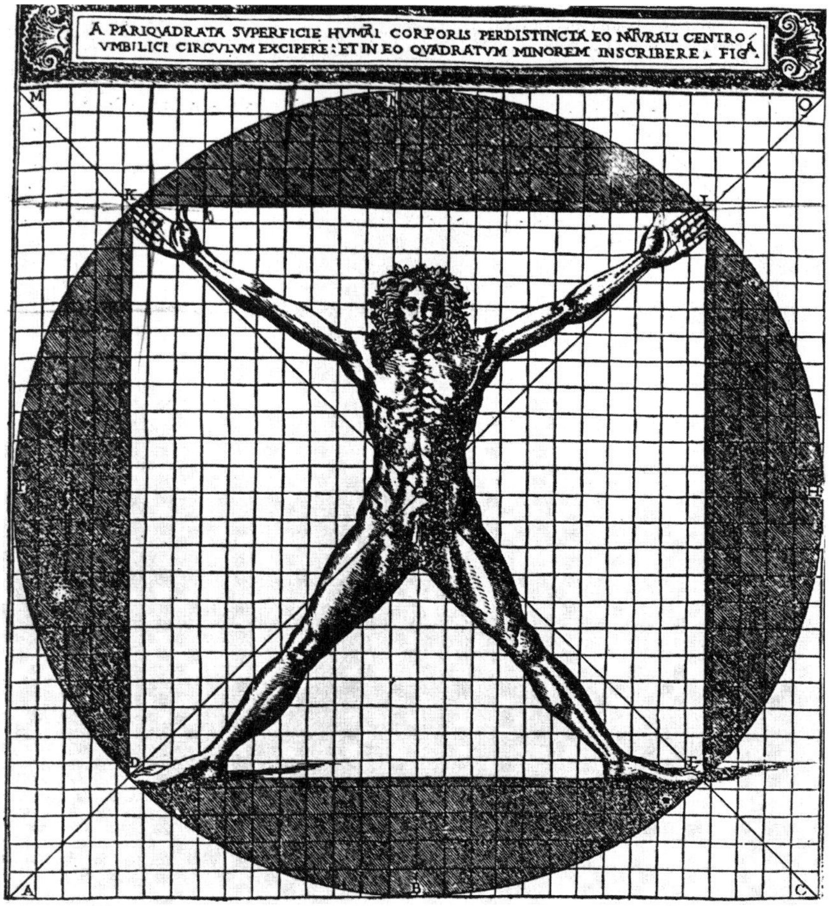

Figure 19
The Vitruvian man in Cesariano's Vitruvius edition (Como, 1521).

divisions and derived, composite and mixed disciplines, while the *"orbis disciplinarum"* thus enriched and articulated, careful not to exclude anything, aspired to being *completum* and *universum* to the point of reflecting the image of the cosmos itself. Rather than covering *argumenta* in succession, the encyclopaedia had claimed the right to enumerate the elements of reality and organise their articulations.

It had thus sacrificed its original identity as a *course of studies* to new aspirations as a *system of the universe.* Such a mutation forced the encyclopaedia to address a previously extraneous question: how to adapt the ambition to contain the universe *"de omni scibile,"* as vast as the perennial matter of reality itself, to the limited human resources. In other words, how to turn the macrocosm/microcosm relationship on its head in such a way that the latter could contain the former.

In substance, the problem was how to summarise and abridge the *"orbis rerum"* in a more constricted but no less universal *"orbis scientiarum."* And the problem involved both space and time. Human life was infinitely shorter than the time of nature, the encyclopaedic space considerably more restricted than the actual one.

The chronological order which had represented the framework of the encyclopaedia as *curriculum* had to be replaced by a criterion of order appropriate to the notion of encyclopaedia as *universe.* "The order of time," wrote Pierre de la Ramée, "must give way to the order of reason."

Irrespective of the choices made by the individual "cosmographers of knowledge," most agreed with de la Ramée's definition. It was clear to all that if you wanted to widen the encyclopaedic sphere and also include those branches of practical and operational knowledge excluded by tradition, you had to move from the various objects of nature towards the unity of the operations involved in understanding the centre of gravity of the *orbis* of knowledge and make human nature the criterion for organising that knowledge, the very condition of its universality. The disciplines must also be reclassified as languages, products and constructions of human activity rather than as parts of an external, objective reality, and arranged in the encyclopaedic space according to the criterion of maximum economy. In other words, starting from the common mental and human origin which obliged all the arts to share the same principles and the same objects, to intersect their orbs, each to cross over its own boundaries, and which made knowledge an *orbis* but an *artificial* orbis, a complete and self-sufficient universe which contains within itself and its author the rules for its functioning and within its functioning, its very legality.

The Vitruvian figure of *homo ad circulum* (figs. 18-19) thus became the symbolic representation of an idea of encyclopaedia considered, in the words of Alsted, *"in mentem ut habitus, extra mentem ut systema"* (Alsted [1620] 2). An idea, bluntly reproposed by Robert Fludd in his encyclopaedic work (fig. 20), which was, so to speak, reflected years before in the way Ringelberg (fig. 8), Scalichius (fig. 7), de la Ramée (fig. 21), Nizolio (fig. 22), Savigny (fig. 16), Barbaro and some others divided knowledge. All very different, but all putting man, his faculties, his skills, his *habitus* at the origin of any classification or ramification of the disciplines.

Constructed by man and therefore *artificial*, the encyclopaedic orb could thus contain all things, words, and concepts, not according to their essential being, but adapted to the human *ratio* and *usus*.

A mean world, the artificial circle of knowledge, as universal as the intelligible world, but at the same time as divisible and mobile as the elementary world; a mean term in the relationship which links the unity of the divine design to the multiplicity of natural objects and orders.

This is the message of Achille Bocchi's symbol of *medietas* of knowledge communicated through the female figure drawing geometric shapes on the ground, similar to those which God holds immobile in the sky (fig. 6).

The *artes*, not as isolated specializations but as the articulations of a universal and coherent system consisting of signs, symbols, languages and human constructions, aspire to reflect and interpret not the act of the divine architect, but his rational design. Nature is no longer the limit before which the mind and hand of man can only retreat, and humanity – liberated by its own knowledge and made eternal by the most perfect of its works, the circle of arts and sciences – is authorised to shape the multiplicity of orders in the tangible world in order to bring them back to a rational model which is presumed to be analogous to the designing rationality of divine mind.

The form of knowledge, the encyclopaedia, is equivalent to the space between the arms of Bocchi's *medietas* who, in the gesture of drawing in the sky and earth, reveals her own universalising *ratio* to herself and adapts the *infinite* universe of concepts and things to it. A *ratio* which is *neither* in things *nor* intelligible, but which, despite this, can impose itself *as a measure* of things and *as the shape* of the intelligible.

Thus a mean cosmos, halfway between the sensible and the intelligible, but capable of expanding to the point of becoming a hypercosmos. As Robert Fludd represents it in his *Utriusque Cosmi Historia*, such a cosmos is vast enough to contain all things and all forms, not according to their essential nature, but adapted and presented according to the criteria of understanding (fig. 23).

Positions such as these expressed a number of perceptions which were becoming established in the mid-sixteenth century.

First of all, the centrality and authority of man in the universe of knowledge and the liberty attributed to the scholar in constructing by himself – as did Bocchi's *medietas* with her compasses – the object of his own knowledge, through the *inventio* of the smallest elements (signs, alphabetic characters, symbols, etc) of a cognitive language and the identification of "universal" laws which regulate its *dispositio*.

The result was the self-referential nature of that knowledge – not of the individual dis-

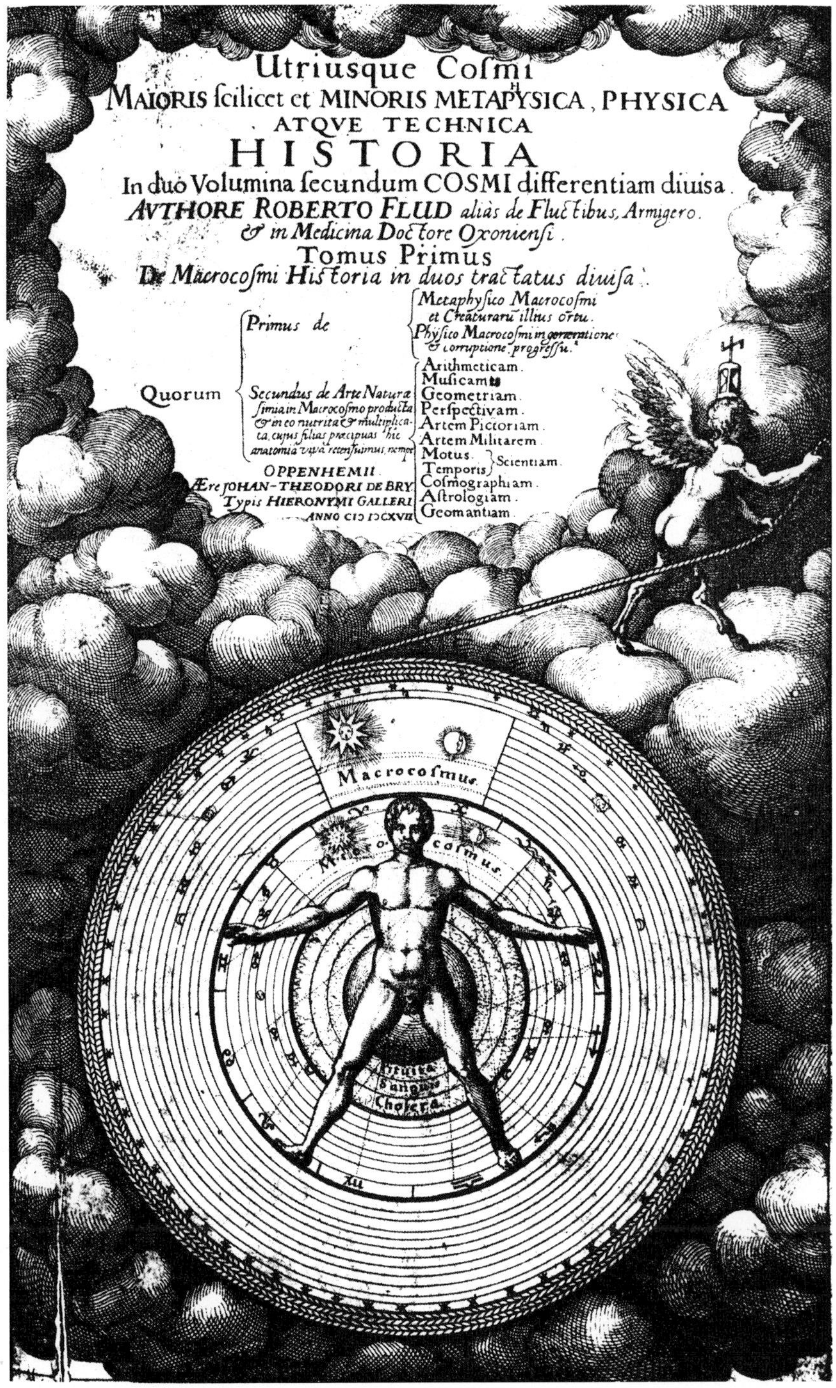

Figure 20
The human knowledge (metaphysics, physics, and technology) in the framework of Macrocosm and Microcosm. Title-page of Fludd: *Utriusque Cosmi ... Historia* (Oppenheim 1617).

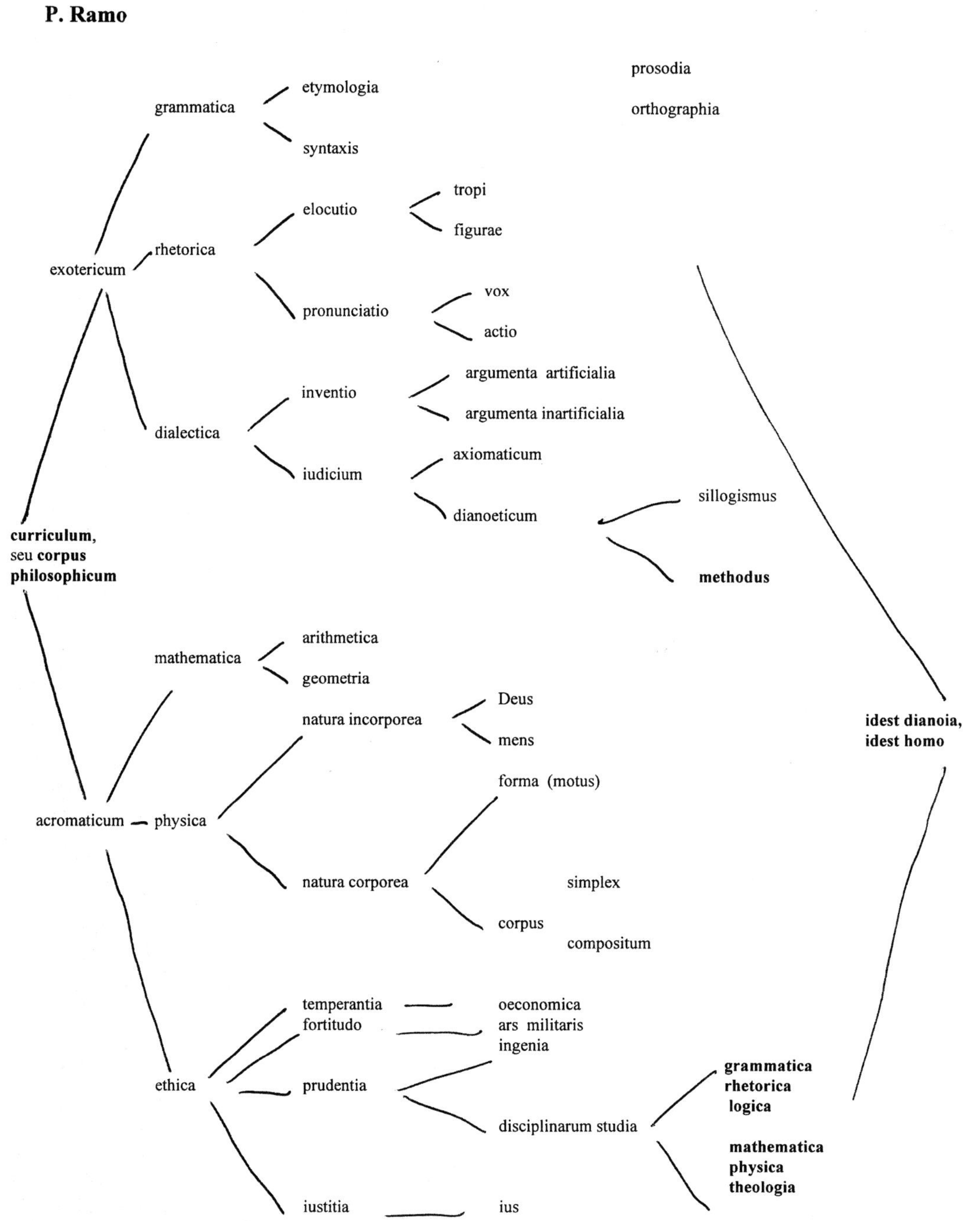

Figure 21
The main classes of the *Corpus philosophicum*. From Ramus' *Professio Regia* (Basel 1576).

ciplines, but of the system which contained them – obliged to renounce both the guarantee of reflecting the *"ordo naturae"* and faith in a trascendent principle, outside the system of knowledge, but capable in itself of guaranteeing certainty and universality. But the universality which had once derived from the *universitas mundi* and which knowledge was now called upon to find within itself, obliged it above all to be *encyclopaedic*: total and unifying, inclusive and not exclusive. It had to be at the same time, in everything – in the framework of the system – and in its parts – the disciplines – speculative and operative or rather, *resolving* and *composing*, analytical and synthetic, theoric and practic. And finally, it had to find in itself a cohesive force sufficient to counter the tendency for rapidly increasing knowledge to explode and shatter into pieces, without natural boundaries and without theological barriers. Among its levels and languages – the disciplines and their

M. NIZOLIO

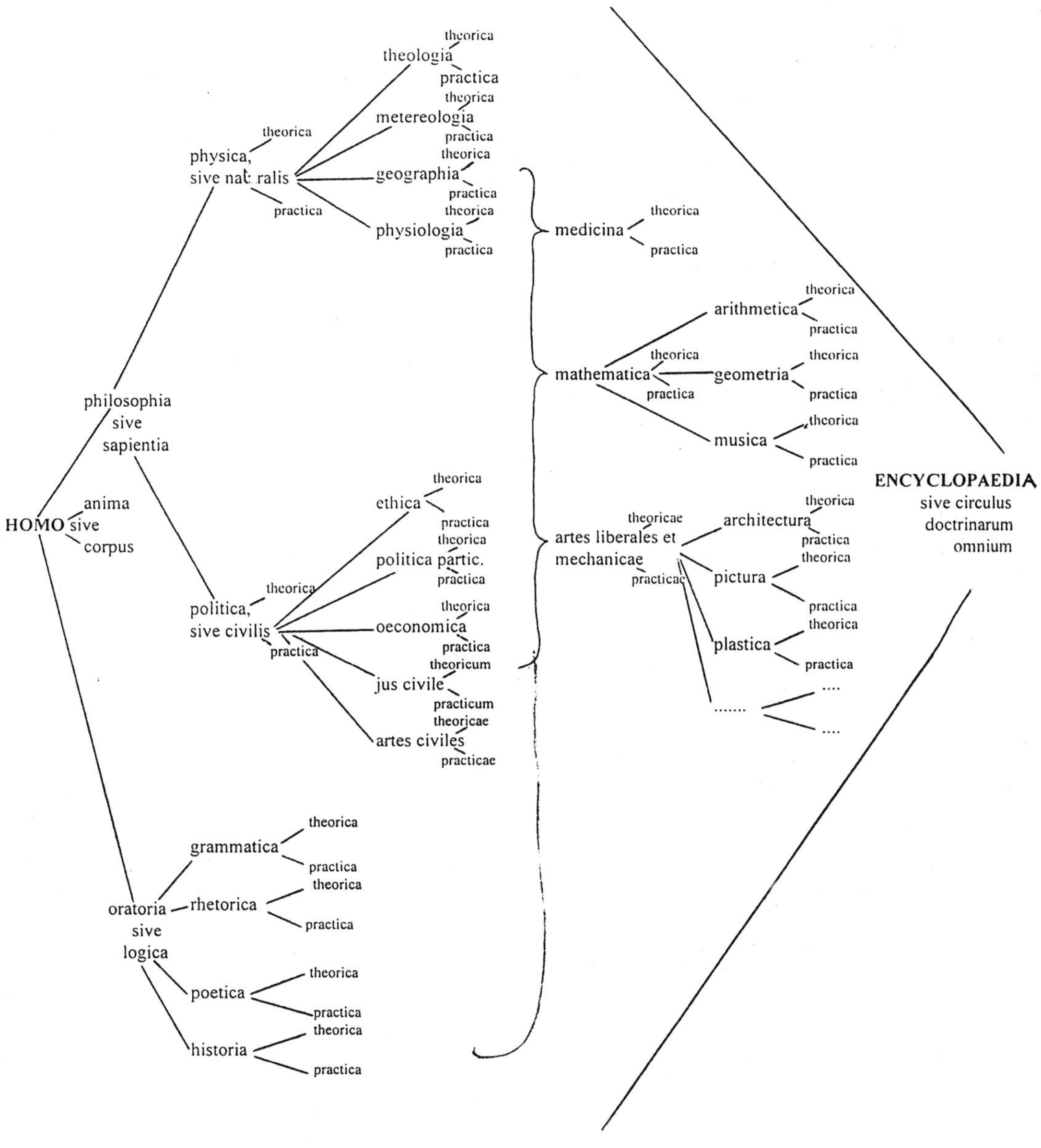

Figure 22
Encyclopaedical scheme. From Nizolio: *De veris principiis et de vera ratione philosophandi* (Parma 1553).

divisions – one had to be identified to *mediate* and *unify* the others. Barbaro called it architecture, Scalichius symbolic philosophy, Bocchi symbolic philology, Nizolio rhetoric, Ramo dialectic, many others mathematics, but for all, it was an encyclopaedic science – better, a science-encyclopaedia – at the beginning and end of all others because it in fact represented nothing other than the *humana ratio ad cognoscendum*.

This need for omni-comprehension explains both the historic limit and the innovative nature of such structures of knowledge. The limit lies in the fact that it was then impossible to conceive of the independent development of a discipline as an attempt to understand and formalise the function of the universe without concentrating on the conception of an all-inclusive world as would be possible within just a few decades. This is because it was inconceivable that individual branches of the tree of knowledge could grow in order to gather and catalogue new facts, or that the encyclopaedia as *systema systematum* could accommodate relatively autonomous individual microcosmos, characterised by their own principles and criteria.

This is confirmed by the plates of Savigny's encyclopaedia (figs. 15–17) where the trees with their numerous, disparate and intricate

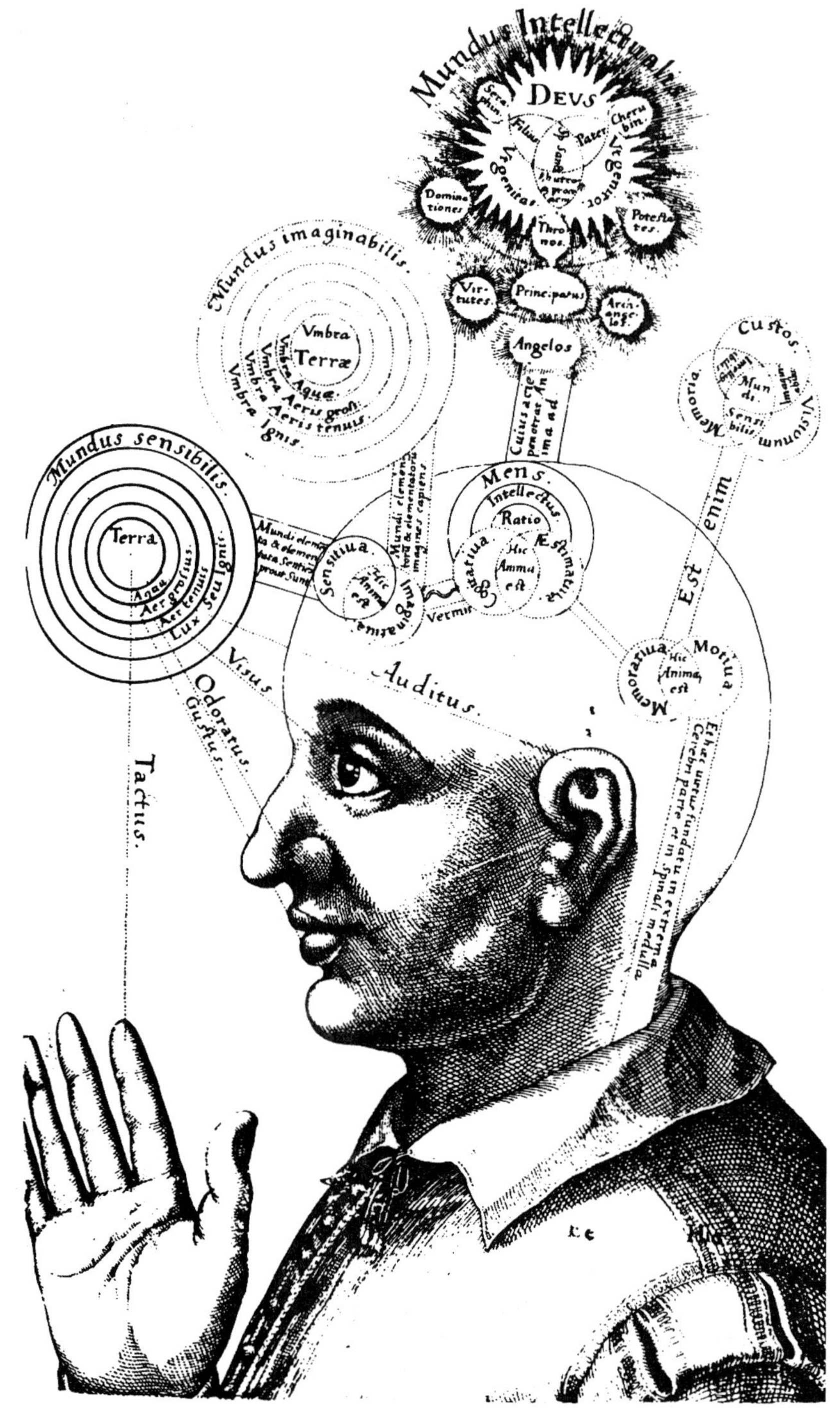

Figure 23
The "spiritual brain," which composes all things of the upper and the lower world according to the criteria of human understanding. From Fludd: *Utriusque Cosmi ... Historia* (Oppenheim 1617).

branches and fruits appearing through the foliage are all included within the closed, circular *"orbis"* of a knowledge which, although allowing itself to be represented by such a naturalistic and irregular image as a tree, places above this image, the geometric, regular, symbolic image of the circle.

Despite this, structures of knowledge such as these implied adhesion to a criterion, if not of investigation, at least of organisation and scientific verification, which turns its back decisively on demonstrative traditional canons. The guarantee of the truth of the conclusion, or the deduction, did not lie in the truth of the premise or the principle. Quite the opposite, it was the conclusions which validated and verified principles and premises, initially assumed as "convenient," "judicious" or simply possible. It was the whole, the totality, the circumference which enclosed and protected the orbis which guaranteed the truth and stability of the centre or the *principium* around which it rotated.

Università di Bologna

Notes

[1] Among the many see: Zwinger [1565]; Bodin [1590]; Camillo [1550]; Garzoni [1583], [1589a], [1586], [1613] and [1589b]; Galluccio [1588]; Bry [1596]; Khunrath [1609]; Patrizi [1561]; Delbene [1609]; Andreae [1619]; Campanella [1623]; Lomazzo [1584]; Recorde [1556], [1542] and [1551].

[2] *"Ordinatio, dispositio, symmetria, eurithmia, decor, distributio"* were the six categories of classical architecture, see Vitruvius, I, 2.

[3] "Fabrica ita habet. Fiat volumen grande, in folio, et illud depescantur in decem partes principales, pro numero decem subjectorum. Singulae partes sibi vendicent bis novena folia; quibus inscribantur 18 praedicata singulis singula. Singuli folia dividantur in 10 particulas quibus inscribantur 10 quaestiones, singulis item singula [...]," Alsted [1610] 16.

[4] You can compare the symbol of humane knowledge (the female figure) in an emblem of Bocchi [1555] (fig. 6) to the image of God the Creator in a thirteenth-century miniature (from *Bible moralisée*) (fig. 5).

[5] Quintilianus, I, 10, 1: "Nunc de caeteris artibus quibus instituendos priusquam rhetori tradantur pueros, existimo strictim subiungam, ut efficiatur orbis ille doctrinae, quem graeci encyclion paedian vocant."

[6] Beroaldus [1521] c. 91r: "[...] orbis ille doctrinae quem graeci *εγκυκλοπαιδιαν* nominaverunt."

[7] Goclenius [1609] 2: "Philosophia est systema disciplinarum liberalium. Alias orbis doctrinarum: Encyclopaedia. Orbis doctrinae liberalis syntagma"; Goclenius [1613] 363, df. 3: "Circulus, Circularis Scientia, seu orbis doctrinarum est encyclopaedia, id est absoluta cognitio omnium scientiarum."

[8] Alsted [1612] 4: "[...] dicitur alias *εγκυκλοπαεδια* circulus omnium disciplinarum."

[9] Ramée [1577] 279: "nonnulli completum quendam velut orbem doctrinae, *εγκυκλοπαιδειαν* vocant."

[10] Scalichius de Lika [1559]; see figure 7 for Scalichius' encyclopaedic scheme.

[11] Ringelberg [1538]. See the scheme (fig. 8) obtained from Ringelberg's text.

[12] On the notion of *encyclopaedia* in ancient and medieval texts see Marrou [1958] 211–235; Col-

lison [1966²]; Rijk [1965]; Auroux [1979]; Simone [1976] 234–243; Levi [1976] 170–184; Freedman [1994]; Porset [1987] 719–726; Cherchi [1993], [1985]; Bos [1989].

13 Le Fevre de la Boderie [1578b]; see also Le Fevre de la Boderie [1571] and the scheme (fig. 11) obtained from Le Fevre de la Boderie's translation of Giorgi's *De Harmonia Mundi*, see Le Fevre de la Boderie [1578a]; the same scheme was employed in organising sciences in *La Galliade*.

14 It might be interesting to observe the title-page of two different editions of *Margarita Philosophica*. The first (fig. 12) was printed at the end of the fifteenth century, the other (fig. 13) in the middle of the sixteenth century. In the first, the allegory of philosophy surrounded by the seven arts of Trivium and Quadrivium is recognisable in the centre. In the other, philosophy is represented as the trunk of liberal arts' tree, whose branches are interlaced with one another. It could mean that even old text books were adapted to the modern notion of encyclopaedia as an interweaving of disciplines.

References

Alsted, J. H.
1610 *Panacea Philosophica, id est, nova et accurata methodus docendi et discendi universam Encyclopaediam*, Sectio II: *De Arte magna, generali et ultima, seu lexico philosophico.* Herborn.

1612 *Philosophia digne restituta.* Herborn.

1620 *Cursus philosophici enciclopaedia.* Herborn.

1630 *Encyclopaedia.* Herborn.

Andreae, J. V.
1619 *Reipublicae Christianopolitanae Descriptio.* Strasbourg.

Auroux, S.
1979 *Sémiotique des encyclopédistes.* Paris.

Badoer, F.
1560 *Instrumento di deputatione.* Venezia.

Barbaro, D.
1567 *I dieci libri dell'Architettura di M. Vitruvio, Tradotti et commentati* [...] *da lui riveduti et ampliati; et hora in più commoda forma ridotti.* Venezia. (First ed. 1556).

Beroaldus, Ph.
1521 *Oratio habita pro aede Divi Petronii.* Bononia.

Bocchi , A.
1555 *Symbolicarum Quaestionum.* Bononia.

Bodin, J.
1590 *Universae naturae theatrum.* Lugdunum.

1984 *Colloque entre sept Scavans qui sont des differens sentimens. Des secrets cachez. Des choses revées* (anonymous translation of *Colloquium Heptaplomeres de Rerum Sublimium Arcanis*), ed. F. Berriot (Paris, BN, MS. français 1923). Genève.

Bos, A.P.
1989 "*Εξοτερικοι λογοι ανδ εγκυκλιοι λογοι* in the Corpus Aristotelicum and the Origin of the Idea of *εγκυκλιο παιδεια*." In: *Journal of the History of Ideas* 50: 179–198.

Bry, T. de
1596 *The Theatre of Human Life.* Oppenheim.

Budé, G.
1966 *De l'Institution du Prince* (reprint ed. Paris 1547). Farnborough.

Camillo, G.
1550 *Idea del Theatro.* Firenze.

Campanella, T.
1623 *Civitas Solis.* Frankfurt.

Cherchi, G.
1985 "Momenti e modelli nella storia dell'Enciclopedia. Il mondo musulmano, ebraico e latino a confronto sul tema dell'organizzazione del sapere." In: *Rivista di Storia della Filosofia* XL/1, ed. M. T. Fumagalli.

1993 "Encyclopaedias y organizacion del saber de la Antiguedad al Rinascimento." In: *De Las Academias a las Enciclopedia. El discurso del saber en la modernidad*, ed. E. Rodriguez Cuadros, 71-94. Valencia.

Cicero
Pro Archia.

Collison, R.
1966² *Encyclopaedias: their history throughout the Ages.* New York.

Delbene, B.
1609 *Civitas veri.* Paris.

Fludd, R.
1617 *Utriusque Cosmi ... Historia.* Oppenheim.

Freedman, J. S.
1994 "Encycloedic philosopical writings in Central Europe during the high and late Renaissance (ca. 1500 – ca. 1700)." In: *Archiv für Begriffsgeschichte* XXXVII: 212-255.

Galluccio, G. P.
1588 *Theatrum mundi et temporis.*Venezia.

Garzoni, T.
1583 *Theatro dei vari e diversi cervelli.* Venezia.

1586 *Hospitale de' Pazzi.* Venezia.

1589a *Piazza universale di tutte le professioni del mondo.* Venezia.

1589b *Sinagoga degli Ignoranti.* Venezia.

1613 *Serraglio delle meraviglie del mondo.* Venezia.

Goclenius, R.
1609 *Conciliator Philosophicus.* Cassellis.

1613 *Lexicon philosophicum.* Frankfurt.

Khunrath, H.
1609 *l'Amphitheatrum sapientiae aeternae.* Hannover.

Le Fevre de la Boderie, G.
1571 *Encyclie des sécrets de l'eternité.* Antwerpia.

1578a "Epistre en forme de Preface" In: *L'Harmonie du Monde, divisée en trois cantiques. Oeuvre singulier et plein d'admirable erudition. Premierement composé en Latin par Francois Georges Venetien, et depuis traduict et illustré par Guy Le Fevre de la Boderie.* Paris.

1578b *La Galliade ou de la Révolution des artes et des sciences*. Paris.

Levi, A. H. T.
1976 *Ethics and the Encyclopaedia in the Sixteenth Century.* Edinburgh.

Lomazzo, G. P.
1584 *Idea del tempio della pittura*. Milano.

Marrou, H. I.
1958 (repr.1983) *Saint Augustin et la fin de la culture antique* (1938), ed. E. De Boccard. Paris.

Nizolio, M.
1553 *De veris principiis et de vera ratione philosophandi*. Parma.

1952 *De veris principiis et de vera ratione philosophandi (first ed. Parma, apud Septium Viottum,* 1553), ed. Q. Breen. Roma.

Ong, W.
1956 "System, Space, and Intellect in Renaissance Symbolism." In *Bibliothèque d'Humanisme et Renaissance* XVIII: 229–235.

Patrizi, F.
1561 *La città felice*. Venezia.

Porset, C.
1987 "Figures de l'encyclopédie." In: *Le Siècle de Voltaire*, ed. C. Mervaud-S. Menant, 719–733. Oxford.

Quintilianus
Institutio Oratoria.

Rabelais, F.
1979 *Gargantua et Pantagruel*, tr. it. Einaudi. Torino.

Ramus, P.
1576 *Professio Regia*. Basel.

1577 "Pro philosophica Parisiensis Academia Disciplina, Oratio (1550)." In: *Collectanae Praefationes, Epistolae, Orationes*. Paris.

Record, R.
1542 *The Grounde of Artes*. London.

1551 *The Pathway of Knowledge*. London.

1556 *The Castle of Knowledge*. London.

Reisch, G.
1496 *Margarita Philosophica*. Heidelberg.

Rijk, L. M. de
1965 "*εγκυκλιοσ παιδεια* A Study of its Original Meaning." In: *Vivarium*, III: 24–93.

Ringelberg, I. F. von
1538 *Lucubrationes, vel potius absolutissima κυκλοπαεδια* [...] *nempe Liber de Ratione Studii*. Basel.

Savigny, Ch. de
1587 *Tableaux accomplis de tous les arts libéraux*. Paris.

Scalichius de Lika, P.
1559 *Encyclopaediae, seu Orbis disciplinarum, tam sacrarum quam prophanarum Epistemon*. Basel.

Simone, F.
1976 "La Notion d'Encyclopédie: Elément caractéristique de la Renaissance française." In: *French Renaissance Studies. 1540–1570. Humanism and Encyclopaedia*, ed. P. Sharratt. Edinburgh.

Valeriano, P.
1556 *Hieroglyphica*. Basel.

Vasoli, C.
1990 "Note sul 'Theatrum Naturae' di Jean Bodin." In: *Rivista di Storia della Filosofia*, NS, 45: 475–537.

Vitruvius
De Architectura.

Zwinger, T.
1565 *Theatrum vitae humanae*. Basel.

The Mathematical Sciences in Raphael's *School of Athens*

Matthias Winner

Who is not familiar with Raphael's *School of Athens*? (Figs. 1–2) It is a mural adorning a room of the papal palace on the Vatican Hill in Rome. Ever since it was painted almost half a millennium ago it has belonged to the collective philosophical and scientific consciousness of the Western world. Pope Julius II (1503–1513) commissioned the fresco from the 25-year-old Raphael in 1508. The date 1511 is recorded in one of the window embrasures as the date of completion.

The painter and architect Giorgio Vasari published his *Lives of the Artists* in 1550; it marked the birth of art history as a separate branch of historical enquiry. Almost two decades previously, in 1531, Vasari had made drawings after Raphael's frescoes in the Stanza della Segnatura, as he called the room. Vasari celebrates Raphael as *pittore universale*, as universal painter and a new Apelles. That Raphael had been called from Florence to Rome by the Pope, on the recommendation of his kinsman Bramante, we know from Vasari. From him we also learn that the gifted architect Bramante had, shortly before, modernised the papal suite of rooms as part of his enlargement of the Vatican Palace, while simultaneously planning and initiating the gigantic building project of the new St. Peter's for the Pope. Vasari calls the room the Stanza della Segnatura (or Grazia), because it was the practice during the pontificate of Paul III to sign general pardons in this room. But the signing of such pardons in the Stanza during the pontificate of the Pope who commissioned the frescoes, Julius II, is not documented. That is why discussion has raged for over a century about the room's original function. There is no doubt that the mural of the *School of Athens* can be classified among the late-medieval encyclopaedic decoration of *studioli*, or, as we would say, studies. Already by the Renaissance a *studiolo* had become, by tradition, a regular feature in every princely palace. Pope Julius II had book cupboards installed in this room, presumably to contain his private library, which in large part survives. So Raphael's decoration of the whole room, reflecting its use as a library, could refer to the whole horizon of humanistic knowledge: philosophy, theology, poetry and jurisprudence. To each of these four faculties one wall is in fact dedicated. Thus the mural that faces the *School of Athens*, the so-called Disputa (figs. 4 and 6), shows a gathering of theologians around an altar, placed under the Holy Trinity in the heaven above. On one of the two window-walls, the one to the right of the *School of Athens*, Raphael painted, in the upper lunette, two groups of poets under laurel trees on Parnassus. Between them, at the centre of the fresco, are grouped the Muses, who are intent on listening to the god of poetry, Apollo, playing his *lira da braccio*. The window to the left of the *School of Athens* is flanked on the one side by the emperor and on the other by the Pope, both sitting against painted stone niches. To the left, the emperor Justinian is being presented with the *Pandects*, the code of Roman law, which to a large extent still provides the basis of our jurisprudence. To the right, in a similar scene, Pope Gregory IX, who is given the features of the reigning Pope Julius II, is accepting the *Decretals*, the medieval com-

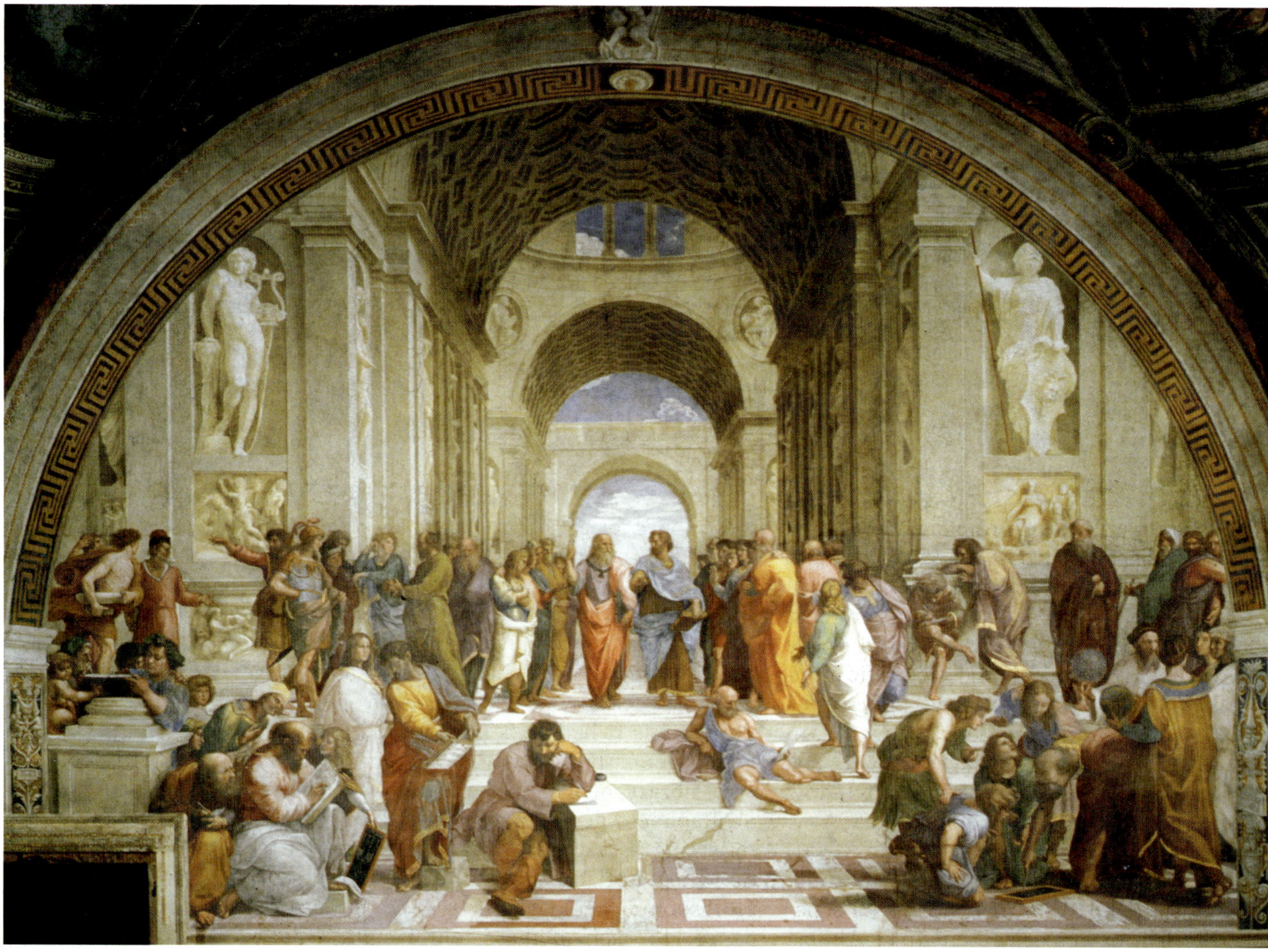

Figure 1
Raphael: *The School of Athens*. Fresco. (Rome, Vatican Palace, Stanza della Segnatura.)

pilation of the Code of Canon Law which still remained in force after 1500. Pope and Emperor are placed below an allegory of *Justitia and Jurisprudentia*.

Vasari describes the murals of the Stanza della Segnatura in the chronological sequence in which, in his view, they had been painted, as he had been able to reconstruct it from artist friends and members of the Curia he had met, some of whom had known Raphael while he was still at work in the Stanza della Segnatura.

According to Vasari, the first fresco to be painted in the room was the *School of Athens*. To some extent it was a trial piece. The Pope was reportedly so struck by it that he ordered all the frescoes already completed by other painters in the room to be destroyed, so that Raphael alone could complete the whole decoration of its walls.

But Vasari's reconstruction of the chronology of the frescoes was already disputed towards the end of the seventeenth century by the learned secretary of the Roman Accademia di San Luca, Gian Pietro Bellori. Bellori argued that the gold ground behind the Trinity in the fresco of the *Disputa* could only have been painted by a still immature Raphael, a Raphael still not emancipated from the late-medieval tradition of painting. So, in his view, the *Disputa* was the first fresco to be painted in the room; only later, when he was at the height of his mastery as an artist, did Raphael paint the *School of Athens*. The critical art history of the nineteenth and twentieth century accepted, without demur, this stylistic judgement of the late-seicento, and followed it in its interpretation of the Stanza della Segnatura. It was the recent restoration and cleaning of the frescoes in the room (still in progress) that confirmed, on grounds of painting technique, my early dating of the *School of Athens*, which I had reproposed on the basis of iconographic

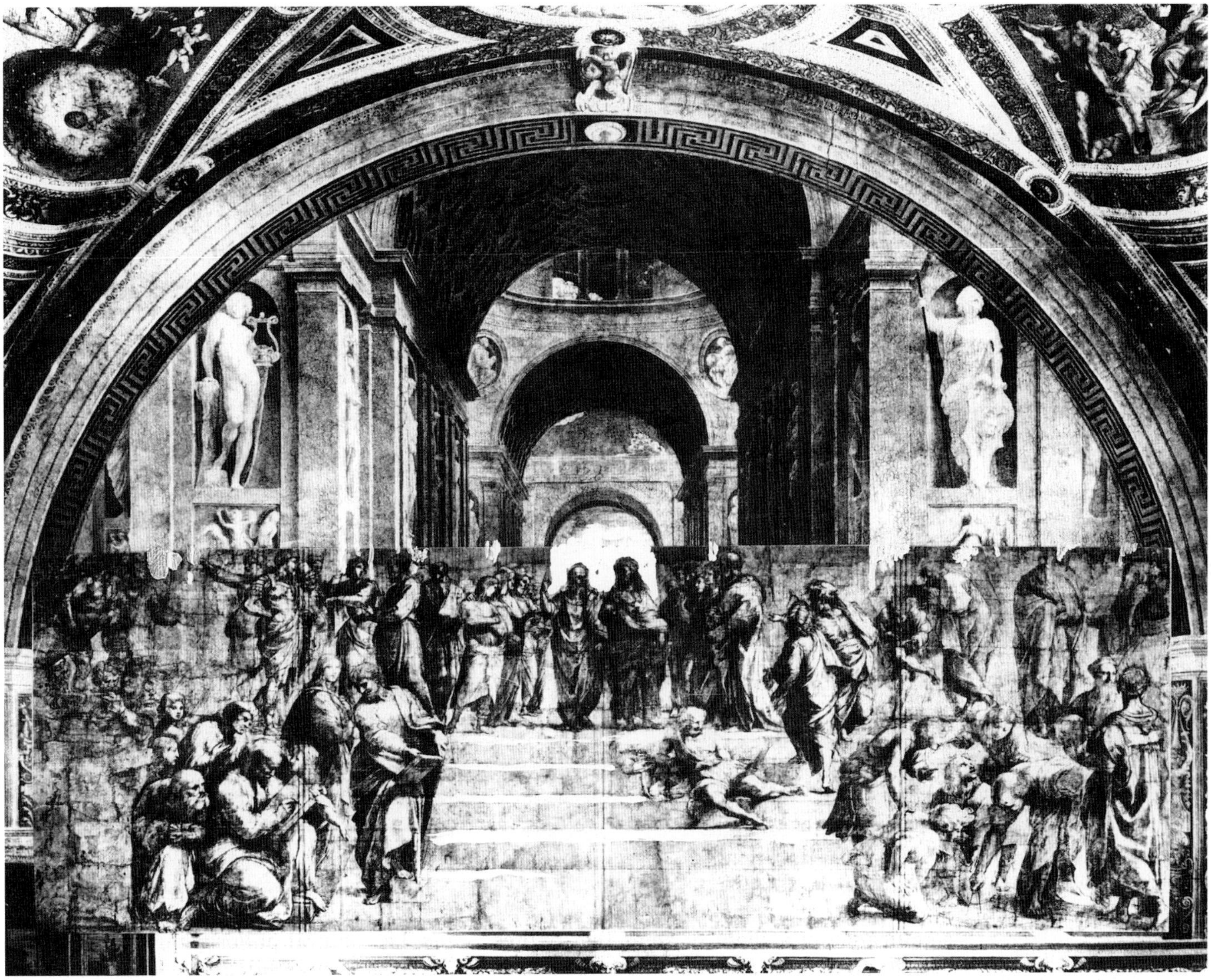

Figure 2
Raphael: *The School of Athens*. Photomontage of cartoon (Ambrosiana) and executed fresco.

considerations fifteen years previously, thus endorsing Vasari's version of the chronology.

Raphael's original full-scale cartoon for the *School of Athens* (fig. 2) is preserved in the Biblioteca Ambrosiana in Milan. In this cartoon the figure of the so-called *pensieroso*, the seated figure of man in the pose of a thinker hunched in thought beside a large block of stone before the flight of steps, is not yet present: he was not added until a later stage. In the cartoon, as later in the finished fresco, the Greek philosophers Plato and Aristotle stand at the centre of the fresco at the top of the flight of steps. They can be clearly identified by the books they each are carrying in their hands: Plato is holding the *Timaios*, his main work on the origin of the cosmos; Aristotle is supporting his *Nicomachean Ethics* against his left thigh. So, in the first version of the fresco, following the Ambrosiana cartoon, the flight of steps in front of Plato and Aristotle was unencumbered; one could, so to speak, mount them without impediment as I have reconstructed the original state of the fresco in a montage illustrated here. Only later did Raphael intervene in his completed fresco, destroying an area measuring roughly two square metres; the plaster was broken with a chisel and a new layer of *intonaco* laid over it. In this area Raphael then added, in fresco, the cube-shaped block of stone and the *pensieroso* which now bar the way to the flight of steps (fig. 3). By so doing the artist did not merely add an additional repoussoir to his painting. His papal patron would scarcely have allowed him to do that. Rather, the addition of the *pensieroso* meant that the

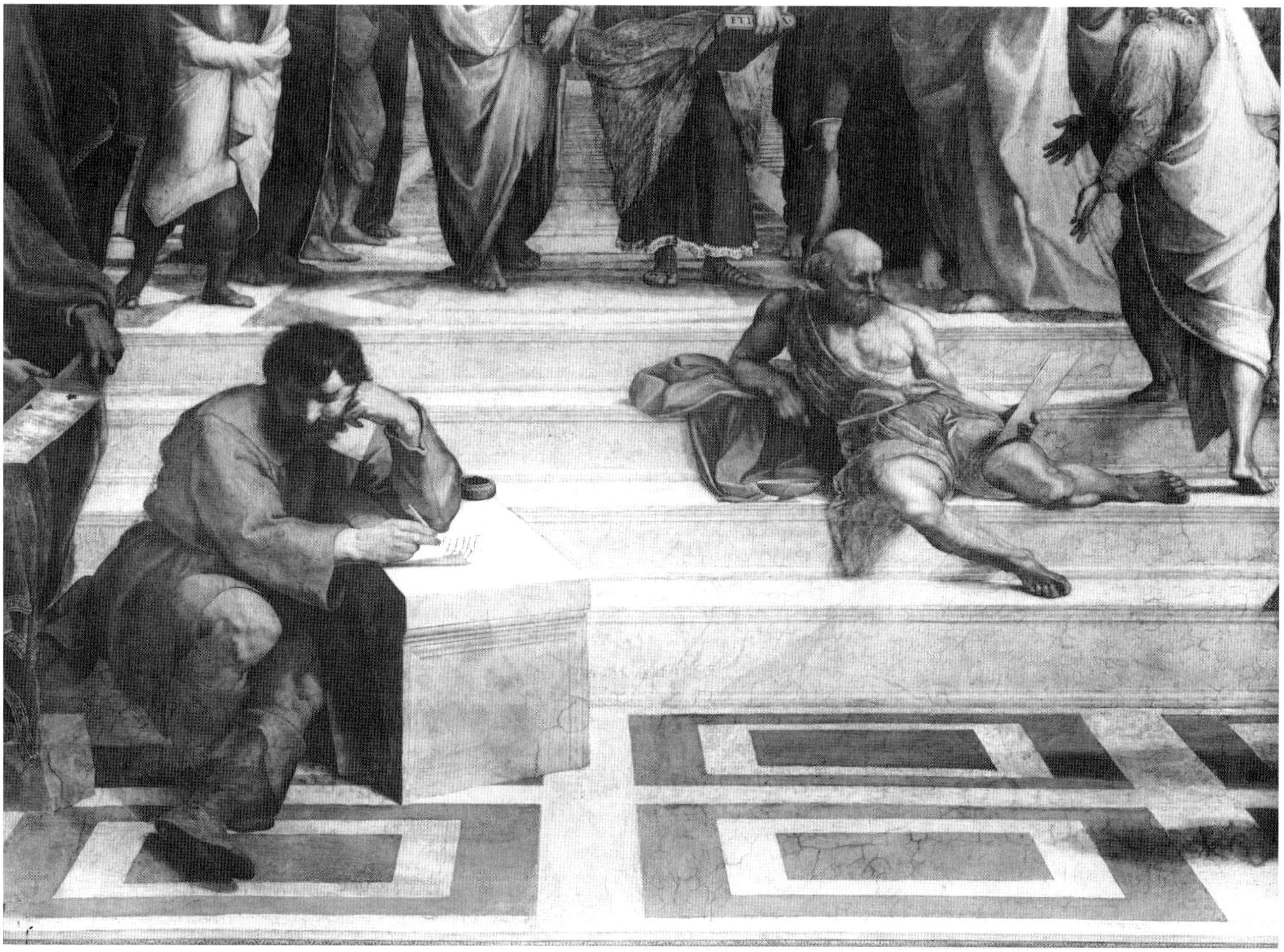

Figure 3
Raphael: *The School of Athens*. Detail with cornerstone.

significance of a flight of steps that invites ascent, and, by implication, the original significance of the *School of Athens* as a whole, was altered.

In the following I will try to relate the block of stone, the flight of steps and the remarkable arcaded building behind Plato and Aristotle to our understanding of Raphael's *School of Athens*.

Vasari describes the fresco as a pictorial history (*storia*) in which all the wise men of the world were represented; they are shown debating among each other in various ways. Vasari also speaks of a large "school of philosophers." On the vaulted ceiling of the room, above his pictorial history with the debating philosophers, Raphael also painted an abstract allegory of philosophy. *Philosophia* appears in the form of the beautiful figure of an enthroned woman, supporting two books on her lap; the lower of the two books is inscribed with the adjective *naturalis*, i.e., natural philosophy, the upper one with the adjective *moralis*, i.e., moral philosophy or ethics; to these adjectives the visible figure of the personified *Philosophia* should be added as substantive. She is flanked by two putti holding tablets bearing the inscription *Causarum cognitio*, i.e., the knowledge of the first causes.

In the mural below the allegory of Philosophy in the ceiling vault, the wise men of the world of various periods converse and gesticulate among themselves. This scene can be immediately recognised by any observer without any particular interpretation as a dramatic scene of philosophic dialogue in action: as a painted history, *storia*, to use the term in use in Raphael's and Vasari's day. Vasari also specified the content of their discussions. The philosophers, he said, were trying to reconcile (*concordare*) theology with philosophy and astrology. On the two slate tablets close to the lower corners of the fresco, at the foot

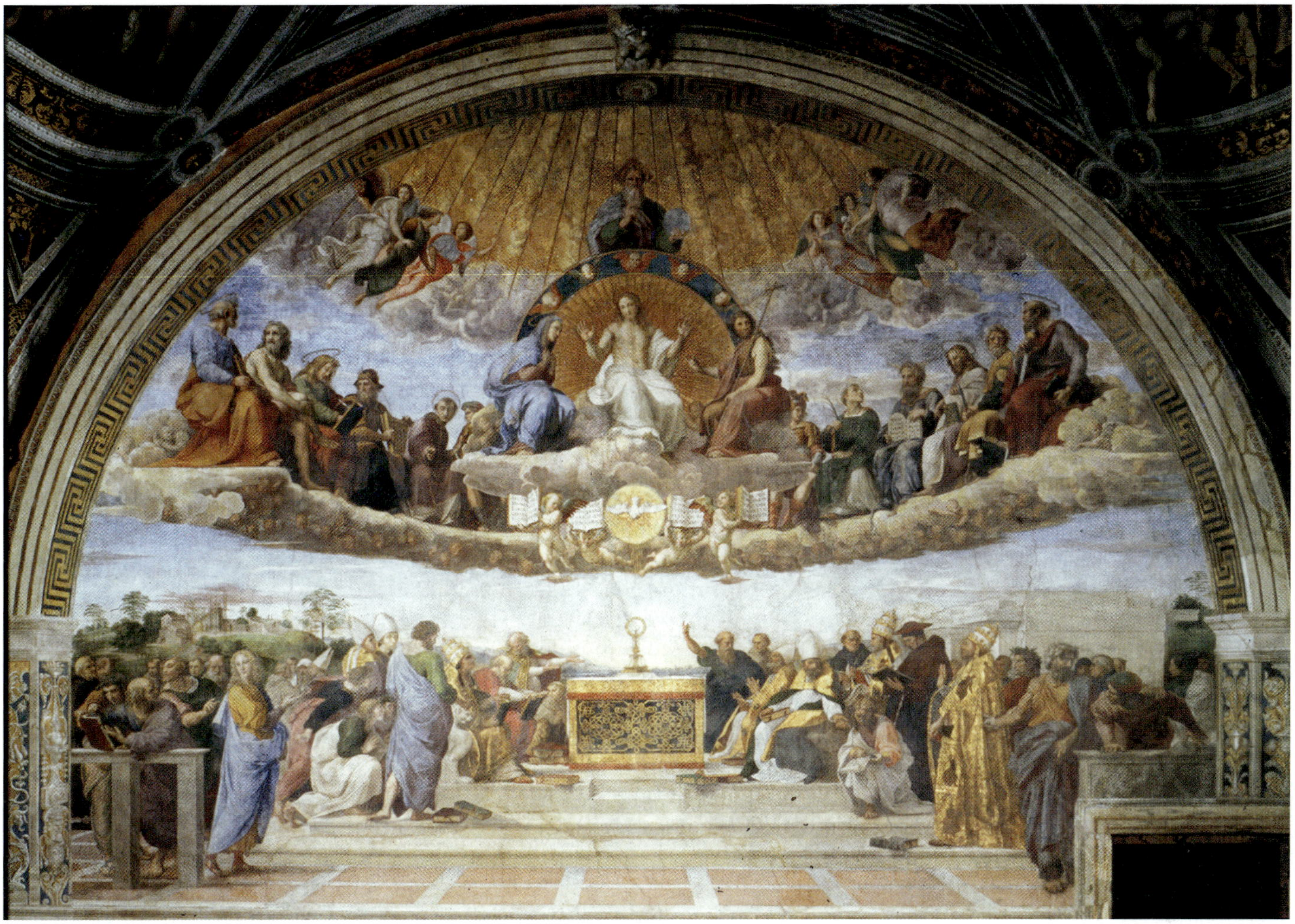

Figure 4
Raphael: *The Disputa.* Fresco. (Rome, Vatican Palace, Stanza della Segnatura.)

of two of the principal figures in this story, both of them seated, Vasari erroneously read "mantic" signs (figs. 7-8). Bellori was the first to identify the right names, which still remain valid today, for the two most important philosophers placed at ground level in front of the flight of steps: to the left Pythagoras and to the right Archimedes; the former as the inventor of arithmetic; the latter as exponent of geometry, as is demonstrated by the harmonious arithmetic proportions on the tablet to the left, and the symmetrical star-shaped figure formed of two intersecting isosceles triangles in the tablet to the right. But Bellori, in the final years of the seventeenth century, already contradicted Vasari's interpretation of the essence of what it is that the philosophers are discussing in Raphael's fresco. How were they supposed to be reconciling theology and philosophy with astrology, if only philosophers were to be seen in the mural? So the difficulties of an all-embracing understanding of the *School of Athens* began already with Vasari, although he was rightly informed that philosophy and theology are in essence reconciled in this picture. But since Vasari described each mural separately in the Stanza, considering each in isolation without completing its meaning by relating it to the mural by which it is faced on the opposite wall, he could not explain the meaning of the picture in the right way. For opposite the mural of the Greek philosophers is Raphael's mural of theology: the *Disputa*. Christian theologians are shown disputing among themselves, below the three-in-one Christian God in the heavens above. There we see the Church father St. Augustine sitting in the circle of the four Doctors of the Latin Church – the *Doctores Ecclesiae*. They flank the central altar on which is placed, in a circular monstrance, the host, the Corpus Domini, forming the centre-point of this lunette. St. Augustine is dictating to an amanuensis. His theological *magnum opus*, the *City of God*, the *Civitas Dei*, lies at his feet.

In Book VII (1–2) of the *City of God* St. Augustine discusses natural theology (*theologia naturalis*), as he calls it in Latin. He equates it with philosophy. We cannot help being reminded of the adjective *naturalis* on the book of the allegory of *Philosophia* opposite. This personified female figure of a woman, placed over the Greek philosophers, herself bears no inscription identifying her name. Only the adjectives *naturalis* and *moralis* are visible on the two books she holds in her hands; from this it can be inferred that she is a personification not only of philosophy, but also of *Theologia naturalis* in the way St. Augustine understood it. So Vasari interpreted correctly the original pictorial sense of the learned dispute of the philosophers. In the following Book of his *Civitas Dei* (VIII, 2) St. Augustine also gives an historical survey of Greek philosophy. It had split, he reports, into two schools, the Italic and the Ionian. The former had been founded by Pythagoras, whereas Socrates had been the first philosopher of the Ionian school "to divert the whole of philosophy towards the improvement and regulation of morals." Only a mind that had been cleansed could, Socrates thought, comprehend the first and highest causes; and

> *therefore he held that men ought to devote their efforts to the purification of life by sound morals, so that the mind, thus freed from the oppressive weight of lower appetites, might rise by the natural impulse to the realm of eternal things and behold with unmixed intelligence the essence of incorporeal and unchangeable light, in which the causes of all created things have their firm abode* (*Civitas Dei* VIII, 3).

In saying so, Augustine is recalling, at least implicitly, the metaphor of the cave that Plato puts into Socrates' mouth in his dialogue on the ideal state (*Politeia*). But, Augustine continues, Plato who had been the pupil of Socrates, and the philosophers who derived from Plato, had taught us to recognise in natural science (*physica*) that God is no material body (*Nullum corpus esse deum*, Civ. Dei VIII, 6). And therefore in seeking God they had transcended all material bodies (*corpora*). They also saw that anything susceptible to change could not be God; and therefore in seeking God they transcended every living soul and all disembodied spirits that are susceptible to change. In his discussion of ethics Augustine again refers to Plato, who, he says, had identified a virtuous life as the supreme good for man; but only those who recognise God, and seek to imitate him, can attain virtue.

Augustine, in his survey of Greek philosophy, as we have said, named Pythagoras as founder of the Italic school of philosophy, and Socrates as leading exponent of the Ionian school, together with his last pupil Plato. All three can be seen occupying prominent positions in Raphael's *School of Athens*. Socrates, who is recognisable due to his silenus-like facial features, stands on the top step to the left of Plato. Vasari made no mention of Socrates, although it cannot have escaped him that Raphael had here, for the first time, furnished Socrates with the silenus-like portrait features handed down by ancient pictorial tradition. When Augustine says that the philosophers dependent on Plato, so not least his pupil Aristotle, had transcended all material bodies – *corpora* – to seek God, who was incorporeal, then Raphael's flight of steps in the *School of Athens*, on which the philosophers move, could be a way of illustrating Augustine's argument of the need to transcend all material bodies. But what could better express a material body, which all philosophers need to transcend, than the massive cube-shaped stone block that Raphael placed below the lowest step? Plato's gesture, pointing up to heaven, underlines the transcendental in his philosophy (to put it in Kantian terms).

In chapter 9 of Book VIII of the *Civitas Dei* Augustine finally sums up the philosophers who had in his view come closest to the Christian faith. Whenever these philosophers recognised the true God as the effective cause (*causa*) of all created things, as the light by which we attain knowledge of all things and as the goal of all our actions – writes Augustine -, whether they be Plato himself as the leading philosopher of Ionia or the Platonists, or Pythagoras or the Pythagoreans, or whether they belong to other nations, such as the wise men among the Egyptians, the Indians, the Persians or the Chaldaeans – if they had this view of the true God and taught it, then Augustine prefers them to all the rest, since they stand closest to the Christian faith. It is with these philosophers that he wants to discuss the questions of faith.

At this point Augustine touches on a basic question of medieval theology: namely, to what extent the philosophy of the Greeks could be reconciled with the revealed truth of the Christian faith. If Aristotle had determined the basic presuppositions of scholastic theology right down to the fifteenth century, it was Plato as philosopher and *natural theologian* who was pushed to the forefront of theological debate ever since Gemistos Plethon had arrived together with Bessarion from Constantinople to attend the Council of Ferrara and Florence in 1438. Marsiglio Ficino's translation of the Platonic writings and his *Theologia Platonica*, platonic theology, as well as the bitter conflict between Aristotelians and Platonists during the conciliar period, also provided the source material for the uncompleted life-work of Giovanni Pico della Mirandola, *De concordia Platonis et Aristotelis*, the concord or reconciliation between Plato and Aristotle.

That it is only these two, Plato and his pupil Aristotle, who are given so prominent a place among all the other heads of the schools of ancient philosophy in the *School of Athens*, and in some sense stand out as *principes*, reflects the intellectual situation of "natural theology" on the eve of the Reformation.

In chapter 12 of Book VIII of the *Civitas Dei* Augustine enquires into the origin of the knowledge of the Platonists, and how they had been able to acquire the understanding by which, like Plato himself, they had come closest to Christian beliefs. In trying to answer this question he refers tellingly to *Romans* 1:19:

> *For what can be known about God is perfectly plain to them, since God has made it plain to them: ever since the creation of the world, the invisible existence of God and his everlasting power have been clearly seen by the mind's understanding of created things.*

Plato himself had arrived at this understanding. His disciple Aristotle, "a man", says Augustine, "of outstanding ability, and in literary style easily the superior of many, although certainly no match for Plato," had, like his master, founded a school that came to be known as the School of the Peripatetics (because it was Aristotle's habit to walk about as he lectured). But all these philosophers believed in polytheistic worship; they were of the view, as was Plato himself, that sacrifice should be offered to many gods, although they would have done better to speak of the one God who had created heaven and earth. So much for St. Augustine.

It must already be clear that Raphael in this fresco neither painted the school of Aristotle alone, nor that of the Academy of Plato, although the teacher-pupil relationship leading from Socrates to Plato to Aristotle in generational sequence can be read from left to right on the topmost step. The facade of a four-way arch of Janus Quadrifons in the form of two barrel-vaulted porticoes behind them can hardly allude to the grove of the Akademos, where Plato met with his pupils, nor to the Lykaion of Aristotle. So the name *School of Athens*, which was not given to the fresco until the seventeenth century, is inappropriate. No doubt this name was chosen because only Apollo and Athena, the civic goddess of Athens, can be recognised among the statues of gods in the niches behind the philosophers. But Pythagoras never taught in Athens: he came from Samos, whence he travelled to Egypt, where he gained the confidence of the priests and learnt from them their mathematical wisdom. He later travelled to Magna Graecia, where he founded a school at Crotona and gathered a sect round him; his pupils included King Numa, who gave Rome its first code of law, based on peace and harmony, and introduced the cult of the god of peace Janus. A turbaned figure among the pupils behind Pythagoras can hardly be a Greek philosopher.

But let us now return to the painted architecture of the *Philosophers' School* (as we might, more appropriately, call Raphael's fresco). Over the crossing of the intersecting barrel-vaults rises an invisible dome. Plato and Aristotle are distinguished from the other groups of teaching and debating philosophers of various age groups, not only because of their central position, but also because a stone triumphal arch, at the far end of the arch of Janus Quadrifons, frames them in such a way that only these two philosopher princes appear before the blue sky, whereas all the others are placed against stone architecture.

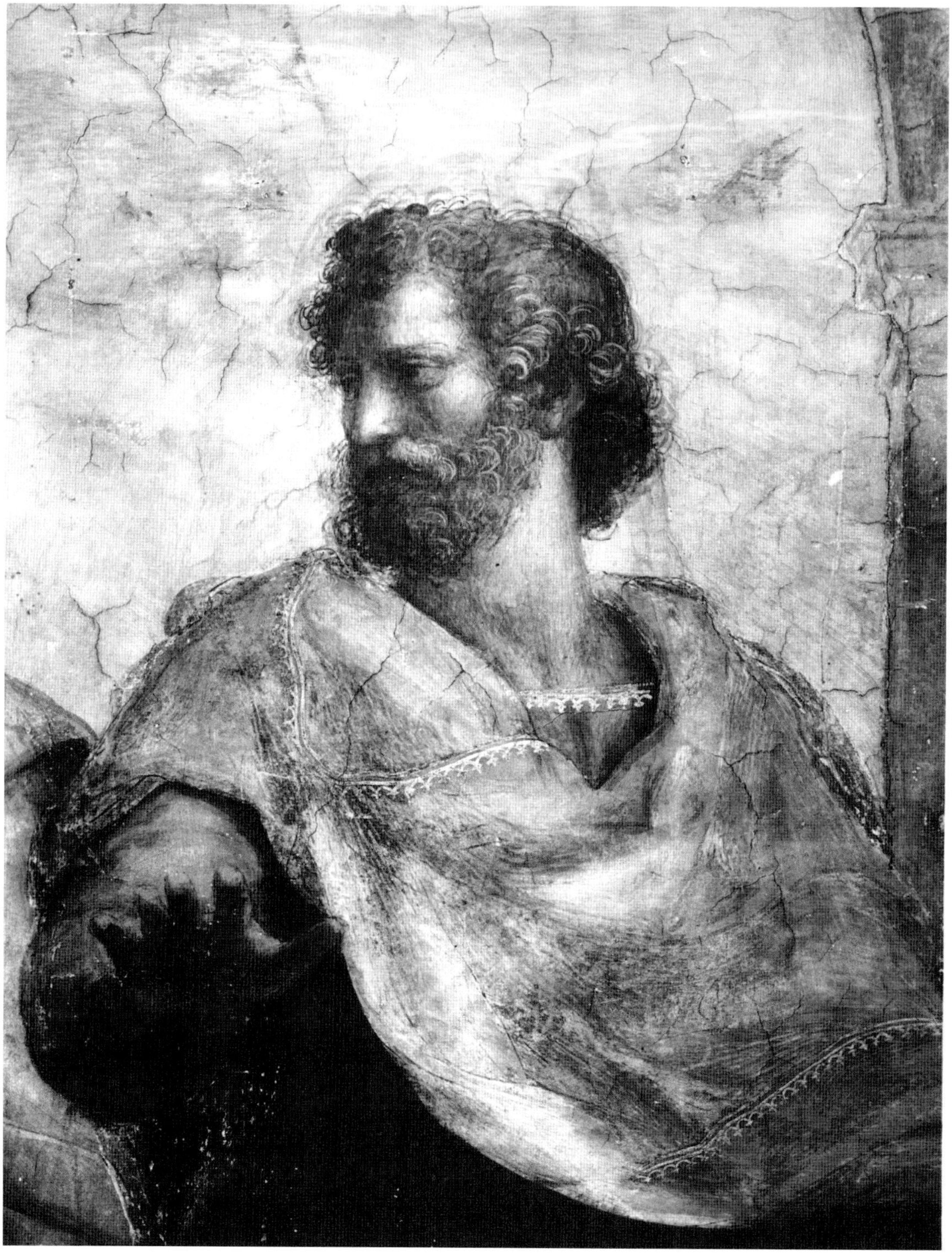

Figure 5
Raphael: *The School of Athens*. Detail with Aristotle.

The coffering of the barrel-vaults is modelled on that of the Basilica of Maxentius on the Forum Romanum. Raphael and his contemporaries confused the Basilica of Maxentius with the Temple of Peace (*Templum pacis*) erected by the emperors Vespasian and Titus after the destruction of Jerusalem. Gian Pietro Bellori already recognised in the late seventeenth century that the horizontally extended right hand of Raphael's Aristotle was the gesture of peace (fig. 5). This discovery of Bellori is now forgotten. But in his accurate interpretation of such gestures Bellori was only following the acute Politian. It was this humanist who, towards the end of the quattrocento, had correctly read the similarly outstretched right hand of the equestrian statue of Marcus Aurelius, which

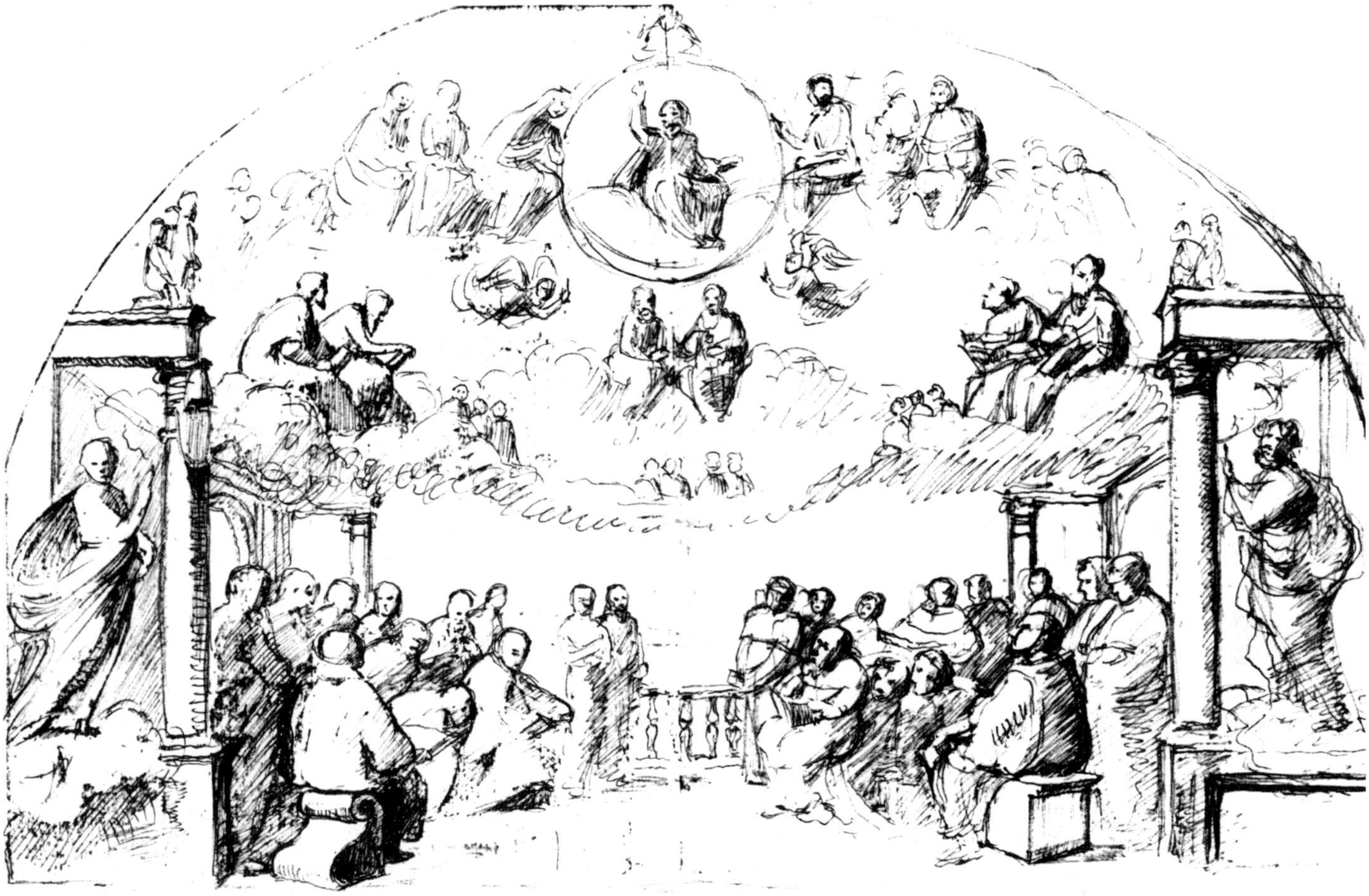

Figure 6
Raphael: *The Disputa*. First (?) modello – synopsis of Raphaels drawings in Windsor Castle, Oxford and Chantilly (M. Winner).

then stood in front of the papal palace at the Lateran, as a gesture of peace; it conformed to the orator's gesture of peace, as Quintilian had described it in his book on oratory (the *Institutiones oratoriae*). That is why the symbolism of peace in Raphael's framing architecture of the *Philosophers' School* should be taken seriously; for the building type of an arch of Janus Quadrifons bears the name of the god of peace, under whose aegis were placed doors, gates, arches and bridges, everything that expresses a passage or thoroughfare, everything that has an opening at both ends. In Roman state religion Janus was also the god of all beginnings, as well as the god of the arch of heaven and the god of chaos. His name was invoked during sacrifice before that of any other god. He was also nicknamed the key-holder, *claviger*, the same name adopted by the pope in the succession of Peter. The fact that the supreme priestly office in the Roman Catholic Church perpetuated the title of the supreme priestly office of the pagan Roman state religion, Pontifex Maximus, meant that the original sacred meaning of Pontifex as bridge-builder had never been forgotten. That explains why the name of Pope Julius with his title Pontifex Maximus can be read on the altar of the *Disputa*. Janus was a Roman god whom one may seek in vain in the Greek Pantheon. In times of peace his temple was closed; for Janus was synonymous with peace. In the two arched doorways behind Plato and Aristotle, Raphael painted for us an image of how access could be gained to the real knowledge of created things, since only these philosophers stand before the infinity of the heavens. He has painted, in effect, what is tantamount to a visual pun on what Christ, the son of God who became man, said of himself: "I am the door" (*John* 10:9). *Sum ostium*, or *Sum Janua*. But if the whole arch of Janus Quadrifons in the *Philosophers' School* represents a self-testimony of the living Son of God, then we must recognise the architecture as an additional bearer of meaning and feel ourselves once again invited to see stairway and stone in a double sense.

What now appears in Raphael's opposite fresco of the *Disputa* in the Stanza della Segnatura was originally planned in a different way. On the basis of three Raphael drawings respectively in Windsor, Chantilly and Oxford, we are able to reconstruct the original pictorial idea with some degree of precision (fig. 6). The altar with the Corpus Christi at the centre is absent from this reconstructed original design for the *Disputa*. In its stead, a balustrade in the middle distance closes the gap at the centre between the two groups of theologians to the right and to the left. In the foreground a single column on a stylobate stands to the right and to the left. Sections of entablature rest on their capitals. To the two columns in the foreground correspond two columns in the background. Their entablatures are in part concealed by banks of cloud on which two apostles are seated on either side. Christ, sitting in Judgement at the centre, is surrounded by a circular aureole. Below Christ are sketched two figures of apostles, framed by the four Evangelists sitting on clouds. Given their central position below the Trinity, the apostles in question can only be the princes of the apostles Peter and Paul. So these two apostles, seated on clouds at the centre of the *Disputa*, were originally intended to correspond to the two central figures of the *Philosophers' School*.

In the foreground, to the left, behind one of the two columns, a female figure in a billowing mantle is standing. With the index finger of her right hand she is pointing to a stone coat of arms, that is being raised to the height of the capital by two putti. Since the papal tiara appears over the coat of arms, it can only be the coat of arms of the reigning pope Julius II Della Rovere. If Raphael had actually executed this original design, the golden oak-tree against an azure ground would surely have been visible on this coat of arms; for the family name of Pope Julius, Della Rovere, means oak-tree in English. The female figure is pointing to his coat of arms. She can therefore be identified as the Sibyl of Cumae. For a verse in Vergil's fourth Eclogue had been given a Christian (as well as a Julian) interpretation: a verse in which Sibyl prophesied that the great sequence of the ages was about to start afresh, and that the firstborn of the new age was already on his way from high heaven down to earth ("*nova progenies demittitur ab caelo alto*").

In Raphael's *modello* for the *Disputa*, the coat of arms of the Pope, like the capital and the column to which it is raised, were of stone. The two theologians with their backs turned to the viewer, one on either side, sit on blocks of stone. In the foreground to the right the block of stone is part of an entablature with a moulded cornice, but polished smooth on the highlight side. In the foreground to the left, by contrast, the block of stone consists of an upside-down console with a scrolled base. In this original design for the *Disputa*, the *Doctores Ecclesiae* do not yet sit on stone *cathedrae*, as they do in Raphael's final painted version of the fresco. They sit on masonry blocks, in an open-air column-supported colonnade over a stone foundation. This is not the place to review the whole corpus of building and masonry metaphors in the Old and New Testament, in order to elucidate the sense of the stone blocks on which the theologians are sitting. But the *Doctores Ecclesiae*, in accordance with Pauline building metaphors, have always been understood as columns, *columnae* of the Church. Paul himself, in the first of his *Letters to the Corinthians* (3:10), likens himself to a trained architect (*sapiens architectus*) who had, by the grace of God, laid the foundations. And on this foundation, which is Christ, anyone could build. In the *Letter to the Ephesians* (2:19–20) Paul speaks of the foundations of the Church built by the prophets and the apostles. In this passage Paul addresses himself to the gentiles in contrast to the Jewish Christians:

> *So you are no longer aliens or foreign visitors; you are fellow-citizens with the holy people of God and part of God's household. You are built upon the foundations of the apostles and prophets, and Christ Jesus himself is the cornerstone.*

And lastly the apostle Peter, writing to his brothers in his first Letter, declares:

> *He* [the Lord] *is the living stone, rejected by human beings but chosen by God and precious to him; set yourself close to him so that you, too, may be living stones (lapides vivi) making a spiritual house as a holy priesthood to offer the spiritual sacrifices made acceptable to God through Jesus Christ. As scripture*

says: 'Now I am laying a stone in Zion, a chosen, precious cornerstone (lapis summus angularis) and no one who relies on this will be brought to disgrace.' To you believers it brings honour. But for unbelievers, it is rather a stone which the builders rejected that became a cornerstone, a stumbling stone, a rock to trip people up. They stumble over it because they do not believe in the Word (1 Peter 2:4–8).

The hewn cube-shaped block of stone painted into Raphael's *Philosophers' School* in a second moment is just such a cornerstone (fig. 3). Its foremost corner in front of the flight of steps articulates the central vertical axis of the whole lunette. Anyone who wants to mount the steps to the height of knowledge represented by Plato and Aristotle must now stumble over the stone. It is the stumbling stone that the builders rejected.

The first *modello* for the *Disputa* thus helps to explain the living stones from which the spiritual house of the Church with its cornerstone Christ is built. The cloud zone with the four Evangelists to the sides of the two princes of the apostles extends horizontally at a level corresponding to that of the entablatures placed above the stone columns to the rear of the colonnade. Living stones: in other words, the new *Civitas Dei*, as St. Augustine would put it. Elsewhere Augustine says that citizens (*cives*) live in each city built by man, whereas in the *Civitas Dei* the citizens (*cives*) are themselves the stones of which the city is built. It was not by accident that Raphael, in his *Philosophers' School* placed *dead* stone statues of gods in stone niches over the heads of the princes of the philosophers standing on the topmost step – at precisely the same height as that which, in the opposite fresco, the spiritual house built of *living* stones is placed over a bank of clouds.

We may further note that the load-bearing structure of the arch of Janus Quadrifrons in the *Philosophers' School* consists merely of pilasters; by contrast, the original *modello* of the *Disputa* is articulated with columns. Only over the paired philosophers Plato and Aristotle can one glimpse a window divided into three parts by two column shafts in the tambour of the dome over the crossing. The capitals of these columns are invisible; they are cut off by the barrel vault in front. But these two columns, the only ones in the whole architecture of the arch of Janus Quadrifons, are located precisely opposite the position of the two princes of the apostles below the Trinity in Raphael's original *modello* for the *Disputa*. A clear analogy is thus drawn between the dead stones of the *Philosophers' School* and the living stones of the *Civitas Dei* opposite; for the column shafts in the tambour window over Plato and Aristotle reflect the theological metaphor *Doctores Ecclesiae columnae Ecclesiae sunt*. The theologians sit as living stones on the flat ground of the foundation that is Christ; the foundation that the trained architect had laid. They need no flight of steps to attain a knowledge of God in the heavens above. They sit *in humilitate* on the ground below.

In the *Philosophers' School* Plato and Aristotle have attained a high degree of knowledge of God's works. But their *theologia naturalis* does not lead to any knowledge of the true God who became man.

Let us now turn to the two philosophers, Pythagoras to the left, Archimedes to the right, who each demonstrate something to their pupils with the help of slate tablets (figs. 7-8).

Pythagoras is sitting in the foreground to the left. Before him a boy is holding a slate tablet upright, supporting it on the ground. Numbers and schematic figures are inscribed on it. In the lower part of the slate tablet Roman numbers from I to IIII are arranged in the form of the Pythagorean triangle with the number I forming the apex and IIII the base line. These numbers, when added together, produce a total of ten, which is shown in the form of a Greek X at the foot of the tablet. The number four enjoyed particular veneration among Pythagoras and his pupils as the "holy *tetraktys*."

In the upper half of the tablet appears the arithmetical Pythagorean scheme of musical proportions. According to Boethius, it was Pythagoras who invented it. It was also known in the Middle Ages. It is based on the numbers six, eight, nine and twelve (inscribed at the top of the tablet). The lower half-circle joins the six with the twelve and illustrates the musical octave in its proportion 1:2. It is called in Greek *diapason*. The other proportions are called *diapente*, or fifth, and *diatessaron*, or fourth. The upper segment joins the eight with the nine and

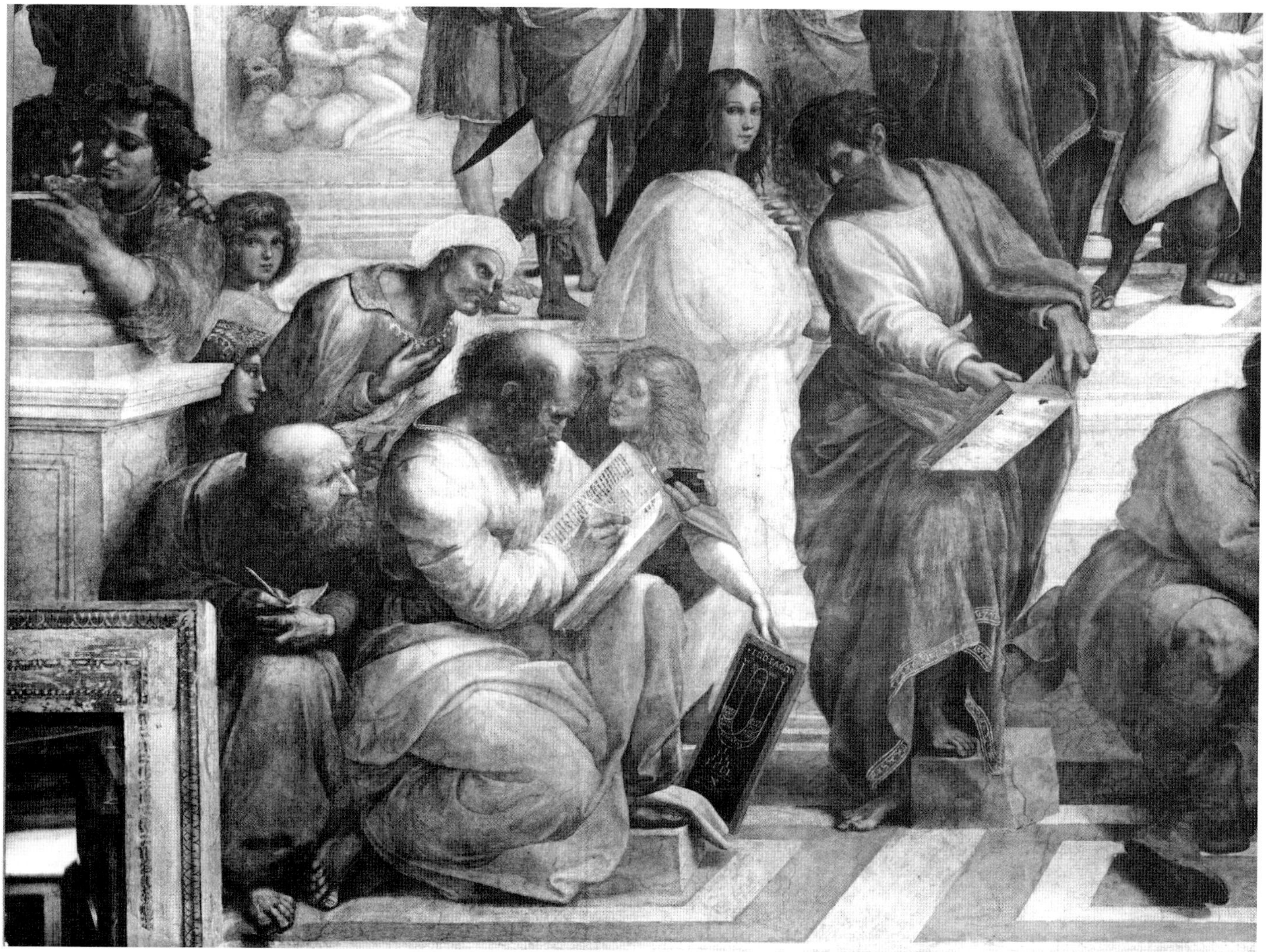

Figure 7
Raphael: *The School of Athens.* Detail with Pythagoras.

thus illustrates the musical second, which expresses a disharmonious proportion and was called *epogdoon* by the Greeks. In Boethius' *Institutio arithmetica*, Pythagoras is described as the inventor of arithmetic and philosophy. The humanist Raffaello Volterrano dedicated his encyclopaedic magnum opus *Philologia* to Pope Julius II in 1508. In this work he calls Pythagoras the inventor of the science of numbers. He also says that Pythagoras was the first to teach his pupils to write on the abacus, or reckoning tablet, as we see it in the *School of Athens*. From the biography of Pythagoras written in Greek by Iamblichus the humanists could discover yet another quality that Raphael has expressed in his picture. Pythagoras, namely, was the first who refused to be called wise. He was the first to call himself merely a lover of wisdom, *philosophus*, instead. That is why he sits to the left of the *Philosophers' School* over a door through which anyone who enters the room has to pass. Pythagoras and his pupils dressed in white. Consequently Raphael clad not only him, but also the youth who looks out of the picture to the right of Pythagoras, in a white mantle. Another man who is ascending the stairs, with his back turned, is also dressed in the Pythagorean white mantle. The unidentified friend standing to the right next to the self-portrait of Raphael, to the far right of the fresco, is also wearing a white mantle. Not only Iamblichus, but also Plato and many other ancient sources transmitted the legend that Pythagoras claimed to be able to hear the harmony of the universe, otherwise inaudible to human ears, and the music of the moving spheres of heaven and their stars. Pythagoras considered numbers to be the essences of things. The abacus of Pythagoras thus has to do with arithmetic and musical harmony, insofar as it can be expressed in arithmetic proportions. Friendship (*amicitia*) was held by Pythagoras to be the basis for any human community, also that of the state. That is why we see so many embracing each other in friendship in Raphael's *Philosophers' School*.

It is hardly by accident that Socrates, standing above Pythagoras on the uppermost step, should be making a gesture of enumeration to his pupils; for with two fingers of his right hand he is holding the index finger of his left hand. Among his audience is Alcibiades; he can be recognised by his armour. We cannot but be reminded here of Plato's dialogue *The Symposium*, in which Socrates says much about friendship and love in conversation with Alcibiades. But Socrates' gesture of ticking off his fingers could also be a way of illustrating the introductory words "One, two, three" of Plato's *Timaios*; for at the beginning of that book Plato represents his teacher Socrates as counting those of his partners in dialogue who were already present: One, two, three. What had happened to the fourth interlocutor of the previous day? The others call him. In fact the man to the rear of the group of listeners gathered round Socrates is beckoning with his outstretched arm to the left, as if calling to someone to join the group. Marsiglio Ficino, in the printing of his Latin translation of Plato's dialogues, placed the *Politeia*, the dialogue of the state, not without reason, before the *Timaios*, which Plato carries under his arm in Raphael's fresco. For, after the partners in dialogue of the *Timaios* had fully assembled, Socrates reminds his interlocutors of what they had discussed about the ideal "state" on the previous day. Socrates then, at the beginning of the *Timaios*, repeats significant propositions about the "state" to his friends.

In the *Politeia* Plato emphasises that only philosophically educated citizens were capable of directing the ideal state; for only they could form it according to their philosophical knowledge of the supreme good. So if kings wanted to govern the state in the right way, they must first become philosophers. Aristotle, by contrast, at the beginning of his *Nicomachean Ethics*, which in Raphael's fresco he is supporting against his thigh with his left hand, posits the highest goal of various forms of conduct which must be striven toward for their own sake, and which therefore also are to be called the "supreme good" in the fields of the empirical sciences and practical arts. According to Aristotle, the "supreme good," in its most authentic and most sovereign sense, belongs to the sphere of the arts. The "art of the state" [Staatskunst] was in this sense the most comprehensive art of all. Thus Raphael's *Philosophers' School* is also, if you like, a political allegory.

According to Plato's *Politeia*, the citizens, who ought to preside over the state as guardians or kings, should only be instructed in philosophy. For only philosophers could turn away from the deceptive world of appearance and shadow, in order to recognise the true world of ideas and of the "supreme good." Socrates argues in Plato's *Politeia* that only the mathematical sciences can purify the mind from the illusions of the senses. Socrates is counting off with his fingers in Raphael's fresco. The inconsequential enumeration one, two, three, with which Plato's *Timaios* opens, could hardly justify Raphael's precise representation of a Socrates who grasps the index finger of his left hand with the thumb and index finger of his right hand. Rather, Socrates seems here to be performing the gesture that he makes to his partners in dialogue in Book 7 of Plato's *Politeia*, in order to explain to them the principle of numbers. In other words, Socrates shows thumb, index finger and middle finger of one hand to his listeners to demonstrate the impossibility of singling out one finger from another by sensory experience alone. Each of the three fingers in question shows infinite qualities, argues Socrates. But if we grasp the unity of only one finger among three other similar ones, we understand it as *unitas*. This concept of unity is, in Socrates' view, the basis of number and counting. Moreover, that one plus one makes two was the basis of philosophical learning. The gesture of Raphael's Socrates must therefore be understood in a twofold sense. He stands on the topmost step over the slate tablet of Pythagoras, the legendary inventor of arithmetic. But at the same time Socrates illustrates the philosophic concept of unity, in Greek *monas*, what so to say expresses the sensory experience of the movement of *one* finger among many. The future guardians of the ideal state should therefore, in Socrates' view, begin with arithmetic, in order to be educated as philosophers. In the second place they should learn geometry, because geometry leads the soul in the direc-

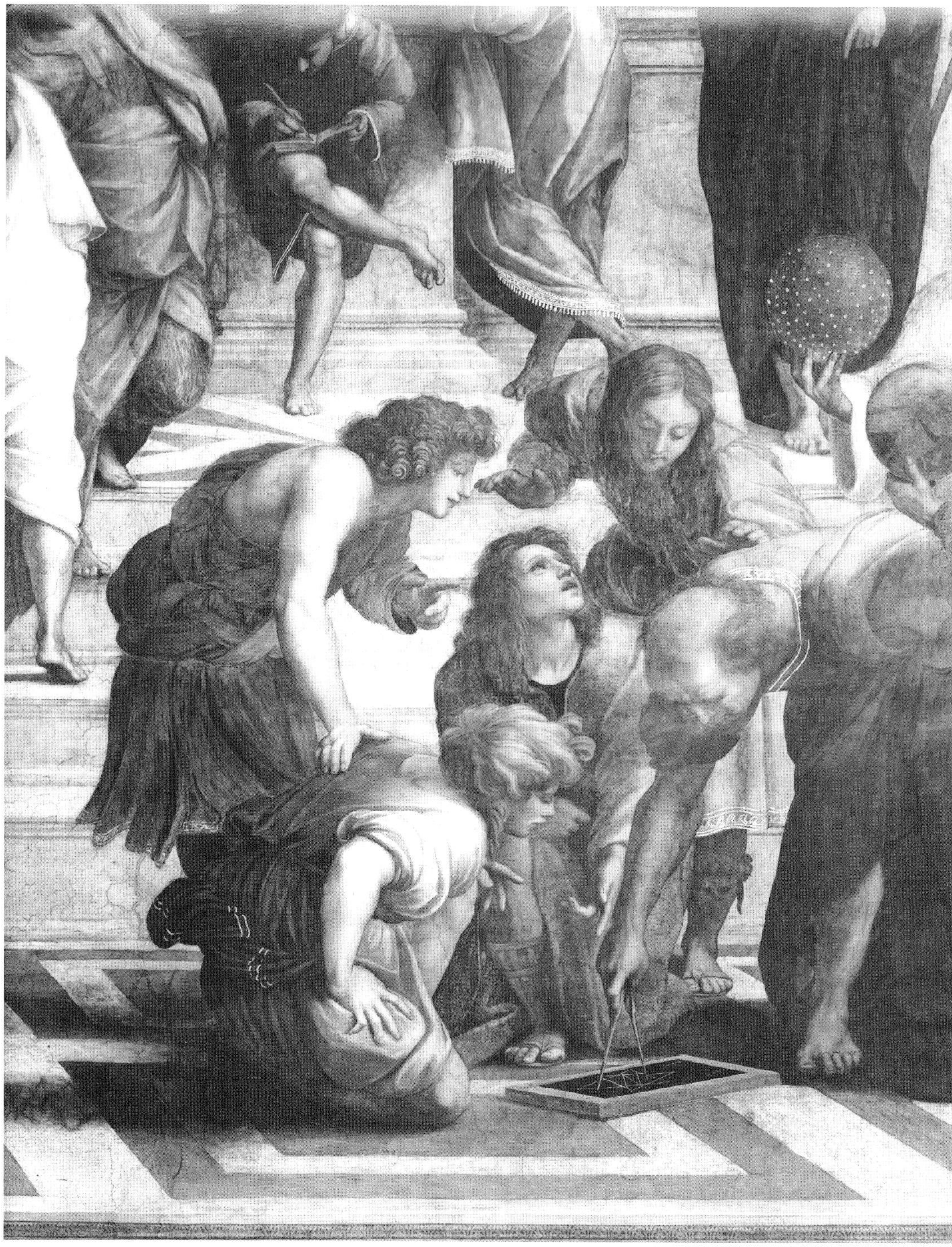

Figure 8
Raphael: *The School of Athens*. Detail with Archimedes.

tion where the supreme good of all being exists. Geometry directs our soul upwards, explains Socrates, although we draw her figures physically bent down to earth. And it is bent exactly like this that Archimedes is placed as the representative of geometry on the right side of Raphael's fresco (fig. 8). He bends down to the ground to measure with his dividers the proportions of a star-shaped figure on the slate tablet at his feet. Bellori rightly identified him as Archimedes because Raphael's pupil Perino del Vaga painted in grisaille a scene of the *Murder of Archimedes* in the dado immediately below. The Roman general Marcellus was forced into a long drawn-out siege of Syracuse as a result of the military machines invented by Archimedes. When the city finally fell, the Roman general ordered his troops to capture the mathematician and philosopher Archimedes alive, so that he could be led in triumph through the city of Rome. A soldier found Archimedes in his own house, as he was bending down to draw geometric figures on the floor. So absorbed was Archimedes in his geometric problem that he could not even remember his own name when he was asked to give it. "Do not disturb my circle" (*Noli disturbare circulos meos*), he is reported to have replied to the Roman soldier, who thereupon stabbed him to death. The philosopher's blood is said to have flown over his geometric figures.

Vasari had heard that Raphael had given the features of the papal architect Bramante to the bent-over geometrician. Since Archimedes, as Vitruvius reports, had pursued the profession of architect before devoting the rest of his life to his studies as a philosopher of mathematics, Vasari's report does conceal a kernel of truth. Unfortunately the geometric star-shaped figure on the slate tablet in front of Archimedes-Bramante has no ancient model, neither in Archimedes' writings nor in Euclid's ancient textbook of geometry. That it has to do with a problem of proportions of squares and sections, just as in the tablet of Pythagoras on the opposite side of the fresco, seems clear; for Archimedes is seeking a point with his dividers, without having yet begun to draw a circle: "Give me a point outside the earth, and I will turn the world upside down," Archimedes is said to have boasted, after his invention of the law of the lever. Does the star-shaped figure on his tablet perhaps allude to one of the many spangled stars that appear on the model globe of the eighth sphere of heaven over Archimedes' head?

According to the pre-Copernican world system, the eighth sphere of heaven with its fixed stars circles the motionless earth at the centre of the cosmos faster than all the others. According to the definition of Boethius, geometry is the science of motionless size.

The pupils of Archimedes follow his geometrical demonstration with gestures of wonder. One of them even points to the spherical form of his bald head. It is as if our attention were being drawn to the stereometric form of three spheres or globes, one placed over the other, on this side of the *Philosophers' School*. The bearded man standing behind Archimedes is wearing a beret adorned with a circlet of gold. That it is Zoroaster, or rather Zarathustra, the legendary king of Bactria, who holds the model of the eighth sphere of heaven, the heaven of the fixed stars, was already recognised by Vasari. Zarathustra was venerated as the legendary inventor of astronomy. Facing him, with his back turned to the viewer, is King Ptolemy of Egypt, wearing a hellenistic radiate crown and a royal mantle, who was identified in the Middle Ages with the mathematician and astronomer of the same name. Ptolemy is holding a small model of the earth and is turning his head to look at the self-portrait of Raphael. Raphael himself seems as if he is being instructed by King Zarathustra and King Ptolemy. Ptolemy's *Geographia* was translated into Latin as early as 1407, but was not printed until 1475. In the introduction to his book Ptolemy draws a distinction between two methods of cartography. The first was geography, literally, the drawn picture of the whole earth. By this Ptolemy understood a planimetric method of projection that would determine the distances between localities and plot general contours such as those of coast-lines and harbours. Its most important auxiliary was geometry, which means, in its prime etymological sense, the science of measuring the earth. But Ptolemy also explains another cartographical science in addition to geography: this he calls *chorographia*; it was the science of mapping a region in which painters represented in as great a detail as possible everything that appeared to the eye within the visible horizon, such as harbours, farms, villages, towns and rivers: the panorama as *vedute*, as Christiane Hessler has put it. Ptolemy then continues: it is as if *geographia* were to represent the whole of a human head, whereas *chorographia* limited itself to the detail of an ear or an eye of this head. The Ptolomaic world map had already been disseminated from Constantinople through Florence to the whole of Europe by the early Quattrocento. When Ptolemy places this world map in the form of a globe before Raphael's eyes, this was undoubtedly one way of spelling out what Ptolemy regarded as the painter's task, namely chorography.

On the right side of the *Philosophers' School* we thus see not only spherical models of the celestial and earthly globes, but are also led to compare them with the spherical form of the human head of Archimedes. In

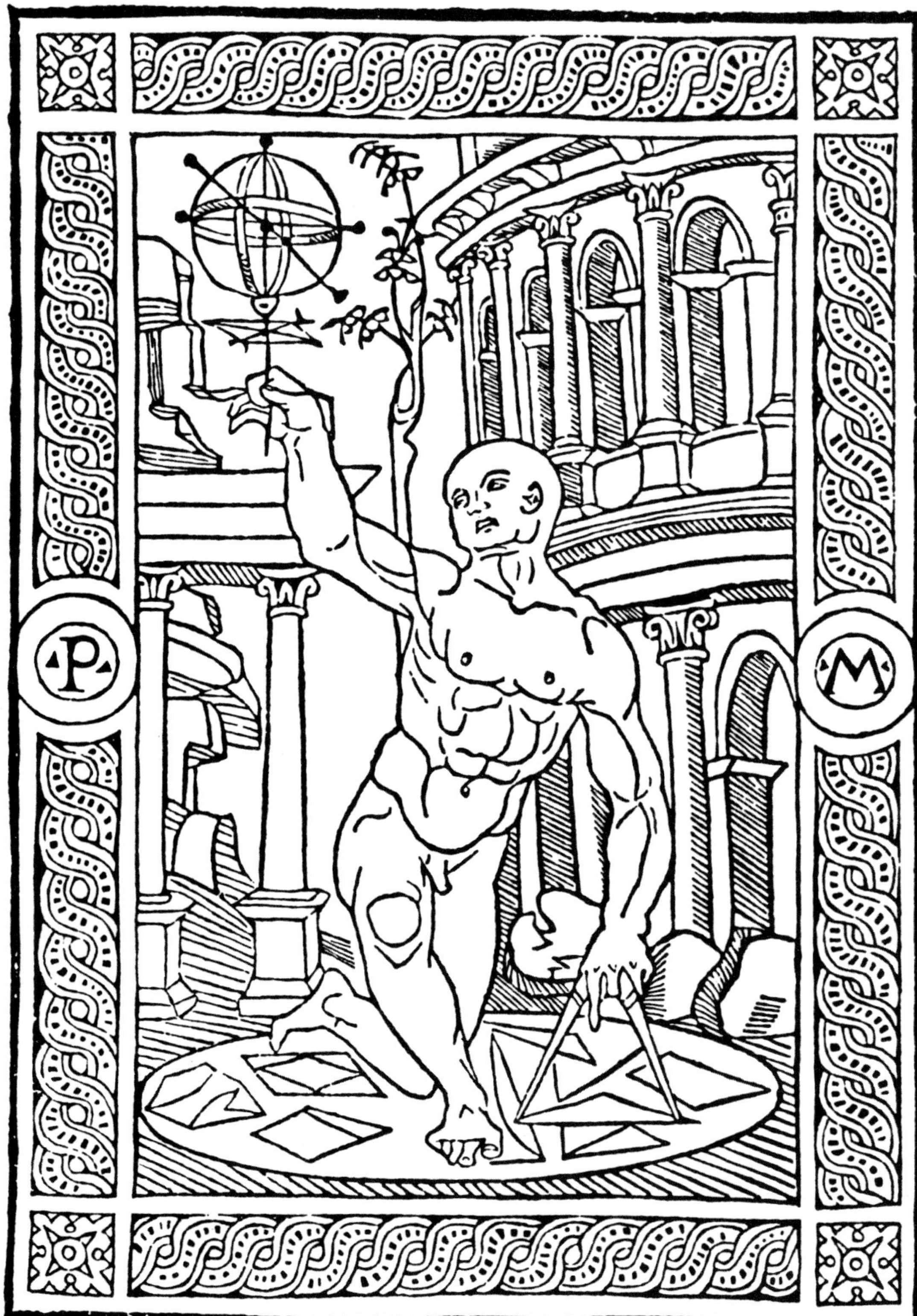

Figure 9
Prospettivo Milanese. Frontispiece of book title, woodcut.

a frontispiece woodcut of a tractate on the ancient works of art of Rome, published ten years before Raphael's arrival in the Eternal City, we find illustrated just this comparison between the microcosmos of the human skull and the macrocosmos of the heavenly sphere in the form of an armillary sphere, as recently explained to us by Hans Ost (fig. 9). The anonymous Milanese artist, author of this woodcut and of the tractate it embellishes, called himself the *Prospettivo Milanese*, the Milanese perspective painter. In Raphael's fresco we find, placed flat on the ground, the tablet with which Archimedes is projecting the geometric figure of a circle with his dividers, and over it the spherical forms of the skull and the celestial globe. It was just this comparison that Plato made in his *Timaios* (44a):

> *In order to imitate the spherical form of the cosmos, they (the gods) bound the divine revolutions, which are two, within a sphere-shaped body, the same body that we now call the head, it being the most divine part and reigning over all the parts within us.*

And in a slightly later passage of his *Timaios* (47a) Plato discusses the sense of sight, which he took to be the prime cause of reason and of philosophy itself. Why, Plato asks, did God bestow us with eyes?

> *Vision, in my view, is the cause of the greatest benefit to us, inasmuch as none of the accounts now given concerning the universe would ever have been given if men had not seen the stars or the sun or the heaven. But as it is, the vision of day and night and of months and circling years has created the art of number and has given us not only the notion of time but also means of research into the nature of the universe. From these we have procured philosophy in all its range, than which no greater boon has ever come or will come, by divine bestowal, into the race of mortals. This I affirm to be the greatest good of eyesight.*

Ptolemy and Zoroaster, their heads turned, seem to be explaining to Raphael the revolution of the heaven round the earth with their hands. But Raphael's gaze is not fixed on them: he is looking instead out of the painting. This was one way of demonstrating the importance of sight not just for the art of painting but as the human faculty *par excellence* that is needed to gain access to philosophy. The science of vision is called optics or perspective, *prospettiva*, which is, as already Dante affirmed in his *Convivio*, a handmaid of geometry. *Logos opticus*, as Vitruvius said in Book I of his architectural treatise.

But let us now return to Socrates, who, in Book VII of Plato's *Politeia*, called for arithmetic and geometry to be the basic disciplines for any philosophic instruction. Socrates also added astronomy to these two sciences, because astronomy is the science of the revolution and motion of the celestial bodies. At this point Socrates' interlocutors interrupt his argument. Why, they ask, does he treat astronomy immediately after geometry? In the last analysis geometry was only aimed at the measurement of surfaces and planes and hence was known as planimetry. The measurement of solid three-dimensional bodies was not the business of geometry. Conversely, astronomy tackled the movement of bodies and could not therefore be solved with geometric knowledge alone. After all, it was not yet known what a solid body – *corpus solidum* – was. Socrates admits that the problem of the third dimension could not be solved by geometry, for that was a science limited to two dimensions. The discipline of solid bodies (*corpora solida*) or physical forms had to do with length, breadth and height or depth. Socrates calls this mathematical discipline stereometry. He cites as an example a *corpus solidum* in the form of a cube, which it was the job of stereometry to measure and elucidate like all other stereometric bodies. We would not be mistaken if we were to find this Platonic example for stereometry, the cube, at the centre of the *Philosophers' School* – placed between arithmetic and geometry. But we also know that this cornerstone in the shape of a cube was only painted in front of the flight of steps by Raphael after the rest of the fresco had been completed. We therefore construed it as the cornerstone metaphor of Christ in the building of the *Civitas Dei*. The mediating discipline of stereometry must, however, also have been implicit in the original state of the fresco, between geometric planimetry and astronomy, if Raphael's *Philosophers' School*, as we believe, was intended to illustrate Plato's mathematical disciplines of philosophy. In fact, a small stone block in the form of a cube lies under Pythagoras' foot. Its edges express length, breadth and depth. But a stone under Pythagoras' foot can only refer to the cubic *numbers*.

According to Socrates, stereometry was a particularly difficult field and was not properly investigated in any state (*civitas*). There were, he says, two reasons for this: first, no state held such inquiries into three-dimensional objects in honour, and hence they were languidly pursued owing to their difficulty; second, the philosophers would

> *need a director, who is indispensable for success and who, to begin with, is not easy to find, and then, even if he could be found, ... the investigators in this field would be too arrogant to submit to his guidance. But if the state as a whole should join in superintending these studies and honour them, these specialists would accept advice, and continuous and strenuous investigation would bring out the truth* (Corpus verum, corpus solidum).

As far as I am aware, there is no interest in this remarkable textual passage in the contemporary philosophical interpretation of Plato's *Republic*. Even Marsiglio Ficino did not comment on it, although it cannot be doubted that the metaphorical sense of the flight of steps in Raphael's *Philosophers' School* was suggested by Ficino. Ficino's commentary on Plato's *Politeia* had been printed fully two decades before Raphael painted his fresco. In it Ficino repeats a stairway metaphor for the mathematical sciences that had first been used by Boethius. The mathematical disciplines led, he suggests, as on a flight of steps, to a higher stage of intellectual knowledge, removed from the senses. Raphael's flight of steps is admittedly also a compositional device, but it must have been dictated primarily by the patron or his theological-philosophical advisers for iconographic reasons. The original sense of the stairway was in any case correctly understood by Raphael's perceptive interpreter Gian Pietro Bellori in 1695. Bellori described the young philosopher, with his back turned, who is ascending the staircase, as follows (fig. 10): Raphael painted a pupil of Aristotle, who has just left the group of pupils being instructed in geometry. He looks as if he had completed the mathematical disciplines and now turns his steps towards philosophy; for another person shows him the way to Plato and Aristotle. When Raphael designed this youth, he was undoubtedly – Bellori suggests in conclusion – following the practice of the ancient Greeks, who had ascended the flight of steps from the mathematical disciplines to the speculative sciences. The purpose of learning the mathematical sciences was to escape from the deceptive world of the senses and so attain intellectual contemplation of things not in their real but in their ideal form.

So Bellori's sharp-sightedness did decipher the meaning of the painted stairway. He did not remark, however, that the youth in question is distinguished by the white mantle of the Pythagoreans. Among the painted philosophers and their pupils he is alone in *ascending* the stairway, in order to attain the height of knowledge of Plato and Aristotle. At the same time, however, he is gesticulating with both hands backwards at the cynic Diogenes, who because of his attribute, the bowl (standing on the step beside him), as well as his disorderly attire, was correctly identified by Vasari. The shadow of the young philosopher falls over the outstretched left leg of Diogenes. He has luxuriant blond hair, held together by a golden diadem. He must be a king. In the first state of the fresco, as can be seen in a drawing copied after it by Parmigianino, executed before the later addition of the cube-shaped block of stone and the brooding philosopher below Plato, this group with a young king beside Diogenes appeared in an even more central position in the whole composition. His hands pointing backwards to Diogenes help us to identify the young philosopher and king as the world conqueror Alexander.

In Plutarch, Aelian, Seneca and other ancient authors the famous answer of Diogenes to the king's question as to what he would like as a present, was transmitted as follows: "Get out of my sunshine!"

The point of this anecdote is enriched with an addendum in most sources. Alexander, namely, was so struck by this reply, that he is reported to have said to his entourage: "Were I not King Alexander, I should wish to be the philosopher Diogenes!" Since Raphael's Alexander is turning to the right to a colleague on a higher step, while at the same time pointing backwards to Diogenes, the king's words could not be more aptly expressed with gestures. Alexander's philosophical teacher was Aristotle. The king is reported to have later confessed that his life he owed to his father Philip, but the ability to live "uprightly" (*eu*) to his teacher Aristotle. The index finger of

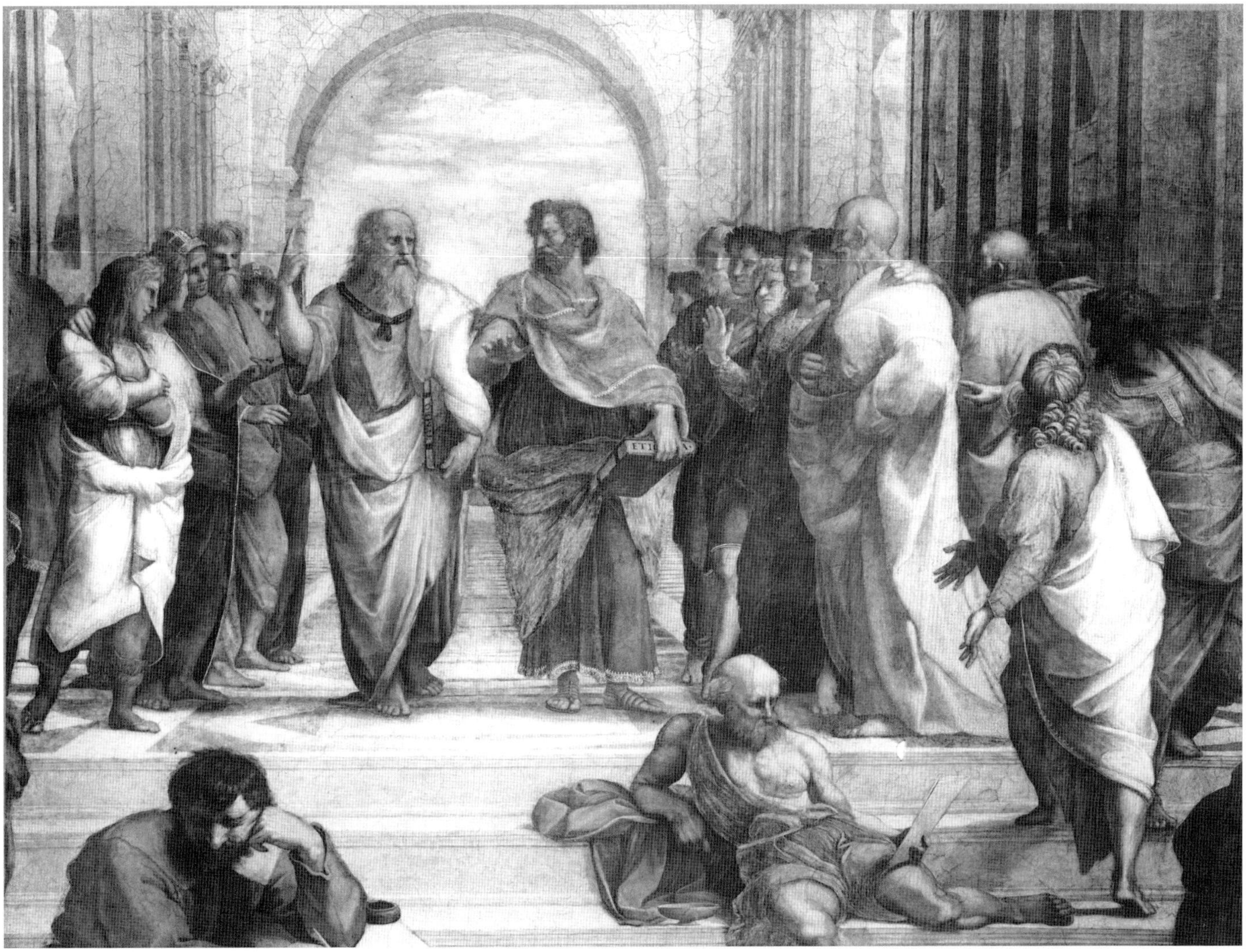

Figure 10
Raphael: *The School of Athens*. Detail with Plato and Aristotle.

his companion seems to be pointing at his teacher Aristotle. But the pointing gesture might equally be intended to indicate the bearded, bald-headed man strikingly dressed in a yellow mantle, who is standing to the right of Aristotle and towards whom Aristotle seems to be stepping. Plutarch reports in his treatise *De Alexandri Magni Fortuna* that King Alexander as a true philosopher had turned into a reality the idea or image of a cosmopolitan world state, which the founder of the Stoa, the philosopher Zeno has only sketched out in theory in his *Politeia*. Alexander, Plutarch says, in this had not followed his teacher Aristotle, who had advised him that he should show himself to the Greeks as their leader and to the other peoples as their lord. But Alexander had felt himself to be the heaven-sent ruler of all men and mediator for the whole world: his mission was to unify all people from wherever they came into a single "body" (*corpus*) and merge them together as if in a large mixing bowl of friendship. This, according to Plutarch, was precisely what the basic idea in Zeno's *Politeia* had expressed. All the inhabitants of this world should not remain divided by different legal systems and religious cults in many states and communities. Instead, all men should be seen as forming a single community and a single state. They should live together in only one community of life and with only one legal system for everyone, in some sense like a herd that pastures together in the same field. So Alexander, in Plutarch's view, had fulfilled the idea of the philosopher Zeno by unifying the world of his day as a philosopher-king.

This utopia of a cosmopolitan world state, in which everyone would live together in peace and friendship under a King Alexander, was therefore a political demand before it would become, in Augustine's *City of God*, a political consequence of belief in Christ, king of kings, the true son of God and at the same time true son of man, and lead to the theological concept of a *Corpus mysticum* of the whole Church. In his *Disputa* Raphael painted an altar-table, on which the Eucharist appears only in the form of the Body of Christ as *corpus verum*. It stands for the real presence of the three-in-one God here on earth. According to St. Augustine, the blessed citizens of the City of God, redeemed by Christ's sacrificial death, would be immersed in the eternal contemplation of God.

This Christian contemplation of the divine light in Augustine's City of God corresponds to Plato's philosophical contemplation of the divine idea of the supreme good, according to which the philosophers should be instructed in the mathematical sciences of arithmetic, the arithmetic doctrine of harmony, geometry, stereometry and astronomy, in order to be able to recognise the kingdom of the light of ideas without the sensory deceptions of their bodies.

This highest stage of philosophical learning and knowledge is called dialectic by Socrates in Book 7 of Plato's *Republic*. So when King Alexander, who had been educated in the mathematical sciences, is on the point of ascending the last step of Raphael's *Philosophers' School*, he is then ready to turn to the true philosophy that is called dialectic in Plato's doctrine of the state.

According to the late-medieval tradition, dialectic was expressed by two disputing philosophers, as it is for example in a relief at the base of the Campanile of the Cathedral of Florence. So in Raphael's *Philosophers' School* Plato and Aristotle embody the Platonic dialectic, in the light of which King Alexander is now in a position to recognise the supreme good for the formation of his cosmopolitan world state.

Bibliotheca Hertziana, Rome

References

Most recent bibliographical references in:

Hall, M. (ed.)
1997 *Raphael's "School of Athens"*. Cambridge.

Nesselrath, A.
1996 *Raphael's School of Athens*, photographs by F. Bono and P. Zigrossi. Vatican City State.

The above-mentioned issues have been dealt with in earlier publications:

Winner, M.
1984 "Il giudizio di Vasari sulle prime tre stanze di Raffaello in Vaticano." In: *Braccio di Carlo Magno Raffaello in Vaticano* (exhib.cat. 16.X.1984 – 16.I.1985), ed. Città del Vaticano, 179–193. Rome.

1986 "Disputa und Schule von Athen." In: *Raffaello a Roma – Il Convegno del 1983*, eds. C. L. Frommel and M. Winner, 29–47. Rome.

1993 "Progetti ed esecuzione nella Stanza della Segnatura." In: *Raffaello nell'appartamento di Giulio II. e Leone X.* Edizioni Musei Vaticani, 247–291. Rome

1994 "Ekphrasis bei Vasari." In: *Beschreibungskunst – Kunstbeschreibung; Ekphrasis von der Antike bis zur Gegenwart*, eds. G. Boehm and H. Pfotenhauer, 259–278. Munster.

1996 "Der Architekt in Raffaels Schule von Athen." In: *Architectural Studies in Memory of Richard Krautheimer*, ed. C. L. Striker, 180–185. Mainz.

1997 "Raffaels Schule von Athen (Philosophischer Unterricht zur Erkenntnis des Summum bonum)." In: *Jahrbuch der Max-Planck-Gesellschaft*, 652–657. Munich.